HANDLING HAZARDOUS MATERIALS

Edited by
Paul M. Bomgardner
American Trucking Associations

Developed under the direction of
The Hazardous Materials Committee
ATA Safety Management Council

Chairman
Michael R. Dean
Carolina Freight Carriers

Hazardous Materials Committee Members
Carl Ayres, CCC Express, Inc.
Dennis Barrett, Miller Transporters, Inc.
Jim Bucko, Rogers Cartage Company
Ed Carnahan, TNT Red Star Express
John Combs, Jr. Consolidated Freightways
Ken Dobisch, Chemical Leaman Tank Lines
Max Drauschak, Preston Trucking Co., Inc.
Ike Gabbard, Propane Transportation, Inc.
Fred Hager, Transport Insurance Company
Don Hagerman, American Tank Transport
Mark Johnson, Roadway Express, Inc.
Wayne King, St. Johnsbury Trucking Co.
Chuck Lovinski, Frederal Express Corp.
Gene Meenan, Matlack, Inc.
Harry Montgomery, Highway Transport, Inc.
Dick O'Boyle, National Continental Insurance
Gary Reid, Tri-State Motor Transit Co.
Paul Tamburelli, American President Trucking Co.
Rainey Turpin, AIG Transport Division
Tom Wilson, Yellow Freight System, Inc.

FOREWORD

The transportation of hazardous materials continues to increase in complexity.

Each year, several thousand new chemicals are introduced into commerce. Many of them are classified as hazardous materials. From time to time, commodities already in commerce are found to create hazards in addition to those originally noted and provided for.

Public interest and pressure for more stringent controls for hazardous materials transportation have also had an impact on government and industry safety programming. A few serious hazardous materials accidents coupled with worldwide publicized accounts of life threatening, environmentally hazardous and costly industrial accidents such as the methylisocyanate spill at Bohpal, India, the gas explosion at Mexico City, and the uranium hexafluoride release at Gore, Oklahoma have produced an atmosphere of public concern about the potential for catastrophic accidents resulting from hazardous materials transportation.

Congressional interest in hazardous materials has been heightened by constituent demands. Recent legislation has promoted an increase in regulatory enforcement by Federal and State agencies.

The Transportation Safety Act of 1974 extended the Department of Transportation's authority to regulate hazardous materials. The Act established the Materials Transportation Bureau within D.O.T. to identify materials moving in commerce that may pose unreasonable risks to health and safety, and to promulgate regulations for the safe transportation of hazardous materials. Sections of the Act establish stringent civil penalties for violations of hazardous materials regulations, increase criminal penalties for willful violations of the regulations and generally provides for extended coverage of the Department's authority over hazardous materials.

The Environmental Protection Agency has developed controls for the proper management and transportation of hazardous waste from "cradle to grave." EPA has established regulations for generators, transporters and treatment, storage, or disposal (TSD) facilities, including significant changes in recordkeeping, reporting, and manifesting requirements.

The Clean Water Act was adopted to minimize the adverse affects on the environment by spills of materials identified as hazardous substances. Both D.O.T. and EPA have promulgated requirements under the Act for notification and clean-up of hazardous substance spills.

Regulatory agencies, primarily the Department of Transportation, and the Environmental Protection Agency, are continuing to develop, implement and enforce regulations under the authority of these Acts. Many state and local jurisdictions are also developing and enforcing restrictions over the transportation of hazardous materials.

In addition to civil penalties imposed on carriers by the U.S. Department of Transportation (USDOT), state penalties, criminal and civil, may be imposed for violation of the regulations contained in Title 49 Code of Federal Regulations (49 CFR) which have now been adopted and incorporated by many states as law.

There has also been a significant increase in enforcement activity with respect to motor carriers by personnel of the Office of Motor Carriers. Numerous violations are found, and while many of them are technical, and have no immediate adverse effect on safety, they are nonetheless a black mark on the fleet's compliance record and result in stiff fines. Civil penalties can be imposed up to a maximum fine of $10,000. Each day is considered a separate violation. The maximum criminal penalty for knowing violations can be as high as a $25,000 fine, imprisonment for a term not to exceed 5 years, or both. Violations can be used as evidence of lack of fitness in connection with renewals of authority, efforts to secure additional temporary or permanent authority, and, in extreme cases, could be the basis for efforts to revoke the carrier's operating authority. Therefore, compliance with the regulations is important to the fleet's future economic well-being, as well to the safety of its operations.

Section 177.800 of the Hazardous Materials Regulations requires that motor carriers make the applicable regulations effective within their operations and provide necessary training for employees. Because of the extensive revisions to the regulations, retraining of employees is essential, and there is a continuing need for follow-up action and further retraining. In its monitoring of compliance with the regulations, the Department of Transportation is successfully prosecuting motor carriers for failure to train their employees. It is the recommendation of the Hazardous Materials Committee of the ATA Safety Management Council that motor carriers record the names of all employees given training in these regulations. This can be done by maintaining signed attendance lists of those attending each session, by having each employee sign a slip indicating that he has received training, or by issuing each employee a certificate upon completion of training.

Handling Hazardous Materials has been developed to help fleets comply with the regulations of the Department of Transportation, and to assist carriers in improving their hazardous materials safety programs. This guidebook provides updated information on the regulations in simplified language, and sets forth the changes in the internal programs and procedures in fleets made necessary by the recent amendments by the Department of Transportation and the Environmental Protection Agency. This guidebook does not cover out-of-the-ordinary situations or shipper responsibilities and shipping container specifications, except as they pertain directly to the responsibilities of the carrier. It also does not address state and local requirements, however a list of state agencies regulating the movement of hazardous waste is provided at page 179.

Handling Hazardous Materials is a guide to the federal requirements. It should not be used as a substitute. The full text of the hazardous materials regulations is contained in Title 49, Code of Federal Regulations, Parts 100-180. A copy of CFR 49 should be on hand in every terminal where hazardous materials are handled. Reprints of CFR 49, 100-180 are available from American Trucking Associations. For more information call toll free 1 800 ATA-LINE.

The development of this guidebook, and other activities of American Trucking Associations concerning hazardous materials, are carried on with the guidance and assistance of the Hazardous Materials Committee of the ATA Safety Management Council.

CONTENTS

Classification Of Hazardous Materials

The Department of Transportation (DOT) regulations governing the transportation of hazardous materials define and classify these commodities according to their nature and the hazard presented in transportation. The definition of each class is referenced in Part 171 of the regulations and the detailed definitions are set forth in Part 173. The definitions below are condensed versions which are intended to provide motor carrier employees with a basic indication of the nature of each class.

The Environmental Protection Agency (EPA), in conjunction with DOT, has promulgated regulations governing the transportation of hazardous wastes and hazardous substances. These terms are also defined below.

HAZARDOUS MATERIAL—A substance or material, including a hazardous substance, which has been determined by the Secretary of Transportation to be capable of posing an unreasonable risk to health, safety, and property when transported in commerce and which has been so designated.

A hazardous substance is further defined as a material, including its mixtures and solutions that —

1) Is listed in Appendix to 172.101, Hazardous Materials Table;

2) Is in a quantity in one package which equals or exceeds the reportable quantity (RQ) listed in the Appendix to the Hazardous Materials Table; and,

3) When in a mixture or solution, is in a concentration by weight corresponding to the RQ of the material as shown in the following table:

RQ pounds	RQ Kilograms	Concentration by Weight %	Concentration by Weight PPM
5,000	2270	10	100,000
1,000	454	2	20,000
100	45.4	0.2	2,000
10	4.54	0.02	200
1	0.45	0.002	20

HAZARDOUS WASTE—A solid waste which has been found by the Administrator of EPA to exhibit one or more of the following characteristics: ignitability, corrosivity, reactivity, or toxicity. *(40 CFR, 261.11). (NOTE: for the purpose of transportation a material is a hazardous waste if it is subject to the manifest requirements of EPA (40 CFR, 262, Subpart B) or any similar requirements of a state having an EPA approved hazardous waste control program.)*

BLASTING AGENT—A material designated for blasting which has undergone prescribed tests and been found to have little probability of initiating an explosion or of burning and then exploding; e.g. "Blasting Agent, n.o.s." *(49 CFR, 173.114a).*

CLASS A EXPLOSIVES—Solid or liquid explosives which function by detonation or otherwise of maximum hazard; e.g. Ammunition for cannon with explosive projectile; Black powder; Nitroglycerin. *(49 CFR, 173.53).*

CLASS B EXPLOSIVES—Explosives which generally function by rapid combustion rather than detonation; e.g. Firewords, special; Igniter; Rocket motor; Jet thrust unit. *(49 CFR, 173.88).*

CLASS C EXPLOSIVES—Manufactured articles which contain restricted quantities of class A and/or B explosives as components, and certain types of fireworks; e.g. Small arms ammunition; Common fireworks; Cigarette loads. *(49 CFR, 173.100).*

COMBUSTIBLE LIQUID—A liquid not meeting the definition of any other hazardous materials class having a flashpoint of 100 degrees F. (37.8 degrees C.) or more and less than 200 degrees F. (93.3 degrees C.); e.g. Crude oil, petroleum; Formaldehyde; Fuel oil. *(49 CFR, 173.115). (NOTE: Crude oil, petroleum is a commodity that can fall within the definition of more than one hazard class. Such may be the case with other commodities.) (NOTE 2: "Flash point" means the minimum temperature at which a liquid gives off vapor within a test vessel in sufficient concentration to form an ignitable mixture with the air near the surface of the liquid.)*

COMPRESSED GAS—Material having an absolute pressure exceeding 40 psi at 70 degrees F., an absolute pressure exceeding 104 psi at 130 degrees F., or any liquid flammable material having a vapor pressure exceeding 40 psi absolute at 100 degrees F. See subclassifications below. *(49 CFR, 173.300).*

a) **Flammable Gas**—Gas which is flammable or explosive under prescribed test procedures; e.g. Acetylene; Hydrogen; Propane.

b) **Non-Flammable Gas**—Gas which is hazardous by virtue of being under pressure, but which does not burn or explode under prescribed tests; e.g. Helium; Oxygen; Sulfur dioxide.

CORROSIVE MATERIAL—A liquid or solid that causes visible destruction or irreversible damage to skin tissue at the point of contact, or that has a severe corrosion rate on steel; e.g. Phosphorus pentachloride, solid; Potassium flouride solution; Sulfuric acid. *(49 CFR, 173.240).*

CRYOGENIC LIQUID—A flammable or nonflammable gas which has been converted to a liquid at super-cold temperature (below -130 F) at normal atmospheric pressure; e.g. Liquified Natural Gas, refrigerated liquid; Nitrogen, refrigerated liquid. *(49 CFR, 173.300(f)).*

ETIOLOGIC AGENTS—Substances which may cause human disease and which are specifically listed in regulations of the Department of Health, Education and Welfare (42 CFR, 72.25(c)); e.g. Biological products; Diagnostic specimens. *(49 CFR, 173.386).*

FLAMMABLE LIQUID—Any liquid which, under specified test procedures, has a flashpoint of less than 100 degrees F. (37.8 degrees C.); e.g. Crude oil, petroleum; Ether; Gasoline. *(49 CFR, 173.115). (NOTE: Crude oil, petroleum is a commodity that can fall within the definition of more than one hazard class. Such may be the case with other commodities.)*

FLAMMABLE SOLID—Solids, other than explosives, which are likely to cause fire through absorption of moisture, through spontaneous chemical changes, or as a result of retained heat from the manufacturing or processing of the commodity, including spontaneously combustible or water-reactive materials; e.g. Charcoal briquettes; Lithium metal; Matches, strike anywhere. *(49 CFR 173.150).*

IRRITATING MATERIALS—Liquids or solids which give off dangerous or intensely irritating fumes when exposed to air or upon contact with fire; e.g. Grenades; Tear gas; Xylyl bromide. *(49 CFR, 173.381).*

ORGANIC PEROXIDE—A derivative of hydrogen peroxide in which part of the hydrogen has been replaced by an organic material; e.g. Acetyl benzoyl peroxide; Benzoyl peroxide; Lauroyl peroxide. *(49 CFR, 173.151).*

OTHER REGULATED MATERIALS (ORM)—ORM A, B or C is a material that does not meet the definition of a hazardous material which is listed as an ORM in the Hazardous Materials Table (Sec. 171.101), or which meets one or more of the characteristics described below. *(49 CFR, 173.500).*

A commodity classed as an ORM-A, B, or C is not subject to the hazardous materials regulations for transportation by highway unless the commodity is a hazardous waste or is a hazardous substance identified by an "E" in Column 1, of the Hazardous Materials Table. If an ORM-A, B or C material is to be interlined with an air or water carrier, the packagings must be marked, and the shipment must be accompanied by a properly prepared shipping paper. Transportation of ORM commodities does not require placarding of the vehicle.

ORM-A—A substance which has an irritating, noxious, anesthetic or similar property which would cause extreme annoyance or discomfort to passengers and crew in the event of leakage during transportation; e.g. Bone oil; Carbon tetrachloride; Dry ice.

ORM-B—A substance which, when wet, would cause significant damage to a transport vehicle and is so designated in the Hazardous Materials Table; e.g. Calcium oxide; Chloroplatinic acid, solid; Lead chloride.

ORM-C—A substance listed in the Hazardous Materials Table not classed as ORM-A or ORM-B, but which has characteristics warranting special treatment during transportation. Examples: Battery parts; Bleaching powder; Castor beans.

ORM-D—A substance which meets the definition of a consumer commodity which would ordinarily be regulated as a hazardous material but which presents a limited hazard in transportation due to its form, quantity, or packaging, and for which a "limited quantity" exception is provided in the regulations. Applicable quantity limitations and packaging requirements for ORM-D shipments are set forth in

Section 173.1200 of the regulations. Under present regulations, the only designated proper shipping name for ORM-D is "Consumer Commodity."

For truck transportation of ORM-D the shipping paper, labeling, and placard rules do not apply. Packagings must be marked "Consumer Commodity" and also with the marking "ORM-D" (or ORM-D AIR) within a rectangular border.

If the packaging is to be interlined for air shipment, it must be marked "ORM-D AIR," and a properly prepared shipping paper must be tendered to the air carrier. The shipping paper tendered to the air carrier may be given by the motor carrier, or it may be provided directly to the air carrier by the shipper.

ORM-E—A material which does not meet the definition of any other hazard class, but which is defined as a "hazardous substance" or a "hazardous waste" by the Environmental Protection Agency (EPA), and is subject to the Hazardous Materials Regulations of the U.S. Department of Transportation.

OXIDIZER—A substance that yields oxygen readily to stimulate the combustion of certain other substances; e.g. Ammonium nitrate mixed fertilizer; Lead nitrate; Lead peroxide. *(49 CFR, 173.151)*

POISON A—Specifically enumerated gases or liquids of such nature that a small quantity of gas or the vapor of the liquid when mixed with air is dangerous to life; e.g. Cyanogen; Phosgene. *(49 CFR, 173.326)*.

POISON B—Liquids or solids (including pastes and semi-solids) known to be so toxic to man as to create a health hazard during transportation; or which are presumed to be toxic to man because of the effects on laboratory animals; e.g. Arsenic; Some Insecticides; Carbolic acid. *(49 CFR, 173.343)*.

RADIOACTIVE MATERIALS—Substances which spontaneously emit radiation capable of penetrating and damaging living tissue and undeveloped film. Fissile radioactive materials are classified according to the controls required for nuclear criticality safety; e.g. Cobalt 60; Gold 198; Iridium 192. *(49 CFR, 173.389)*.

TECHNICAL NAME—The scientific designation of a chemical in accordance with the nomenclature system developed by the International Union of Pure and Applied Chemistry (IUPAC) or the Chemical Abstracts Service (CAS) rules of nomenclature, or a name currently recognized in the Registry of Toxic Effects of Chemical Substances (RTECS). The term DOES NOT INCLUDE trade names.

TOXIC BY INHALATION—A liquid, other than a liquified compressed gas, which gives off toxic vapors at normal temperature (68 F), and which may, or may not, also meet the definition of another class of hazardous material: e.g. Methyl Isocyanate, flammable liquid. *(49 CFR, 173.3a)*.

Hazardous Materials Table

This table is a basic reference in dealing with the transportation of hazardous materials. It is useful to the motor carrier in determining whether or not a shipment contains a hazardous material. It is useful in checking to determine whether information on the shipping paper, and the markings and label(s) on packagings are in agreement. The table includes references to packaging requirements which may be of assistance by referencing a specification number which can be checked against what is on the packaging.

The items of concern to motor carriers are:

Column 1—Shows the applicability of the regulations in special situations as follows *(49 CFR, 171.101)*:

a) A plus sign (+) indicates that the designated proper shipping name and hazard class of the material must always be shown whether or not the material or its mixtures or solution meet the definition of the class.

b) A letter "A" or "W" signifies that the material is subject to the regulations only when transported by air (A) or by water (W) as indicated, unless the material is a hazardous substance or a hazardous waste. A material that is a hazardous substance or hazardous waste is subject to the regulations regardless of the mode of transport.

Column 2—Lists the proper shipping name of hazardous materials. It is the responsibility of the shipper to determine the proper shipping name and provide it where required on shipping papers and the markings on packagings. If wording in italics is shown adjacent to a proper shipping name, it may be included as additional information but is not required.

a) To verify the proper shipping name of a hazardous substance, check the Appendix to §172.101. If the Hazardous Substance (column 1 of the Appendix) is marked with "*" or "@," it indicates that the material is listed in the Hazardous Materials table by the same proper shipping name which must be used. For any other hazardous substance, the proper shipping name is, "hazardous substance, liquid or solid, n.o.s., NA9188."

b) If the word "waste" is not a part of the proper shipping name of a hazardous material that is a hazardous waste, the word "waste" must be entered preceding the proper shipping name; e.g., "Waste acetone."

c) For a hazardous waste that is not listed in the hazardous material table, the proper shipping name is, "Hazardous waste, liquid or solid, n.o.s., NA9189." When this proper shipping name is used, supplemental identification must be shown in parentheses to indicate the nature of the waste or waste stream being transported. See below - Additional Description Requirements, Hazardous Substances (page 154, 1(a)).

Column 3—Shows the hazard class of the material that must be shown on the shipping paper. *NOTE: A "Forbidden" material may not be transported.*

Column 3A—Shows the hazard identification number for the material which must be shown on the shipping paper, on portable tanks and cargo tanks, and on the exterior of packages.

a) A number with the prefix "UN" is authorized for all international shipments.

b) A number with the prefix "NA" is not recognized for international shipments, except between the United States and Canada.

Column 4—Shows the hazardous materials label(s) that must be applied to a packaging. For some commodities, the table specifies that more than one hazardous materials label be affixed to a packaging; e.g. Arsine, which requires both Poison Gas and Flammable Gas labels, but which is assigned the single hazard class of Poison A.

Column 5—This two-part column (a & b) shows the section numbers covering the packaging requirements for each hazardous material. Compliance with packaging requirements is the responsibility of the shipper. The references are of value to the motor carrier in verifying packaging specification numbers of a regulation against the specification number on any packaging.

Column 5a shows section numbers which provide for exceptions to packaging requirements under specified conditions and which qualify the packaging for a "Limited Quantity" (Ltd. Qty.) exception. If the notation "None" appears in Column 5(a) there are no conditions under which the material is entitled to a "Limited Quantity" exception.

Column 5(b) lists the section numbers for all packaging requirements other than "Limited Quantity" exceptions.

Placards Required On Trucks - The official printing of the Hazardous Materials Table provides information on the special restrictions and precautions which apply to the shipment of hazardous materials by air and water. American Trucking Associations, Inc. has replaced this information with a listing of the placard, if any, required for shipment by truck for each commodity.

The placard shown for each material is the one required (or authorized) for that material, alone. See Placarding section and the Hazardous Materials Guide (inside back cover) for detailed information on mixed loads and other requirements.

Materials which have been identified as toxic by inhalation, and a shipping paper entry states "POISON - INHALATION HAZARD," when transported by highway require a poison placard on each side and each end in addition to any other required placards. Duplication of poison placard or display of class numbers on additional placard are not required. The 1000 lb. exception for placards in 172.504(c) does not apply. E.G. Methyl Isocyanate, flammable liquid, UN2480, 500 lbs. would require a flammable placard on all 4 sides plus a poison placard on all four sides.

PLACARDING NOTES

In determining the placards to be used, the following special situations must be considered as indicated in the "Placards required" column.

NONE - No placard required for the material.

* - An asterisk after the placard name indicates that placarding is required for any quantity of material on board the vehicle.

1 - The COMBUSTIBLE placard is not required except for packagings larger than 110 gallons capacity. On cargo tanks or portable tanks, the FLAMMABLE placard may be used in lieu of COMBUSTIBLE.

2 - Indicates a material specifically identified as "Toxic By Inhalation." For packagings larger than 1 liter, the POISON placard must be displayed in addition to any other placard which is prescribed. It is not necessary to display two POISON placards for a material otherwise classed as a Poison. Materials in addition to those indicated by this footnote may be Toxic By Inhalation as indicated by the notation "Poison - Inhalation Hazard" on the shipping paper, as well as a marking, and a Poison label as a supplemental label on a packaging.

3 - The RADIOACTIVE placard is prescribed only when transporting shipments bearing the "Radioactive Yellow III" label, or when Radioactive Materials, Low Specific Activity, are transported in a vehicle on an "exclusive use" basis.

4 - The CHLORINE placard is required for shipments of chlorine in packagings larger than 110 gallons. The NONFLAMMABLE GAS placard is required for chlorine shipments in packagings of 110 gallons or less.

5 - The use of the FLAMMABLE placard is an authorized alternative to the FLAMMABLE SOLID placard. The FLAMMABLE SOLID placard must be used for shipments by water.

6 - DOT does not prescribe a placard for the transportation of polychlorinated biphenlys (PCB's). Regulations of the Environmental Protection Agency require the display of the large PCB label on vehicles in the manner prescribed for placards.

7 - Placarding shipments of Uranium Hexaflouride is governed by the following provisions:

a) RADIOACTIVE placard required for any quantity bearing a Yellow III label and
b) CORROSIVE placard must also be displayed if the shipment is 1000 lbs. or more.

§172.101 HAZARDOUS MATERIALS TABLE

(1) +/ A/ W	(2) Hazardous materials descriptions and proper shipping names	(3) Hazard Class	(3A) Identi- fication number	(4) Label(s) required (if not excepted)	(5) Packaging (a) Exceptions	(5) Packaging (b) Specific require- ments	Placards required (*denotes placard for any quantity)
	ACCUMULATOR, PRESSURIZED *(PNEUMATIC OR HYDRAULIC), CONTAINING NONFLAMMABLE GAS*	NONFLAMMABLE GAS	NA1956	NONFLAMMABLE GAS	173.306		NONFLAMMABLE GAS
	ACETAL	FLAMMABLE LIQUID	UN1088	FLAMMABLE LIQUID	173.118	173.119	FLAMMABLE
	ACETALDEHYDE (*ETHYL ALDEHYDE*)	FLAMMABLE LIQUID	UN1089	FLAMMABLE LIQUID	NONE	173.119	FLAMMABLE
A	ACETALDEHYDE AMMONIA	ORM-A	UN1841	NONE	173.505	173.510	NONE
	ACETIC ACID (*AQUEOUS SOLUTION*)	CORROSIVE MATERIAL	UN2790	CORROSIVE	173.244	173.245	CORROSIVE
	ACETIC ACID, GLACIAL	CORROSIVE MATERIAL	UN2789	CORROSIVE	173.244	173.245	CORROSIVE
	ACETIC ANHYDRIDE	CORROSIVE MATERIAL	UN1715	CORROSIVE	173.244	173.245	CORROSIVE
	ACETONE	FLAMMABLE LIQUID	UN1090	FLAMMABLE LIQUID	173.118	173.119	FLAMMABLE
	ACETONE CYANOHYDRIN	POISON B	UN1541	POISON	NONE	173.346	POISON[2]
	ACETONE OIL	FLAMMABLE LIQUID	UN1091	FLAMMABLE LIQUID	173.118	173.119	FLAMMABLE
	ACETONITRILE	FLAMMABLE LIQUID	NA1648	FLAMMABLE LIQUID	173.118	173.119	FLAMMABLE
	ACETYL ACETONE PEROXIDE, *IN SOLUTION WITH NOT MORE THAN 9% BY WEIGHT ACTIVE OXYGEN. SEE* ORGANIC PEROXIDE, LIQUID *OR* SOLUTION, N.O.S.		UN2080				
	ACETYL ACETONE PEROXIDE WITH MORE THAN 9% BY WEIGHT ACTIVE OXYGEN	FORBIDDEN					
	ACETYL BENZOYL PEROXIDE, *NOT MORE THAN 40% IN SOLUTION. SEE* ACETYL BENZOYL PEROXIDE SOLUTION, *NOT OVER 40% PEROXIDE*		UN2081				
	ACETYL BENZOYL PEROXIDE, SOLID, OR MORE THAN 40% IN SOLUTION	FORBIDDEN					
	ACETYL BENZOYL PEROXIDE SOLUTION, *NOT OVER 40% PEROXIDE*	ORGANIC PEROXIDE	UN2081	ORGANIC PEROXIDE	NONE	173.222	ORGANIC PEROXIDE
	ACETYL BROMIDE	CORROSIVE MATERIAL	UN1716	CORROSIVE	173.244	173.247	CORROSIVE
	ACETYL CHLORIDE	FLAMMABLE LIQUID	UN1717	FLAMMABLE LIQUID	173.244	173.247	FLAMMABLE
	ACETYL CYCLOHEXANESULFONYL PEROXIDE, MORE THAN 82%, WETTED WITH LESS THAN 12% WATER	FORBIDDEN					
	ACETYL CYCLOHEXANESULPHONYL PEROXIDE, *NOT MORE THAN 82%, WETTED WITH NOT LESS THAN 12% WATER. SEE* ORGANIC PEROXIDE, SOLID, N.O.S.		UN2082				
	ACETYL CYCLOHEXANESULPHONYL PEROXIDE, *NOT MORE THAN 32% IN SOLUTION. SEE* ORGANIC PEROXIDE, LIQUID *OR* SOLUTION, N.O.S.		UN2083				
	ACETYLENE	FLAMMABLE GAS	UN1001	FLAMMABLE GAS	NONE	173.303	FLAMMABLE GAS
	ACETYLENE (LIQUID)	FORBIDDEN					
	ACETYLENE SILVER NITRATE	FORBIDDEN					
A	ACETYLENE TETRABROMIDE	ORM-A	UN2504	NONE	173.505	173.510	NONE
	ACETYL IODIDE	CORROSIVE MATERIAL	UN1898	CORROSIVE	173.244	173.247	CORROSIVE
	ACETYL PEROXIDE, *NOT MORE THAN 25% IN SOLUTION. SEE* ACETYL PEROXIDE SOLUTION, *NOT OVER 25% PEROXIDE*		UN2084				

2-SEE PLACARDING NOTES, PAGE 5

§172.101 HAZARDOUS MATERIALS TABLE

(1) +/ A/ W	(2) Hazardous materials descriptions and proper shipping names	(3) Hazard Class	(3A) Identi- fication number	(4) Label(s) required (if not excepted)	(5) Packaging (a) Exceptions	(5) Packaging (b) Specific require- ments	Placards required (*denotes placard for any quantity)
	ACETYL PEROXIDE, SOLID, OR MORE THAN 25% IN SOLUTION	FORBIDDEN					
	ACETYL PEROXIDE SOLUTION, *NOT OVER 25% PEROXIDE*.	ORGANIC PEROXIDE	UN2084	ORGANIC PEROXIDE	173.153	173.222	ORGANIC PEROXIDE
	ACID BUTYL PHOSPHATE	CORROSIVE MATERIAL	UN1718	CORROSIVE	173.244	173.245	CORROSIVE
	ACID CARBOY, EMPTY. *SEE* CARBOY, EMPTY						
	ACID, LIQUID, N.O.S.	CORROSIVE MATERIAL	NA1760	CORROSIVE	173.244	173.245	CORROSIVE
+	ACROLEIN, INHIBITED	FLAMMABLE LIQUID	UN1092	FLAMMABLE LIQUID AND POISON	NONE	173.122	FLAMMABLE[2]
	ACRYLIC ACID	CORROSIVE MATERIAL	UN2218	CORROSIVE	173.244	173.245	CORROSIVE
	ACRYLONITRILE	FLAMMABLE LIQUID	UN1093	FLAMMABLE LIQUID AND POISON	NONE	173.119	FLAMMABLE
	ACTUATING CARTRIDGE, EXPLOSIVE(*FIRE EXTINGUISHER, OR VALVE*)	CLASS C EXPLOSIVE		EXPLOSIVE C	173.114		DANGEROUS
	ADHESIVE	COMBUSTIBLE LIQUID	UN1133	NONE	173.118a	NONE	COMBUSTIBLE[1]
	ADHESIVE	FLAMMABLE LIQUID	UN1133	FLAMMABLE LIQUID	173.118	173.132	FLAMMABLE
	AEROSOL PRODUCT. SEE COMPRESSED GAS, N.O.S.						
	AIR, COMPRESSED	NONFLAMMABLE GAS	UN1002	NONFLAMMABLE GAS	173.306	173.302	NONFLAMMABLE GAS
	AIR, REFRIGERATED LIQUID *(CRYOGENIC LIQUID)*	NONFLAMMABLE GAS	UN1003	NONFLAMMABLE GAS	173.320	173.316	NONFLAMMABLE GAS
	AIR CONDITIONING MACHINE. *SEE* REFRIGERATING MACHINE						
	AIRPLANE FLARE. SEE FIREWORKS, SPECIAL						
	ALCOHOLIC BEVERAGE	FLAMMABLE LIQUID	UN1170	FLAMMABLE LIQUID	173.118	173.125	FLAMMABLE
	ALCOHOLIC BEVERAGE	COMBUSTIBLE LIQUID	UN1170	NONE	173.118a	NONE	COMBUSTIBLE[1]
	ALCOHOL, N.O.S.	FLAMMABLE LIQUID	UN1987	FLAMMABLE LIQUID	173.118	173.125	FLAMMABLE
	ALCOHOL, N.O.S.	COMBUSTIBLE LIQUID	UN1987	NONE	173.118a	NONE	COMBUSTIBLE
	ALDRIN	POISON B	NA2761	POISON	173.364	173.376	POISON
A	ALDRIN, CAST SOLID	ORM-A	NA2761	NONE	173.505	173.510	NONE
	ALDRIN MIXTURE, DRY (*WITH MORE THAN 65% ALDRIN*).	POISON B	NA2761	POISON	173.364	173.376	POISON
A	ALDRIN MIXTURE, DRY, WITH 65% OR LESS ALDRIN	ORM-A	NA2761	NONE	173.505	173.510	NONE
	ALDRIN MIXTURE, LIQUID (*WITH MORE THAN 60% ALDRIN*)	POISON B	NA2762	POISON	173.345	173.361	POISON
A	ALDRIN MIXTURE, LIQUID, WITH 60% OR LESS ALDRIN	ORM-A	NA2762	NONE	173.505	173.510	NONE
	ALKALINE *(CORROSIVE)* LIQUID, N.O.S.	CORROSIVE MATERIAL	NA1719	CORROSIVE	173.244	173.249	CORROSIVE
	ALKANESULFONIC ACID	CORROSIVE MATERIAL	UN2584	CORROSIVE	173.244	173.245	CORROSIVE
	ALKYL ALUMINUM HALIDES. SEE PYROPHORIC LIQUID, N.O.S.						
A	ALLETHRIN	ORM-A	NA2902	NONE	173.505	173.510	NONE

1,2-SEE PLACARDING NOTES, PAGE 5

§172.101 HAZARDOUS MATERIALS TABLE

(1) +/ A/ W	(2) Hazardous materials descriptions and proper shipping names	(3) Hazard Class	(3A) Identi-fication number	(4) Label(s) required (if not excepted)	(5) Packaging (a) Exceptions	(5) Packaging (b) Specific require-ments	Placards required (*denotes placard for any quantity)
	ALLYL ALCOHOL	FLAMMABLE LIQUID	UN1098	FLAMMABLE LIQUID AND POISON	NONE	173.119 173.3a	FLAMMABLE[2]
	ALLYL BROMIDE	FLAMMABLE LIQUID	UN1099	FLAMMABLE LIQUID	173.118	173.119	FLAMMABLE
	ALLYL CHLORIDE	FLAMMABLE LIQUID	UN1100	FLAMMABLE LIQUID	NONE	173.119	FLAMMABLE
	ALLYL CHLOROCARBONATE	FLAMMABLE LIQUID	UN1722	FLAMMABLE LIQUID	NONE	173.288	FLAMMABLE
	ALLYL CHLOROFORMATE. *SEE* ALLYL CHLOROCARBONATE						
	ALLYL TRICHLOROSILANE	CORROSIVE MATERIAL	UN1724	CORROSIVE	NONE	173.280	CORROSIVE
	ALUMINUM ALKYL. *SEE* PYROPHORIC LIQUID, N.O.S.						
	ALUMINUM BROMIDE, ANHYDROUS	CORROSIVE MATERIAL	UN1725	CORROSIVE	173.244	173.245b	CORROSIVE
	ALUMINUM CHLORIDE, ANHYDROUS	CORROSIVE MATERIAL	UN1726	CORROSIVE	173.244	173.245b	CORROSIVE
	ALUMINUM DROSS, WET OR HOT. SEE 173.173	FORBIDDEN					
	ALUMINUM HYDRIDE	FLAMMABLE SOLID	UN2463	FLAMMABLE SOLID AND DANGEROUS WHEN WET	NONE	173.206	FLAMMABLE SOLID -W-*
	ALUMINUM, METALLIC, POWDER	FLAMMABLE SOLID	UN1396	FLAMMABLE SOLID	173.232	173.232	FLAMMABLE SOLID[5]
	ALUMINUM NITRATE	OXIDIZER	UN1438	OXIDIZER	173.153	173.182	OXIDIZER
	ALUMINUM PHOSPHATE SOLUTION	CORROSIVE MATERIAL	NA1760	CORROSIVE	173.244	173.245	CORROSIVE
	ALUMINUM PHOSPHIDE	FLAMMABLE SOLID	UN1397	FLAMMABLE SOLID AND DANGEROUS WHEN WET	NONE	173.154	FLAMMABLE SOLID -W-*
A	ALUMINUM SULFATE SOLUTION	ORM-B	NA1760	NONE	173.505	173.510	NONE
	AMATOL. SEE HIGH EXPLOSIVE						
	2-(2-AMINOETHOXY) ETHANOL	CORROSIVE MATERIAL	NA1760	CORROSIVE	173.244	173.245	CORROSIVE
	N-AMINOETHYLPIPERAZINE	CORROSIVE MATERIAL	UN2815	CORROSIVE	173.244	173.245	CORROSIVE
	AMINOPROPYLDIETHANOLAMINE	CORROSIVE MATERIAL	NA1760	CORROSIVE	173.244	173.245	CORROSIVE
	N-AMINOPROPYLMORPHOLINE	CORROSIVE MATERIAL	NA1760	CORROSIVE	173.244	173.245	CORROSIVE
	BIS (AMINOPROPYL) PIPERAZINE	CORROSIVE MATERIAL	NA1760	CORROSIVE	173.244	173.245	CORROSIVE
	AMMONIA, ANHYDROUS	NONFLAMM-ABLE GAS	UN1005	NONFLAMM-ABLE GAS	173.306	173.304 173.314 173.315	NONFLAMM-ABLE GAS
	AMMONIA SOLUTION (*CONTAINING MORE THAN 44% AMMONIA*)	NONFLAMM-ABLE GAS	UN2073	NONFLAMM-ABLE GAS	173.306	173.304 173.314 173.315	NONFLAMM-ABLE GAS
	AMMONIA SOLUTION (CONTAINING 44% OR LESS AMMONIA IN WATER). SEE AMMONIUM HYDROXIDE						
	AMMONIUM ARSENATE, SOLID	POISON B	UN1546	POISON	173.364	173.365	POISON
	AMMONIUM AZIDE	FORBIDDEN					
	AMMONIUM BIFLUORIDE, SOLID *OR* SOLUTION. *SEE* AMMONIUM HYDROGEN FLUORIDE, SOLID *OR* SOLUTION						

2, 5 - SEE PLACARDING NOTES, PAGE 5

§172.101 HAZARDOUS MATERIALS TABLE

(1) +/ A/ W	(2) Hazardous materials descriptions and proper shipping names	(3) Hazard Class	(3A) Identi-fication number	(4) Label(s) required (if not excepted)	(5) Packaging (a) Exceptions	(5) Packaging (b) Specific require-ments	Placards required (*denotes placard for any quantity)
A	AMMONIUM BISULFITE, SOLID	ORM-B	NA2693	NONE	173.505	173.510	NONE
	AMMONIUM BISULFITE SOLUTION	CORROSIVE MATERIAL	NA2693	CORROSIVE	173.244	173.245	CORROSIVE
	AMMONIUM BROMATE	FORBIDDEN					
A	AMMONIUM CARBAMATE	ORM-A	NA9083	NONE	173.505	173.510	NONE
A	AMMONIUM CARBONATE	ORM-A	NA9084	NONE	173.505	173.510	NONE
	AMMONIUM CHLORATE	FORBIDDEN					
	AMMONIUM DICHROMATE (*AMMONIUM BICHROMATE*)	OXIDIZER	UN1439	OXIDIZER	173.153	173.154 173.235	OXIDIZER
A	AMMONIUM FLUOBORATE	ORM-B	NA9088	NONE	NONE	173.510	NONE
A	AMMONIUM FLUORIDE	ORM-B	UN2505	NONE	173.505	173.800	NONE
	AMMONIUM FULMINATE	FORBIDDEN					
	AMMONIUM HYDROGEN FLUORIDE, SOLID	CORROSIVE MATERIAL	UN1727	CORROSIVE	173.244	173.245b	CORROSIVE
	AMMONIUM HYDROGEN FLUORIDE SOLUTION	CORROSIVE MATERIAL	UN2817	CORROSIVE	173.244	173.245	CORROSIVE
A	AMMONIUM HYDROGEN SULFATE	ORM-B	UN2506	NONE	173.505	173.800	NONE
A	AMMONIUM HYDROSULFIDE SOLUTION	ORM-A	NA2683	NONE	173.505	173.605	NONE
	AMMONIUM HYDROXIDE (*CONTAINING NOT LESS THAN 12% BUT NOT MORE THAN 44% AMMONIA*)	CORROSIVE MATERIAL	NA2672	CORROSIVE	173.244	173.245	CORROSIVE
AW	AMMONIUM HYDROXIDE (*CONTAINING LESS THAN 12% AMMONIA*)	ORM-A	NA2672	NONE	173.505	173.510	NONE
	AMMONIUM NITRATE-CARBONATE MIXTURE	OXIDIZER	UN2068	OXIDIZER	173.153	173.182	OXIDIZER
	AMMONIUM NITRATE FERTILIZER, *CONTAINING NO MORE THAN 0.2% CARBON*	OXIDIZER	UN2067	OXIDIZER	173.153	173.182	OXIDIZER
	AMMONIUM NITRATE - FUEL OIL MIXTURE. SEE HIGH EXPLOSIVE						
	AMMONIUM NITRATE - FUEL OIL MIXTURE (*CONTAINING ONLY PRILLED AMMONIUM NITRATE AND FUEL OIL*)	BLASTING AGENT		BLASTING AGENT	NONE	173.114a	BLASTING AGENTS
	AMMONIUM NITRATE MIXED FERTILIZER	OXIDIZER	UN2069	OXIDIZER	173.153	173.182	OXIDIZER
	AMMONIUM NITRATE (NO ORGANIC COATING)	OXIDIZER	UN1942	OXIDIZER	173.153	173.182	OXIDIZER
	AMMONIUM NITRATE (ORGANIC COATING)	OXIDIZER	NA1942	OXIDIZER	173.153	173.182	OXIDIZER
	AMMONIUM NITRATE-PHOSPHATE	OXIDIZER	UN2070	OXIDIZER	173.153	173.182	OXIDIZER
	AMMONIUM NITRATE, SOLUTION (*CONTAINING NOT LESS THAN 15% WATER*). *SEE 173.154(A)(17) AND 173.154(A)(18)*	OXIDIZER	UN2426	OXIDIZER			OXIDIZER
	AMMONIUM NITRITE	FORBIDDEN					
	AMMONIUM OXALATE	ORM-A	NA2449	NONE	173.505	173.510	NONE
	AMMONIUM PERCHLORATE	OXIDIZER	UN1442	OXIDIZER	173.153	173.239a	OXIDIZER
	AMMONIUM PERCHLORATE. *SEE* HIGH EXPLOSIVE						
	AMMONIUM PERMANGANATE	OXIDIZER	NA9190	OXIDIZER	NONE	173.154	OXIDIZER

§172.101 HAZARDOUS MATERIALS TABLE

(1) +/ A/ W	(2) Hazardous materials descriptions and proper shipping names	(3) Hazard Class	(3A) Identi-fication number	(4) Label(s) required (if not excepted)	(5) Packaging (a) Exceptions	(5) Packaging (b) Specific require-ments	Placards required (*denotes placard for any quantity)
	AMMONIUM PERSULFATE	OXIDIZER	UN1444	OXIDIZER	173.153	173.154	OXIDIZER
	AMMONIUM PICRATE, DRY. SEE HIGH EXPLOSIVE						
	AMMONIUM PICRATE, WET (*WITH 10% OR MORE WATER*)	FLAMMABLE SOLID	UN1310	FLAMMABLE SOLID	173.192		FLAMMABLE SOLID[5]
	AMMONIUM PICRATE, WET, WITH 10% OR MORE WATER, OVER 16 OUNCES IN ONE OUTSIDE PACKAGING. SEE HIGH EXPLOSIVE						
	AMMONIUM POLYSULFIDE SOLUTION	ORM-A	UN2818	NONE	173.505	173.605	NONE
A	AMMONIUM SILICOFLUORIDE	ORM-B	UN2854	NONE	173.505	173.510	NONE
W	AMMONIUM SULFATE NITRATE	ORM-C	NA1477	NONE	173.505	173.910	NONE
	AMMONIUM SULFIDE SOLUTION	FLAMMABLE LIQUID	UN2683	FLAMMABLE LIQUID	173.118	173.119	FLAMMABLE
	AMMUNITION, CHEMICAL (CONTAINING A POISON A LIQUID OR GAS). SEE CHEMICAL AMMUNITION, NONEXPLOSIVE (*CONTAINING A POISON A MATERIAL)*						
	AMMUNITION, CHEMICAL (CONTAINING A POISON B MATERIAL). SEE CHEMICAL AMMUNITION, NONEXPLOSIVE (*CONTAINING A POISON B MATERIAL)*						
	AMMUNITION, CHEMICAL (CONTAINING AN IRRITATING LIQUID OR SOLID). SEE CHEMICAL AMMUNITION, NONEXPLOSIVE (*CONTAINING AN IRRITATING MATERIAL)*						
	AMMUNITION, CHEMICAL, EXPLOSIVE, WITH POISON A MATERIAL	CLASS A EXPLOSIVE		EXPLOSIVE A AND POISON GAS	NONE	173.59	EXPLOSIVES A* AND POISON GAS*
	AMMUNITION, CHEMICAL, EXPLOSIVE, WITH POISON B MATERIAL	CLASS A EXPLOSIVE		EXPLOSIVE A AND POISON	NONE	173.59	EXPLOSIVES A*
	AMMUNITION, CHEMICAL, EXPLOSIVE, WITH IRRITANT	CLASS A EXPLOSIVE		EXPLOSIVE A AND IRRITANT	NONE	173.59	EXPLOSIVES A*
	AMMUNITION FOR CANNON WITH EMPTY PROJECTILE	CLASS B EXPLOSIVE		EXPLOSIVE B	NONE	173.89	EXPLOSIVES B*
	AMMUNITION FOR CANNON WITH EXPLOSIVE PROJECTILE	CLASS A EXPLOSIVE		EXPLOSIVE A	NONE	173.54	EXPLOSIVES A*
	AMMUNITION FOR CANNON WITH GAS PROJECTILE	CLASS A EXPLOSIVE		EXPLOSIVE A	NONE	173.54	EXPLOSIVES A*
	AMMUNITION FOR CANNON WITH ILLUMINATING PROJECTILE	CLASS A EXPLOSIVE		EXPLOSIVE A	NONE	173.54	EXPLOSIVES A*
	AMMUNITION FOR CANNON WITH INCENDIARY PROJECTILE	CLASS A EXPLOSIVE		EXPLOSIVE A	NONE	173.54	EXPLOSIVES A*
	AMMUNITION FOR CANNON WITH INERT LOADED PROJECTILE	CLASS B EXPLOSIVE		EXPLOSIVE B	NONE	173.89	EXPLOSIVES B*
	AMMUNITION FOR CANNON WITHOUT PROJECTILE	CLASS B EXPLOSIVE		EXPLOSIVE B	NONE	173.89	EXPLOSIVES B*
	AMMUNITION FOR CANNON WITH SMOKE PROJECTILE	CLASS A EXPLOSIVE		EXPLOSIVE A	NONE	173.54	EXPLOSIVES A*
	AMMUNITION FOR CANNON WITH SOLID PROJECTILE	CLASS B EXPLOSIVE		EXPLOSIVE B	NONE	173.89	EXPLOSIVES B*
	AMMUNITION FOR CANNON WITH TEAR GAS PROJECTILE	CLASS A EXPLOSIVE		EXPLOSIVE A	NONE	173.54	EXPLOSIVES A*
	AMMUNITION FOR CANNON WITH TEAR GAS PROJECTILE	CLASS B EXPLOSIVE		EXPLOSIVE B	NONE	173.89	EXPLOSIVES B*
	AMMUNITION FOR SMALL-ARMS WITH EXPLOSIVE PROJECTILE	CLASS A EXPLOSIVE		EXPLOSIVE A	NONE	173.58	EXPLOSIVES A*

5 - SEE PLACARDING NOTES, PAGE 5

§172.101 HAZARDOUS MATERIALS TABLE

(1) +/ A/ W	(2) Hazardous materials descriptions and proper shipping names	(3) Hazard Class	(3A) Identi- fication number	(4) Label(s) required (if not excepted)	(5) Packaging (a) Exceptions	(5) Packaging (b) Specific require- ments	Placards required (*denotes placard for any quantity)
	AMMUNITION FOR SMALL-ARMS WITH INCENDIARY PROJECTILE	CLASS A EXPLOSIVE		EXPLOSIVE A	NONE	173.58	EXPLOSIVES A*
	AMMUNITION, NON-EXPLOSIVE. SEE 173.55						
	AMMUNITION, ROCKET. SEE ROCKET AMMUNITION WITH						
	AMMUNITION, SMALL-ARMS. *SEE* SMALL-ARMS AMMUNITION						
	AMYL ACETATE	FLAMMABLE LIQUID	UN1104	FLAMMABLE LIQUID	173.118	173.119	FLAMMABLE
	AMYL ACID PHOSPHATE	CORROSIVE MATERIAL	UN2819	CORROSIVE	173.244	173.245	CORROSIVE
	AMYLAMINE	FLAMMABLE LIQUID	UN1106	FLAMMABLE LIQUID	173.118	173.119	FLAMMABLE
	AMYL CHLORIDE	FLAMMABLE LIQUID	UN1107	FLAMMABLE LIQUID	173.118	173.119	FLAMMABLE
	AMYLENE, *NORMAL*	FLAMMABLE LIQUID	UN1108	FLAMMABLE LIQUID	173.118	173.119	FLAMMABLE
	AMYL FORMATE	FLAMMABLE LIQUID	UN1109	FLAMMABLE LIQUID	173.118	173.119	FLAMMABLE
	AMYL MERCAPTAN	FLAMMABLE LIQUID	UN1111	FLAMMABLE LIQUID	NONE	173.141	FLAMMABLE
	AMYL NITRITE	FLAMMABLE LIQUID	UN1113	FLAMMABLE LIQUID	173.118	173.119	FLAMMABLE
	TERT-AMYL PEROXY-2-ETHYLHEXANOATE, *TECHNICALLY PURE. SEE* ORGANIC PEROXIDE, LIQUID *OR* SOLUTION, N.O.S.		UN2898				
	TERT- AMYL PEROXYNEODECANOATE, *NOT MORE THAN 75% WITH PHLEGMATIZER. SEE* ORGANIC PEROXIDE, LIQUID *OR* SOLUTION, N.O.S.		UN2891				
	AMYL TRICHOROSILANE	CORROSIVE MATERIAL	UN1728	CORROSIVE	NONE	173.280	CORROSIVE
	ANHYDROUS AMMONIA. *SEE* AMMONIA, ANHYDROUS						
	ANHYDROUS HYDRAZINE. *SEE* HYDRAZINE, ANHYDROUS						
	ANHYDROUS HYDROFLUORIC ACID. *SEE* HYDROGEN FLUORIDE						
	ANILINE OIL DRUM, EMPTY *SEE 173.347(D)*	POISON B					
+	ANILINE OIL, LIQUID	POISON B	UN1547	POISON	NONE	173.347	POISON
	ANISOYL CHLORIDE	CORROSIVE MATERIAL	UN1729	CORROSIVE	173.244	173.279	CORROSIVE
	ANTIMONOUS CHLORIDE. *SEE* ANTIMONY TRICHLORIDE						
A	ANTIMONY LACTATE, SOLID	ORM-A	UN1550	NONE	173.505	173.510	NONE
	ANTIMONY PENTACHLORIDE	CORROSIVE MATERIAL	UN1730	CORROSIVE	NONE	173.247	CORROSIVE
	ANTIMONY PENTACHLORIDE SOLUTION	CORROSIVE MATERIAL	UN1731	CORROSIVE	173.244	173.245	CORROSIVE
	ANTIMONY PENTAFLUORIDE	CORROSIVE MATERIAL	UN1732	CORROSIVE	NONE	173.246	CORROSIVE
A	ANTIMONY POTASSIUM TARTRATE, SOLID	ORM-A	UN1551	NONE	173.505	173.510	NONE
	ANTIMONY SULFIDE AND A CHLORATE, MIXTURES OF	FORBIDDEN					
A	ANTIMONY SULFIDE, SOLID	ORM-A	NA1325	NONE	173.505	173.510	NONE

§172.101 HAZARDOUS MATERIALS TABLE

(1) +/ A/ W	(2) Hazardous materials descriptions and proper shipping names	(3) Hazard Class	(3A) Identification number	(4) Label(s) required (if not excepted)	(5) Packaging (a) Exceptions	(5) Packaging (b) Specific requirements	Placards required (*denotes placard for any quantity)
	ANTIMONY TRIBROMIDE, SOLID	CORROSIVE MATERIAL	NA1549	CORROSIVE	173.244	173.245b	CORROSIVE
	ANTIMONY TRIBROMIDE SOLUTION	CORROSIVE MATERIAL	NA1549	CORROSIVE	173.244	173.245	CORROSIVE
	ANTIMONY TRICHLORIDE, SOLID	CORROSIVE MATERIAL	UN1733	CORROSIVE	173.244	173.245b	CORROSIVE
	ANTIMONY TRICHLORIDE SOLUTION	CORROSIVE MATERIAL	UN1733	CORROSIVE	173.244	173.245	CORROSIVE
	ANTIMONY TRIFLUORIDE, SOLID	CORROSIVE MATERIAL	NA1549	CORROSIVE	173.244	173.245b	CORROSIVE
	ANTIMONY TRIFLUORIDE SOLUTION	CORROSIVE MATERIAL	NA1549	CORROSIVE	173.244	173.245	CORROSIVE
	AQUA AMMONIA SOLUTION *(CONTAINING 44% OR LESS AMMONIA). SEE* AMMONIUM HYDROXIDE						
	ARGON *OR* ARGON, COMPRESSED	NONFLAMMABLE GAS	UN1006	NONFLAMMABLE GAS	173.306	173.302 173.314	NONFLAMMABLE GAS
	ARGON, REFRIGERATED LIQUID *(CRYOGENIC LIQUID)*	NONFLAMMABLE GAS	UN1951	NONFLAMMABLE GAS	173.320	173.316 173.318	NONFLAMMABLE GAS
	ARSENIC ACID, SOLID	POISON B	UN1554	POISON	173.364	173.366	POISON
	ARSENIC ACID SOLUTION	POISON B	UN1553	POISON	173.345	173.348	POISON
	ARSENICAL COMPOUND, LIQUID, N.O.S., *OR* ARSENICAL MIXTURE, LIQUID, N.O.S.	POISON B	UN1556	POISON	173.345	173.346	POISON
	ARSENICAL COMPOUND, SOLID, N.O.S., *OR* ARSENICAL MIXTURE, SOLID, N.O.S.	POISON B	UN1557	POISON	173.364	173.367	POISON
	ARSENICAL DIP, LIQUID *(SHEEP DIP)*	POISON B	NA1557	POISON	173.345	173.346	POISON
	ARSENICAL DUST	POISON B	UN1562	POISON	173.364	173.368	POISON
	ARSENICAL PESTICIDE, LIQUID, N.O.S. (*COMPOUNDS AND PREPARATIONS*)	FLAMMABLE LIQUID	UN2760	FLAMMABLE LIQUID	173.118	173.119	FLAMMABLE
	ARSENICAL PESTICIDE, LIQUID, N.O.S. (*COMPOUNDS AND PREPARATIONS*)	POISON B	UN2759	POISON	173.345	173.346	POISON
	ARSENICAL PESTICIDE, SOLID, N.O.S. (*COMPOUNDS AND PREPARATIONS*)	POISON B	UN2759	POISON	173.364	173.367	POISON
	ARSENIC BROMIDE, SOLID	POISON B	UN1555	POISON	173.364	173.365	POISON
	ARSENIC CHLORIDE, LIQUID. *SEE* ARSENIC TRICHLORIDE						
	ARSENIC DISULFIDE. *SEE* ARSENIC SULFIDE, SOLID						
	ARSENIC IODIDE, SOLID	POISON B	NA1557	POISON	173.364	173.365	POISON
	ARSENIC PENTOXIDE, SOLID	POISON B	UN1559	POISON	173.364	173.365	POISON
	ARSENIC, SOLID	POISON B	UN1558	POISON	173.364	173.366	POISON
	ARSENIC SULFIDE AND A CHLORATE, MIXTURES OF	FORBIDDEN					
	ARSENIC SULFIDE, SOLID	POISON B	NA1557	POISON	173.364	173.365	POISON
	ARSENIC TRICHLORIDE, LIQUID	POISON B	UN1560	POISON	173.345	173.346	POISON
	ARSENIC TRIOXIDE, SOLID	POISON B	UN1561	POISON	173.364	173.366 173.368	POISON
	ARSENIC TRISULFIDE	POISON B	NA1557	POISON	173.364	173.365	POISON
	ARSENIC, WHITE, SOLID. *SEE* ARSENIC TRIOXIDE, SOLID						

§172.101 HAZARDOUS MATERIALS TABLE

(1) +/ A/ W	(2) Hazardous materials descriptions and proper shipping names	(3) Hazard Class	(3A) Identi-fication number	(4) Label(s) required (if not excepted)	(5) Packaging (a) Exceptions	(5) Packaging (b) Specific require-ments	Placards required (*denotes placard for any quantity)
	ARSENIOUS ACID, SOLID. *SEE* ARSENIC TRIOXIDE, SOLID.						
	ARSENIOUS AND MERCURIC IODIDE SOLUTION	POISON B	NA2810	POISON	173.345	173.346	POISON
	ARSINE	POISON A	UN2188	POISON GAS AND FLAMMABLE GAS	NONE	173.328	POISON GAS*
	ASBESTOS	ORM-C		NONE	173.1090	173.1090	NONE
	ASCARIDOLE (ORGANIC PEROXIDE)	FORBIDDEN					
W	ASPHALT, *AT OR ABOVE ITS FLASHPOINT*	ORM-C	NA1999	NONE	NONE	NONE	NONE
	ASPHALT, CUT BACK	FLAMMABLE LIQUID	NA1999	FLAMMABLE LIQUID	173.118	173.131	FLAMMABLE
	ASPHALT, CUT BACK	COMBUSTIBLE LIQUID	NA1999	NONE	173.118a	NONE	COMBUSTIBLE[1]
	AUTOMOBILE, MOTORCYCLE, TRACTOR, OR OTHER SELF-PROPELLED VEHICLE. SEE MOTOR VEHICLE						
	AUTOMOBILE, MOTORCYCLE, TRACTOR, OR OTHER SELF-PROPELLED VEHICLE, ENGINE, OR OTHER MECHANICAL APPARATUS, WITH CHARGED ELECTRIC STORAGE BATTERY, WET. SEE MOTOR VEHICLE						
	AZAUROLIC ACID (SALT OF), (DRY)	FORBIDDEN					
	3-AZIDO-1,2-PROPYLENE GLYCOL DINITRATE	FORBIDDEN					
	5-AZIDO-1-HYDROXY TETRAZOLE	FORBIDDEN					
	AZIDODITHIOCARBONIC ACID	FORBIDDEN					
	AZIDOETHYL NITRATE	FORBIDDEN					
	AZIDO GUANIDINE PICRATE (DRY)	FORBIDDEN					
	AZIDO HYDROXY TETRAZOLE (MERCURY AND SILVER SALTS)	FORBIDDEN					
	AZINPHOS METHYL	POISON B	NA2783	POISON	173.364	173.365	POISON
	AZINPHOS METHYL MIXTURE, LIQUID	POISON B	NA2783	POISON	173.345	173.346	POISON
	1-AZIRIDINYL PHOSPHINE OXIDE (TRIS). *SEE* TRIS-(1-AZIRIDINYL) PHOSPHINE OXIDE						
	AZOTETRAZOLE (DRY)	FORBIDDEN					
	BAGS, BURLAP, USED, MUST BE CLASSED FOR THE HAZARDOUS MATERIAL PREVIOUSLY CONTAINED IN BAG. SEE 173.28, 173.29						
	BARIUM AZIDE, WET, *50% OR MORE WATER*	FLAMMABLE SOLID	UN1571	FLAMMABLE SOLID	NONE	173.239	FLAMMABLE SOLID[5]
	BARIUM CHLORATE	OXIDIZER	UN1445	OXIDIZER	173.153	173.163	OXIDIZER
	BARIUM CHLORATE, WET	OXIDIZER	NA1445	OXIDIZER	173.153	173.163	OXIDIZER
	BARIUM CYANIDE, SOLID	POISON B	UN1565	POISON	173.370	173.370	POISON
	BARIUM NITRATE	OXIDIZER	UN1446	OXIDIZER	173.153	173.182	OXIDIZER
A	BARIUM OXIDE	ORM-B	UN1884	NONE	173.505	173.800	NONE
	BARIUM PERCHLORATE	OXIDIZER	UN1447	OXIDIZER	173.153	173.154	OXIDIZER
	BARIUM PERMANGANATE	OXIDIZER	UN1448	OXIDIZER	173.153	172.154	OXIDIZER

1,5-SEE PLACARDING NOTES, PAGE 5

§172.101 HAZARDOUS MATERIALS TABLE

(1) +/ A/ W	(2) Hazardous materials descriptions and proper shipping names	(3) Hazard Class	(3A) Identification number	(4) Label(s) required (if not excepted)	(5) Packaging (a) Exceptions	(5) Packaging (b) Specific requirements	Placards required (*denotes placard for any quantity)
	BARIUM PEROXIDE	OXIDIZER	UN1449	OXIDIZER	173.153	173.156	OXIDIZER
	BARIUM STYPHNATE, MONOHYDRATE. SEE INITIATING EXPLOSIVE						
	BARREL, EMPTY. SEE DRUM, EMPTY						
	BATTERY, DRY. NOT SUBJECT TO PARTS 170-189 OF THIS SUBCHAPTER						
	BATTERY, *ELECTRIC STORAGE,* DRY (CONTAINING POTASSIUM HYDROXIDE, DRY, *SOLID, FLAKE, BEAD, OR GRANULAR)*	CORROSIVE MATERIAL	NA1813	CORROSIVE	173.244	173.245b	CORROSIVE
	BATTERY, *ELECTRIC STORAGE,* WET, FILLED WITH ACID.	CORROSIVE MATERIAL	UN2794	CORROSIVE	173.260	173.260	CORROSIVE
	BATTERY, *ELECTRIC STORAGE,* WET, *FILLED WITH* ACID, WITH AUTOMOBILE (*OR SPECIFICALLY NAMED SELF-PROPELLED VEHICLE OR MECHANICAL APPARATUS*)	CORROSIVE MATERIAL	NA2794	CORROSIVE	173.250	173.260	CORROSIVE
	BATTERY, *ELECTRIC STORAGE, WET,* FILLED WITH ALKALI, WITH AUTOMOBILE *(OR SPECIFICALLY NAMED SELF-PROPELLED VEHICLE OR MECHANICAL APPARATUS)*	CORROSIVE MATERIAL	NA2795	CORROSIVE	173.250	173.260	CORROSIVE
	BATTERY, *ELECTRIC STORAGE,* WET, WITH WHEELCHAIR	CORROSIVE MATERIAL		CORROSIVE	173.250	173.250 175.10	CORROSIVE
	BATTERY, ELECTRIC STORAGE, WET, NONSPILLABLE. SEE 173.260(D)						
	BATTERY, ELECTRIC STORAGE, WET, FILLED WITH ALKALI	CORROSIVE MATERIAL	UN2795	CORROSIVE	173.260	173.260	CORROSIVE
	BATTERY FLUID, ACID	CORROSIVE MATERIAL	UN2796	CORROSIVE	173.244	173.257	CORROSIVE
	BATTERY FLUID, ACID, WITH ELECTRONIC EQUIPMENT *OR* ACTUATING DEVICE	CORROSIVE MATERIAL	NA2796	CORROSIVE	NONE	173.259	CORROSIVE
	BATTERY FLUID, ACID, WITH BATTERY, *ELECTRIC STORAGE, WET, EMPTY, OR DRY*	CORROSIVE MATERIAL	NA2796	CORROSIVE	NONE	173.258	CORROSIVE
	BATTERY FLUID, ALKALI	CORROSIVE MATERIAL	UN2797	CORROSIVE	173.244	173.257	CORROSIVE
	BATTERY FLUID, ALKALI, WITH ELECTRONIC EQUIPMENT *OR* ACTUATING DEVICE	CORROSIVE MATERIAL	NA2797	CORROSIVE	NONE	173.259	CORROSIVE
	BATTERY FLUID, ALKALI, WITH BATTERY, *ELECTRIC STORAGE, WET, EMPTY OR DRY*	CORROSIVE MATERIAL	UN2797	CORROSIVE	NONE	173.258	CORROSIVE
	BATTERY, LITHIUM. SEE 173.206(F)						
W	BATTERY PARTS *(PLATES, GRIDS, ETC. UNWASHED, EXHAUSTED)*	ORM-C		NONE	173.505	173.915	NONE
	BENZALDEHYDE	COMBUSTIBLE LIQUID	NA1989	NONE	173.118a	NONE	COMBUSTIBLE[1]
	BENZENE (*BENZOL*)	FLAMMABLE LIQUID	UN1114	FLAMMABLE LIQUID	173.118	173.119	FLAMMABLE
	BENZENE DIAZONIUM CHLORIDE (DRY)	FORBIDDEN					
	BENZENE DIAZONIUM NITRATE (DRY)	FORBIDDEN					
	BENZENE PHOSPHORUS DICHLORIDE	CORROSIVE MATERIAL	UN2798	CORROSIVE	NONE	173.250a	CORROSIVE
	BENZENE PHOSPHORUS THIODICHLORIDE	CORROSIVE MATERIAL	UN2799	CORROSIVE	NONE	173.250a	CORROSIVE
	BENZENETHIOL. SEE PHENYL MERCAPTAN						

1-SEE PLACARDING NOTES, PAGE 5

§172.101 HAZARDOUS MATERIALS TABLE

(1) +/ A/ W	(2) Hazardous materials descriptions and proper shipping names	(3) Hazard Class	(3A) Identi- fication number	(4) Label(s) required (if not excepted)	(5) Packaging (a) Exceptions	(5) Packaging (b) Specific require- ments	Placards required (*denotes placard for any quantity)
	BENZENE TRIOZONIDE	FORBIDDEN					
	BENZIDINE	POISON B	UN1885	POISON	173.364	173.365	POISON
	BENZINE	FLAMMABLE LIQUID	UN1115	FLAMMABLE LIQUID	173.118	173.119	FLAMMABLE
	BENZOIC DERIVATIVE PESTICIDE, LIQUID, N.O.S. (*COMPOUNDS AND PREPARATIONS*)	FLAMMABLE LIQUID	UN2770	FLAMMABLE LIQUID	173.118	173.119	FLAMMABLE
	BENZOIC DERIVATIVE PESTICIDE, LIQUID, N.O.S. (*COMPOUNDS AND PREPARATIONS*)	POISON B	UN2769	POISON	173.345	173.346	POISON
	BENZOIC DERIVATIVE PESTICIDE, SOLID, N.O.S. (*COMPOUNDS AND PREPARATIONS*)	POISON B	UN2769	POISON	173.364	173.365	POISON
	BENZONITRILE	COMBUSTIBLE LIQUID	UN2224	NONE	173.118b	NONE	COMBUSTIBLE[1]
	BENZOXIDIAZOLES (DRY)	FORBIDDEN					
	BENZOYL AZIDE	FORBIDDEN					
	BENZOYL CHLORIDE	CORROSIVE MATERIAL	UN1736	CORROSIVE	173.244	173.247	CORROSIVE
	BENZOYL PEROXIDE	ORGANIC PEROXIDE	NA2085	ORGANIC PEROXIDE	NONE	173.157 173.158	ORGANIC PEROXIDE
	BENZOYL PEROXIDE, *MORE THAN 77% BUT LESS THAN 95% WITH WATER. SEE* BENZOYL PEROXIDE.		UN2088				
	BENZOYL PEROXIDE, *NOT LESS THAN 30% BUT NOT MORE THAN 52% WITH INERT SOLID. SEE* ORGANIC PEROXIDE, SOLID, N.O.S.		UN2089				
	BENZOYL PEROXIDE, *NOT MORE THAN 72% AS A PASTE. SEE* ORGANIC PEROXIDE, SOLID, N.O.S.		UN2087				
	BENZOYL PEROXIDE, *NOT MORE THAN 77% WITH WATER. SEE* BENZOYL PEROXIDE.		UN2090				
	BENZOYL PEROXIDE, *TECHNICALLY PURE OR* BENZOYL PEROXIDE, *MORE THAN 52% WITH INERT SOLID. SEE* BENZOYL PEROXIDE		UN2085				
	BENZYL BROMIDE (*BROMOTOLUENE, ALPHA*)	CORROSIVE MATERIAL	UN1737	CORROSIVE	NONE	173.281	CORROSIVE
	BENZYL CHLORIDE	CORROSIVE MATERIAL	UN1738	CORROSIVE	173.244	173.295	CORROSIVE
	BENZYL CHLOROFORMATE (*BENZYL CHLOROCARBONATE*)	CORROSIVE MATERIAL	UN1739	CORROSIVE	NONE	173.288	CORROSIVE
	BERYLLIUM CHLORIDE	POISON B	NA1566	POISON	173.364	173.365	POISON
	BERYLLIUM COMPOUND, N.O.S.	POISON B	UN1566	POISON	173.364	173.365	POISON
	BERYLLIUM FLUORIDE	POISON B	NA1566	POISON	173.364	173.365	POISON
	BERYLLIUM NITRATE	OXIDIZER	UN2464	OXIDIZER	173.153	173.182	OXIDIZER
	BIPHENYL TRIOZONIDE	FORBIDDEN					
	BIPYRIDILIUM PESTICIDE, LIQUID, N.O.S. (*COMPOUNDS AND PREPARATIONS*)	FLAMMABLE LIQUID	UN2782	FLAMMABLE LIQUID	173.118	173.119	FLAMMABLE
	BIPYRIDILIUM PESTICIDE, LIQUID, N.O.S. (*COMPOUNDS AND PREPARATIONS*)	POISON B	UN2781	POISON	173.345	173.346	POISON
	BIPYRIDILIUM PESTICIDE, SOLID, N.O.S. (*COMPOUNDS AND PREPARATIONS*)	POISON B	UN2781	POISON	173.364	173.365	POISON

1 - SEE PLACARDING NOTES, PAGE 5

§172.101 HAZARDOUS MATERIALS TABLE

(1) +/ A/ W	(2) Hazardous materials descriptions and proper shipping names	(3) Hazard Class	(3A) Identi-fication number	(4) Label(s) required (if not excepted)	(5) Packaging (a) Exceptions	(5) Packaging (b) Specific require-ments	Placards required (*denotes placard for any quantity)
	BLACK POWDER	CLASS A EXPLOSIVE		EXPLOSIVE A	NONE	173.60	EXPLOSIVE A*
	BLACK POWDER IGNITER WITH EMPTY CARTRIDGE BAG.	CLASS C EXPLOSIVE		EXPLOSIVE C	NONE	173.106	DANGEROUS
	BLASTING AGENT, N.O.S.	BLASTING AGENT		BLASTING AGENT	NONE	173.114a	BLASTING AGENTS
	BLASTING CAPS. SEE DETONATORS, CLASS A *OR* CLASS C EXPLOSIVES						
	BLASTING CAPS, ELECTRIC. SEE DETONATORS, CLASS A *OR* CLASS C EXPLOSIVES						
	BLASTING CAPS, PERCUSSION ACTIVATED. SEE DETONATORS, CLASS A *OR* CLASS C EXPLOSIVES						
	BLASTING CAPS WITH DETONATING CORD. SEE DETONATORS, CLASS A *OR* CLASS C EXPLOSIVES *AND* DETONATING PRIMERS, CLASS A *OR* CLASS C EXPLOSIVES						
	BLASTING CAPS WITH METAL CLAD MILD DETONATING FUZE. SEE DETONATORS, CLASS A *OR* CLASS C EXPLOSIVES						
	BLASTING CAPS WITH SAFETY FUSE. SEE DETONATORS, CLASS A *OR* CLASS C EXPLOSIVES						
	BLASTING CAPS WITH SHOCK TUBES. SEE DETONATORS, CLASS A *OR* CLASS C EXPLOSIVES						
	BLASTING GELATIN. SEE HIGH EXPLOSIVE						
	BLASTING POWDER. SEE BLACK POWDER						
W	BLEACHING POWDER, *CONTAINING 39% OR LESS AVAILABLE CHLORINE*	ORM-C	UN2208	NONE	173.505	173.920	NONE
	BOMB, EXPLOSIVE. SEE EXPLOSIVE BOMB						
	BOMB, EXPLOSIVE WITH GAS, SMOKE, OR INCENDIARY MATERIAL. SEE EXPLOSIVE BOMB						
	BOMB, FIREWORKS. SEE FIREWORKS, SPECIAL						
	BOMB, GAS, SMOKE, OR INCENDIARY, NONEXPLOSIVE. SEE CHEMICAL AMMUNITION, NONEXPLOSIVE						
	BOMB, INCENDIARY, OR SMOKE WITHOUT BURSTING CHARGE. SEE FIREWORKS, SPECIAL						
	BOMB, PRACTICE, WITH ELECTRIC PRIMER OR ELECTRIC SQUIB (NON-EXPLOSIVE). SEE 173.55						
	BOMB, SAND-LOADED OR EMPTY (NON-EXPLOSIVE). SEE 173.55.						
	BOOSTER, EXPLOSIVE	CLASS A EXPLOSIVE		EXPLOSIVE A	NONE	173.69	EXPLOSIVES A*
	BORDEAUX ARSENITE, LIQUID	POISON B	NA2759	POISON	173.345	173.346	POISON
	BORDEAUX ARSENITE, SOLID	POISON B	NA2759	POISON	173.364	173.365	POISON
	BORON TRIBROMIDE	CORROSIVE MATERIAL	UN2692	CORROSIVE	NONE	173.251	CORROSIVE
	BORON TRICHLORIDE	CORROSIVE MATERIAL	UN1741	CORROSIVE	NONE	173.251	CORROSIVE
+	BORON TRIFLUORIDE	NONFLAMMABLE GAS	UN1008	NONFLAMMABLE GAS AND POISON	NONE	173.302	NONFLAMMABLE GAS

§172.101 HAZARDOUS MATERIALS TABLE

(1) +/ A/ W	(2) Hazardous materials descriptions and proper shipping names	(3) Hazard Class	(3A) Identification number	(4) Label(s) required (if not excepted)	(5) Packaging (a) Exceptions	(5) Packaging (b) Specific requirements	Placards required (*denotes placard for any quantity)
	BORON TRIFLUORIDE-ACETIC ACID COMPLEX	CORROSIVE MATERIAL	UN1742	CORROSIVE	173.244	173.247	CORROSIVE
	BROMINE	CORROSIVE MATERIAL	UN1744	CORROSIVE	NONE	173.252	CORROSIVE
	BROMINE AZIDE	FORBIDDEN					
	BROMINE PENTAFLUORIDE	OXIDIZER	UN1745	OXIDIZER	NONE	173.246	OXIDIZER
	BROMINE TRIFLUORIDE	OXIDIZER	UN1746	OXIDIZER AND POISON	NONE	173.246	OXIDIZER[2]
	4-BROMO-1,2-DINITROBENZENE (UNSTABLE AT 59 DEG C.)	FORBIDDEN					
	BROMOACETIC ACID, SOLID	CORROSIVE MATERIAL	UN1938	CORROSIVE	173.244	173.245b	CORROSIVE
	BROMOACETIC ACID SOLUTION	CORROSIVE MATERIAL	UN1938	CORROSIVE	173.244	173.245	CORROSIVE
	BROMOACETONE, LIQUID	POISON A	UN1569	POISON GAS	NONE	173.329	POISON GAS*
	BROMOBENZENE	COMBUSTIBLE LIQUID	UN2514	NONE	173.118a	NONE	COMBUSTIBLE[1]
A	BROMOCHLOROMETHANE	ORM-A	UN1887	NONE	173.505	173.605	NONE
	BROMOSILANE	FORBIDDEN					
	BROMOTOLUENE, ALPHA. SEE BENZYL BROMIDE						
	BROMOTRIFLUOROMETHANE (*R-13B1 OR H-1301*)	NONFLAMMABLE GAS	UN1009	NONFLAMMABLE GAS	173.306	173.304 173.314 173.315	NONFLAMMABLE GAS
	BRUCINE, SOLID (*DIMETHOXY STRYCHNINE*)	POISON B	UN1570	POISON	173.364	173.365	POISON
	BURNT COTTON, NOT REPICKED	FLAMMABLE SOLID	NA1325	FLAMMABLE SOLID	NONE	173.159	FLAMMABLE SOLID[5]
	BURSTER, EXPLOSIVE	CLASS A EXPLOSIVE		EXPLOSIVE A	NONE	173.69	EXPLOSIVE A*
+	BUTADIENE, INHIBITED	FLAMMABLE GAS	UN1010	FLAMMABLE GAS	173.306	173.304 173.314 173.315	FLAMMABLE GAS
	BUTANE *OR* LIQUEFIED PETROLEUM GAS. *SEE* LIQUEFIED PETROLEUM GAS						
	1,2,4-BUTANETRIOL TRINITRATE	FORBIDDEN					
	TERT-BUTOXYCARBONYL AZIDE	FORBIDDEN					
	N-BUTYL-4,4-DI-(TERT-BUTYLPEROXY)VALERATE, *TECHNICALLY PURE. SEE* ORGANIC PEROXIDE, LIQUID *OR* SOLUTION, N.O.S.		UN2140				
	N-BUTYL-4,4-DI-(TERT-BUTYLPEROXY)VALERATE, *NOT MORE THAN 52% WITH INERT SOLID. SEE* ORGANIC PEROXIDE, SOLID, N.O.S.		UN2141				
	BUTYL ACETATE	FLAMMABLE LIQUID	UN1123	FLAMMABLE LIQUID	173.118	173.119	FLAMMABLE
	N-BUTYL ACID PHOSPHATE. *SEE* ACID BUTYL PHOSPHATE						
	BUTYL ALCOHOL	FLAMMABLE LIQUID	NA1120	FLAMMABLE LIQUID	173.118	173.125	FLAMMABLE
	BUTYLAMINE	FLAMMABLE LIQUID	UN1125	FLAMMABLE LIQUID	173.118	173.119	FLAMMABLE
	BUTYL BROMIDE, *NORMAL*	FLAMMABLE LIQUID	UN1126	FLAMMABLE LIQUID	173.118	173.119	FLAMMABLE

1, 2 - SEE PLACARDING NOTES, PAGE 5

(1) +/ A/ W	(2) Hazardous materials descriptions and proper shipping names	(3) Hazard Class	(3A) Identi-fication number	(4) Label(s) required (if not excepted)	(5) Packaging (a) Exceptions	(5) Packaging (b) Specific require-ments	Placards required (*denotes placard for any quantity)
	BUTYL CHLORIDE	FLAMMABLE LIQUID	UN1127	FLAMMABLE LIQUID	173.118	173.119	FLAMMABLE
	TERT-BUTYL CUMYL PEROXIDE, *TECHNICALLY PURE OR* TERT-BUTYL CUMENE PEROXIDE, *TECHNICALLY PURE. SEE* ORGANIC PEROXIDE, LIQUID *OR* SOLUTION, N.O.S.		UN2091				
	BUTYL ETHER	FLAMMABLE LIQUID	UN1149	FLAMMABLE LIQUID	173.118	173.119	FLAMMABLE
	BUTYL FORMATE	FLAMMABLE LIQUID	UN1128	FLAMMABLE LIQUID	173.118	173.119	FLAMMABLE
	TERT-BUTYL HYDROPEROXIDE, *MORE THAN 72% BUT NOT MORE THAN 90% WITH WATER. SEE* ORGANIC PEROXIDE, LIQUID *OR* SOLUTION, N.O.S.		UN2094				
	TERT-BUTYL HYDROPEROXIDE, *NOT MORE THAN 72% WITH WATER. SEE* ORGANIC PEROXIDE, LIQUID *OR* SOLUTION, N.O.S.		UN2093				
	TERT-BUTYL HYDROPEROXIDE, *NOT MORE THAN 80% IN DI-TERT-BUTYL PEROXIDE AND SOLVENT. SEE* ORGANIC PEROXIDE, LIQUID *OR* SOLUTION, N.O.S.		UN2092				
	TERT-BUTYL HYDROPEROXIDE, *NOT MORE THAN 80% IN DI-TERT-BUTYL PEROXIDE OR SOLVENT. SEE* ORGANIC PEROXIDE, LIQUID *OR* SOLUTION, N.O.S.		UN2092				
	TERT-BUTYL HYDROPEROXIDE, MORE THAN 90% WITH WATER	FORBIDDEN					
	N-BUTYL ISOCYANATE	FLAMMABLE LIQUID	UN2485	FLAMMABLE LIQUID AND POISON	NONE	173.119 173.3a	FLAMMABLE[2]
	TERT-BUTYL ISOPROPYL BENZENE HYDROPEROXIDE	ORGANIC PEROXIDE	NA2091	ORGANIC PEROXIDE	173.153	173.224	ORGANIC PEROXIDE
	BUTYL MERCAPTAN	FLAMMABLE LIQUID	UN2347	FLAMMABLE LIQUID	NONE	173.141	FLAMMABLE
	TERT-BUTYL PEROXY-2-ETHYLHEXANOATE, *TECHNICALLY PURE. SEE* ORGANIC PEROXIDE, LIQUID *OR* SOLUTION, N.O.S.		UN2143				
	TERT-BUTYL PEROXY-2-ETHYLHEXANOATE, *NOT MORE THAN 30%* WITH 2,2-DI-(TERT-BUTYLPEROXY)BUTANE, *NOT MORE THAN 35%, WITH NOT LESS THAN 35% PHLEGMATIZER. SEE* ORGANIC PEROXIDE, LIQUID *OR* SOLUTION, N.O.S.		UN2886				
	TERT-BUTYL PEROXY-2-ETHYLHEXANOATE, *NOT MORE THAN 12%* WITH 2,2-DI-(TERT-BUTYLPEROXY)BUTANE, *NOT MORE THAN 14% WITH NOT LESS THAN 14% PHLEGMATIZER AND 60% INERT INORGANIC SOLID. SEE* ORGANIC PEROXIDE, SOLID, N.O.S.		UN2887				
	TERT-BUTYL PEROXY-2-ETHYLHEXANOATE, *NOT MORE THAN 50% WITH PHLEGMATIZER. SEE* ORGANIC PEROXIDE, LIQUID *OR* SOLUTION, N.O.S.		UN2888				
	TERT-BUTYL PEROXY-3,5,5-TRIMETHYLHEXANOATE *OR* TERT-BUTYL PEROXYISONONANOATE, *TECHNICALLY PURE. SEE* ORGANIC PEROXIDE, LIQUID *OR* SOLUTION, N.O.S.		UN2104				
	3-TERT-BUTYL PEROXY-3-PHENYLPHTHALIDE, *TECHNICALLY PURE. SEE* ORGANIC PEROXIDE, SOLID, N.O.S.		UN2596				
	TERT-BUTYL PEROXYACETATE, *NOT MORE THAN 76% IN SOLUTION. SEE* ORGANIC PEROXIDE, LIQUID *OR* SOLUTION, N.O.S.		UN2095				

2-SEE PLACARDING NOTES, PAGE 5

(1) +/ A/ W/	(2) Hazardous materials descriptions and proper shipping names	(3) Hazard Class	(3A) Identi- fication number	(4) Label(s) required (if not excepted)	(5) Packaging (a) Exceptions	(5) Packaging (b) Specific require- ments	Placards required (*denotes placard for any quantity)
	TERT-BUTYL PEROXYACETATE, *NOT MORE THAN 52% IN SOLUTION. SEE* ORGANIC PEROXIDE, LIQUID *OR* SOLUTION, N.O.S.		UN2096				
	TERT-BUTYL PEROXYACETATE, MORE THAN 76% IN SOLUTION	FORBIDDEN					
	TERT-BUTYL PEROXYBENZOATE, *NOT MORE THAN 75% IN SOLUTION. , SEE* ORGANIC PEROXIDE, LIQUID *OR* SOLUTION, N.O.S.		UN2098				
	TERT-BUTYL PEROXYBENZOATE, *NOT MORE THAN 50% WITH INERT IN ORGANIC SOLID. SEE* ORGANIC PEROXIDE, SOLID, N.O.S.		UN2890				
	TERT-BUTYL PEROXYBENZOATE, *TECHNICALLY PURE OR* TERT-BUTYL PEROXYBENZOATE, *MORE THAN 75% IN SOLUTION. SEE* ORGANIC PEROXIDE, LIQUID *OR* SOLUTION, N.O.S.		UN2097				
	TERT-BUTYL PEROXYCROTONATE, *NOT MORE THAN 76% IN SOLUTION. SEE* ORGANIC PEROXIDE, LIQUID *OR* SOLUTION, N.O.S.		UN2183				
	N-BUTYL PEROXYDICARBONATE, *NOT MORE THAN 52% IN SOLUTION. , SEE* ORGANIC PEROXIDE, LIQUID *OR* SOLUTION, N.O.S.		UN2169				
	N-BUTYL PEROXYDICARBONATE, *NOT MORE THAN 27% IN SOLUTION. SEE* ORGANIC PEROXIDE, LIQUID *OR* SOLUTION, N.O.S.		UN2170				
	N-BUTYL PEROXYDICARBONATE, MORE THAN 52% IN SOLUTION.	*FORBIDDEN*					
	TERT-BUTYL PEROXYDIETHYLACETATE, *33%* WITH TERT-BUTYL PEROXYBENZOATE, *33%, AND SOLVENT. SEE* ORGANIC PEROXIDE, LIQUID *OR* SOLUTION, N.O.S.		UN2551				
	TERT-BUTYL PEROXYDIETHYLACETATE, *TECHNICALLY PURE. SEE* ORGANIC PEROXIDE, LIQUID *OR* SOLUTION, N.O.S.		UN2144				
	TERT-BUTYL PEROXYISOBUTYRATE, *MORE THAN 52% BUT NOT MORE THAN 77% IN SOLUTION. SEE* ORGANIC PEROXIDE, LIQUID *OR* SOLUTION, N.O.S.		UN2142				
	TERT-BUTYL PEROXYISOBUTYRATE, *NOT MORE THAN 52% IN SOLUTION. SEE* ORGANIC PEROXIDE, LIQUID *OR* SOLUTION, N.O.S.		UN2562				
	TERT-BUTYL PEROXYISOBUTYRATE, MORE THAN 77% IN SOLUTION	FORBIDDEN					
	TERT-BUTYL PEROXYISOPROPYL CARBONATE, *TECHNICALLY PURE. SEE* ORGANIC PEROXIDE, LIQUID *OR* SOLUTION, N.O.S.		UN2103				
	TERT-BUTYL PEROXYMALEATE, *NOT MORE THAN 55% IN SOLUTION. SEE* ORGANIC PEROXIDE, LIQUID *OR* SOLUTION, N.O.S.		UN2100				
	TERT-BUTYL PEROXYMALEATE, *NOT MORE THAN 55% AS A PASTE. SEE* ORGANIC PEROXIDE, SOLID, N.O.S.		UN2101				
	TERT-BUTYL PEROXYMALEATE, *TECHNICALLY PURE. SEE* ORGANIC PEROXIDE, SOLID, N.O.S.		UN2099				
	TERT-BUTYL PEROXYNEODECANOATE, *NOT MORE THAN 77% IN SOLUTION. SEE* ORGANIC PEROXIDE, LIQUID *OR* SOLUTION, N.O.S.		UN2177				

§172.101 HAZARDOUS MATERIALS TABLE

(1) +/ A/ W	(2) Hazardous materials descriptions and proper shipping names	(3) Hazard Class	(3A) Identi-fication number	(4) Label(s) required (if not excepted)	(5) Packaging (a) Exceptions	 (b) Specific require-ments	Placards required (*denotes placard for any quantity)
	TERT-BUTYL PEROXYNEODECANOATE, *TECHNICALLY PURE. SEE* ORGANIC PEROXIDE, LIQUID *OR* SOLUTION, N.O.S.		UN2594				
	TERT-BUTYL PEROXYPHTHALATE, *TECHNICALLY PURE. SEE* ORGANIC PEROXIDE, SOLID, N.O.S.		UN2105				
	TERT-BUTYL PEROXYPIVALATE, *NOT MORE THAN 77% IN SOLUTION.* , *SEE* ORGANIC PEROXIDE, LIQUID *OR* SOLUTION, N.O.S.		UN2110				
	BUTYL PHOSPHORIC ACID. *SEE* ACID BUTYL PHOSPHATE						
	BUTYL TRICHLOROSILANE	CORROSIVE MATERIAL	UN1747	CORROSIVE	NONE	173.280	CORROSIVE
	BUTYRALDEHYDE	FLAMMABLE LIQUID	UN1129	FLAMMABLE LIQUID	173.118	173.119	FLAMMABLE
	BUTYRIC ACID	CORROSIVE MATERIAL	UN2820	CORROSIVE	173.244	173.245	CORROSIVE
	CABAZIDE	FORBIDDEN					
	CALCIUM ARSENATE, SOLID	POISON B	UN1573	POISON	173.364	173.367 173.368	POISON
	CALCIUM ARSENITE, SOLID	POISON B	NA1574	POISON	173.364	173.365	POISON
	CALCIUM BISULFITE SOLUTION. *SEE* CALCIUM HYDROGEN SULFITE SOLUTION						
	CALCIUM CARBIDE	FLAMMABLE SOLID	UN1402	FLAMMABLE SOLID AND DANGEROUS WHEN WET	NONE	173.178	FLAMMABLE SOLID -W-*
	CALCIUM CHLORATE	OXIDIZER	UN1452	OXIDIZER	173.153	173.163	OXIDIZER
	CALCIUM CHLORITE	OXIDIZER	UN1453	OXIDIZER	NONE	173.160	OXIDIZER
AW	CALCIUM CYANAMIDE, NOT HYDRATED (*CONTAINING MORE THAN 0.1% CALCIUM CARBIDE*)	ORM-C	UN1403	NONE	NONE	173.945	NONE
	CALCIUM CYANIDE, SOLID *OR* CALCIUM CYANIDE MIXTURE, SOLID	POISON B	UN1575	POISON	173.370	173.370	POISON
	CALCIUM HYDROGEN SULFITE SOLUTION	CORROSIVE MATERIAL	NA2693	CORROSIVE	173.244	173.245	CORROSIVE
	CALCIUM HYPOCHLORITE, HYDRATED (*MINIMUM 5.5% BUT NOT MORE THAN 10% WATER, AND CONTAINING MORE THAN 39% AVAILABLE CHLORINE*)	OXIDIZER	UN2880	OXIDIZER	173.153	173.217	OXIDIZER
	CALCIUM HYPOCHLORITE MIXTURE, *DRY. (CONTAINING MORE THAN 39% AVAILABLE CHLORINE)*	OXIDIZER	UN1748	OXIDIZER	173.153	173.217	OXIDIZER
	CALCIUM, METAL	FLAMMABLE SOLID	UN1401	FLAMMABLE SOLID AND DANGEROUS WHEN WET	173.153	173.154	FLAMMABLE SOLID -W-*
	CALCIUM, METAL, CRYSTALLINE	FLAMMABLE SOLID	NA1401	FLAMMABLE SOLID AND DANGEROUS WHEN WET	NONE	173.231	FLAMMALE SOLID -W-*
	CALCIUM NITRATE (*SEE 173.182(A) NOTE*)	OXIDIZER	UN1454	OXIDIZER	173.153	173.182	OXIDIZER
AW	CALCIUM OXIDE	ORM-B	UN1910	NONE	173.505	173.850	NONE
	CALCIUM PERMANGANATE	OXIDIZER	UN1456	OXIDIZER	173.153	173.154	OXIDIZER
	CALCIUM PEROXIDE	OXIDIZER	UN1457	OXIDIZER	173.153	173.156	OXIDIZER

§172.101 HAZARDOUS MATERIALS TABLE

(1) +/ A/ W	(2) Hazardous materials descriptions and proper shipping names	(3) Hazard Class	(3A) Identi-fication number	(4) Label(s) required (if not excepted)	(5) Packaging (a) Exceptions	 (b) Specific require-ments	Placards required (*denotes placard for any quantity)
	CALCIUM PHOSPHIDE	FLAMMABLE SOLID	UN1360	FLAMMABLE SOLID AND DANGEROUS WHEN WET	NONE	173.161	FLAMMABLE SOLID -W-*
	CALCIUM RESINATE	FLAMMABLE SOLID	UN1313	FLAMMABLE SOLID	NONE	173.166	FLAMMABLE SOLID[5]
	CALCIUM RESINATE, FUSED	FLAMMABLE SOLID	UN1314	FLAMMABLE SOLID	NONE	173.166	FLAMMABLE SOLID[5]
	CALCIUM SILICON *(POWDER)*	FLAMMABLE SOLID	UN1406	FLAMMABLE SOLID AND DANGEROUS WHEN WET	173.153	173.178	FLAMMABLE SOLID -W-*
AW	CAMPHENE	ORM-A	NA9011	NONE	173.505	173.610	NONE
	CAMPHOR OIL	COMBUSTIBLE LIQUID	UN1130	NONE	173.118a	NONE	COMBUSTIBLE[1]
	CANNON PRIMER	CLASS C EXPLOSIVE		NONE	NONE	173.107	NONE
	CAPRYLYL PEROXIDE SOLUTION	ORGANIC PEROXIDE	NA2129	ORGANIC PEROXIDE	173.153	173.221	ORGANIC PEROXIDE
	CAPS, BLASTING. SEE DETONATORS, CLASS A *OR* CLASS C EXPLOSIVES						
	CAPS, TOY. SEE TOY CAPS						
	CARBAMATE PESTICIDE, LIQUID, N.O.S. *(COMPOUNDS AND PREPARATIONS)*	FLAMMABLE LIQUID	UN2758	FLAMMABLE LIQUID	173.118	173.119	FLAMMABLE
	CARBAMATE PESTICIDE, LIQUID, N.O.S. *(COMPOUNDS AND PREPARATIONS)*	POISON B	UN2757	POISON	173.345	173.346	POISON
	CARBAMATE PESTICIDE, SOLID, N.O.S. *(COMPOUNDS AND PREPARATIONS)*	POISON B	UN2757	POISON	173.364	173.365	POISON
A	CARBARYL	ORM-A	NA2757	NONE	173.505	173.510	NONE
	CARBOFURAN	POISON B	NA2757	POISON	173.364	173.365	POISON
	CARBOFURAN MIXTURE, LIQUID	POISON B	NA2757	POISON	173.345	173.346	POISON
+	CARBOLIC ACID. *SEE* PHENOL						
	CARBOLIC ACID, LIQUID *(LIQUID TAR ACID CONTAINING OVER 50% PHENOL). SEE* PHENOL, LIQUID						
	CARBON BISULFIDE, *OR* CARBON DISULFIDE	FLAMMABLE LIQUID	UN1131	FLAMMABLE LIQUID	NONE	173.121	FLAMMABLE
	CARBON DIOXIDE	NONFLAMM-ABLE GAS	UN1013	NONFLAMM-ABLE GAS	173.306	173.302 173.304	NONFLAMM-ABLE GAS
	CARBON DIOXIDE-NITROUS OXIDE MIXTURE	NONFLAMM-ABLE GAS	UN1015	NONFLAMM-ABLE GAS	173.306	173..304	NONFLAMM-ABLE GAS
	CARBON DIOXIDE-OXYGEN MIXTURE	NONFLAMM-ABLE GAS	UN1014	NONFLAMM-ABLE GAS	173.306	173.304	NONFLAMM-ABLE GAS
	CARBON DIOXIDE, REFRIGERATED LIQUID	NONFLAMM-ABLE GAS	UN2187	NONFLAMM-ABLE GAS	173.306	173.314 173.315	NONFLAMM-ABLE GAS
AW	CARBON DIOXIDE, SOLID, *OR* DRY ICE, *OR* CARBONICE	ORM-A	UN1845	NONE	NONE	173.615	NONE
+	CARBON MONOXIDE	FLAMMABLE GAS	UN1016	FLAMMABLE GAS	173.306	173.302	FLAMMABLE GAS
	CARBON MONOXIDE, CRYOGENIC LIQUID	FLAMMABLE GAS	NA9202	FLAMMABLE GAS	NONE	173.318	FLAMMABLE GAS
AW	CARBON TETRACHLORIDE	ORM-A	UN1846	NONE	173.505	173.620	NONE
	CARBONYL CHLORIDE. SEE PHOSGENE						

1,5-SEE PLACARDING NOTES, PAGE 5

§172.101 HAZARDOUS MATERIALS TABLE

(1) +/ A/ W	(2) Hazardous materials descriptions and proper shipping names	(3) Hazard Class	(3A) Identi-fication number	(4) Label(s) required (if not excepted)	(5) Packaging (a) Exceptions	(5) Packaging (b) Specific require-ments	Placards required (*denotes placard for any quantity)
	CARBOYS, EMPTY, MUST BE CLASSED FOR THE HAZARDOUS MATERIAL PREVIOUSLY CONTAINED IN CARBOY. SEE 173.29						
	CARTRIDGE BAGS, EMPTY, WITH BLACK POWDER IGNITER	CLASS C EXPLOSIVE		EXPLOSIVE C	NONE	173.106	DANGEROUS
	CARTRIDGE, PRACTICE AMMUNITION	CLASS C EXPLOSIVE		EXPLOSIVE C	NONE	173.101a	DANGEROUS
	CASE OIL. SEE GASOLINE *OR* NAPTHA						
	CASINGHEAD GASOLINE. SEE GASOLINE						
W	CASTOR BEANS	ORM-C		NONE	173.505	173.952	NONE
W	CASTOR POMACE. *SEE* CASTOR BEANS						
	CAUSTIC, POTASH, DRY, SOLID, FLAKE, BEAD, *OR* GRANULAR. *SEE* POTASSIUM HYDROXIDE, DRY, *ETC.*						
	CAUSTIC POTASH, LIQUID *OR* SOLUTION. *SEE* POTASSIUM HYDROXIDE SOLUTION						
	CAUSTIC SODA, DRY, SOLID, FLAKE, BEAD, *OR* GRANULAR. *SEE* SODIUM HYDROXIDE, DRY, *ETC.*						
	CAUSTIC SODA, LIQUID *OR* SOLUTION. *SEE* SODIUM HYDROXIDE SOLUTION						
W	*CELLOSOLVE. SEE* ETHYLENE GLYCOL MONOETHYL ETHER						
W	*CELLOSOLVE ACETATE. SEE* ETHYLENE GLYCOL MONOETHYL ETHER , ACETATE						
	CEMENT	COMBUSTIBLE LIQUID	NA1133	NONE	173.118a	NONE	COMBUSTIBLE[1]
	CEMENT	FLAMMABLE LIQUID	NA1133	FLAMMABLE LIQUID	173.118	173.132	FLAMMABLE
	CEMENT, ADHESIVE, N.O.S. *SEE* CEMENT						
	CEMENT, CONTAINER, LINOLEUM, TILE, *OR* WALLBOARD, LIQUID	FLAMMABLE LIQUID	NA1133	FLAMMABLE LIQUID	173.118	173.132	FLAMMABLE
	CEMENT, LEATHER	FLAMMABLE LIQUID	NA1133	FLAMMABLE LIQUID	173.118	173.119	FLAMMABLE
	CEMENT, PYROXYLIN	FLAMMABLE LIQUID	NA1133	FLAMMABLE LIQUID	173.118	173.132	FLAMMABLE
	CEMENT, ROOFING, LIQUID	FLAMMABLE LIQUID	NA1133	FLAMMABLE LIQUID	173.118	173.119	FLAMMABLE
	CEMENT, RUBBER	FLAMMABLE LIQUID	NA1133	FLAMMABLE LIQUID	173.118	173.132	FLAMMABLE
	CESIUM METAL	FLAMMABLE SOLID	UN1407	FLAMMABLE SOLID AND DANGEROUS WHEN WET	NONE	173.206	FLAMMABLE SOLID -W-*
	CHARCOAL BRIQUETTES *OR* BRIQUETS	FLAMMABLE SOLID	NA1361	FLAMMABLE SOLID	173.162	173.162	FLAMMABLE SOLID[5]
	CHARCOAL SCREENINGS, MADE FROM 'PINON' WOOD	FLAMMABLE SOLID	NA1361	FLAMMABLE SOLID	173.162	173.162	FLAMMABLE SOLID[5]
	CHARCOAL, SHELL	FLAMMABLE SOLID	NA1361	FLAMMABLE SOLID	173.162	173.162	FLAMMABLE SOLID[5]
	CHARCOAL, WOOD, GROUND, CRUSHED, GRANULATED, *OR* PULVERIZED	FLAMMABLE SOLID	NA1361	FLAMMABLE SOLID	173.162	173.162	FLAMMABLE SOLID[5]
	CHARCOAL, WOOD, LUMP	FLAMMABLE SOLID	NA1361	FLAMMABLE SOLID	173.162	173.162	FLAMMABLE SOLID[5]
	CHARCOAL WOOD SCREENINGS, OTHER THAN 'PINON' WOOD SCREENINGS	FLAMMABLE SOLID	NA1361	FLAMMABLE SOLID	NONE	173.162	FLAMMABLE SOLID[5]

1, 5 - SEE PLACARDING NOTES, PAGE 5

§172.101 HAZARDOUS MATERIALS TABLE

(1) +/ A/ W	(2) Hazardous materials descriptions and proper shipping names	(3) Hazard Class	(3A) Identification number	(4) Label(s) required (if not excepted)	(5) Packaging (a) Exceptions	(5) Packaging (b) Specific requirements	Placards required (*denotes placard for any quantity)
	CHARGED WELL CASING JET PERFORATING GUN (*TOTAL EXPLOSIVE CONTENTS IN GUNS 20 POUNDS OR MORE PER MOTOR VEHICLE*)	CLASS A EXPLOSIVE		EXPLOSIVE A	NONE	173.53 173.80	EXPLOSIVE A
	CHARGED WELL CASING JET PERFORATING GUN (*TOTAL EXPLOSIVE CONTENTS IN GUNS NOT EXCEEDING 20 POUNDS PER MOTOR VEHICLE OR SPECIAL OFFSHORE DOWN HOLE TOOL PALLET*)	CLASS C EXPLOSIVE		EXPLOSIVE C	NONE	173.53 173.110	DANGEROUS
	CHEMICAL AMMUNITION, EXPLOSIVE. SEE AMMUNITION, CHEMICAL, EXPLOSIVE, WITH						
	CHEMICAL AMMUNITION, NONEXPLOSIVE (*CONTAINING A POISON B MATERIAL*)	POISON B	UN2016	POISON	173.345	173.350	POISON
	CHEMICAL AMMUNITION, NONEXPLOSIVE (*CONTAINING AN IRRITATING MATERIAL*)	IRRITATING MATERIAL	UN2017	IRRITANT	NONE	173.383	DANGEROUS
	CHEMICAL AMMUNITION, NONEXPLOSIVE (*CONTAINING A POISON A MATERIAL*)	POISON A	UN2016	POISON GAS	NONE	173.330	POISON GAS*
	CHEMICAL KIT	CORROSIVE MATERIAL	NA1760	CORROSIVE	173.286		CORROSIVE
	CHLORATE AND BORATE MIXTURE (*CONTAINING MORE THAN 28% CHLORATE*)	OXIDIZER	UN1458	OXIDIZER	173.153	173.229	OXIDIZER
	CHLORATE AND MAGNESIUM CHLORIDE MIXTURE (*CONTAINING MORE THAN 28% CHLORATE*)	OXIDIZER	UN1459	OXIDIZER	173.153	173.229	OXIDIZER
	CHLORATE EXPLOSIVE, DRY. SEE HIGH EXPLOSIVE						
	CHLORATE, N.O.S.	OXIDIZER	UN1461	OXIDIZER	173.153	173.163	OXIDIZER
	CHLORATE, N.O.S., WET	OXIDIZER	NA1461	OXIDIZER	173.153	173.163	OXIDIZER
	CHLORATE OF POTASH. *SEE* POTASSIUM CHLORATE						
	CHLORATE OF SODA. *SEE* SODIUM CHLORATE						
	CHLORATE POWDER. *SEE* HIGH EXPLOSIVE						
	CHLORDANE, LIQUID	FLAMMABLE LIQUID	NA2762	FLAMMABLE LIQUID	173.118	173.119	FLAMMABLE
	CHLORDANE, LIQUID	COMBUSTIBLE LIQUID	NA2762	NONE	173.118a	NONE	COMBUSTIBLE[1]
	CHLORIC ACID	OXIDIZER	NA2626	OXIDIZER AND POISON	NONE	173.237	OZIDIXER
	CHLORIDE OF PHOSPHORUS. *SEE* PHOSPHORUS TRICHLORIDE						
	CHLORIDE OF SULFUR. *SEE* SULFUR CHLORIDE						
W	CHLORINATED LIME (*CHLORIDE OF LIME.) SEE* BLEACHING POWDER						
+	CHLORINE	NONFLAMMABLE GAS	UN1017	NONFLAMMABLE GAS AND POISON	NONE	173.304 173.314 173.315	NONFLAMMABLE GAS[4] OR CHLORINE[4]
	CHLORINE AZIDE	FORBIDDEN					
	CHLORINE DIOXIDE HYDRATE, FROZEN	OXIDIZER	NA9191	OXIDIZER AND POISON	NONE	173.237	OXIDIZER
	CHLORINE DIOXIDE (NOT HYDRATE)	FORBIDDEN					
	CHLORINE TRIFLUORIDE	OXIDIZER	UN1749	OXIDIZER AND POISON	NONE	173.246	OXIDIZER[2]
	CHLOROACETIC ACID, LIQUID *OR* SOLUTION	CORROSIVE MATERIAL	UN1750	CORROSIVE	173.244	173.294	CORROSIVE

1,2,4,5-SEE PLACARDING NOTES, PAGE 5

§172.101 HAZARDOUS MATERIALS TABLE

(1) +/ A/ W/	(2) Hazardous materials descriptions and proper shipping names	(3) Hazard Class	(3A) Identi-fication number	(4) Label(s) required (if not excepted)	(5) Packaging (a) Exceptions	(5) Packaging (b) Specific require-ments	Placards required (*denotes placard for any quantity)
	CHLOROACETIC ACID, SOLID	CORROSIVE MATERIAL	UN1751	CORROSIVE	173.244	173.245b	CORROSIVE
	CHLOROACETOPHENONE, GAS, LIQUID, *OR* SOLID (*CN*)	IRRITATING MATERIAL	UN1697	IRRITANT	NONE	173.382	DANGEROUS
	CHLOROACETYL CHLORIDE	CORROSIVE MATERIAL	UN1752	CORROSIVE	NONE	173.253	CORROSIVE
	CHLOROBENZENE	FLAMMABLE LIQUID	UN1134	FLAMMABLE LIQUID	173.118	173.119	FLAMMABLE
	CHLOROBENZOL. *SEE* CHLOROBENZENE						
	P-CHLOROBENZOYL PEROXIDE	ORGANIC PEROXIDE	UN2113	ORGANIC PEROXIDE	NONE	173.157 173.158	ORGANIC PEROXIDE
	P-CHLOROBENZOYL PEROXIDE, *NOT MORE THAN 75% WITH WATER. SEE* P-CHLOROBENZOYL PEROXIDE		UN2113				
	P-CHLOROBENZOYL PEROXIDE, *NOT MORE THAN 52% AS A PASTE. SEE* ORGANIC PEROXIDE, SOLID, N.O.S.		UN2114				
	P-CHLOROBENZOYL PEROXIDE, *NOT MORE THAN 52% IN SOLUTION. SEE* ORGANIC PEROXIDE, LIQUID *OR* SOLUTION, N.O.S.		UN2115				
	CHLORODINITROBENZENE. SEE DINITROCHLOROBENZENE						
	1-CHLORO-1,1-DIFLUOROETHANE. *SEE* CHLORODIFLUOROETHANE (*R-142*)						
+	CHLORODIFLUOROETHANE (*R-142B*) *OR* (1-CHLORO-1,1-DIFLUOROETHANE)	FLAMMABLE GAS	UN2517	FLAMMABLE GAS	173.306	173.304 173.314 173.315	FLAMMABLE GAS
	CHLORODIFLUOROMETHANE (*R-22*)	NONFLAMMABLE GAS	UN1018	NONFLAMMABLE GAS	173.306	173.304 173.314 173.315	NONFLAMMABLE GAS
	CHLORODIFLUOROMETHANE AND CHLOROPENTAFLUOROETHANE MIXTURE (*CONSTANT BOILING MIXTURE*) (*R-502*). *SEE* REFRIGERANT GAS, N.O.S.						
AW	CHLOROFORM	ORM-A	UN1888	NONE	173.505	173.630	NONE
+	4-CHLORO-O-TOLUIDINE HYDROCHLORIDE	POISON B	UN1579	POISON	NONE	173.362	POISON
	CHLOROPENTAFLUOROETHANE (*R-115*)	NONFLAMMABLE GAS	UN1020	NONFLAMMABLE GAS	173.306	173.304 173.314 173.315	NONFLAMMABLE GAS
	3-CHLOROPEROXYBENZOIC ACID, *NOT MORE THAN 86% WITH 3-CHLOROBENZOIC ACID SEE* ORGANIC PEROXIDE, SOLID, N.O.S.		UN2755				
	CHLOROPHENYLTRICHLOROSILANE	CORROSIVE MATERIAL	UN1753	CORROSIVE	NONE	173.280	CORROSIVE
+	CHLOROPICRIN, ABSORBED	POISON B	NA1583	POISON	NONE	173.357	POISON[2]
+	CHLOROPICRIN AND METHYL CHLORIDE MIXTURE	POISON A	UN1582	POISON GAS AND FLAMMABLE GAS	NONE	173.329	POISON GAS*
+	CHLOROPICRIN AND NONFLAMMABLE, NONLIQUEFIED COMPRESSED GAS MIXTURE	POISON A	NA1955	POISON GAS AND NONFLAMMABLE GAS	NONE	173.329	POISON GAS*
	CHLOROPICRIN MIXTURE, FLAMMABLE (*PRESSURE NOT EXCEEDING 14.7 PSIA; FLASH POINT BELOW 100° F*)	POISON B	NA2929	POISON AND FLAMMABLE LIQUID	NONE	173.357	POISON[2]
+	CHLOROPICRIN, LIQUID	POISON B	UN1580	POISON	NONE	173.357	POISON[2]
+	CHLOROPICRIN MIXTURE (*CONTAINING NO COMPRESSED GAS OR POISON A LIQUID*)	POISON B	UN1583	POISON	NONE	173.357	POISON[2]

2 - SEE PLACARDING NOTES, PAGE 5

§172.101 HAZARDOUS MATERIALS TABLE

(1) +/ A/ W	(2) Hazardous materials descriptions and proper shipping names	(3) Hazard Class	(3A) Identi- fication number	(4) Label(s) required (if not excepted)	(5) Packaging (a) Exceptions	(5) Packaging (b) Specific require- ments	Placards required (*denotes placard for any quantity)
A	CHLOROPLATINIC ACID, SOLID	ORM-B	UN2507	NONE	173.505	173.800	NONE
	CHLOROPRENE, INHIBITED	FLAMMABLE LIQUID	UN1991	FLAMMABLE LIQUID	173.118	173.119	FLAMMABLE
	CHLOROPRENE, UNINHIBITED	FORBIDDEN					
	2-CHLOROPROPENE	FLAMMABLE LIQUID	UN2456	FLAMMABLE LIQUID	NONE	173.119	FLAMMABLE
	CHLOROSULFONIC ACID	CORROSIVE MATERIAL	UN1754	CORROSIVE	173.244	173.254	CORROSIVE
	CHLOROSULFONIC ACID-SULFUR TRIOXIDE MIXTURE	CORROSIVE MATERIAL	UN1754	CORROSIVE	173.144	173.254	CORROSIVE
	CHLOROTETRAFLUOROETHANE (*R-124*)	NONFLAMM- ABLE GAS	UN1021	NONFLAMM- ABLE GAS	173.306	173.304 173.314	NONFLAMM- ABLE GAS
	CHLOROTRIFLUOROMETHANE (*R-13*)	NONFLAMM- ABLE GAS	UN1022	NONFLAMM- ABLE GAS	173.306	173.304 173.314 173.315	NONFLAMM- ABLE GAS
A	CHLORPYRIFOS	ORM-A	NA2783	NONE	173.505	173.510	NONE
	CHROMIC ACID MIXTURE, DRY	OXIDIZER	NA1463	OXIDIZER	173.153	173.164	OXIDIZER
	CHROMIC ACID, SOLID	OXIDIZER	NA1463	OXIDIZER	173.153	173.164	OXIDIZER
	CHROMIC ACID SOLUTION	CORROSIVE MATERIAL	UN1755	CORROSIVE	173.244	173.287	CORROSIVE
	CHROMIC ANHYDRIDE. *SEE* CHROMIC ACID, SOLID						
	CHROMIC FLUORIDE, SOLID	CORROSIVE MATERIAL	UN1756	CORROSIVE	173.244	173.245b	CORROSIVE
	CHROMIC FLUORIDE SOLUTION	CORROSIVE MATERIAL	UN1757	CORROSIVE	173.244	173.245	CORROSIVE
	CHROMIC TRIOXIDE. *SEE* CHROMIC ACID, SOLID						
	CHROMIUM OXYCHLORIDE *OR* CHROMYL CHLORIDE	CORROSIVE MATERIAL	UN1758	CORROSIVE	NONE	173.247	CORROSIVE
	CIGARETTE LIGHTER (*OR OTHER SIMILAR IGNITION DEVICE*)	FLAMMABLE GAS	UN1057	FLAMMABLE GAS	173.21 175.10	173.308	FLAMMABLE GAS
	CIGARETTE LIGHTER (*OR OTHER SIMILAR IGNITION DEVICE*)	FLAMMABLE LIQUID	UN1226	FLAMMABLE LIQUID	173.21 175.10	173.118	FLAMMABLE
	CIGARETTE LOAD	CLASS C EXPLOSIVE		EXPLOSIVE C	NONE	173.111	DANGEROUS
	CLOUD GAS CYLINDER. SEE CHEMICAL AMMUNITION, NONEXPLOSIVE						
	COAL BRIQUETTES, HOT	FORBIDDEN					
	COAL FACINGS. *SEE* COAL GROUND BITUMINOUS, *ETC.*						
	COAL GAS. SEE HYDROCARBON GAS, NONLIQUEFIED						
	COAL, GROUND BITUMINOUS, SEA COAL, COAL FACINGS, *ETC.*	FLAMMABLE SOLID	NA1361	FLAMMABLE SOLID	173.165		FLAMMABLE SOLID[5]
	COAL OIL (*EXPORT SHIPMENT ONLY*). *SEE* KEROSENE						
	COAL TAR DISTILLATE	COMBUSTIBLE LIQUID	UN1137	NONE	173.118a	NONE	COMBUSTIBLE[1]
	COAL TAR DISTILLATE	FLAMMABLE LIQUID	UN1136	FLAMMABLE LIQUID	173.118	173.119	FLAMMABLE
	COAL TAR DYE, LIQUID (*NOT OTHERWISE SPECIFICALLY NAMED IN 172.101*)	CORROSIVE MATERIAL	NA2801	CORROSIVE	173.244	173.249a	CORROSIVE
	COATING SOLUTION	FLAMMABLE LIQUID	UN1139	FLAMMABLE LIQUID	173.118	173.132	FLAMMABLE
	COBALT RESINATE, PRECIPITATED	FLAMMABLE SOLID	UN1318	FLAMMABLE SOLID	NONE	173.166	FLAMMABLE SOLID[5]

1,5-SEE PLACARDING NOTES, PAGE 5

§172.101 HAZARDOUS MATERIALS TABLE

(1) +/ A/ W	(2) Hazardous materials descriptions and proper shipping names	(3) Hazard Class	(3A) Identification number	(4) Label(s) required (if not excepted)	(5) Packaging (a) Exceptions	(5) Packaging (b) Specific requirements	Placards required (*denotes placard for any quantity)
	COCCULUS, SOLID (*FISHBERRY*)	POISON B	UN1584	POISON	173.364	173.365	POISON
W	COCONUT MEAL PELLETS *CONTAINING AT LEAST 6% AND NOT MORE THAN 13% MOISTURE AND NOT MORE THAN 10% RESIDUAL FAT CONTENT*	ORM-C		NONE	173.505	173.955	NONE
	COKE, HOT	FORBIDDEN					
	COLLODION	FLAMMABLE LIQUID	NA2059	FLAMMABLE LIQUID	173.118	173.119	FLAMMABLE
	COMBINATION FUZE	CLASS C EXPLOSIVE		EXPLOSIVE C	NONE	173.105	DANGEROUS
	COMBINATION PRIMER	CLASS C EXPLOSIVE		NONE	NONE	173.107	DANGEROUS
	COMBUSTIBLE LIQUID, N.O.S.	COMBUSTIBLE LIQUID	NA1993	NONE	173.118a	NONE	COMBUSTIBLE[1]
	COMMERCIAL SHAPED CHARGE. SEE HIGH EXPLOSIVE						
	COMMON FIREWORKS. *SEE* FIREWORKS, COMMON						
	COMPOUND, CLEANING, LIQUID	FLAMMABLE LIQUID	NA1993	FLAMMABLE LIQUID	173.118	173.119	FLAMMABLE
	COMPOUND, CLEANING, LIQUID	CORROSIVE MATERIAL	NA1760	CORROSIVE	173.244	173.245	CORROSIVE
	COMPOUND, CLEANING, LIQUID	COMBUSTIBLE LIQUID	NA1993	NONE	173.118a	NONE	COMBUSTIBLE[1]
	COMPOUND, CLEANING, LIQUID (*CONTAINING PHOSPHORIC ACID, ACETIC ACID, SODIUM HYDROXIDE OR POTASSIUM HYDROXIDE*)	CORROSIVE MATERIAL	NA1760	CORROSIVE	173.244	173.249a	CORROSIVE
	COMPOUND, CLEANING, LIQUID (CONTAINING HYDROCHLORIC (MURIATIC) ACID)	CORROSIVE MATERIAL	NA1789	CORROSIVE	173.244	173.263	CORROSIVE
	COMPOUND, CLEANING, LIQUID (CONTAINING HYDROFLUORIC ACID)	CORROSIVE MATERIAL	NA1790	CORROSIVE	172.244	173.256	CORROSIVE
	COMPOUND, POLISHING, LIQUID	FLAMMABLE LIQUID	NA1142	FLAMMABLE LIQUID	173.118	173.119	FLAMMABLE
	COMPOUND, RUST PREVENTING *OR* COMPOUND, RUST REMOVING	CORROSIVE MATERIAL	NA1760	CORROSIVE	173.244	173.245	CORROSIVE
	COMPOUND, TREE *OR* WEED KILLING, LIQUID	COMBUSTIBLE LIQUID	NA1993	NONE	173.118a	NONE	COMBUSTIBLE[1]
	COMPOUND, TREE *OR* WEED KILLING, LIQUID	CORROSIVE MATERIAL	NA1760	CORROSIVE	173.244	173.245	CORROSIVE
	COMPOUND, TREE *OR* WEED KILLING, LIQUID	FLAMMABLE LIQUID	NA1993	FLAMMABLE LIQUID	173.118	173.119	FLAMMABLE
	COMPOUND, TREE *OR* WEED KILLING, LIQUID	POISON B	NA2810	POISON	173.345	173.346	POISON
	COMPOUND, TREE *OR* WEED KILLING, SOLID	OXIDIZER	NA1479	OXIDIZER	173.153	173.154 173.229	OXIDIZER
	COMPOUND, VULCANIZING, LIQUID	CORROSIVE MATERIAL	NA1760	CORROSIVE	173.244	173.245	CORROSIVE
	COMPOUND, VULCANIZING, LIQUID	FLAMMABLE LIQUID	NA1142	FLAMMABLE LIQUID	173.118	172.119	FLAMMABLE
	COMPRESSED GAS, N.O.S.	FLAMMABLE GAS	UN1954	FLAMMABLE GAS	173.306	173.302 173.304 173.305	FLAMMABLE GAS
	COMPRESSED GAS, N.O.S.	NONFLAMMABLE GAS	UN1956	NONFLAMMABLE GAS	173.306 173.307	173.302 173.304 173.305	NONFLAMMABLE GAS
	CONSUMER COMMODITY	ORM-D		NONE	173.505(B)	173.510 173.1200	NONE
	CONTAINER, REUSED OR EMPTY, MUST BE CLASSED FOR THE HAZARDOUS MATERIAL PREVIOUSLY CONTAINED. SEE 173.28, 173.29						

1-SEE PLACARDING NOTES, PAGE 5

§172.101 HAZARDOUS MATERIALS TABLE

(1) +/ A/ W	(2) Hazardous materials descriptions and proper shipping names	(3) Hazard Class	(3A) Identi-fication number	(4) Label(s) required (if not excepted)	(5) Packaging (a) Exceptions	(5) Packaging (b) Specific require-ments	Placards required (*denotes placard for any quantity)
	COPPER ACETOARSENITE, SOLID	POISON B	UN1585	POISON	173.364	173.367	POISON
	COPPER ACETYLIDE	FORBIDDEN					
	COPPER AMINE AZIDE	FORBIDDEN					
	COPPER ARSENITE, SOLID	POISON B	UN1586	POISON	173.364	173.365	POISON
	COPPER BASED PESTICIDE, LIQUID, N.O.S. (*COMPOUNDS AND PREPARATIONS*)	FLAMMABLE LIQUID	UN2776	FLAMMABLE LIQUID	173.118	173.119	FLAMMABLE
	COPPER BASED PESTICIDE, LIQUID, N.O.S. (*COMPOUNDS AND PREPARATIONS*)	POISON B	UN2775	POISON	173.345	173.346	POISON
	COPPER BASED PESTICIDE, SOLID, N.O.S. (*COMPOUNDS AND PREPARATIONS*)	POISON B	UN2775	POISON	173.364	173.365	POISON
A	COPPER CHLORIDE	ORM-B	UN2802	NONE	173.505	173.800	NONE
	COPPER CYANIDE	POISON B	UN1587	POISON	173.370	173.370	POISON
	COPPER TETRAMINE NITRATE	FORBIDDEN					
W	COPRA	ORM-C	UN1363	NONE	173.505	173.960	NONE
	COPRA PELLETS. SEE COCONUT MEAL PELLETS						
	CORD, DETONATING *FLEXIBLE*	CLASS A EXPLOSIVE		EXPLOSIVE A	173.81	173.81	EXPLOSIVE A*
	CORD, DETONATING *FLEXIBLE*	CLASS C EXPLOSIVE		EXPLOSIVE C	NONE	173.104	DANGEROUS
	CORROSIVE LIQUID, N.O.S.	CORROSIVE MATERIAL	UN1760	CORROSIVE	173.244	173.245 173.245a	CORROSIVE
	CORROSIVE LIQUID, POISONOUS, N.O.S.	CORROSIVE MATERIAL	UN2922	CORROSIVE AND POISON	173.244	173.245	CORROSIVE
	CORROSIVE SOLID, N.O.S.	CORROSIVE MATERIAL	UN1759	CORROSIVE	173.244	173.245b	CORROSIVE
	COSMETICS, LIQUID, N.O.S.	CORROSIVE MATERIAL	NA1760	CORROSIVE	173.244	173.245	CORROSIVE
	COSMETICS, N.O.S.	COMBUSTIBLE LIQUID	NA1993	NONE	173.118a	NONE	COMBUSTIBLE[1]
	COSMETICS, N.O.S.	FLAMMABLE LIQUID	NA1993	FLAMMABLE LIQUID	173.118	173.119	FLAMMABLE
	COSMETICS, N.O.S.	FLAMMABLE SOLID	NA1325	FLAMMABLE SOLID	173.153	173.154	FLAMMABLE SOLID[5]
	COSMETICS, N.O.S.	OXIDIZER	NA1479	OXIDIZER	173.153	173.154	OXIDIZER
	COSMETICS, SOLID, N.O.S.	CORROSIVE MATERIAL	NA1759	CORROSIVE	173.244	173.245b	CORROSIVE
W	COTTON	ORM-C		NONE	173.505	173.965	NONE
	COUMAPHOS	POISON B	NA2783	POISON	173.364	173.365	POISON
	COUMAPHOS MIXTURE, LIQUID	POISON B	NA2783	POISON	173.345	173.346	POISON
	CRESOL	CORROSIVE MATERIAL	UN2076	CORROSIVE	173.244	173.245	CORROSIVE
	CROTONALDEHYDE	FLAMMABLE LIQUID	UN1143	FLAMMABLE LIQUID AND POISON	NONE	173.119 173.3a	FLAMMABLE[2]
	CROTONIC ACID	CORROSIVE MATERIAL	UN2823	CORROSIVE	173.244	173.245	CORROSIVE
	CROTONYLENE	FLAMMABLE LIQUID	UN1144	FLAMMABLE LIQUID	173.118	173.119	FLAMMABLE
	CRUDE OIL, PETROLEUM	COMBUSTIBLE LIQUID	UN1267	NONE	173.118a	NONE	COMBUSTIBLE[1]
	CRUDE OIL, PETROLEUM	FLAMMABLE LIQUID	UN1267	FLAMMABLE LIQUID	173.118	173.119	FLAMMABLE

1,5-SEE PLACARDING NOTES, PAGE 5

§172.101 HAZARDOUS MATERIALS TABLE

(1) +/A/W	(2) Hazardous materials descriptions and proper shipping names	(3) Hazard Class	(3A) Identification number	(4) Label(s) required (if not excepted)	(5) Packaging (a) Exceptions	(5) Packaging (b) Specific requirements	Placards required (*denotes placard for any quantity)
	CUMENE HYDROPEROXIDE	ORGANIC PEROXIDE	UN2116	ORGANIC PEROXIDE	173.153	173.224	ORGANIC PEROXIDE
	CUMENE HYDROPEROXIDE, *TECHNICALLY PURE. SEE* CUMENE HYDROPEROXIDE		UN2116				
	CUPRIC CYANIDE. SEE COPPER CYANIDE						
	CUPRIC NITRATE	OXIDIZER	NA1479	OXIDIZER	173.153	173.182	OXIDIZER
	CUPRIETHYLENE-DIAMINE SOLUTION	CORROSIVE MATERIAL	UN1761	CORROSIVE	173.244	173.249	CORROSIVE
	CYANIDE *OR* CYANIDE MIXTURE, DRY	POISON B	UN1588	POISON	173.364	173.370	POISON
	CYANIDE SOLUTION, N.O.S.	POISON B	UN1935	POISON	173.345	173.352	POISON
	CYANOGEN BROMIDE	POISON B	UN1889	POISON	NONE	173.379	POISON
	CYANOGEN CHLORIDE *CONTAINING LESS THAN 0.9% WATER*	POISON A	UN1589	POISON GAS AND FLAMMABLE GAS	NONE	173.328	POISON GAS*
	CYANOGEN GAS	POISON A	UN1026	POISON GAS AND FLAMMABLE GAS	NONE	173.328	POISON GAS*
	CYANURIC TRIAZIDE	FORBIDDEN					
	CYCLOHEXANE	FLAMMABLE LIQUID	UN1145	FLAMMABLE LIQUID	173.118	173.119	FLAMMABLE
	CYCLOHEXANONE PEROXIDE, *50 TO 85% PEROXIDE*	ORGANIC PEROXIDE	UN2119	ORGANIC PEROXIDE	173.157	173.158	ORGANIC PEROXIDE
	CYCLOHEXANONE PEROXIDE, *AS A PASTE WITH NOT MORE THAN 9% BY WEIGHT ACTIVE OXYGEN. SEE* CYCLOHEXANONE PEROXIDE, *50 TO 85% PEROXIDE.*		UN2896				
	CYCLOHEXANONE PEROXIDE, *IN SOLUTION WITH NOT MORE THAN 9% BY WEIGHT ACTIVE OXYGEN. SEE* CYCLOHEXANONE PEROXIDE, *50 TO 85% PEROXIDE*		UN2118				
	CYCLOHEXANONE PEROXIDE, *NOT OVER 50% PEROXIDE*	ORGANIC PEROXIDE	UN2896	ORGANIC PEROXIDE	173.153	173.154	ORGANIC PEROXIDE
	CYCLOHEXANONE PEROXIDE AND DI-(1-HYDROXY CYCLOHEXYL) PEROXIDE MIXTURE. *SEE APPROPRIATE CYCLOHEXANONE PEROXIDE ENTRY*						
	CYCLOHEXENYL TRICHLOROSILANE	CORROSIVE MATERIAL	UN1762	CORROSIVE	NONE	173.280	CORROSIVE
	CYCLOHEXYLAMINE	FLAMMABLE LIQUID	UN2357	FLAMMABLE LIQUID	173.118	173.119	FLAMMABLE
	CYCLOHEXYL TRICHLOROSILANE	CORROSIVE MATERIAL	UN1763	CORROSIVE	NONE	173.280	CORROSIVE
	CYCLOPENTANE	FLAMMABLE LIQUID	UN1146	FLAMMABLE LIQUID	173.118	173.119	FLAMMABLE
	CYCLOPENTANE, METHYL	FLAMMABLE LIQUID	UN2298	FLAMMABLE LIQUID	173.118	173.119	FLAMMABLE
+	CYCLOPROPANE	FLAMMABLE GAS	UN1027	FLAMMABLE GAS	173.306	173.304	FLAMMABLE GAS
	CYCLOTETRAMETHYLENE TETRANITRAMINE (DRY) (HMX)	FORBIDDEN					
	CYCLOTETRAMETHYLENE TETRANITRAMINE, WET WITH NOT LESS THAN 10% WATER. SEE HIGH EXPLOSIVE						
	CYCLOTRIMETHYLENE TRINITRAMINE, DESENSITIZED. SEE HIGH EXPLOSIVE						
	CYCLOTRIMETHYLENE TRINITRAMINE, WET WITH NOT LESS THAN 10% WATER. SEE HIGH EXPLOSIVE						

§172.101 HAZARDOUS MATERIALS TABLE

(1) +/ A/ W/	(2) Hazardous materials descriptions and proper shipping names	(3) Hazard Class	(3A) Identi-fication number	(4) Label(s) required (if not excepted)	(5) Packaging (a) Exceptions	(5) Packaging (b) Specific require-ments	Placards required (*denotes placard for any quantity)
	CYLINDER, EMPTY, INCLUDING TON TANKS, MUST BE CLASSED FOR THE HAZARDOUS MATERIAL PREVIOUSLY CONTAINED IN CYLINDER. SEE 173.29						
A	2,4-D. *SEE* 2,4-DICHLOROPHENOXYACETIC ACID						
A	DDT *OR* DICHLORODIPHENYLTRICHLOROETHANE	ORM-A	NA2761	NONE	173.505	173.510	NONE
	DECABORANE	FLAMMABLE SOLID	UN1868	FLAMMABLE SOLID AND POISON	NONE	173.236	FLAMMABLE SOLID[5]
	DECAHYDRONAPHTHALENE	COMBUSTIBLE LIQUID	UN1147	NONE	173.118a	NONE	COMBUSTIBLE[1]
	DECALIN. SEE DECAHYDRONAPHTHALENE						
	DECANOYL PEROXIDE, *TECHNICALLY PURE. SEE* ORGANIC PEROXIDE, SOLID, N.O.S.		UN2120				
	DELAY CONNECTORS. SEE DETONATORS, CLASS A *OR* CLASS C EXPLOSIVES *AND* DETONATING PRIMERS, CLASS A *OR* CLASS C EXPLOSIVES						
	DELAY ELECTRIC IGNITER	CLASS C EXPLOSIVE		EXPLOSIVE C	NONE	173.106	DANGEROUS
	DENATURED ALCOHOL	FLAMMABLE LIQUID	NA1986	FLAMMABLE LIQUID	173.118	173.125	FLAMMABLE
	DEPTH BOMB. SEE EXPLOSIVE BOMB						
	DETONATING FUZE, CLASS A, *WITH OR WITHOUT RADIOACTIVE COMPONENTS*	CLASS A EXPLOSIVE		EXPLOSIVE A	NONE	173.69	EXPLOSIVES A*
	DETONATING FUZE, CLASS C EXPLOSIVE	CLASS C EXPLOSIVE		EXPLOSIVE C	NONE	173.113	DANGEROUS
	DETONATING PRIMERS, CLASS A EXPLOSIVES. *SEE 173.53.*	CLASS A EXPLOSIVE		EXPLOSIVE A	NONE	173.68	EXPLOSIVES A*
	DETONATING PRIMERS, CLASS C EXPLOSIVES. *SEE 173.100*	CLASS C EXPLOSIVE		EXPLOSIVE C	NONE	173.68	DANGEROUS
	DETONATORS, CLASS A EXPLOSIVES. *SEE 173.53*	CLASS A EXPLOSIVE		EXPLOSIVE A	NONE	173.66	EXPLOSIVES A*
	DETONATORS, CLASS C EXPLOSIVES. *SEE 173.100*	CLASS C EXPLOSIVE		EXPLOSIVE C	NONE	173.66	DANGEROUS
	DETONATORS, COMMERCIAL. SEE DETONATORS, CLASS A *OR* CLASS C EXPLOSIVES						
	DI-(1-HYDROXYCYCLOHEXYL) PEROXIDE, *TECHNICALLY PURE. SEE* ORGANIC PEROXIDE, SOLID, N.O.S.		UN2148				
	DI-(1-HYDROXYTETRAZOLE) (DRY)	FORBIDDEN					
	DI-(1-NAPHTHOYL)PEROXIDE	FORBIDDEN					
	DI-(2-ETHYLHEXYL) PEROXYDICARBONATE, *TECHNICALLY PURE. SEE* ORGANIC PEROXIDE, LIQUID *OR* SOLUTION, N.O.S.		UN2122				
	DI-(2-ETHYLHEXYL) PEROXYDICARBONATE, *NOT MORE THAN 77% IN SOLUTION. SEE* ORGANIC PEROXIDE, LIQUID *OR* SOLUTION, N.O.S.		UN2123				
	DI-(2-ETHYLHEXYL) PHOSPHORIC ACID	CORROSIVE MATERIAL	NA1902	CORROSIVE	173.244	173.245	CORROSIVE
	DI-(2-METHYLBENZOYL)PEROXIDE, *NOT MORE THAN 85% WITH WATER. SEE* ORGANIC PEROXIDE, SOLID, N.O.S.		UN2593				

1,5-SEE PLACARDING NOTES, PAGE 5

(1) +/ A/ W	(2) Hazardous materials descriptions and proper shipping names	(3) Hazard Class	(3A) Identi- fication number	(4) Label(s) required (if not excepted)	(5) Packaging (a) Exceptions	(5) Packaging (b) Specific require- ments	Placards required (*denotes placard for any quantity)
	1,3-DI-(2-TERT-BUTYLPEROXYISOPROPYL) BENZENE, *TECHNICALLY PURE OR MORE THAN 40% WITH INERT SOLID. SEE* ORGANIC PEROXIDE, SOLID, N.O.S....		UN2112				
	1,3-DI-(2-TERT-BUTYLPEROXYISOPROPYL) BENZENE AND 1,4-DI-(2-TERT-BUTYLPEROXYISOPROPYL) BENZENE MIXTURE, *TECHNICALLY PURE OR MORE THAN 40% WITH INERT SOLID. SEE* ORGANIC PEROXIDE, SOLID, N.O.S.		UN2112				
	1,4-DI-(2-TERT-BUTYLPEROXYISOPROPYL) BENZENE, *TECHNICALLY PURE OR MORE THAN 40% WITH INERT SOLID. SEE* ORGANIC PEROXIDE, SOLID, N.O.S....		UN2112				
	DI-(3,5,5-TRIMETHYL-1,2-DIOXOLANYL-3)PEROXIDE, *NOT MORE THAN 50% AS A PASTE, WITH PHLEGMATIZER. SEE* ORGANIC PEROXIDE, SOLID, N.O.S.		UN2597				
	2,2-DI-(4,4-DI-TERT-BUTYLPEROXYCYCLO-HEXYL)PROPANE, *NOT MORE THAN 42% WITH INERT SOLID. SEE* ORGANIC PEROXIDE, SOLID, N.O.S.		UN2168				
	2,2-DI-(4,4-DI-TERT-BUTYLPEROXYCYCLOHEXYL) PROPANE, MORE THAN 42% WITH INERT SOLID	FORBIDDEN					
	DI-(4-TERT-BUTYLCYCLOHEXYL)PEROXYDICARBONATE, *TECHNICALLY PURE. SEE* ORGANIC PEROXIDE, SOLID, N.O.S.		UN2154				
	DI-(4-TERT-BUTYLCYCLOHEXYL)PEROXYDICARBONATE, *NOT MORE THAN 42%, STABLE DISPERSION, IN WATER. SEE* ORGANIC PEROXIDE, LIQUID *OR*, SOLUTION, N.O.S.		UN2894				
	DIACETONE ALCOHOL	COMBUSTIBLE LIQUID	UN1148	NONE	173.118a	NONE	COMBUSTIBLE
	DIACETONE ALCOHOL	FLAMMABLE LIQUID	UN1148	FLAMMABLE LIQUID	173.118	173.119	FLAMMABLE
	DIACETONE ALCOHOL PEROXIDE, *NOT MORE THAN 57% IN SOLUTION WITH NOT MORE THAN 9% HYDROGEN PEROXIDE, NOT LESS THAN 26% DIACETONE ALCOHOL AND NOT LESS THAN 9% WATER; TOTAL ACTIVE OXYGEN CONTENT NOT MORE THAN 9%. SEE* ORGANIC PEROXIDE, LIQUID *OR* SOLUTION, N.O.S.		UN2163				
	DIACETONE ALCOHOL PEROXIDES, MORE THAN 57% IN SOLUTION WITH MORE THAN 9% HYDROGEN PEROXIDE, LESS THAN 26% DIACETONE ALCOHOL AND LESS THAN 9% WATER; TOTAL ACTIVE OXYGEN CONTENT MORE THAN 9% BY WEIGHT	FORBIDDEN					
	DIACETYL	FLAMMABLE LIQUID	UN2346	FLAMMABLE LIQUID	173.118	173.119	FLAMMABLE
	P-DIAZIDOBENZENE	FORBIDDEN					
	1,2-DIAZIDOETHANE	FORBIDDEN					
A	DIAZINON	ORM-A	NA2783	NONE	173.505	173.510	NONE
	1,1'-DIAZOAMINONAPHTHALENE	FORBIDDEN					
	DIAZOAMINOTETRAZOLE (DRY)	FORBIDDEN					
	DIAZODINITROPHENOL. *SEE* INITIATING EXPLOSIVE						
	DIAZODINITROPHENOL (DRY)	FORBIDDEN					
	DIAZODIPHENYLMETHANE	FORBIDDEN					
	DIAZONIUM NITRATES (DRY)	FORBIDDEN					

§172.101 HAZARDOUS MATERIALS TABLE

(1) +/ A/ W/	(2) Hazardous materials descriptions and proper shipping names	(3) Hazard Class	(3A) Identi-fication number	(4) Label(s) required (if not excepted)	(5) Packaging (a) Exceptions	(5) Packaging (b) Specific require-ments	Placards required (*denotes placard for any quantity)
	DIAZONIUM PERCHLORATES (DRY)	FORBIDDEN					
	1,3-DIAZOPROPANE	FORBIDDEN					
	DIBENZYL PEROXYDICARBONATE, *NOT MORE THAN 87% WITH WATER. SEE* ORGANIC PEROXIDE, SOLID, N.O.S.		UN2149				
	DIBENZYL PEROXYDICARBONATE, MORE THAN 87% WITH WATER	FORBIDDEN					
	DI-(BETA-NITROXYETHYL) AMMONIUM NITRATE	*FORBIDDEN*					
+	DIBORANE *OR* DIBORANE MIXTURES	FLAMMABLE GAS	UN1911	FLAMMABLE GAS AND POISON	NONE	173.302	FLAMMABLE GAS
	DIBROMOACETYLENE	FORBIDDEN					
A	DIBROMODIFLUOROMETHANE	ORM-A	UN1941	NONE	173.505	173.605	NONE
AW	1,2-DIBROMOETHANE. *SEE* ETHYLENE DIBROMIDE						
	DICETYL PEROXYDICARBONATE, *NOT MORE THAN 42%, STABLE DISPERSION, IN WATER. SEE* ORGANIC PEROXIDE, LIQUID *OR* SOLUTION, N.O.S.		UN2895				
	DICETYL PEROXYDICARBONATE, *TECHNICALLY PURE. SEE* ORGANIC PEROXIDE, SOLID, N.O.S.		UN2164				
	N, N'-DICHLORAZODICARBONAMIDINE (SALT OF), (DRY)	FORBIDDEN					
	1, 1-DICHLORO -2, 2 -BIS (PARACHLOROPHENYL) ETHANE. *SEE TDE*						
	DICHLOROACETIC ACID	CORROSIVE MATERIAL	UN1764	CORROSIVE	173.244	173.245	CORROSIVE
	DICHLOROACETYL CHLORIDE	CORROSIVE MATERIAL	UN1765	CORROSIVE	173.244	173.247	CORROSIVE
	DICHLOROACETYLENE	FORBIDDEN					
A	DICHLOROBENZENE, ORTHO, LIQUID	ORM-A	UN1591	NONE	173.505	173.510	NONE
A	DICHLOROBENZENE, PARA, SOLID	ORM-A	UN1592	NONE	173.505	173.510	NONE
	2,4-DICHLOROBENZOYL PEROXIDE, *NOT MORE THAN 75% WITH WATER. SEE* ORGANIC PEROXIDE, SOLID, N.O.S.		UN2137				
	2,4-DICHLOROBENZOYL PEROXIDE, *NOT MORE THAN 52% AS A PASTE. SEE* ORGANIC PEROXIDE, SOLID, N.O.S.		UN2138				
	2,4-DICHLOROBENZOYL PEROXIDE, *NOT MORE THAN 52% IN SOLUTION. SEE* ORGANIC PEROXIDE, LIQUID *OR* SOLUTION, N.O.S.		UN2139				
	2,4-DICHLOROBENZOYL PEROXIDE, MORE THAN 75% WITH WATER	FORBIDDEN					
	DICHLOROBUTENE	FLAMMABLE LIQUID	NA2924	FLAMMABLE LIQUID	173.118	173.119	FLAMMABLE
	DICHLOROBUTENE	CORROSIVE MATERIAL	NA2924	CORROSIVE	173.244	173.245 173.245a	CORROSIVE
A	DICHLORODIFLUOROETHYLENE	ORM-A	NA9018	NONE	173.505	173.605	NONE
	DICHLORODIFLUOROMETHANE (*R-12*)	NONFLAMM-ABLE GAS	UN1028	NONFLAMM-ABLE GAS	173.306	173.304 173.314 173.315	NONFLAMM-ABLE GAS
	DICHLORODIFLUOROMETHANE (*R-12*) AND CHLORODI-FLUOROMETHANE (*R-22*) MIXTURE. *SEE* REFRIGER-ANT GAS, N.O.S. *OR* DISPERSANT GAS, N.O.S.						

§172.101 HAZARDOUS MATERIALS TABLE

(1) +/ A/ W	(2) Hazardous materials descriptions and proper shipping names	(3) Hazard Class	(3A) Identi-fication number	(4) Label(s) required (if not excepted)	(5) Packaging (a) Exceptions	(5) Packaging (b) Specific require-ments	Placards required (*denotes placard for any quantity)
	DICHLORODIFLUOROMETHANE (*R-12*) AND DICHLOROTETRAFLUOROETHANE (*R-114*) MIXTURE. *SEE* REFRIGERANT GAS, N.O.S. *OR* DISPERSANT GAS, N.O.S.						
	DICHLORODIFLUOROMETHANE AND DIFLUOROETHANE MIXTURE (*CONSTANT BOILING MIXTURE) (R-500). SEE* REFRIGERANT GAS, N.O.S. *OR* DISPERSANT GAS, N.O.S.						
	DICHLORODIFLUOROMETHANE (*R-12*) AND TRICHLOROFLUOROMETHANE (*R-11*) MIXTURE. *SEE* REFRIGERANT GAS, N.O.S. *OR* DISPERSANT GAS, N.O.S.						
	DICHLORODIFLUOROMETHANE (*R-12*) AND TRICHLOROTRIFLUOROETHANE (*R-113*) MIXTURE. *SEE* REFRIGERANT GAS, N.O.S. *OR* DISPERSANT GAS, N.O.S.						
	DICHLORODIFLUOROMETHANE (*R-12*), TRICHLOROFLUOROMETHANE (*R-11*) AND CHLORODIFLUOROMETHANE (*R-22*) MIXTURE. *SEE* REFRIGERANT GAS, N.O.S. *OR* DISPERSANT GAS, N.O.S.						
A	DICHLORODIPHENYLTRICHLOROETHANE. *SEE* DDT						
	DICHLOROETHYLENE	FLAMMABLE LIQUID	UN1150	FLAMMABLE LIQUID	173.118	173.119	FLAMMABLE
	DICHLOROISOPROPYL ETHER	CORROSIVE MATERIAL	UN2490	CORROSIVE	173.244	173.245	CORROSIVE
A	DICHLOROMETHANE *OR* METHYLENE CHLORIDE	ORM-A	UN1593	NONE	173.505	173.605	NONE
	DICHLOROPENTANE	FLAMMABLE LIQUID	UN1152	FLAMMABLE LIQUID	173.118	173.119	FLAMMABLE
A	2,4-DICHLOROPHENOXYACETIC ACID	ORM-A	NA2765	NONE	173.505	173.510	NONE
	DICHLOROPHENYLTRICHLOROSILANE	CORROSIVE MATERIAL	UN1766	CORROSIVE	NONE	173.280	CORROSIVE
	DICHLOROPROPANE. *SEE* PROPYLENE DICHLORIDE						
	DICHLOROPROPENE	FLAMMABLE LIQUID	UN2047	FLAMMABLE LIQUID	173.118	173.119	FLAMMABLE
	DICHLOROPROPENE AND PROPYLENE DICHLORIDE MIXTURE	FLAMMABLE LIQUID	NA2047	FLAMMABLE LIQUID	173.118	173.119	FLAMMABLE
	2,2-DICHLOROPROPIONIC ACID	CORROSIVE MATERIAL	NA1760	CORROSIVE	173.244	173.245	CORROSIVE
	DICHLORVOS	POISON B	NA2783	POISON	173.345	173.346	POISON
	DICHLORVOS MIXTURE, DRY	POISON B	NA2783	POISON	173.364	173.365	POISON
	DICUMYL PEROXIDE *50% SOLUTION*	ORGANIC PEROXIDE	NA2121	ORGANIC PEROXIDE	173.153	173.224	ORGANIC PEROXIDE
	DICUMYL PEROXIDE, *TECHNICALLY PURE OR* DICUMYL PEROXIDE, *WITH INERT SOLID. SEE* DICUMYL PEROXIDE, DRY.		UN2121				
	DICUMYL PEROXIDE, DRY	ORGANIC PEROXIDE	UN2121	ORGANIC PEROXIDE	173.153	173.154	ORGANIC PEROXIDE
	DICYCLOHEXYL PEROXYDICARBONATE, *TECHNICALLY PURE. SEE* ORGANIC PEROXIDE, SOLID, N.O.S.		UN2152				
	DICYCLOHEXYL PEROXYDICARBONATE, *NOT MORE THAN 91% WITH WATER. SEE* ORGANIC PEROXIDE, SOLID, N.O.S.		UN2153				
A	DIELDRIN	ORM-A	NA2761	NONE	173.505	173.510	NONE
	DIESEL FUEL. *SEE* FUEL OIL						
	DIETHANOL NITROSAMINE DINITRATE (DRY)	FORBIDDEN					
	DIETHYLAMINE	FLAMMABLE LIQUID	UN1154	FLAMMABLE LIQUID	173.118	173.119	FLAMMABLE

§172.101 HAZARDOUS MATERIALS TABLE

(1) +/ A/ W/	(2) Hazardous materials descriptions and proper shipping names	(3) Hazard Class	(3A) Identi-fication number	(4) Label(s) required (if not excepted)	(5) Packaging (a) Exceptions	(5) Packaging (b) Specific require-ments	Placards required (*denotes placard for any quantity)
	DIETHYL CELLOSOLVE. *SEE* ETHYLENE GLYCOL DIETHYL ETHER						
	DIETHYL DICHLOROSILANE	FLAMMABLE LIQUID	UN1767	FLAMMABLE LIQUID	NONE	173.135	FLAMMABLE
	DIETHYLENE GLYCOL DINITRATE. SEE 173.51	FORBIDDEN					
	DIETHYLGOLD BROMIDE	FORBIDDEN					
	DIETHYL KETONE	FLAMMABLE LIQUID	UN1156	FLAMMABLE LIQUID	173.118	173.119	FLAMMABLE
	DIETHYL PEROXYDICARBONATE, *NOT MORE THAN 27% IN SOLUTION.* , *SEE* ORGANIC PEROXIDE, LIQUID *OR* SOLUTION, N.O.S.		UN2175				
	DIETHYL PEROXYDICARBONATE, MORE THAN 27% IN SOLUTION	FORBIDDEN					
	DIETHYLZINC. SEE PYROPHORIC LIQUID, N.O.S						
	DIFLUOROETHANE *(R-152A)*	FLAMMABLE GAS	UN1030	FLAMMABLE GAS	173.306	173.304 173.314 173.315	FLAMMABLE GAS
A	1,1-DIFLOUROETHYLENE *(R-1132A)*	FLAMMABLE GAS	UN1959	FLAMMABLE GAS	173.306	173.304	FLAMMABLE GAS
	DIFLUOROPHOSPHORIC ACID, ANHYDROUS	CORROSIVE MATERIAL	UN1768	CORROSIVE	NONE	173.275	CORROSIVE
	2,2-DIHYDROPEROXY PROPANE, *NOT MORE THAN 25% WITH INERT ORGANIC SOLID. SEE* ORGANIC PEROXIDE, SOLID, N.O.S.		UN2178				
	DIHYDROPYRAN	FLAMMABLE LIQUID	UN2376	FLAMMABLE LIQUID	173.118	173.119	FLAMMABLE
	1,8-DIHYDROXY-2,4,5,7-TETRANITROANTHRAQUINONE (CHRYSAMMINIC ACID)	FORBIDDEN					
	DIIODOACETYLENE	FORBIDDEN					
	DIISOBUTYL KETONE	COMBUSTIBLE LIQUID	UN1157	NONE	173.118a	NONE	COMBUSTIBLE[1]
	DIISOBUTYRYL PEROXIDE, *NOT MORE THAN 52% IN SOLUTION. SEE* ORGANIC PEROXIDE, LIQUID *OR* SOLUTION, N.O.S.		UN2182				
	DIISOOCTYL ACID PHOSPHATE	CORROSIVE MATERIAL	UN1902	CORROSIVE	173.244	173.296	CORROSIVE
	DIISOPROPYLAMINE	FLAMMABLE LIQUID	UN1158	FLAMMABLE LIQUID	173.118	173.119	FLAMMABLE
	DIISOPROPYLBENZENE HYDROPEROXIDE, *NOT MORE THAN 72% IN SOLUTION. SEE* DIISOPROPYLBENZENE HYDROPEROXIDE SOLUTION, *NOT MORE THAN 72% PEROXIDE.*		UN2171				
	DIISOPROPYLBENZENE HYDROPEROXIDE, MORE THAN 72% IN SOLUTION	FORBIDDEN					
	DIISOPROPYLBENZENE HYDROPEROXIDE SOLUTION, *NOT OVER 72% PEROXIDE*	ORGANIC PEROXIDE	UN2171	ORGANIC PEROXIDE	173.153	173.224	ORGANIC PEROXIDE
	DIISOPROPYL ETHER	FLAMMABLE LIQUID	UN1159	FLAMMABLE LIQUID	173.118	173.119	FLAMMABLE
	DIISOTRIDECYL PEROXYDICARBONATE, *TECHNICALLY PURE. SEE* ORGANIC PEROXIDE, LIQUID *OR* SOLUTION, N.O.S.		UN2889				
	2,5-DIMETHYL-2,5-DI-(2-ETHYLHEXANOYL-PEROXY)HEXANE, *TECHNICALLY PURE. SEE* ORGANIC PEROXIDE, LIQUID *OR* SOLUTION, N.O.S.		UN2157				

1-SEE PLACARDING NOTES, PAGE 5

§172.101 HAZARDOUS MATERIALS TABLE

(1) +/ A/ W/	(2) Hazardous materials descriptions and proper shipping names	(3) Hazard Class	(3A) Identi- fication number	(4) Label(s) required (if not excepted)	(5) Packaging (a) Exceptions	(5) Packaging (b) Specific require- ments	Placards required (*denotes placard for any quantity)
	2,5-DIMETHYL-2,5-DI-(BENZOYLPEROXY)HEXANE, *TECHNICALLY PURE. SEE* ORGANIC PEROXIDE, SOLID, N.O.S.		UN2172				
	2,5-DIMETHYL-2,5-DI-(BENZOYLPEROXY)HEXANE, *NOT MORE THAN 82% WITH INERT SOLID. SEE* ORGANIC PEROXIDE, SOLID, N.O.S.		UN2173				
	2,5-DIMETHYL-2,5-DIHYDROPEROXY HEXANE, *NOT MORE THAN 82% WITH WATER. SEE* DIMETHYLHEXANE DIHYDROPEROXIDE, *WITH 18% OR MORE WATER*		UN2174				
	2,5-DIMETHYL-2,5-DIHYDROPEROXY HEXANE, MORE THAN 82% WITH WATER	FORBIDDEN					
	2,5-DIMETHYL-2,5-DI-(TERT-BUTYLPEROXY)HEXANE, *TECHNICALLY PURE. SEE* ORGANIC PEROXIDE, LIQUID *OR* SOLUTION, N.O.S.		UN2155				
	2,5-DIMETHYL-2,5-DI-(TERT-BUTYLPEROXY)HEXANE, *NOT MORE THAN 52% WITH INERT SOLID. SEE* ORGANIC PEROXIDE, SOLID N.O.S.		UN2156				
	2,5-DIMETHYL-2,5-DI-(TERT-BUTYLPEROXY)HEXYNE-3, *TECHNICALLY PURE. SEE* ORGANIC PEROXIDE, LIQUID *OR* SOLUTION, N.O.S.		UN2158				
	2,5-DIMETHYL-2,5-DI-(TERT-BUTYLPEROXY)HEXYNE-3, *NOT MORE THAN 52% WITH INERT SOLID. SEE* ORGANIC PEROXIDE, SOLID, N.O.S.		UN2159				
+	DIMETHYLAMINE, ANHYDROUS	FLAMMABLE GAS	UN1032	FLAMMABLE GAS	173.306	173.304 173.314 173.315	FLAMMABLE GAS
	DIMETHYLAMINE, AQUEOUS SOLUTION	FLAMMABLE LIQUID	UN1160	FLAMMABLE LIQUID	173.118	173.119	FLAMMABLE
	2,3-DIMETHYLBUTANE	FLAMMABLE LIQUID	UN2457	FLAMMABLE LIQUID	173.118	173.119	FLAMMABLE
	DIMETHYL CARBONATE	FLAMMABLE LIQUID	UN1161	FLAMMABLE LIQUID	173.118	173.119	FLAMMABLE
	DIMETHYL CHLOROTHIOPHOSPHATE, *SEE* DIMETHYL PHOSPHOROCHLORIDOTHIOATE						
	1,4-DIMETHYLCYCLOHEXANE	FLAMMABLE LIQUID	UN2263	FLAMMABLE LIQUID	173.118	173.119	FLAMMABLE
	DIMETHYLDICHLOROSILANE	FLAMMABLE LIQUID	UN1162	FLAMMABLE LIQUID	NONE	173.135	FLAMMABLE
+	DIMETHYL ETHER	FLAMMABLE GAS	UN1033	FLAMMABLE GAS	173.306	173.304 173.314 173.315	FLAMMABLE GAS
	DIMETHYLHEXANE DIHYDROPEROXIDE (DRY)	FORBIDDEN					
	DIMETHYLHEXANE DIHYDROPEROXIDE, (WITH 18% OR MORE WATER)	ORGANIC PEROXIDE	UN2174	ORGANIC PEROXIDE	NONE	173.157	ORGANIC PEROXIDE
	DIMETHYLHYDRAZINE, UNSYMMETRICAL (*UDMH*)	FLAMMABLE LIQUID	UN1163	FLAMMABLE LIQUID AND POISON	NONE	173.145	FLAMMABLE[2]
	DIMETHYL PHOSPHOROCHLORIDOTHIOATE	CORROSIVE MATERIAL	NA2267	CORROSIVE AND POISON	173.244	173.245	CORROSIVE[2]
	DIMETHYL SULFATE	CORROSIVE MATERIAL	UN1595	CORROSIVE	NONE	173.255	CORROSIVE
	DIMETHYL SULFIDE	FLAMMABLE LIQUID	UN1164	FLAMMABLE LIQUID	NONE	173.119	FLAMMABLE
	DIMYRISTYL PEROXYDICARBONATE, *TECHNICALLY PURE. SEE* ORGANIC PEROXIDE, SOLID, N.O.S.		UN2595				
	DIMYRISTYL PEROXYDICARBONATE, *NOT MORE THAN 22%, STABLE DISPERSION, IN WATER. SEE* ORGANIC PEROXIDE, LIQUID *OR* SOLUTION, N.O.S.		UN2892				

2-SEE PLACARDING NOTES, PAGE 5

§172.101 HAZARDOUS MATERIALS TABLE

(1) +/ A/ W/	(2) Hazardous materials descriptions and proper shipping names	(3) Hazard Class	(3A) Identi-fication number	(4) Label(s) required (if not excepted)	(5) Packaging (a) Exceptions	(5) Packaging (b) Specific require-ments	Placards required (*denotes placard for any quantity)
	1,4-DINITRO-1,1,4,4-TETRAMETHYLOLBUTANETETRANITRATE (DRY)	FORBIDDEN					
	2,4-DINITRO-1,3,5-TRIMETHYLBENZENE	FORBIDDEN					
	1,3-DINITRO-4,5-DINITROSOBENZENE	FORBIDDEN					
	1,3-DINITRO-5,5-DIMETHYL HYDANTOIN	FORBIDDEN					
	DINITRO-7,8-DIMETHYLGLYCOLURIL (DRY)	FORBIDDEN					
	DINITROBENZENE, SOLID, *OR* DINITROBENZOL, SOLID	POISON B	UN1597	POISON	173.364	173.371	POISON
	DINITROBENZENE SOLUTION	POISON B	UN1597	POISON	173.345	173.346	POISON
	DINITROCHLOROBENZENE	POISON B	UN1577	POISON	173.364	173.365	POISON
A	DINITROCYCLOHEXYLPHENOL	ORM-A	NA9026	NONE	173.505	173.510	NONE
	1,2-DINITROETHANE	FORBIDDEN					
	1,1-DINITROETHANE (DRY)	FORBIDDEN					
	DINITROGLYCOLURIL	FORBIDDEN					
	DINITROMETHANE	FORBIDDEN					
	DINITROPHENOL SOLUTION	POISON B	UN1599	POISON	173.345	173.362a	POISON
	DINITROPROPYLENE GLYCOL	FORBIDDEN					
	4,6-DINITRORESORCINOL (HEAVY METAL SALTS OF), (DRY)	FORBIDDEN					
	2,4-DINITRORESORCINOL (HEAVY METAL SALTS OF), (DRY)	FORBIDDEN					
	3,5-DINITROSALICYLIC ACID (LEAD SALT), (DRY)	FORBIDDEN					
	DINITROSOBENZYLAMIDINE AND SALTS OF (DRY)	FORBIDDEN					
	2,2-DINITROSTILBENE	FORBIDDEN					
	A,A'-DI-(NITROXY)METHYLETHER	FORBIDDEN					
	1,9-DINITROXY PENTAMETHYLENE-2,4,6,8-TETRAMINE (DRY)	FORBIDDEN					
	DI-N-PROPYL PEROXYDICARBONATE, TECHNICALLY PURE. SEE ORGANIC PEROXIDE, LIQUID, N.O.S.		UN2176				
	DIOXANE	FLAMMABLE LIQUID	UN1165	FLAMMABLE LIQUID	173.118	173.119	FLAMMABLE
	DIOXOLANE	FLAMMABLE LIQUID	UN1166	FLAMMABLE LIQUID	173.118	173.119	FLAMMABLE
	DIPHENYLAMINECHLOROARSINE (*DM*)	IRRITATING MATERIAL	UN1698	IRRITANT	NONE	173.382	DANGEROUS
	DIPHENYL DICHLOROSILANE	CORROSIVE MATERIAL	UN1769	CORROSIVE	NONE	173.280	CORROSIVE
	DIPHENYL METHYL BROMIDE, SOLID	CORROSIVE MATERIAL	UN1770	CORROSIVE	173.244	173.245b	CORROSIVE
	DIPHENYL METHYL BROMIDE SOLUTION	CORROSIVE MATERIAL	UN1770	CORROSIVE	173.244	173.247	CORROSIVE
	DIPHOSGENE. SEE PHOSGENE						

§172.101 HAZARDOUS MATERIALS TABLE

(1) +/ A/ W	(2) Hazardous materials descriptions and proper shipping names	(3) Hazard Class	(3A) Identi-fication number	(4) Label(s) required (if not excepted)	(5) Packaging (a) Exceptions	(5) Packaging (b) Specific require-ments	Placards required (*denotes placard for any quantity)
	DI-SEC-BUTYL PEROXYDICARBONATE, *TECHNICALLY PURE. SEE* ORGANIC PEROXIDE, LIQUID *OR* SOLUTION, N.O.S.		UN2150				
	DI-SEC-BUTYL PEROXYDICARBONATE, *NOT MORE THAN 52% IN SOLUTION. SEE* ORGANIC PEROXIDE, LIQUID *OR* SOLUTION, N.O.S.		UN2151				
	DISINFECTANT, LIQUID	CORROSIVE MATERIAL	UN1903	CORROSIVE	173.244	173.245	CORROSIVE
	DISINFECTANT, LIQUID	POISON B	UN1601	POISON	173.345	173.346	POISON
	DISINFECTANT, LIQUID, N.O.S.	COMBUSTIBLE LIQUID	NA1993	NONE	173.118a	NONE	COMBUSTIBLE[1]
	DISINFECTANT, SOLID	POISON B	UN1601	POISON	173.364	173.365	POISON
	DISPERSANT GAS, N.O.S. *SEE* REFRIGERANT GAS, N.O.S.						
	DISTEARYL PEROXYDICARBONATE, *NOT MORE THAN 85% WITH STEARYL ALCOHOL. SEE* ORGANIC PEROXIDE, SOLID, N.O.S.		UN2592				
	DISULFOTON	POISON B	NA2783	POISON	NONE	173.358	POISON
	DISULFOTON MIXTURE, DRY	POISON B	NA2783	POISON	173.377	173.377	POISON
	DISULFOTON MIXTURE, LIQUID	POISON B	NA2783	POISON	173.359	173.359	POISON
	DI-TERT-BUTYL PEROXIDE, *TECHNICALLY PURE. SEE* ORGANIC PEROXIDE, LIQUID *OR* SOLUTION, N.O.S.		UN2102				
	1,1-DI-(TERT-BUTYLPEROXY)-3,3,5-TRIMETHYL CYCLOHEXANE, *TECHNICALLY PURE. SEE* ORGANIC PEROXIDE, LIQUID *OR* SOLUTION N.O.S.		UN2145				
	1,1-DI-(TERT-BUTYLPEROXY)-3,3,5-TRIMETHYL CYCLOHEXANE, *NOT MORE THAN 57% IN SOLUTION. SEE* ORGANIC PEROXIDE, LIQUID *OR* SOLUTION, N.O.S.		UN2146				
	1,1-DI-(TERT-BUTYLPEROXY)-3,3,5-TRIMETHYL CYCLOHEXANE, *NOT MORE THAN 58% WITH INERT SOLID. SEE* ORGANIC PEROXIDE, SOLID, N.O.S.		UN2147				
	2,2-DI-(TERT-BUTYLPEROXY)BUTANE, *NOT MORE THAN 55% IN SOLUTION. SEE* ORGANIC PEROXIDE, LIQUID *OR* SOLUTION, N.O.S.		UN2111				
	2,2-DI-(TERT-BUTYLPEROXY) BUTANE, MORE THAN 55% IN SOLUTION	FORBIDDEN					
	1,1-DI-(TERT-BUTYLPEROXY)CYCLOHEXANE, *TECHNICALLY PURE. SEE* ORGANIC PEROXIDE, LIQUID *OR* SOLUTION, N.O.S.		UN2179				
	1,1-DI-(TERT-BUTYLPEROXY)CYCLOHEXANE, *NOT MORE THAN 77% IN SOLUTION. SEE* ORGANIC PEROXIDE, LIQUID *OR* SOLUTION, N.O.S.		UN2180				
	1,2-DI-(TERT-BUTYLPEROXY)CYCLOHEXANE, *NOT MORE THAN 77% IN SOLUTION. SEE* ORGANIC PEROXIDE, LIQUID *OR* SOLUTION, N.O.S.		UN2181				
	1,1-DI-(TERT-BUTYLPEROXY)CYCLOHEXANE, *NOT MORE THAN 40% WITH INERT INORGANIC SOLID, WITH NOT LESS THAN 13% PHLEGMATIZER. SEE* ORGANIC PEROXIDE, SOLID, N.O.S.		UN2885				

1 - SEE PLACARDING NOTES, PAGE 5

(1) +/ A/ W	(2) Hazardous materials descriptions and proper shipping names	(3) Hazard Class	(3A) Identi- fication number	(4) Label(s) required (if not excepted)	(5) Packaging (a) Exceptions	(5) Packaging (b) Specific require- ments	Placards required (*denotes placard for any quantity)
	1,1-DI-(TERT-BUTYLPEROXY)CYCLOHEXANE, *NOT MORE THAN 50% WITH PHLEGMATIZER. SEE* ORGANIC PEROXIDE, LIQUID *OR* SOLUTION, N.O.S.		UN2897				
	DI-(TERT-BUTYLPEROXY) PHTHALATE, MORE THAN 55% IN SOLUTION	FORBIDDEN					
	DI-(TERT-BUTYLPEROXY)PHTHALATE, *TECHNICALLY PURE. SEE* ORGANIC PEROXIDE, SOLID, N.O.S.		UN2106				
	DI-(TERT-BUTYLPEROXY)PHTHALATE, *NOT MORE THAN 55% IN SOLUTION. SEE* ORGANIC PEROXIDE, LIQUID *OR* SOLUTION, N.O.S.		UN2107				
	DI-(TERT-BUTYLPEROXY)PHTHALATE, *NOT MORE THAN 55% AS A PASTE. SEE* ORGANIC PEROXIDE, SOLID, N.O.S.		UN2108				
	2,2-DI-(TERT-BUTYLPEROXY)PROPANE, *NOT MORE THAN 50% WITH PHLEGMATIZER. SEE* ORGANIC PEROXIDE, LIQUID *OR* SOLUTION, N.O.S.		UN2883				
	2,2-DI-(TERT-BUTYLPEROXY)PROPANE, *NOT MORE THAN 40% WITH INERT INORGANIC SOLID WITH NOT LESS THAN 13% PHLEGMATIZER. SEE* ORGANIC PEROXIDE, SOLID, N.O.S.		UN2884				
	DITHIOCARBAMATE PESTICIDE, LIQUID, N.O.S. *(COMPOUNDS AND PREPARATIONS)*	FLAMMABLE LIQUID	UN2772	FLAMMABLE LIQUID	173.118	173.119	FLAMMABLE
	DITHIOCARBAMATE PESTICIDE, LIQUID, N.O.S. *(COMPOUNDS AND PREPARATIONS)*	POISON B	UN2771	POISON	173.345	173.346	POISON
	DITHIOCARBAMATE PESTICIDE, SOLID, N.O.S. *(COMPOUNDS AND PREPARATIONS)*	POISON B	UN2771	POISON	173.364	173.365	POISON
	DIVINYL ETHER	FLAMMABLE LIQUID	UN1167	FLAMMABLE LIQUID	NONE	173.119	FLAMMABLE
	DODECYLBENZENESULFONIC ACID	CORROSIVE MATERIAL	NA2584	CORROSIVE	173.244	173.245	CORROSIVE
	DODECYL TRICHLOROSILANE	CORROSIVE MATERIAL	UN1771	CORROSIVE	NONE	173.280	CORROSIVE
	DRIERS, PAINT *OR* VARNISH, LIQUID, N.O.S.	COMBUSTIBLE LIQUID	UN1168	NONE	173.118a	NONE	COMBUSTIBLE
	DRIERS, PAINT *OR* VARNISH, LIQUID, N.O.S.	FLAMMABLE LIQUID	UN1168	FLAMMABLE LIQUID	173.118 173.128	173.128	FLAMMABLE
	DRILL CARTRIDGE. SEE 173.55						
	DRUGS, LIQUID, N.O.S.	CORROSIVE MATERIAL	NA1760	CORROSIVE	173.244	173.245	CORROSIVE
	DRUGS, LIQUID, N.O.S.	POISON B	NA2810	POISON	173.345	173.346	POISON
	DRUGS, N.O.S.	COMBUSTIBLE LIQUID	NA1993	NONE	173.118a	NONE	COMBUSTIBLE[1]
	DRUGS, N.O.S.	FLAMMABLE SOLID	NA1325	FLAMMABLE SOLID	173.153	173.154	FLAMMABLE SOLID[5]
	DRUGS, N.O.S.	OXIDIZER	NA1479	OXIDIZER	173.153	173.154	OXIDIZER
	DRUGS, N.O.S.	FLAMMABLE LIQUID	NA1993	FLAMMABLE LIQUID	173.118	173.119	FLAMMABLE
	DRUGS, SOLID, N.O.S.	CORROSIVE MATERIAL	NA1759	CORROSIVE	173.244	173.245b	CORROSIVE
	DRUGS, SOLID, N.O.S.	POISON B	NA2811	POISON	173.364	173.365	POISON
	DRUMS, EMPTY, MUST BE CLASSED FOR THE HAZARDOUS MATERIAL PREVIOUSLY CONTAINED IN DRUM. SEE 173.29						
	DRY ICE. *SEE* CARBON DIOXIDE, SOLID						

5-SEE PLACARDING NOTES, PAGE 5

§172.101 HAZARDOUS MATERIALS TABLE

(1) +/ A/ W/	(2) Hazardous materials descriptions and proper shipping names	(3) Hazard Class	(3A) Identi-fication number	(4) Label(s) required (if not excepted)	(5) Packaging (a) Exceptions	(5) Packaging (b) Specific require-ments	Placards required (*denotes placard for any quantity)
	DYE INTERMEDIATE, LIQUID	CORROSIVE MATERIAL	UN2801	CORROSIVE	173.244	173.249a	CORROSIVE
	DYNAMITE. SEE HIGH EXPLOSIVE						
	ELECTRIC BLASTING CAPS. SEE DETONATORS, CLASS A *OR* CLASS C EXPLOSIVES						
	ELECTRIC SQUIB	CLASS C EXPLOSIVE		EXPLOSIVE C	NONE	173.106	DANGEROUS
	ELECTROLYTE (ACID) BATTERY FLUID (*NOT OVER 47% ACID*). *SEE* BATTERY FLUID, ACID						
	EMPTY CARTRIDGE BAG WITH BLACK POWDER IGNITER.	CLASS C EXPLOSIVE		EXPLOSIVE C	NONE	173.106	DANGEROUS
	ENDOSULFAN	POISON B	NA2761	POISON	173.364	173.365	POISON
	ENDOSULFAN MIXTURE, LIQUID	POISON B	NA2761	POISON	173.345	173.346	POISON
	ENDRIN	POISON B	NA2761	POISON	173.364	173.365	POISON
	ENDRIN MIXTURE, LIQUID	POISON B	NA2761	POISON	173.345	173.346	POISON
	ENGINE, INTERNAL COMBUSTION				173.120		
	ENGINE STARTING FLUID	FLAMMABLE GAS	UN1960	FLAMMABLE GAS	NONE	173.304	FLAMMABLE GAS
	EPICHLOROHYDRIN	FLAMMABLE LIQUID	UN2023	FLAMMABLE LIQUID	173.118	173.119	FLAMMABLE
	ESCAPE *OR* EVACUATION SLIDE, INFLATABLE. *SEE* LIFE RAFTS, INFLATABLE						
	ETCHING ACID, LIQUID, N.O.S.	CORROSIVE MATERIAL	NA1790	CORROSIVE	NONE	173.299	CORROSIVE
+	ETHANE *OR* ETHANE, COMPRESSED	FLAMMABLE GAS	UN1035	FLAMMABLE GAS	173.306	173.304	FLAMMABLE GAS
	ETHANE-PROPANE MIXTURE, REFRIGERATED LIQUID	FLAMMABLE GAS	UN1961	FLAMMABLE GAS	NONE	173.315	FLAMMABLE GAS
	ETHANE, REFRIGERATED LIQUID	FLAMMABLE GAS	UN1961	FLAMMABLE GAS	NONE	173.315	FLAMMABLE GAS
	ETHANOL AMINE DINITRATE	FORBIDDEN					
	ETHANOL. *SEE* ETHYL ALCOHOL						
	ETHION	POISON B	NA2783	POISON	173.345	173.346	POISON
	ETHION MIXTURE, DRY	POISON B	NA2783	POISON	173.364	173.365	POISON
	ETHYL-3,3-DI-(TERT-BUTYLPEROXY)BUTYRATE, *TECHNICALLY PURE. SEE* ORGANIC PEROXIDE, LIQUID *OR* SOLUTION, N.O.S.		UN2184				
	ETHYL-3,3-DI-(TERT-BUTYLPEROXY)BUTYRATE, *NOT MORE THAN 77% IN SOLUTION. SEE* ORGANIC PEROXIDE, LIQUID *OR* SOLUTION, N.O.S.		UN2185				
	ETHYL-3,3-DI-(TERT-BUTYLPEROXY)BUTYRATE, *NOT MORE THAN 50% WITH INERT INORGANIC SOLID. SEE* ORGANIC PEROXIDE, SOLID, N.O.S.		UN2598				
	ETHYL ACETATE	FLAMMABLE LIQUID	UN1173	FLAMMABLE LIQUID	173.118	173.119	FLAMMABLE
	ETHYL ACRYLATE, INHIBITED	FLAMMABLE LIQUID	UN1917	FLAMMABLE LIQUID	173.118	173.119	FLAMMABLE
	ETHYL ALCOHOL	FLAMMABLE LIQUID	UN1170	FLAMMABLE LIQUID	173.118	173.125	FLAMMABLE
	ETHYL ALDEHYDE. *SEE* ACETALDEHYDE						

§172.101 HAZARDOUS MATERIALS TABLE

(1) +/ A/ W	(2) Hazardous materials descriptions and proper shipping names	(3) Hazard Class	(3A) Identi- fication number	(4) Label(s) required (if not excepted)	(5) Packaging (a) Exceptions	(5) Packaging (b) Specific require- ments	Placards required (*denotes placard for any quantity)
	ETHYL BENZENE	FLAMMABLE LIQUID	UN1175	FLAMMABLE LIQUID	173.118	173.119	FLAMMABLE
	ETHYL BORATE	FLAMMABLE LIQUID	UN1176	FLAMMABLE LIQUID	173.118	173.119	FLAMMABLE
	ETHYL BUTYL ACETATE	COMBUSTIBLE LIQUID	UN1177	NONE	173.118a	NONE	COMBUSTIBLE[1]
	ETHYL BUTYL ETHER	FLAMMABLE LIQUID	UN1179	FLAMMABLE LIQUID	173.118	173.119	FLAMMABLE
	ETHYL BUTYRALDEHYDE	FLAMMABLE LIQUID	UN1178	FLAMMABLE LIQUID	173.118	173.119	FLAMMABLE
	ETHYL BUTYRATE	FLAMMABLE LIQUID	UN1180	FLAMMABLE LIQUID	173.118	173.119	FLAMMABLE
	ETHYL CHLORIDE	FLAMMABLE LIQUID	UN1037	FLAMMABLE LIQUID	NONE	173.123	FLAMMABLE
	ETHYL CHLOROACETATE	COMBUSTIBLE LIQUID	UN1181	NONE	173.118a	NONE	COMBUSTIBLE[1]
	ETHYL CHLOROFORMATE (*CHLOROCARBONATE*)	FLAMMABLE LIQUID	UN1182	FLAMMABLE LIQUID AND POISON	NONE	173.288	FLAMMABLE[2]
	ETHYL CHLOROTHIOFORMATE	CORROSIVE MATERIAL	UN2826	CORROSIVE	173.244	173.245 173.245a	CORROSIVE
	ETHYL CROTONATE	FLAMMABLE LIQUID	UN1862	FLAMMABLE LIQUID	173.118	173.119	FLAMMABLE
	ETHYL DICHLOROSILANE	FLAMMABLE LIQUID	UN1183	FLAMMABLE LIQUID	NONE	173.135	FLAMMABLE
	ETHYLENE CHLOROHYDRIN	POISON B	UN1135	POISON	173.345 173.3a	173.346	POISON[2]
	ETHYLENEDIAMINE	CORROSIVE MATERIAL	UN1604	CORROSIVE	173.244	173.245	CORROSIVE
	ETHYLENE DIAMINE DIPERCHLORATE	FORBIDDEN					
	ETHYLENE DIBROMIDE	POISON B	UN1605	POISON	173.505	173.620	POISON
	ETHYLENE DICHLORIDE	FLAMMABLE LIQUID	UN1184	FLAMMABLE LIQUID	173.118	173.119	FLAMMABLE
	ETHYLENE GLYCOL DIETHYL ETHER (*DIETHYL 'CELLOSOLVE'*)	FLAMMABLE LIQUID	UN1153	FLAMMABLE LIQUID	173.118a	NONE	FLAMMABLE
	ETHYLENE GLYCOL DINITRATE	FORBIDDEN					
	ETHYLENE GLYCOL MONOETHYL ETHER (*'CELLOSOLVE'*)	COMBUSTIBLE LIQUID	UN1171	NONE	173.118a	NONE	COMBUSTIBLE[1]
	ETHYLENE GLYCOL MONOETHYL ETHER ACETATE (*'CELLOSOLVE ACETATE'*)	COMBUSTIBLE LIQUID	UN1172	NONE	173.118a	NONE	COMBUSTIBLE[1]
	ETHYLENE GLYCOL MONOMETHYL ETHER (*METHYL 'CELLOSOLVE'*)	COMBUSTIBLE LIQUID	UN1188	NONE	173.118a	NONE	COMBUSTIBLE[1]
	ETHYLENE GLYCOL MONOMETHYL ETHER ACETATE (*METHYL 'CELLOSOLVE ACETATE'*)	COMBUSTIBLE LIQUID	UN1189	NONE	173.118a	NONE	COMBUSTIBLE[1]
	ETHYLENE IMINE, INHIBITED	FLAMMABLE LIQUID	UN1185	FLAMMABLE LIQUID AND POISON	NONE	173.139	FLAMMABLE[2]
+	ETHYLENE *OR* ETHYLENE, COMPRESSED	FLAMMABLE GAS	UN1962	FLAMMABLE GAS	173.306	173.304	FLAMMABLE GAS
	ETHYLENE OXIDE	FLAMMABLE LIQUID	UN1040	FLAMMABLE LIQUID	NONE	173.124	FLAMMABLE
	ETHYLENE, REFRIGERATED LIQUID (*CRYOGENIC LIQUID*)	FLAMMABLE GAS	UN1038	FLAMMABLE GAS	NONE	173.318 173.319	FLAMMABLE GAS
	ETHYL ETHER	FLAMMABLE LIQUID	UN1155	FLAMMABLE LIQUID	NONE	173.119	FLAMMABLE
	ETHYL FORMATE	FLAMMABLE LIQUID	UN1190	FLAMMABLE LIQUID	173.118	173.119	FLAMMABLE
	ETHYLHEXALDEHYDE	COMBUSTIBLE LIQUID	UN1191	NONE	173.118a	NONE	COMBUSTIBLE[1]
	ETHYL HYDROPEROXIDE (EXPLODES ABOVE 100 DEG C)	FORBIDDEN					

1,2-SEE PLACARDING NOTES, PAGE 5

(1) +/ A/ W/	(2) Hazardous materials descriptions and proper shipping names	(3) Hazard Class	(3A) Identi-fication number	(4) Label(s) required (if not excepted)	(5) Packaging (a) Exceptions	(5) Packaging (b) Specific require-ments	Placards required (*denotes placard for any quantity)
	ETHYL LACTATE	COMBUSTIBLE LIQUID	UN1192	NONE	173.118a	NONE	COMBUSTIBLE[1]
	ETHYL MERCAPTAN	FLAMMABLE LIQUID	UN2363	FLAMMABLE LIQUID	NONE	173.141	FLAMMABLE
	ETHYL METHYL ETHER	FLAMMABLE LIQUID	UN1039	FLAMMABLE LIQUID	NONE	173.119	FLAMMABLE
	ETHYL METHYL KETONE	FLAMMABLE LIQUID	UN1193	FLAMMABLE LIQUID	173.118	173.119	FLAMMABLE
	ETHYL NITRATE (*NITRIC ETHER*)	FLAMMABLE LIQUID	NA1993	FLAMMABLE LIQUID	173.118	173.119	FLAMMABLE
	ETHYL NITRITE (NITROUS ETHER)	FLAMMABLE LIQUID	UN1194	FLAMMABLE LIQUID	NONE	173.119	FLAMMABLE
	ETHYL PERCHLORATE	FORBIDDEN					
	ETHYL PHENYL DICHLOROSILANE	CORROSIVE MATERIAL	UN2435	CORROSIVE	NONE	173.280	CORROSIVE
	ETHYL PHOSPHONOUS DICHLORIDE. *SEE* PYROFORIC LIQUID, N.O.S.						
	ETHYL PHOSPHORODICHLORIDATE	CORROSIVE MATERIAL	NA1760	CORROSIVE	173.244	173.245 173.245a	CORROSIVE
	ETHYL PROPIONATE	FLAMMABLE LIQUID	UN1195	FLAMMABLE LIQUID	173.118	173.119	FLAMMABLE
	ETHYL SILICATE (*TETRAETHYL ORTHOSILICATE*)	COMBUSTIBLE LIQUID	UN1292	NONE	173.118a	NONE	COMBUSTIBLE[1]
	ETHYL TRICHLOROSILANE	FLAMMABLE LIQUID	UN1196	FLAMMABLE LIQUID	NONE	173.135	FLAMMABLE
	ETIOLOGIC AGENT, N.O.S.	ETIOLOGIC AGENT	NA2814	ETIOLOGIC AGENT	173.386	173.387	NONE
	EXPLOSIVE AUTO ALARM	CLASS C EXPLOSIVE		EXPLOSIVE C	NONE	173.111	DANGEROUS
	EXPLOSIVE BOMB	CLASS A EXPLOSIVE		EXPLOSIVE A	NONE	173.56	EXPLOSIVES A*
	EXPLOSIVE CABLE CUTTER	CLASS C EXPLOSIVE		EXPLOSIVE C	NONE	173.102	DANGEROUS
	EXPLOSIVE, FORBIDDEN. SEE SEC. 173.51	FORBIDDEN					
	EXPLOSIVE MINE	CLASS A EXPLOSIVE		EXPLOSIVE A	NONE	173.56	EXPLOSIVES A*
	EXPLOSIVE, NEW APPROVAL, AND EVALUATION. SEE 173.86						
	EXPLOSIVE PEST CONTROL DEVICES	CLASS C EXPLOSIVE		EXPLOSIVE C	NONE	173.100	DANGEROUS
	EXPLOSIVE POWER DEVICE, CLASS B	CLASS B EXPLOSIVE		EXPLOSIVE B	NONE	173.94	EXPLOSIVES B*
	EXPLOSIVE POWER DEVICE, CLASS C	CLASS C EXPLOSIVE		EXPLOSIVE C	NONE	173.102	DANGEROUS
	EXPLOSIVE PROJECTILE	CLASS A EXPLOSIVE		EXPLOSIVE A	NONE	173.56	EXPLOSIVES A*
	EXPLOSIVE RELEASE DEVICE	CLASS C EXPLOSIVE		EXPLOSIVE C	NONE	173.102	DANGEROUS
	EXPLOSIVE RIVET	CLASS C EXPLOSIVE		EXPLOSIVE C	NONE	173.100	DANGEROUS
	EXPLOSIVE, SAMPLE FOR LABORATORY EXAMINATION				173.86		
	EXPLOSIVE TORPEDO	CLASS A EXPLOSIVE		EXPLOSIVE A	NONE	173.56	EXPLOSIVES A*
	EXTRACT, LIQUID, FLAVORING	FLAMMABLE LIQUID	UN1197	FLAMMABLE LIQUID	173.118	173.119	FLAMMABLE
	FERRIC ARSENATE, SOLID	POISON B	UN1606	POISON	173.364	173.365	POISON
	FERRIC ARSENITE, SOLID	POISON B	UN1607	POISON	173.364	173.365	POISON
A	FERRIC CHLORIDE, SOLID, *ANHYDROUS*	ORM-B	UN1773	NONE	173.505	173.510	NONE

1-SEE PLACARDING NOTES, PAGE 5

§172.101 HAZARDOUS MATERIALS TABLE

(1) +/ A/ W/	(2) Hazardous materials descriptions and proper shipping names	(3) Hazard Class	(3A) Identi- fication number	(4) Label(s) required (if not excepted)	(5) Packaging (a) Exceptions	(5) Packaging (b) Specific require- ments	Placards required (*denotes placard for any quantity)
	FERRIC CHLORIDE SOLUTION	CORROSIVE MATERIAL	UN2582	CORROSIVE	173.244	173.245	CORROSIVE
	FERRIC NITRATE	OXIDIZER	UN1466	OXIDIZER	173.153	173.182	OXIDIZER
W	FERROCHROME, EXOTHERMIC	ORM-C		NONE	173.505	173.985	NONE
W	FERROMANGANESE, EXOTHERMIC. *SEE* FERROCHROME, EXOTHERMIC						
W	FERROPHOSPHORUS	ORM-A		NONE	173.505	173.635	NONE
AW	FERROSILICON, *CONTAINING 30% OR MORE BUT NOT MORE THAN 70% SILICON*	ORM-A	UN1408	NONE	173.505	173.645	NONE
	FERROUS ARSENATE, SOLID	POISON B	UN1608	POISON	173.364	173.365	POISON
A	FERROUS CHLORIDE, SOLID	ORM-B	NA1759	NONE	173.505	173.510	NONE
	FERROUS CHLORIDE, SOLUTION	CORROSIVE MATERIAL	NA1760	CORROSIVE	173.244	173.245	CORROSIVE
	FERTILIZER AMMONIATING SOLUTION *CONTAINING FREE AMMONIA (MORE THAN 25.3 P.S.I.G.)*	NONFLAMM-ABLE GAS	UN1043	NONFLAMM-ABLE GAS	173.306	173.304 173.314	NONFLAMM-ABLE GAS
W	FIBERS (*JUTE, HEMP, FLAX, SISAL, COIR, KAPOK, AND SIMILAR VEGETABLE FIBERS*)	ORM-C	NA1372	NONE	173.505	173.965	NONE
	FILM (*NITROCELLULOSE*)	FLAMMABLE SOLID	NA1324	FLAMMABLE SOLID	NONE	173.177	FLAMMABLE SOLID[5]
	FIRECRACKER. SEE FIREWORKS, COMMON *OR* SPECIAL						
	FIRECRACKER SALUTE. SEE FIREWORKS, COMMON *OR* SPECIAL						
	FIRE EXTINGUISHER	NONFLAMM-ABLE GAS	UN1044	NONFLAMM-ABLE GAS	173.306		NONFLAMM-ABLE GAS
	FIRE EXTINGUISHER CHARGE CONTAINING NOT MORE THAN 50 GRAINS OF , PROPELLANT EXPLOSIVE PER UNIT. NOT SUBJECT TO REQUIREMENTS OF THIS SUBCHAPTER						
	FIRE EXTINGUISHER CHARGE CONTAINING SULFURIC ACID	CORROSIVE MATERIAL	UN1774	CORROSIVE	173.261		CORROSIVE
	FIREWORKS, COMMON	CLASS C EXPLOSIVE		EXPLOSIVE C	NONE	173.100 173.108	DANGEROUS
	FIREWORKS, EXHIBITION DISPLAY PIECE. SEE FIREWORKS, SPECIAL						
	FIREWORKS, SPECIAL	CLASS B EXPLOSIVE		EXPLOSIVE B	NONE	173.88 173.91	EXPLOSIVES B*
W	FISH MEAL *OR* FISH SCRAP CONTAINING 6% TO 12% WATER	ORM-C	NA2216	NONE	173.505	173.995	NONE
	FISH MEAL *OR* FISH SCRAP CONTAINING LESS THAN 6% OR MORE THAN 12% WATER	FLAMMABLE SOLID	NA1374	FLAMMABLE SOLID	NONE	173.171	FLAMMABLE SOLID[5]
	FISSILE RADIOACTIVE MATERIAL. *SEE* RADIOACTIVE MATERIAL, FISSILE						
	FLAME RETARDANT COMPOUND LIQUID	CORROSIVE MATERIAL	NA1760	CORROSIVE	173.244	173.291	CORROSIVE
+	FLAMMABLE GAS N.O.S. *SEE* COMPRESSED GAS, N.O.S						
	FLAMMABLE LIQUID, CORROSIVE, N.O.S.	FLAMMABLE LIQUID	UN2924	FLAMMABLE LIQUID AND CORROSIVE	NONE	173.119	FLAMMABLE
	FLAMMABLE LIQUID, N.O.S.	FLAMMABLE LIQUID	UN1993	FLAMMABLE LIQUID	173.118	173.119	FLAMMABLE

2,5-SEE PLACARDING NOTES, PAGE 5

§172.101 HAZARDOUS MATERIALS TABLE

(1) +/ A/ W	(2) Hazardous materials descriptions and proper shipping names	(3) Hazard Class	(3A) Identification number	(4) Label(s) required (if not excepted)	(5) Packaging (a) Exceptions	(5) Packaging (b) Specific requirements	Placards required (*denotes placard for any quantity)
	FLAMMABLE LIQUID, POISONOUS, N.O.S.	FLAMMABLE LIQUID	UN1992	FLAMMABLE LIQUID AND POISON	NONE	173.119	FLAMMABLE[2]
	FLAMMABLE SOLID, CORROSIVE, N.O.S.	FLAMMABLE SOLID	UN2925	FLAMMABLE SOLID AND CORROSIVE	173.153	173.154	FLAMMABLE SOLID[5]
	FLAMMABLE SOLID, N.O.S.	FLAMMABLE SOLID	UN1325	FLAMMABLE SOLID	173.153	173.154	FLAMMABLE SOLID[5]
	FLAMMABLE SOLID, POISONOUS, N.O.S.	FLAMMABLE SOLID	UN2926	FLAMMABLE SOLID AND POISON	173.153	173.154	FLAMMABLE SOLID[5]
	FLARE. SEE FIREWORKS, COMMON						
	FLARE, AIRPLANE. SEE FIREWORKS, SPECIAL						
	FLASH CARTRIDGE. SEE FIREWORKS, SPECIAL *OR* LOW EXPLOSIVES						
	FLASH CRACKER. SEE FIREWORKS, COMMON *OR* SPECIAL						
	FLASH POWDER. SEE FIREWORKS, SPECIAL *OR* LOW EXPLOSIVES						
	FLEXIBLE LINEAR SHAPED CHARGE, METAL CLAD	CLASS C EXPLOSIVE		EXPLOSIVE C	NONE	173.104	DANGEROUS
	FLOWERS OF SULFUR. *SEE* SULFUR						
	FLUE DUST, POISONOUS	POISON B	NA2811	POISON	173.364	173.368	POISON
	FLUOBORIC ACID	CORROSIVE MATERIAL	UN1775	CORROSIVE	173.244	173.283	CORROSIVE
	FLUORIC ACID. *SEE* HYDROFLUORIC ACID						
	FLUORINE	NONFLAMMABLE GAS	UN1045	POISON AND OXIDIZER	NONE	173.302	NONFLAMMABLE GAS
	FLUOROPHOSPHORIC ACID, ANHYDROUS. *SEE* MONOFLUOROPHOSPHORIC ACID, ANHYDROUS						
	FLUOROSILICIC ACID. *SEE* HYDROFLUOROSILICIC ACID						
	FLUOROSULFONIC ACID *OR* FLUOSULFONIC ACID	CORROSIVE MATERIAL	UN1777	CORROSIVE	NONE	173.274	CORROSIVE
	FLUOSILICIC ACID	CORROSIVE MATERIAL	UN1778	CORROSIVE	NONE	173.265	CORROSIVE
	FORBIDDEN EXPLOSIVES. SEE 173.51	FORBIDDEN					
	FORBIDDEN MATERIALS. SEE 173.21	FORBIDDEN					
AW	FORMALDEHYDE SOLUTION (*FLASH POINT MORE THAN 141 DEG F.; IN CONTAINERS OF 110 GALLONS OR LESS*)	ORM-A	UN2209	NONE	173.505	173.510	NONE
	FORMALDEHYDE SOLUTION (*FLASH POINT NOT MORE THAN 141 DEG F.; IN CONTAINERS OVER 110 GALLONS*)	COMBUSTIBLE LIQUID	UN1198	NONE	173.118a	NONE	COMBUSTIBLE[1]
AW	FORMALDEHYDE SOLUTION (*FLASH POINT NOT MORE THAN 141 DEG F.; IN CONTAINERS OF 110 GALLONS OR LESS*)	ORM-A	UN1198	NONE	173.505	173.510	NONE
	FORMALDEHYDE SOLUTION (*FLASH POINT MORE THAN 141 DEG F.; IN CONTAINERS OVER 110 GALLONS*)	COMBUSTIBLE LIQUID	UN2209	NONE	173.118a	NONE	COMBUSTIBLE[1]
	FORMALIN. *SEE* FORMALDEHYDE SOLUTION						
	FORMIC ACID	CORROSIVE MATERIAL	UN1779	CORROSIVE	173.244	173.289	CORROSIVE
	FORMIC ACID SOLUTION	CORROSIVE MATERIAL	UN1779	CORROSIVE	173.244	173.289	CORROSIVE

1,2,5-SEE PLACARDING NOTES, PAGE 5

§172.101 HAZARDOUS MATERIALS TABLE

(1) +/ A/ W	(2) Hazardous materials descriptions and proper shipping names	(3) Hazard Class	(3A) Identification number	(4) Label(s) required (if not excepted)	(5) Packaging (a) Exceptions	(5) Packaging (b) Specific requirements	Placards required (*denotes placard for any quantity)
	FUEL, AVIATION, TURBINE ENGINE	FLAMMABLE LIQUID	UN1863	FLAMMABLE LIQUID	173.118	173.119	FLAMMABLE
	FUEL, AVIATION, TURBINE ENGINE	COMBUSTIBLE LIQUID	UN1863	NONE	173.118a	NONE	COMBUSTIBLE[1]
	FUEL OIL	COMBUSTIBLE LIQUID	NA1993	NONE	173.118a	NONE	COMBUSTIBLE[1]
	FUEL OIL, DIESEL. *SEE* FUEL OIL						
	FUEL OIL, NO. 1, 2, 4, 5 *OR* 6	COMBUSTIBLE LIQUID	NA1993	NONE	173.118a	NONE	COMBUSTIBLE[1]
	FULMINATE OF MERCURY (DRY)	FORBIDDEN					
	FULMINATE OF MERCURY, WET. SEE INITIATING EXPLOSIVE						
	FULMINATING GOLD	FORBIDDEN					
	FULMINATING MERCURY	FORBIDDEN					
	FULMINATING PLATINUM	FORBIDDEN					
	FULMINATING SILVER	FORBIDDEN					
	FULMINIC ACID	FORBIDDEN					
	FUMARYL CHLORIDE	CORROSIVE MATERIAL	UN1780	CORROSIVE	173.244	173.245	CORROSIVE
	FUMIGANT. SEE 173.152(A) NOTE 1						
	FURAN	FLAMMABLE LIQUID	UN2389	FLAMMABLE LIQUID	173.118	173.119	FLAMMABLE
	FURFURAL	COMBUSTIBLE LIQUID	UN1199	NONE	173.118a	NONE	COMBUSTIBLE[1]
	FUSEE (*RAILWAY OR HIGHWAY*)	FLAMMABLE SOLID	NA1325	FLAMMABLE SOLID	NONE	173.154a	FLAMMABLE SOLID
	FUSE IGNITER	CLASS C EXPLOSIVE		EXPLOSIVE C	NONE	173.106	DANGEROUS
	FUSE, INSTANTANEOUS	CLASS C EXPLOSIVE		EXPLOSIVE C	173.100		DANGEROUS
	FUSE LIGHTER	CLASS C EXPLOSIVE		EXPLOSIVE C	NONE	173.106	DANGEROUS
	FUSEL OIL	COMBUSTIBLE LIQUID	UN1201	NONE	173.118a	NONE	COMBUSTIBLE[1]
	FUSE, MILD DETONATING, METAL CLAD	CLASS C EXPLOSIVE		EXPLOSIVE C	NONE	173.104	DANGEROUS
	FUSE, SAFETY	CLASS C EXPLOSIVE		EXPLOSIVE C	173.100	173.100	DANGEROUS
	FUZE, COMBINATION	CLASS C EXPLOSIVE		EXPLOSIVE C	NONE	173.105	DANGEROUS
	FUZE, DETONATING	CLASS A EXPLOSIVE		EXPLOSIVE A	NONE	173.69	EXPLOSIVES A*
	FUZE, DETONATING, CLASS C EXPLOSIVE	CLASS C EXPLOSIVE		EXPLOSIVE C	NONE	173.113	DANGEROUS
	FUZE, DETONATING, RADIOACTIVE	CLASS A EXPLOSIVE		EXPLOSIVE A	NONE	173.69	EXPLOSIVES A*
	FUZE, PERCUSSION	CLASS C EXPLOSIVE		EXPLOSIVE C	NONE	173.105	DANGEROUS
	FUZE, TIME	CLASS C EXPLOSIVE		EXPLOSIVE C	NONE	173.105	DANGEROUS
	FUZE, TRACER	CLASS C EXPLOSIVE		EXPLOSIVE C	NONE	173.105	DANGEROUS
	GALACTSAN TRINITRATE	FORBIDDEN					
	GALLIUM METAL, LIQUID	ORM-B	UN2803	NONE	NONE	173.861	NONE
	GALLIUM METAL, SOLID	ORM-B	UN2803	NONE	NONE	173.862	NONE

1-SEE PLACARDING NOTES, PAGE 5

§172.101 HAZARDOUS MATERIALS TABLE

(1) +/ A/ W	(2) Hazardous materials descriptions and proper shipping names	(3) Hazard Class	(3A) Identi- fication number	(4) Label(s) required (if not excepted)	(5) Packaging (a) Exceptions	(5) Packaging (b) Specific require- ments	Placards required (*denotes placard for any quantity)
	GAS CYLINDER, EMPTY. SEE CYLINDER, EMPTY						
	GAS DRIPS, HYDROCARBON	COMBUSTIBLE LIQUID	UN1864	NONE	173.118a	NONE	COMBUSTIBLE[1]
	GAS DRIPS, HYDROCARBON	FLAMMABLE LIQUID	UN1864	FLAMMABLE LIQUID	173.118	173.119	FLAMMABLE
	GAS IDENTIFICATION SET	POISON A	NA9035	POISON GAS	NONE	173.331	POISON GAS*
	GAS IDENTIFICATION SET	IRRITATING MATERIAL	NA9035	IRRITANT	NONE	173.331	DANGEROUS
	GAS MINE. SEE EXPLOSIVE MINE						
	GASOHOL *(GASOLINE MIXED WITH ETHYL ALCOHOL)* CONTAINING 20% MAXIMUM ALCOHOL. SEE GASOLINE.						
	GASOLINE (*INCLUDING CASING-HEAD AND NATURAL*)	FLAMMABLE LIQUID	UN1203	FLAMMABLE LIQUID	173.118	173.119	FLAMMABLE
	GELATINE DYNAMITE. SEE HIGH EXPLOSIVE						
	GERMANE	POISON A	UN2192	POISON GAS AND FLAMMABLE GAS	NONE	173.328	POISON GAS*
	GLYCEROL-1,3-DINITRATE	FORBIDDEN					
	GLYCEROL MONOGLUCONATE TRINITRATE	FORBIDDEN					
	GLYCEROL MONOLACTATE TRINITRATE	FORBIDDEN					
	GRENADE *WITHOUT BURSTING CHARGE. (WITH INCENDIARY MATERIAL)*	CLASS B EXPLOSIVE		EXPLOSIVE B	NONE	173.91	EXPLOSIVES B*
	GRENADE *WITHOUT BURSTING CHARGE. (WITH SMOKE CHARGE)* (SMOKE GRENADE)	CLASS C EXPLOSIVE		EXPLOSIVE C	NONE	173.108	DANGEROUS
	GRENADE *WITHOUT BURSTING CHARGE. (WITH POISON A GAS CHARGE)*	POISON A	NA2016	POISON GAS	NONE	173.330	POISON GAS*
	GRENADE *WITHOUT BURSTING CHARGE. (WITH POISON B CHARGE)*	POISON B	NA2016	POISON	NONE	173.350	POISON
	GRENADE, EMPTY, PRIMED	CLASS C EXPLOSIVE		NONE	NONE	173.107	NONE
	GRENADE, HAND *OR* RIFLE, EXPLOSIVE (*WITH OR WITHOUT GAS, SMOKE, OR INCENDIARY MATERIAL)*	CLASS A EXPLOSIVE		EXPLOSIVE A	NONE	173.56	EXPLOSIVES A*
	GRENADE, TEAR GAS	IRRITATING MATERIAL	NA2017	IRRITANT	NONE	173.385	DANGEROUS
	GUANIDINE NITRATE	OXIDIZER	UN1467	OXIDIZER	173.153	173.182	OXIDIZER
	GUANYL NITROSAMINO GUANYLIDENE HYDRAZINE. SEE INITIATING EXPLOSIVE						
	GUANYL NITROSAMINO GUANYLIDENE HYDRAZINE (DRY)	FORBIDDEN					
	GUANYL NITROSAMINO GUANYL TETRAZENE. SEE INITIATING EXPLOSIVE						
	GUIDED MISSILE, WITHOUT WARHEAD. SEE ROCKET MOTOR, CLASS A EXPLOSIVE *OR* ROCKET MOTOR, CLASS B EXPLOSIVE						
	GUIDED MISSILE WITH WARHEAD. SEE ROCKET AMMUNITION WITH EXPLOSIVE, ILLUMINATING, GAS, INCENDIARY, *OR* SMOKE PROJECTILE						
	GUNCOTTON. SEE HIGH EXPLOSIVE						

1-SEE PLACARDING NOTES, PAGE 5

(1) +/ A/ W/	(2) Hazardous materials descriptions and proper shipping names	(3) Hazard Class	(3A) Identification number	(4) Label(s) required (if not excepted)	(5) Packaging (a) Exceptions	(5) Packaging (b) Specific requirements	Placards required (*denotes placard for any quantity)
	GUTHION. *SEE* AZINPHOS METHYL						
	GUTHION MIXTURE, LIQUID. *SEE* AZINPHOS METHYL MIXTURE, LIQUID						
	HAFNIUM METAL, DRY. (*SEE 173.214 NOTE 3*)	FLAMMABLE SOLID	UN2545	FLAMMABLE SOLID	NONE	173.214	FLAMMABLE SOLID[5]
	HAFNIUM METAL, WET	FLAMMABLE SOLID	UN1326	FLAMMABLE SOLID	NONE	173.214	FLAMMABLE SOLID[5]
	HAND SIGNAL DEVICE	CLASS C EXPLOSIVE		EXPLOSIVE C	NONE	173.108	DANGEROUS
	HAZARDOUS SUBSTANCE, LIQUID *OR* SOLID, N.O.S., ORM-E, LIQUID OR SOLID, N.O.S.	ORM-E	NA9188	NONE	NONE	173.1300	NONE
	HAZARDOUS WASTE, LIQUID *OR* SOLID, N.O.S.	ORM-E	NA9189	NONE	NONE	173.1300	NONE
	HAZARDOUS WASTE, MEETING THE DEFINITION OF A HAZARD CLASS OTHER THAN ORM-E. SEE 172.101(C)(10)						
	HEATER FOR REFRIGERATOR CAR, LIQUID FUEL TYPE (*CONTAINING FUEL*)	FLAMMABLE LIQUID	NA1993	FLAMMABLE LIQUID	173.146		FLAMMABLE
	HELIUM *OR* HELIUM, COMPRESSED	NONFLAMMABLE GAS	UN1046	NONFLAMMABLE GAS	173.306	173.302 173.314	NONFLAMMABLE GAS
	HELIUM-OXYGEN MIXTURE	NONFLAMMABLE GAS	NA1980	NONFLAMMABLE GAS	173.306	173.302	NONFLAMMABLE GAS
	HELIUM, REFRIGERATED LIQUID (*CRYOGENIC LIQUID*)	NONFLAMMABLE GAS	UN1963	NONFLAMMABLE GAS	173.320	173.316 173.318	NONFLAMMABLE GAS
	HEPTANE	FLAMMABLE LIQUID	UN1206	FLAMMABLE LIQUID	173.118	173.119	FLAMMABLE
	HEXACHLOROCYCLOPENTADIENE	CORROSIVE MATERIAL	UN2646	CORROSIVE	173.244	173.245	CORROSIVE
W	HEXACHLOROETHANE	ORM-A	NA9037	NONE	173.505	173.650	NONE
	HEXADECYLTRICHLOROSILANE	CORROSIVE MATERIAL	UN1781	CORROSIVE	NONE	173.280	CORROSIVE
	HEXADIENE	FLAMMABLE LIQUID	UN2458	FLAMMABLE LIQUID	NONE	173.119	FLAMMABLE
	HEXAETHYL TETRAPHOSPHATE AND COMPRESSED GAS MIXTURE	POISON A	UN1612	POISON GAS	NONE	173.334	POISON GAS*
	HEXAETHYL TETRAPHOSPHATE, LIQUID	POISON B	UN1611	POISON	NONE	173.358	POISON
	HEXAETHYL TETRAPHOSPHATE MIXTURE, DRY (*CONTAINING MORE THAN 2% HEXAETHYL TETRAPHOSPHATE*)	POISON B	NA2783	POISON	NONE	173.377	POISON
	HEXAETHYL TETRAPHOSPHATE MIXTURE, DRY (*CONTAINING NOT MORE THAN 2% HEXAETHYL TETRAPHOSPHATE*)	POISON B	NA2783	POISON	173.377	173.377	POISON
	HEXAETHYL TETRAPHOSPHATE MIXTURE, LIQUID (*CONTAINING MORE THAN 25% HEXAETHYL TETRAPHOSPHATE*)	POISON B	NA2783	POISON	NONE	173.359	POISON
	HEXAETHYL TETRAPHOSPHATE MIXTURE, LIQUID (*CONTAINING NOT MORE THAN 25% HEXAETHYL TETRAPHOSPHATE*)	POISON B	UN2783	POISON	173.359	173.359	POISON
	HEXAFLUOROPHOSPHORIC ACID	CORROSIVE MATERIAL	UN1782	CORROSIVE	NONE	173.275	CORROSIVE
	HEXAFLUOROPROPYLENE	NONFLAMMABLE GAS	UN1858	NONFLAMMABLE GAS	173.306	173.304 173.314 173.315	NONFLAMMABLE GAS
	HEXAFLUOROPROPYLENE OXIDE	NONFLAMMABLE GAS	NA1956	NONFLAMMABLE GAS	173.306	173.304 173.314	NONFLAMMABLE GAS
	HEXALDEHYDE	FLAMMABLE LIQUID	UN1207	FLAMMABLE LIQUID	173.118	173.119	FLAMMABLE

5-SEE PLACARDING NOTES, PAGE 5

§172.101 HAZARDOUS MATERIALS TABLE

(1) +/ A/ W	(2) Hazardous materials descriptions and proper shipping names	(3) Hazard Class	(3A) Identi-fication number	(4) Label(s) required (if not excepted)	(5) Packaging (a) Exceptions	(5) Packaging (b) Specific require-ments	Placards required (*denotes placard for any quantity)
	HEXAMETHYLENEDIAMINE, SOLID	CORROSIVE MATERIAL	UN2280	CORROSIVE	173.244	173.245b	CORROSIVE
	HEXAMETHYLENEDIAMINE, SOLUTION	CORROSIVE MATERIAL	UN1783	CORROSIVE	173.244	173.292	CORROSIVE
	HEXAMETHYLENEIMINE	FLAMMABLE LIQUID	UN2493	FLAMMABLE LIQUID AND CORROSIVE	NONE	173.119	FLAMMABLE
	HEXAMETHYLENE TRIPEROXIDE DIAMINE (DRY)	FORBIDDEN					
	HEXAMETHYLOL BENZENE HEXANITRATE	FORBIDDEN					
	3,3,6,6,9,9-HEXAMETHYL-1,2,4,5-TETRAOXOCYCLONONANE, *TECHNICALLY PURE. SEE* ORGANIC PEROXIDE, SOLID, N.O.S.		UN2165				
	3,3,6,6,9,9-HEXAMETHYL-1,2,4,5-TETRAOXOCYCLONONANE, *NOT MORE THAN 52% WITH INERT SOLID. SEE* ORGANIC PEROXIDE, SOLID, N.O.S.		UN2166				
	3,3,6,6,9,9-HEXAMETHYL-1,2,4,5-TETRAOXOCYCLONONANE, *NOT MORE THAN 52% IN SOLUTION. SEE* ORGANIC PEROXIDE, LIQUID *OR* SOLUTION, N.O.S.		UN2167				
	HEXANE	FLAMMABLE LIQUID	UN1208	FLAMMABLE LIQUID	173.118	173.119	FLAMMABLE
	HEXANITROAZOXY BENZENE	FORBIDDEN					
	2,2',4,4',6,6'-HEXANITRO-3,3'-DIHYDROXYAZOBENZENE (DRY)	FORBIDDEN					
	2,2',3',4,4',6-HEXANITRODIPHENYLAMINE	FORBIDDEN					
	2,3',4,4',6,6'-HEXANITRODIPHENYLETHER	FORBIDDEN					
	N,N'-(HEXANITRODIPHENYL)ETHYLENE DINITRAMINE (DRY)	*FORBIDDEN*					
	HEXANITRODIPHENYL UREA	FORBIDDEN					
	HEXANITROETHANE	FORBIDDEN					
	HEXANITROOXANILIDE	FORBIDDEN					
	HEXANOIC ACID	CORROSIVE MATERIAL	NA1760	CORROSIVE	173.244	173.245	CORROSIVE
	HEXYLTRICHLOROSILANE	CORROSIVE MATERIAL	UN1784	CORROSIVE	NONE	173.280	CORROSIVE
	HIGH EXPLOSIVE	CLASS A EXPLOSIVE		EXPLOSIVE A	173.65	173.61 TO 173.87	EXPLOSIVES A*
	HIGH EXPLOSIVE, LIQUID	CLASS A EXPLOSIVE		EXPLOSIVE A	NONE	173.62	EXPLOSIVES A*
	HYDRAULIC ACCUMULATOR. *SEE* ACCUMULATOR, PRESSURIZED						
	HYDRAZINE, ANHYDROUS	FLAMMABLE LIQUID	UN2029	FLAMMABLE LIQUID AND POISON	NONE	173.276	FLAMMABLE
	HYDRAZINE, AQUEOUS SOLUTION	CORROSIVE MATERIAL	UN2030	CORROSIVE	NONE	173.276	CORROSIVE
	HYDRAZINE AZIDE	FORBIDDEN					
	HYDRAZINE CHLORATE	FORBIDDEN					
	HYDRAZINE DICARBONIC ACID DIAZIDE	FORBIDDEN					
	HYDRAZINE PERCHLORATE	FORBIDDEN					
	HYDRAZINE SELENATE	FORBIDDEN					

§172.101 HAZARDOUS MATERIALS TABLE

(1) +/ A/ W	(2) Hazardous materials descriptions and proper shipping names	(3) Hazard Class	(3A) Identi- fication number	(4) Label(s) required (if not excepted)	(5) Packaging (a) Exceptions	(5) Packaging (b) Specific require- ments	Placards required (*denotes placard for any quantity)
	HYDRIODIC ACID	CORROSIVE MATERIAL	UN1787	CORROSIVE	173.244	173.245	CORROSIVE
	HYDROBROMIC ACID, *MORE THAN 49% STRENGTH*	CORROSIVE MATERIAL	UN1788	CORROSIVE	NONE	173.262	CORROSIVE
	HYDROBROMIC ACID, ANHYDROUS. *SEE* HYDROGEN BROMIDE						
	HYDROBROMIC ACID NOT MORE THAN 49% STRENGTH	CORROSIVE MATERIAL	UN1788	CORROSIVE	173.244	173.262	CORROSIVE
+	HYDROCARBON GAS, LIQUEFIED	FLAMMABLE GAS	UN1965	FLAMMABLE GAS	173.306	173.304 173.314	FLAMMABLE GAS
+	HYDROCARBON GAS, NONLIQUEFIED	FLAMMABLE GAS	UN1964	FLAMMABLE GAS	173.306	173.302	FLAMMABLE GAS
	HYDROCHLORIC ACID	CORROSIVE MATERIAL	UN1789	CORROSIVE	173.244	173.263	CORROSIVE
	HYDROCHLORIC ACID, ANHYDROUS. *SEE* HYDROGEN CHLORIDE						
	HYDROCHLORIC ACID MIXTURE	CORROSIVE MATERIAL	NA1789	CORROSIVE	173.244	173.263	CORROSIVE
	HYDROCHLORIC ACID SOLUTION, INHIBITED	CORROSIVE MATERIAL	UN1789	CORROSIVE	173.244	173.263	CORROSIVE
	HYDROCYANIC ACID *(PRUSSIC)* SOLUTION *(5% OR MORE HYDROCYANIC ACID)*	POISON A	UN1613	POISON GAS AND FLAMM- ABLE GAS	NONE	173.332	POISON GAS*
	HYDROCYANIC ACID, LIQUEFIED	POISON A	NA1051	POISON GAS AND FLAMM- ABLE GAS	NONE	173.332	POISON GAS*
	HYDROCYANIC ACID (PRUSSIC), UNSTABILIZED	FORBIDDEN					
+	HYDROCYANIC ACID SOLUTION, LESS THAN 5% HYDRO- CYANIC ACID	POISON B	UN1613	POISON	NONE	173.351	POISON
	HYDROFLUORIC ACID, ANHYDROUS. *SEE* HYDROGEN FLUORIDE						
	HYDROFLUORIC ACID SOLUTION	CORROSIVE MATERIAL	UN1790	CORROSIVE	173.244	173.264	CORROSIVE
	HYDROFLUORIC AND SULFURIC ACID MIXTURE	CORROSIVE MATERIAL	UN1786	CORROSIVE	NONE	173.290	CORROSIVE
	HYDROFLUOROBORIC ACID. *SEE* FLUOBORIC ACID						
	HYDROFLUOROSILICIC ACID OR HYDROFLUOSILICIC ACID	CORROSIVE MATERIAL	NA1778	CORROSIVE	NONE	173.265	CORROSIVE
+	HYDROGEN *OR* HYDROGEN, COMPRESSED	FLAMMABLE GAS	UN1049	FLAMMABLE GAS	173.306	173.302 173.314	FLAMMABLE GAS
+	HYDROGEN, REFRIGERATED LIQUID (*CRYOGENIC LIQ- UID*)	FLAMMABLE GAS	UN1966	FLAMMABLE GAS	NONE	173.316 173.318 173.319	FLAMMABLE GAS
	HYDROGEN BROMIDE	NONFLAMM- ABLE GAS	UN1048	NONFLAMM- ABLE GAS	173.306	173.304	NONFLAMM- ABLE GAS
	HYDROGEN CHLORIDE *OR* HYDROGEN CHLORIDE, ANHYDROUS	NONFLAMM- ABLE GAS	UN1050	NONFLAMM- ABLE GAS	173.306	173.304	NONFLAMM- ABLE GAS
	HYDROGEN CHLORIDE, REFRIGERATED LIQUID	NON FLAMM- ABLE GAS	UN2186	NONFLAMM- ABLE GAS	NONE	173.314 173.315	NONFLAMM- ABLE GAS
	HYDROGEN FLUORIDE	CORROSIVE MATERIAL	UN1052	CORROSIVE	NONE	173.264	CORROSIVE
	HYDROGEN IODIDE SOLUTION. *SEE* HYDRIODIC ACID						
	HYDROGEN PEROXIDE SOLUTION (40% TO 52% PEROX- IDE)	OXIDIZER	UN2014	OXIDIZER	173.244	173.266	OXIDIZER
	HYDROGEN PEROXIDE SOLUTION (8% TO 40% PEROXIDE)	OXIDIZER	UN2014	OXIDIZER	173.244	173.266	OXIDIZER
	HYDROGEN PEROXIDE SOLUTION (OVER 52% PEROXIDE).	OXIDIZER	UN2015	OXIDIZER AND CORROSIVE	NONE	173.266	OXIDIZER

§172.101 HAZARDOUS MATERIALS TABLE

(1) +/ A/ W/	(2) Hazardous materials descriptions and proper shipping names	(3) Hazard Class	(3A) Identi-fication number	(4) Label(s) required (if not excepted)	(5) Packaging (a) Exceptions	(5) Packaging (b) Specific require-ments	Placards required (*denotes placard for any quantity)
+	HYDROGEN SELENIDE	POISON A	UN2202	FLAMMABLE GAS AND POISON GAS	NONE	173.328	POISON GAS
	HYDROGEN SULFATE. *SEE* SULFURIC ACID						
+	HYDROGEN SULFIDE	FLAMMABLE GAS	UN1053	FLAMMABLE GAS AND POISON	NONE	173.304 173.314	FLAMMABLE GAS
	HYDROSILICOFLUORIC ACID. *SEE* HYDROFLUOROSILICIC ACID						
	HYDROXYL AMINE IODIDE	FORBIDDEN					
	HYPOCHLORITE SOLUTION *CONTAINING MORE THAN 7% AVAILABLE CHLORINE BY WEIGHT*	CORROSIVE MATERIAL	UN1791	CORROSIVE	173.244	173.277	CORROSIVE
A	HYPOCHLORITE SOLUTION CONTAINING NOT MORE THAN 7% AVAILABLE CHLORINE *BY WEIGHT*	ORM-B	NA1791	NONE	173.505	173.510	NONE
	HYPONITROUS ACID	FORBIDDEN					
	IGNITER	CLASS C EXPLOSIVE		EXPLOSIVE C	NONE	173.106	DANGEROUS
	IGNITER CORD	CLASS C EXPLOSIVE		EXPLOSIVE C	NONE	173.100	DANGEROUS
	IGNITER FUSE, METAL CLAD	CLASS C EXPLOSIVE		EXPLOSIVE C	NONE	173.106	DANGEROUS
	IGNITER, JET THRUST (JATO)	CLASS A EXPLOSIVE		EXPLOSIVE A	NONE	173.79	EXPLOSIVES A*
	IGNITER, JET THRUST (JATO)	CLASS B EXPLOSIVE		EXPLOSIVE B	NONE	173.92	EXPLOSIVES B*
	IGNITER, ROCKET MOTOR	CLASS A EXPLOSIVE		EXPLOSIVE A	NONE	173.79	EXPLOSIVES A*
	IGNITER, ROCKET MOTOR	CLASS B EXPLOSIVE		EXPLOSIVE B	NONE	173.92	EXPLOSIVES B*
	ILLUMINATING PROJECTILE. SEE FIREWORKS, SPECIAL.						
	IMINOBISPROPYLAMINE	CORROSIVE MATERIAL	UN2269	CORROSIVE	173.244	173.245	CORROSIVE
	INFECTIOUS SUBSTANCE, HUMAN, N.O.S. *SEE* ETIOLOGIC AGENT, N.O.S.		UN2814				
	INITIATING EXPLOSIVE (*DIAZODINITROPHENOL*)	CLASS A EXPLOSIVE		EXPLOSIVE A	NONE	173.70	EXPLOSIVES A*
	INITIATING EXPLOSIVE (*FULMINATE OF MERCURY*)	CLASS A EXPLOSIVE		EXPLOSIVE A	NONE	173.71	EXPLOSIVES A*
	INITIATING EXPLOSIVE (*GUANYL NITROSAMINO GUANYLIDENE HYDRAZINE*)	CLASS A EXPLOSIVE		EXPLOSIVE A	NONE	173.72	EXPLOSIVES A*
	INITIATING EXPLOSIVE (*LEAD AZIDE, DEXTRINATED TYPE ONLY*)	CLASS A EXPLOSIVE		EXPLOSIVE A	NONE	173.73	EXPLOSIVES A*
	INITIATING EXPLOSIVE (*LEAD MONONITRORESORCINATE*)	CLASS A EXPLOSIVE		EXPLOSIVE A	NONE	173.70	EXPLOSIVES A*
	INITIATING EXPLOSIVE (*BARIUM STYPHNATE, MONOHYDRATE) LEAD STYPHNATE (LEAD TRINITRORESORCINATE*)	CLASS A EXPLOSIVE		EXPLOSIVE A	NONE	173.74	EXPLOSIVES A*
	INITIATING EXPLOSIVE (*NITRO MANNITE*)	CLASS A EXPLOSIVE		EXPLOSIVE A	NONE	173.75	EXPLOSIVES A*
	INITIATING EXPLOSIVE (*NITROSOGUANIDINE*)	CLASS A EXPLOSIVE		EXPLOSIVE A	NONE	173.76	EXPLOSIVES A*
	INITIATING EXPLOSIVE (*PENTAERYTHRITE TETRANITRATE*)	CLASS A EXPLOSIVE		EXPLOSIVE A	NONE	173.77	EXPLOSIVES A*
	INITIATING EXPLOSIVE (*TETRAZENE (GUANYL NITROSAMINE GUANYL TETRAZENE*))	CLASS A EXPLOSIVE		EXPLOSIVE A	NONE	173.78	EXPLOSIVES A*

§172.101 HAZARDOUS MATERIALS TABLE

(1) +/ A/ W	(2) Hazardous materials descriptions and proper shipping names	(3) Hazard Class	(3A) Identi- fication number	(4) Label(s) required (if not excepted)	(5) Packaging (a) Exceptions	(5) Packaging (b) Specific require- ments	Placards required (*denotes placard for any quantity)
	INITIATING EXPLOSIVES (DRY)	FORBIDDEN					
	INK	COMBUSTIBLE LIQUID	UN1210	NONE	173.118a	NONE	COMBUSTIBLE[1]
	INK	FLAMMABLE LIQUID	UN1210	FLAMMABLE LIQUID	173.118	173.144	FLAMMABLE
	INOSITOL HEXANITRATE (DRY)	FORBIDDEN					
	INSECTICIDE, DRY, N.O.S.	POISON B	NA2588	POISON	173.364	173.365	POISON
	INSECTICIDE, LIQUEFIED GAS (*CONTAINING NO POISON A OR B MATERIAL*)	NONFLAMM- ABLE GAS	NA1968	NONFLAMM- ABLE GAS	173.306	173.304	NONFLAMM- ABLE GAS
	INSECTICIDE, LIQUEFIED GAS, CONTAINING POISON A MATERIAL *OR* POISON B MATERIAL	POISON A	NA1967	POISON GAS	NONE	173.329 173.334	POISON GAS
	INSECTICIDE, LIQUID, N.O.S.	COMBUSTIBLE LIQUID	NA1993	NONE	173.118a	NONE	COMBUSTIBLE[1]
	INSECTICIDE, LIQUID, N.O.S.	FLAMMABLE LIQUID	NA1993	FLAMMABLE LIQUID	173.118	173.119	FLAMMABLE
	INSECTICIDE, LIQUID, N.O.S.	POISON B	NA2902	POISON	173.345	173.346	POISON
	INULIN TRINITRATE (DRY)	FORBIDDEN					
	IODINE AZIDE (DRY)	FORBIDDEN					
	IODINE MONOCHLORIDE	CORROSIVE MATERIAL	UN1792	CORROSIVE	NONE	173.293	CORROSIVE
	IODINE PENTAFLUORIDE	OXIDIZER	UN2495	OXIDIZER AND POISON	NONE	173.246	OXIDIZER
	IODOXY COMPOUNDS (DRY)	FORBIDDEN					
	IRIDIUM NITRATOPENTAMINE IRIDIUM NITRATE	FORBIDDEN					
	IRON CHLORIDE, SOLID. *SEE* FERRIC CHLORIDE, SOLID						
	IRON MASS *OR* SPONGE, *NOT PROPERLY OXIDIZED*	FLAMMABLE SOLID	NA1383	FLAMMABLE SOLID	NONE	173.174	FLAMMABLE SOLID[5]
	IRON MASS *OR* SPONGE, SPENT	FLAMMABLE SOLID	UN1376	FLAMMABLE SOLID	NONE	173.174	FLAMMABLE SOLID[5]
	IRON OXIDE, SPENT. *SEE* IRON MASS *OR* SPONGE, SPENT						
	IRON SESQUICHLORIDE, SOLID. *SEE* FERRIC CHLORIDE						
	IRRITATING AGENT, N.O.S.	IRRITATING MATERIAL	NA1693	IRRITANT	NONE	173.382	DANGEROUS
	ISOBUTANE *OR* LIQUEFIED PETROLEUM GAS. *SEE* LIQUEFIED PETROLEUM GAS						
	ISOBUTYL ACETATE	FLAMMABLE LIQUID	UN1213	FLAMMABLE LIQUID	173.118	173.119	FLAMMABLE
	ISOBUTYLAMINE	FLAMMABLE LIQUID	UN1214	FLAMMABLE LIQUID	173.118	173.119	FLAMMABLE
	ISOBUTYLENE *OR* LIQUEFIED PETROLEUM GAS. *SEE* LIQUEFIED PETROLEUM GAS						
	ISOBUTYRIC ACID	CORROSIVE MATERIAL	UN2529	CORROSIVE	173.244	173.245	CORROSIVE
	ISOBUTYRIC ANHYDRIDE	CORROSIVE MATERIAL	UN2530	CORROSIVE	173.244	173.245	CORROSIVE
	ISONONANYL PEROXIDE, *TECHNICALLY PURE OR* ISONONANYL PEROXIDE, *IN SOLUTION. SEE* ORGANIC PEROXIDE, LIQUID *OR* SOLUTION, N.O.S.		UN2128				
	ISOOCTANE	FLAMMABLE LIQUID	UN1262	FLAMMABLE LIQUID	173.118	173.119	FLAMMABLE
	ISOOCTENE	FLAMMABLE LIQUID	UN1216	FLAMMABLE LIQUID	173.118	173.119	FLAMMABLE

1-SEE PLACARDING NOTES, PAGE 5

§172.101 HAZARDOUS MATERIALS TABLE

(1) +/ A/ W/	(2) Hazardous materials descriptions and proper shipping names	(3) Hazard Class	(3A) Identi-fication number	(4) Label(s) required (if not excepted)	(5) Packaging (a) Exceptions	(5) Packaging (b) Specific require-ments	Placards required (*denotes placard for any quantity)
	ISOPENTANE	FLAMMABLE LIQUID	UN1265	FLAMMABLE LIQUID	173.118	173.119	FLAMMABLE
	ISOPENTANOIC ACID	CORROSIVE MATERIAL	NA1760	CORROSIVE	173.244	173.245	CORROSIVE
	ISOPRENE	FLAMMABLE LIQUID	UN1218	FLAMMABLE LIQUID	173.118	173.119	FLAMMABLE
	ISOPROPANOL	FLAMMABLE LIQUID	UN1219	FLAMMABLE LIQUID	173.118	173.125	FLAMMABLE
	ISOPROPYL ACETATE	FLAMMABLE LIQUID	UN1220	FLAMMABLE LIQUID	173.118	173.119	FLAMMABLE
	ISOPROPYL ACID PHOSPHATE, SOLID	CORROSIVE MATERIAL	UN1793	CORROSIVE	173.244	173.245b	CORROSIVE
	ISOPROPYL ALCOHOL. *SEE* ISOPROPANOL						
	ISOPROPYLAMINE	FLAMMABLE LIQUID	UN1221	FLAMMABLE LIQUID	NONE	173.119	FLAMMABLE
	ISOPROPYL MERCAPTAN	FLAMMABLE LIQUID	NA2402	FLAMMABLE LIQUID	NONE	173.141	FLAMMABLE
	ISOPROPYL NITRATE	FLAMMABLE LIQUID	UN1222	FLAMMABLE LIQUID	173.118	173.119	FLAMMABLE
	ISOPROPYL PERCARBONATE, STABILIZED	ORGANIC PEROXIDE	NA2134	ORGANIC PEROXIDE	NONE	173.282	ORGANIC PEROXIDE
	ISOPROPYL PERCARBONATE, UNSTABILIZED	ORGANIC PEROXIDE	NA2133	ORGANIC PEROXIDE	NONE	173.218	ORGANIC PEROXIDE
	ISOPROPYL PEROXYDICARBONATE, *TECHNICALLY PURE. SEE* ISOPROPYL PERCARBONATE, UNSTABILIZED		UN2133				
	ISOPROPYL PEROXYDICARBONATE, *NOT MORE THAN 52% IN SOLUTION. SEE* ORGANIC PEROXIDE, LIQUID *OR* SOLUTION, N.O.S.		UN2134				
	ISOPROPYL PHOSPHORIC ACID, SOLID. *SEE* ISOPROPYL ACID PHOSPHATE, SOLID						
	ISOTHIOCYANIC ACID (POLYMERIZATION HAZARD)	FORBIDDEN					
	JET THRUST IGNITER. *SEE* IGNITER, JET THRUST						
	JET THRUST UNIT (*JATO*)	CLASS A EXPLOSIVE		EXPLOSIVE A	NONE	173.79	EXPLOSIVES A*
	JET THRUST UNIT (*JATO*)	CLASS B EXPLOSIVE		EXPLOSIVE B	NONE	173.92	EXPLOSIVES B*
	KEROSENE	COMBUSTIBLE LIQUID	UN1223	NONE	173.118a	NONE	COMBUSTIBLE[1]
	LACQUER BASE *OR* LACQUER CHIPS, PLASTIC (*WET WITH ALCOHOL OR SOLVENT*)	FLAMMABLE LIQUID	UN1263	FLAMMABLE LIQUID	173.118	173.127	FLAMMABLE
	LACQUER BASE, *OR* LACQUER CHIPS, DRY	FLAMMABLE SOLID	NA2557	FLAMMABLE SOLID	173.153	173.175	FLAMMABLE SOLID[5]
	LAUROYL PEROXIDE	ORGANIC PEROXIDE	UN2124	ORGANIC PEROXIDE	173.153	173.157 173.158	ORGANIC PEROXIDE
	LAUROYL PEROXIDE, *NOT MORE THAN 42%, STABLE DISPERSION, IN WATER. SEE* ORGANIC PEROXIDE, LIQUID *OR* SOLUTION, N.O.S.		UN2893				
	LAUROYL PEROXIDE, *TECHNICALLY PURE. SEE* LAUROYL PEROXIDE		UN2124				
	LEAD ARSENATE, SOLID	POISON B	UN1617	POISON	173.364	173.367	POISON
	LEAD ARSENITE, SOLID	POISON B	UN1618	POISON	173.364	173.365	POISON
	LEAD AZIDE. SEE INITIATING EXPLOSIVE						
	LEAD AZIDE (DRY)	FORBIDDEN					

5-SEE PLACARDING NOTES, PAGE 5

§172.101 HAZARDOUS MATERIALS TABLE

(1) +/ A/ W/	(2) Hazardous materials descriptions and proper shipping names	(3) Hazard Class	(3A) Identi-fication number	(4) Label(s) required (if not excepted)	(5) Packaging (a) Exceptions	(5) Packaging (b) Specific require-ments	Placards required (*denotes placard for any quantity)
A	LEAD CHLORIDE	ORM-B	NA2291	NONE	173.505	173.800	NONE
	LEAD CYANIDE	POISON B	UN1620	POISON	173.370		POISON
W	LEAD DROSS *(CONTAINING 3% OR MORE FREE ACID)*	ORM-C	NA1794	NONE	173.505	173.1010	NONE
A	LEAD FLUOBORATE	ORM-B	NA2291	NONE	173.505	173.510	NONE
A	LEAD FLUORIDE	ORM-B	NA2811	NONE	173.505	173.510	NONE
	LEAD MONONITRORESORCINATE. SEE INITIATING EXPLOSIVE						
	LEAD MONONITRORESORCINATE (DRY)	FORBIDDEN					
	LEAD NITRATE	OXIDIZER	UN1469	OXIDIZER	173.153	173.182	OXIDIZER
	LEAD PEROXIDE	OXIDIZER	UN1872	OXIDIZER	173.153	173.154	OXIDIZER
	LEAD PICRATE (DRY)	FORBIDDEN					
W	LEAD SCRAP. *SEE* LEAD DROSS						
	LEAD STYPHNATE (DRY)	FORBIDDEN					
	LEAD STYPHNATE (LEAD TRINITRORESORCINATE). SEE INITIATING , EXPLOSIVE						
	LEAD SULFATE, SOLID *(CONTAINING MORE THAN 3% FREE ACID)*	CORROSIVE MATERIAL	UN1794	CORROSIVE	173.244	173.245b	CORROSIVE
	LIFESAVING APPLIANCES, SELF INFLATING	ORM-C	UN2990	NONE	NONE	173.906	NONE
	LIME-NITROGEN. SEE CALCIUM CYANAMIDE, NOT HYDRATED						
	LIME, UNSLAKED. *SEE* CALCIUM OXIDE						
A	LINDANE	ORM-A	NA2761	NONE	173.505	173.510	NONE
+	LIQUEFIED HYDROCARBON GAS. *SEE* HYDROCARBON GAS, LIQUEFIED						
	LIQUEFIED NONFLAMMABLE GAS *(CHARGED WITH NITROGEN, CARBON DIOXIDE, OR AIR)*	NONFLAMMABLE GAS	NA1058	NONFLAMMABLE GAS	173.306	173.304	NONFLAMMABLE GAS
+	LIQUEFIED PETROLEUM GAS	FLAMMABLE GAS	UN1075	FLAMMABLE GAS	173.306	173.304 173.314 173.315	FLAMMABLE GAS
	LIQUID OTHER THAN ONE CLASSED AS FLAMMABLE, CORROSIVE, POISON OR IRRITANT, CHARGED WITH NITROGEN, CARBON DIOXIDE, OR AIR. SEE COMPRESSED GAS N.O.S.						
	LITHIUM ACETYLIDE-ETHYLENE DIAMINE COMPLEX	FLAMMABLE SOLID	NA2813	FLAMMABLE SOLID AND DANGEROUS WHEN WET	NONE	173.206	FLAMMABLE SOLID -W-*
	LITHIUM ALUMINUM HYDRIDE	FLAMMABLE SOLID	UN1410	FLAMMABLE SOLID AND DANGEROUS WHEN WET	NONE	173.206	FLAMMABLE SOLID -W-*
	LITHIUM ALUMINUM HYDRIDE, ETHEREAL	FLAMMABLE LIQUID	UN1411	FLAMMABLE LIQUID	NONE	173.137	FLAMMABLE
	LITHIUM AMIDE, POWDERED	FLAMMABLE SOLID	UN1412	FLAMMABLE SOLID	173.153	173.168	FLAMMABLE SOLID[5]
	LITHIUM BATTERIES, FOR DISPOSAL	ORM-C		NONE	NONE	173.1015	NONE
	LITHIUM BATTERY. SEE 173.206(f).						

5-SEE PLACARDING NOTES, PAGE 5

§172.101 HAZARDOUS MATERIALS TABLE

(1) +/ A/ W	(2) Hazardous materials descriptions and proper shipping names	(3) Hazard Class	(3A) Identi- fication number	(4) Label(s) required (if not excepted)	(5) Packaging (a) Exceptions	(5) Packaging (b) Specific require- ments	Placards required (*denotes placard for any quantity)
	LITHIUM BOROHYDRIDE	FLAMMABLE SOLID	UN1413	FLAMMABLE SOLID AND DANGEROUS WHEN WET	NONE	173.206	FLAMMABLE SOLID -W-*
	LITHIUM FERROSILICON	FLAMMABLE SOLID	UN2830	FLAMMABLE SOLID AND DANGEROUS WHEN WET	NONE	173.206	FLAMMABLE SOLID -W-*
	LITHIUM HYDRIDE	FLAMMABLE SOLID	UN1414	FLAMMABLE SOLID AND DANGEROUS WHEN WET	NONE	173.206	FLAMMABLE SOLID -W-*
	LITHIUM HYDRIDE IN FUSED SOLID FORM	FLAMMABLE SOLID	UN2805	FLAMMABLE SOLID AND DANGEROUS WHEN WET	NONE	173.206	FLAMMABLE SOLID -W-*
	LITHIUM HYPOCHLORITE COMPOUND, DRY *(CONTAINING MORE THAN 39% AVAILABLE CHLORINE)*	OXIDIZER	UN1471	OXIDIZER	173.153	173.217	OXIDIZER
	LITHIUM METAL	FLAMMABLE SOLID	UN1415	FLAMMABLE SOLID AND DANGEROUS WHEN WET	NONE	173.206	FLAMMABLE SOLID -W-*
	LITHIUM METAL, IN CARTRIDGES	FLAMMABLE SOLID	UN1415	FLAMMABLE SOLID AND DANGEROUS WHEN WET	173.206	173.206	FLAMMABLE SOLID -W-*
	LITHIUM NITRIDE	FLAMMABLE SOLID	UN2806	FLAMMABLE SOLID AND DANGEROUS WHEN WET	NONE	173.206	FLAMMABLE SOLID -W-*
	LITHIUM PEROXIDE	OXIDIZER	UN1472	OXIDIZER	173.153	173.154	OXIDIZER
	LITHIUM SILICON	FLAMMABLE SOLID	UN1417	FLAMMABLE SOLID AND DANGEROUS WHEN WET	NONE	173.206	FLAMMABLE SOLID -W-*
	LONDON PURPLE, SOLID	POISON B	UN1621	POISON	173.364	173.365	POISON
	LOW BLASTING EXPLOSIVE. SEE LOW EXPLOSIVE						
	LOW EXPLOSIVE	CLASS A EXPLOSIVE		EXPLOSIVE A	NONE	173.60	EXPLOSIVES A*
	LYE. SEE SODIUM HYDROXIDE, SOLID						
	MAGNESIUM ALUMINUM PHOSPHIDE	FLAMMABLE SOLID	UN1419	FLAMMABLE SOLID AND DANGEROUS WHEN WET	NONE	173.206	FLAMMABLE SOLID -W-*
+	MAGNESIUM ARSENATE, SOLID	POISON B	UN1622	POISON	173.364	173.367	POISON
	MAGNESIUM DROSS, WET OR HOT. SEE 173.173	FORBIDDEN					
	MAGNESIUM GRANULES COATED, *PARTICLE SIZE NOT LESS THAN 149 MICRONS*	FLAMMABLE SOLID	UN2950	FLAMMABLE SOLID AND DANGEROUS WHEN WET	173.153	173.178	FLAMMABLE SOLID -W-*
	MAGNESIUM, METAL *(POWDERED, PELLETS, TURNINGS, OR RIBBON) OR* MAGNESIUM ALUMINUM POWDER	FLAMMABLE SOLID	UN1869	FLAMMABLE SOLID AND DANGEROUS WHEN WET	173.153	173.220	FLAMMABLE SOLID -W-*
	MAGNESIUM NITRATE	OXIDIZER	UN1474	OXIDIZER	173.153	173.182	OXIDIZER
	MAGNESIUM PERCHLORATE	OXIDIZER	UN1475	OXIDIZER	173.153	173.154	OXIDIZER
	MAGNESIUM PEROXIDE, SOLID	OXIDIZER	UN1476	OXIDIZER	173.153	173.154	OXIDIZER

§172.101 HAZARDOUS MATERIALS TABLE

(1) +/ A/ W	(2) Hazardous materials descriptions and proper shipping names	(3) Hazard Class	(3A) Identi- fication number	(4) Label(s) required (if not excepted)	(5) Packaging (a) Exceptions	(5) Packaging (b) Specific require- ments	Placards required (*denotes placard for any quantity)
	MAGNESIUM SCRAP (*BORINGS, CLIPPINGS, SHAVINGS, SHEET, TURNINGS, OR SCALPINGS*)	FLAMMABLE SOLID	NA1869	FLAMMABLE SOLID AND DANGEROUS WHEN WET	173.153	173.220	FLAMMABLE SOLID -W-*
A	MAGNETIZED MATERIAL	ORM-C	UN2807	MAGNETIZED MATERIAL	NONE	173.1020	NONE
A	*MAGNETIZED MATERIAL SEE 173.21(F)*						
A	MALATHION	ORM-A	NA2783	NONE	173.505	173.510	NONE
A	MALEIC ACID	ORM-A	NA2215	NONE	173.505	173.510	NONE
AW	MALEIC ANHYDRIDE	ORM-A	UN2215	NONE	173.505	173.510	NONE
	MANNITAN TETRANITRATE	FORBIDDEN					
	MATCHES, BLOCK. SEE MATCHES, STRIKE ANYWHERE						
	MATCHES, SAFETY, *BOOK, CARD, OR STRIKE-ON-BOX*	FLAMMABLE SOLID	UN1944	NONE	173.176	NONE	FLAMMABLE SOLID[5]
	MATCHES, STRIKE ANYWHERE	FLAMMABLE SOLID	UN1331	FLAMMABLE SOLID	NONE	173.176a	FLAMMABLE SOLID[5]
	MATTING ACID. *SEE* SULFURIC ACID						
	MEDICINES, N.O.S.	COMBUSTIBLE LIQUID	UN1851	NONE	173.118a	NONE	COMBUSTIBLE[1]
	MEDICINES, N.O.S.	FLAMMABLE LIQUID	UN1851	FLAMMABLE LIQUID	173.118	173.119	FLAMMABLE
	MEDICINES, N.O.S.	FLAMMABLE SOLID	UN1851	FLAMMABLE SOLID	173.153	173.154	FLAMMABLE SOLID
	MEDICINES, N.O.S.	OXIDIZER	UN1851	OXIDIZER	173.153	173.154	OXIDIZER
	MEDICINES, LIQUID, N.O.S.	CORROSIVE MATERIAL	UN1851	CORROSIVE	173.244	173.245	CORROSIVE
	MEDICINES, LIQUID, N.O.S.	POISON B	UN1851	POISON	173.345	173.346	POISON
	MEDICINES, SOLID, N.O.S.	CORROSIVE MATERIAL	UN1851	CORROSIVE	173.244	173.245b	CORROSIVE
	MEDICINES, SOLID, N.O.S.	POISON B	UN1851	POISON	173.364	173.365	POISON
	P-MENTHANE HYDROPEROXIDE, *TECHNICALLY PURE. SEE* PARAMENTHANE HYDROPEROXIDE		UN2125				
	MERCAPTAN MIXTURE, ALIPHATIC	FLAMMABLE LIQUID	NA1228	FLAMMABLE LIQUID	NONE	173.141	FLAMMABLE
	MERCAPTAN MIXTURE, ALIPHATIC (*IN CONTAINERS OVER 110 GALLONS*)	COMBUSTIBLE LIQUID	NA1228	NONE	173.118a	NONE	COMBUSTIBLE[1]
AW	MERCAPTAN MIXTURE, ALIPHATIC (*IN CONTAINERS OF 110 GALLONS OR LESS). SEE 173.141(B)*	ORM-A	NA1228	NONE	173.505	173.510	NONE
+	MERCURIC ACETATE	POISON B	UN1629	POISON	173.364	173.365	POISON
+	MERCURIC AMMONIUM CHLORIDE, SOLID	POISON B	UN1630	POISON	173.364	173.365	POISON
+	MERCURIC BENZOATE, SOLID	POISON B	UN1631	POISON	173.364	173.365	POISON
+	MERCURIC BROMIDE, SOLID	POISON B	UN1634	POISON	173.364	173.365	POISON
	MERCURIC CHLORIDE, SOLID	POISON B	UN1624	POISON	173.364	173.372	POISON
+	MERCURIC CYANIDE, SOLID	POISON B	UN1636	POISON	173.370	173.370	POISON
	MERCURIC IODIDE, SOLID	POISON B	UN1638	POISON	173.364	173.365	POISON
	MERCURIC IODIDE, SOLUTION	POISON B	UN1638	POISON	173.345	173.346	POISON
	MERCURIC NITRATE	OXIDIZER	UN1625	OXIDIZER	173.153	173.182	OXIDIZER

1,5-SEE PLACARDING NOTES, PAGE 5

§172.101 HAZARDOUS MATERIALS TABLE

(1) +/ A/ W	(2) Hazardous materials descriptions and proper shipping names	(3) Hazard Class	(3A) Identi-fication number	(4) Label(s) required (if not excepted)	(5) Packaging (a) Exceptions	(5) Packaging (b) Specific require-ments	Placards required (*denotes placard for any quantity)
+	MERCURIC OLEATE, SOLID	POISON B	UN1640	POISON	173.364	173.365	POISON
	MERCURIC OXIDE, SOLID	POISON B	UN1641	POISON	173.364	173.365	POISON
+	MERCURIC OXYCYANIDE, SOLID (*DESENSITIZED*)	POISON B	UN1642	POISON	173.364	173.365	POISON
+	MERCURIC POTASSIUM CYANIDE, SOLID	POISON B	UN1626	POISON	173.364	173.365 173.370	POISON
+	MERCURIC POTASSIUM IODIDE, SOLID	POISON B	UN1643	POISON	173.364	173.365	POISON
+	MERCURIC SALICYLATE SOLID	POISON B	UN1644	POISON	173.364	173.365	POISON
+	MERCURIC SUBSULFATE, SOLID	POISON B	NA2025	POISON	173.364	173.365	POISON
+	MERCURIC SULFATE, SOLID	POISON B	UN1645	POISON	173.364	173.365	POISON
+	MERCURIC SULFOCYANATE, SOLID *OR* MERCURIC THIOCYANATE, SOLID	POISON B	UN1646	POISON	173.364	173.365	POISON
+	MERCUROL *OR* MERCURY NUCLEATE, SOLID	POISON B	UN1639	POISON	173.364	173.365	POISON
+	MERCUROUS ACETATE, SOLID	POISON B	UN1629	POISON	173.364	173.365	POISON
	MERCUROUS AZIDE	FORBIDDEN					
+	MERCUROUS BROMIDE, SOLID	POISON B	UN1634	POISON	173.364	173.365	POISON
+	MERCUROUS GLUCONATE, SOLID	POISON B	UN1637	POISON	173.364	173.365	POISON
+	MERCUROUS IODIDE, SOLID	POISON B	UN1638	POISON	173.364	173.365	POISON
	MERCUROUS NITRATE, SOLID	OXIDIZER	UN1627	OXIDIZER	173.153	173.154	OXIDIZER
+	MERCUROUS OXIDE, BLACK, SOLID	POISON B	UN1641	POISON	173.364	173.365	POISON
+	MERCUROUS SULFATE, SOLID	POISON B	UN1628	POISON	173.364	173.365	POISON
	MERCURY ACETYLIDE	FORBIDDEN					
	MERCURY BASED PESTICIDE, LIQUID, N.O.S. (*COMPOUNDS AND PREPARATIONS*)	FLAMMABLE LIQUID	UN2778	FLAMMABLE LIQUID	173.118	173.119	FLAMMABLE
	MERCURY BASED PESTICIDE, LIQUID, N.O.S. (*COMPOUNDS AND PREPARATIONS*)	POISON B	UN2777	POISON	173.345	173.346	POISON
	MERCURY BASED PESTICIDE, SOLID, N.O.S. (*COMPOUNDS AND PREPARATIONS*)	POISON B	UN2777	POISON	173.364	173.365	POISON
	MERCURY COMPOUND, SOLID, N.O.S.	POISON B	UN2025	POISON	173.364	173.365	POISON
	MERCURY FULMINATE. SEE INITIATING EXPLOSIVE						
	MERCURY IODIDE AQUABASIC AMMONOBASIC (IODIDE OF MILLION'S BASE)	FORBIDDEN					
A	MERCURY, METALLIC	ORM-B	NA2809	NONE	NONE	173.860	NONE
	MERCURY NITRIDE	FORBIDDEN					
	MERCURY OXYCYANIDE	FORBIDDEN					
	MESITYL OXIDE	FLAMMABLE LIQUID	UN1229	FLAMMABLE LIQUID	NONE	173.119	FLAMMABLE
	METAL ALKYL, SOLUTION, N.O.S.	FLAMMABLE LIQUID	NA9195	FLAMMABLE LIQUID	173.118	173.119	FLAMMABLE
W	METAL BORINGS, SHAVINGS, TURNINGS, *OR* CUTTINGS (*FERROUS METALS ONLY, EXCEPT STAINLESS STEEL*)	ORM-C	UN2793	NONE	173.505	173.1025	NONE

§172.101 HAZARDOUS MATERIALS TABLE

(1) +/ A/ W	(2) Hazardous materials descriptions and proper shipping names	(3) Hazard Class	(3A) Identi-fication number	(4) Label(s) required (if not excepted)	(5) Packaging (a) Exceptions	(5) Packaging (b) Specific require-ments	Placards required (*denotes placard for any quantity)
	METAL SALTS OF METHYL NITRAMINE (DRY)	FORBIDDEN					
+	METHANE *OR* METHANE, COMPRESSED	FLAMMABLE GAS	UN1971	FLAMMABLE GAS	173.306	173.302	FLAMMABLE GAS
	METHANE, REFRIGERATED LIQUID (*CRYOGENIC LIQUID*)	FLAMMABLE GAS	UN1972	FLAMMABLE GAS	NONE	173.318	FLAMMABLE GAS
	METHANOL. *SEE* METHYL ALCOHOL						
	METHAZOIC ACID	FORBIDDEN					
	METHYL ACETATE	FLAMMABLE LIQUID	UN1231	FLAMMABLE LIQUID	173.118	173.119	FLAMMABLE
	METHYL ACETONE	FLAMMABLE LIQUID	UN1232	FLAMMABLE LIQUID	173.118	173.119	FLAMMABLE
+	METHYLACETYLENE-PROPADIENE, STABILIZED	FLAMMABLE GAS	UN1060	FLAMMABLE GAS	173.306	173.304 173.314 173.315	FLAMMABLE GAS
	METHYL ACRYLATE, INHIBITED	FLAMMABLE LIQUID	UN1919	FLAMMABLE LIQUID	173.118	173.119	FLAMMABLE
	METHYLAL	FLAMMABLE LIQUID	UN1234	FLAMMABLE LIQUID	NONE	173.119	FLAMMABLE
	METHYL ALCOHOL	FLAMMABLE LIQUID	UN1230	FLAMMABLE LIQUID	173.118	173.119	FLAMMABLE
+	METHYLAMINE, ANHYDROUS	FLAMMABLE GAS	UN1061	FLAMMABLE GAS	173.306	173.304 173.314 173.315	FLAMMABLE GAS
	METHYLAMINE, AQUEOUS SOLUTION	FLAMMABLE LIQUID	UN1235	FLAMMABLE LIQUID	173.118	173.119	FLAMMABLE
	METHYLAMINE DINITRAMINE AND DRY SALTS THEREOF	FORBIDDEN					
	METHYLAMINE NITROFORM	FORBIDDEN					
	METHYLAMINE PERCHLORATE (DRY)	FORBIDDEN					
	METHYLAMYL ACETATE	FLAMMABLE LIQUID	UN1233	FLAMMABLE LIQUID	173.118	173.119	FLAMMABLE
	METHYL AMYL KETONE	COMBUSTIBLE LIQUID	UN1110	NONE	173.118a	NONE	COMBUSTIBLE[1]
+	METHYL BROMIDE AND MORE THAN 2% CHLOROPICRIN MIXTURE, LIQUID	POISON B	NA1581	POISON	NONE	173.353	POISON[2]
+	METHYL BROMIDE AND NONFLAMMABLE, NONLIQUEFIED COMPRESSED GAS MIXTURE, LIQUID (*INCLUDING UP TO 2% CHLOROPICRIN*)	POISON B	NA1955	POISON	NONE	173.353a	POISON[2]
+	METHYL BROMIDE - ETHYLENE DIBROMIDE MIXTURE, LIQUID	POISON B	UN1647	POISON	NONE	173.353	POISON[2]
+	METHYL BROMIDE, LIQUID (*INCLUDING UP TO 2% CHLOROPICRIN*)	POISON B	UN1062	POISON	NONE	173.353	POISON[2]
	METHYL BUTENE	FLAMMABLE LIQUID	UN2460	FLAMMABLE LIQUID	NONE	173.119	FLAMMABLE
	METHYL BUTYRATE	FLAMMABLE LIQUID	UN1237	FLAMMABLE LIQUID	173.118	173.119	FLAMMABLE
	METHYL CELLOSOLVE. SEE ETHYLENE GLYCOL MONOMETHYL ETHER						
	METHYL CELLOSOLVE ACETATE. SEE ETHYLENE GLYCOL MONOMETHYL ETHER ACETATE						
+	METHYL CHLORIDE	FLAMMABLE GAS	UN1063	FLAMMABLE GAS	173.306	173.304 173.314 173.315	FLAMMABLE GAS
+	METHYL CHLORIDE-METHYLENE CHLORIDE MIXTURE	FLAMMABLE GAS	UN1912	FLAMMABLE GAS	173.306	173.304 173.314	FLAMMABLE GAS

2-SEE PLACARDING NOTES, PAGE 5

§172.101 HAZARDOUS MATERIALS TABLE

(1) +/A/W	(2) Hazardous materials descriptions and proper shipping names	(3) Hazard Class	(3A) Identi-fication number	(4) Label(s) required (if not excepted)	(5) Packaging (a) Exceptions	(5) Packaging (b) Specific require-ments	Placards required (*denotes placard for any quantity)
	METHYL CHLOROCARBONATE. *SEE* METHYL CHLOROFORMATE						
A	METHYL CHLOROFORM. *SEE* 1,1,1-TRICHLOROETHANE						
	METHYL CHLOROFORMATE	FLAMMABLE LIQUID	UN1238	FLAMMABLE LIQUID AND POISON	NONE	173.288	FLAMMABLE[2]
	METHYLCHLOROMETHYL ETHER, ANHYDROUS	FLAMMABLE LIQUID	UN1239	FLAMMABLE LIQUID AND POISON	NONE	173.143	FLAMMABLE[2]
	METHYL CYANIDE	FLAMMABLE LIQUID	UN1648	FLAMMABLE LIQUID	1783.118	173.119	FLAMMABLE[2]
	METHYLCYCLOHEXANE	FLAMMABLE LIQUID	UN2296	FLAMMABLE LIQUID	173.118	173.119	FLAMMABLE
	METHYLCYCLOPENTANE	FLAMMABLE LIQUID	UN2298	FLAMMABLE LIQUID	173.118	173.119	FLAMMABLE
	METHYL DICHLOROACETATE	CORROSIVE MATERIAL	UN2299	CORROSIVE	173.244	173.245	CORROSIVE
	METHYLDICHLOROARSINE	POISON A	NA1556	POISON GAS	NONE	173.328	POISON GAS*
	METHYL DICHLOROSILANE	FLAMMABLE LIQUID	UN1242	FLAMMABLE LIQUID	NONE	173.136	FLAMMABLE
	METHYLENE CHLORIDE. *SEE* DICHLOROMETHANE						
	METHYLENE GLYCOL DINITRATE	FORBIDDEN					
	METHYL ETHYL ETHER. *SEE* ETHYL METHYL ETHER						
	METHYL ETHYL KETONE	FLAMMABLE LIQUID	UN1193	FLAMMABLE LIQUID	173.118	173.119	FLAMMABLE
	METHYL ETHYL KETONE PEROXIDE, *IN SOLUTION WITH NOT MORE THAN 9% BY WEIGHT ACTIVE OXYGEN. SEE* ORGANIC PEROXIDE, LIQUID, *OR* SOLUTION, N.O.S.		UN2550				
	METHYL ETHYL KETONE PEROXIDE, IN SOLUTION WITH MORE THAN 9% BY WEIGHT ACTIVE OXYGEN	FORBIDDEN					
	METHYL ETHYL PYRIDINE	CORROSIVE MATERIAL	UN2300	CORROSIVE	173.244	173.245	CORROSIVE
	METHYL FORMATE	FLAMMABLE LIQUID	UN1243	FLAMMABLE LIQUID	173.118	173.119	FLAMMABLE
	METHYLFURAN	FLAMMABLE LIQUID	UN2301	FLAMMABLE LIQUID	173.118	173.119	FLAMMABLE
	A-METHYLGLUCOSID TETRANITRATE	FORBIDDEN					
	A-METHYLGLYCEROL TRINITRATE	FORBIDDEN					
	METHYLHYDRAZINE	FLAMMABLE LIQUID	UN1244	FLAMMABLE LIQUID AND POISON	NONE	173.145	FLAMMABLE[2]
	METHYL ISOBUTYL KETONE PEROXIDE, *IN SOLUTION WITH NOT MORE THAN 9% BY WEIGHT ACTIVE OXYGEN. SEE* ORGANIC PEROXIDE, LIQUID *OR* SOLUTION, N.O.S.		UN2126				
	METHYL ISOBUTYL KETONE PEROXIDE, IN SOLUTION WITH MORE THAN 9% BY WEIGHT ACTIVE OXYGEN	FORBIDDEN					
	METHYL ISOCYANATE	FLAMMABLE LIQUID	UN2480	FLAMMABLE LIQUID AND POISON	NONE	173.119 173.3a	FLAMMABLE[2]
	METHYL ISOPROPENYL KETONE, INHIBITED	FLAMMABLE LIQUID	UN1246	FLAMMABLE LIQUID	173.118	173.119	FLAMMABLE
	METHYL MAGNESIUM BROMIDE IN ETHYL ETHER *NOT OVER 40% CONCENTRATION*	FLAMMABLE LIQUID	UN1928	FLAMMABLE LIQUID	NONE	173.149	FLAMMABLE

2-SEE PLACARDING NOTES, PAGE 5

§172.101 HAZARDOUS MATERIALS TABLE

(1) +/ A/ W/	(2) Hazardous materials descriptions and proper shipping names	(3) Hazard Class	(3A) Identi-fication number	(4) Label(s) required (if not excepted)	(5) Packaging (a) Exceptions	(5) Packaging (b) Specific require-ments	Placards required (*denotes placard for any quantity)
+	METHYL MERCAPTAN	FLAMMABLE GAS	UN1064	FLAMMABLE GAS	173.306	173.304 173.314 173.315	FLAMMABLE GAS
	METHYL METHACRYLATE MONOMER, INHIBITED	FLAMMABLE LIQUID	UN1247	FLAMMABLE LIQUID	173.118	173.119	FLAMMABLE
	METHYL METHACRYLATE MONOMER, UNINHIBITED *(HIGH-PURITY, IF ACCEPTABLE UNDER 173.21 OF THIS SUBCHAPTER)*	FLAMMABLE LIQUID	NA1247	FLAMMABLE LIQUID	173.118	173.119	FLAMMABLE
	METHYL NITRATE	FORBIDDEN					
	N-METHYL-N'-NITRO-N-NITROSOGUANIDINE *(NOT EXCEEDING 25 GRAMS IN ONE OUTSIDE PACKAGING).*	FLAMMABLE SOLID	NA1325	FLAMMABLE SOLID	NONE	173.179	FLAMMABLE SOLID[5]
	METHYL PARATHION, LIQUID	POISON B	NA2783	POISON	NONE	173.358	POISON
	METHYL PARATHION MIXTURE, DRY	POISON B	NA2783	POISON	173.377	173.377	POISON
	METHYL PARATHION MIXTURE, LIQUID, *(CONTAINING OVER 25% METHYL PARATHION)*	POISON B	NA2783	POISON	NONE	173.359	POISON
	METHYL PARATHION MIXTURE, LIQUID, (CONTAINING 25% OR LESS METHYL PARATHION)	POISON B	NA2783	POISON	NONE	173.359	POISON
	METHYLPENTADIENE	FLAMMABLE LIQUID	UN2461	FLAMMABLE LIQUID	173.118	173.119	FLAMMABLE
	METHYL PENTANE	FLAMMABLE LIQUID	UN2462	FLAMMABLE LIQUID	173.118	173.119	FLAMMABLE
	METHYL PHOSPHONIC DICHLORIDE	CORROSIVE MATERIAL	NA9206	CORROSIVE AND POISON	NONE	173.271	CORROSIVE
	METHYL PHOSPHONOTHIOIC DICHLORIDE, ANHYDROUS	CORROSIVE MATERIAL	NA1760	CORROSIVE	173.244	173.245 173.245a	CORROSIVE
	METHYL PHOSPHONOUS DICHLORIDE. *SEE* PYROFORIC LIQUID, N.O.S.						
	METHYL PICRIC ACID (HEAVY METAL SALTS OF)	FORBIDDEN					
	METHYL PROPIONATE	FLAMMABLE LIQUID	UN1248	FLAMMABLE LIQUID	173.118	173.119	FLAMMABLE
	METHYL PROPYL KETONE	FLAMMABLE LIQUID	UN1249	FLAMMABLE LIQUID	173.118	173.119	FLAMMABLE
	METHYL SULFATE. *SEE* DIMETHYL SULFATE						
	METHYL SULFIDE. *SEE* DIMETHYL SULFIDE						
	METHYLTRICHLOROSILANE	FLAMMABLE LIQUID	UN1250	FLAMMABLE LIQUID	NONE	173.135	FLAMMABLE
	METHYL TRIMETHYLOL METHANE TRINITRATE	FORBIDDEN					
	METHYL VINYL KETONE, INHIBITED	FLAMMABLE LIQUID	UN1251	FLAMMABLE LIQUID	173.147	173.147	FLAMMABLE
	MEVINPHOS	POISON B	NA2783	POISON	NONE	173.358	POISON
	MEVINPHOS MIXTURE, DRY	POISON B	NA2783	POISON	173.377	173.377	POISON
	MEVINPHOS MIXTURE, LIQUID	POISON B	NA2783	POISON	173.359	173.359	POISON
	MEXACARBATE	POISON B	NA2757	POISON	173.364	173.365	POISON
	MILD DETONATING FUSE, METAL CLAD. SEE FUSE, MILD DETONATING, METAL CLAD						
	MINE, EMPTY. SEE 173.55						

5-SEE PLACARDING NOTES, PAGE 5

§172.101 HAZARDOUS MATERIALS TABLE

(1) +/ A/ W	(2) Hazardous materials descriptions and proper shipping names	(3) Hazard Class	(3A) Identi-fication number	(4) Label(s) required (if not excepted)	(5) Packaging (a) Exceptions	(5) Packaging (b) Specific require-ments	Placards required (*denotes placard for any quantity)
	MINE, EXPLOSIVE, WITH GASEOUS MATERIAL. SEE EXPLOSIVE MINE						
	MINE RESCUE EQUIPMENT CONTAINING CARBON DIOXIDE	NONFLAMMABLE GAS	NA1956	NONFLAMMABLE GAS	173.306		NONFLAMMABLE GAS
	MINING REAGENT, LIQUID *(CONTAINING 20% OR MORE CRESYLIC ACID)*	CORROSIVE MATERIAL	NA2022	CORROSIVE	173.244	173.249a	CORROSIVE
A	MIPAFOX	ORM-A	UN2783	NONE	173.505	173.510	NONE
	MIXED ACID. *SEE* NITRATING ACID						
A	MOLYBDENUM PENTACHLORIDE	ORM-B	UN2508	NONE	173.505	173.800	NONE
	MONOCHLOROACETONE, STABILIZED *OR* INHIBITED	IRRITATING MATERIAL	UN1695	IRRITANT	NONE	173.384	DANGEROUS
	MONOCHLOROACETONE (UNSTABILIZED)	FORBIDDEN					
	MONOCHLOROETHYLENE. *SEE* VINYL CHLORIDE						
	MONOETHANOLAMINE	CORROSIVE MATERIAL	UN2491	CORROSIVE	173.244	173.245	CORROSIVE
	MONOETHYLAMINE	FLAMMABLE LIQUID	UN1036	FLAMMABLE LIQUID	NONE	173.148	FLAMMABLE
	MONOFLUOROPHOSPHORIC ACID, ANHYDROUS	CORROSIVE MATERIAL	UN1776	CORROSIVE	NONE	173.275	CORROSIVE
	MORPHOLINE	FLAMMABLE LIQUID	UN2054	FLAMMABLE LIQUID	173.118	173.119	FLAMMABLE
	MORPHOLINE, AQUEOUS, MIXTURE	FLAMMABLE LIQUID	NA2054	FLAMMABLE LIQUID	173.118	173.119	FLAMMABLE
	MORPHOLINE, AQUEOUS, MIXTURE	CORROSIVE MATERIAL	NA1760	CORROSIVE	173.244	173.245	CORROSIVE
	MOTH BALLS. SEE NAPHTHALENE						
	MOTION PICTURE FILM. SEE FILM						
+	MOTOR FUEL ANTIKNOCK COMPOUND *OR* ANTIKNOCK COMPOUND *(THESE MATERIALS MAY CONTAIN VARIOUS HAZARDOUS SUBSTANCES FOR WHICH THE APPROPRIATE RQ APPLIES)*	POISON B	UN1649	POISON	NONE	173.354	POISON
	MOTOR, INTERNAL COMBUSTION				173.120		
+	MOTOR VEHICLE, *ETC., INCLUDING AUTOMOBILE, MOTORCYCLE, TRUCK, TRACTOR, AND OTHER SELF-PROPELLED VEHICLE OR EQUIPMENT POWERED BY INTERNAL COMBUSTION ENGINE, WHEN OFFERED NEW OR USED FOR TRANSPORTATION AND WHICH CONTAIN FUEL IN THE ENGINE OR FUEL TANK OR THE ELECTRIC STORAGE BATTERY IS CONNECTED TO EITHER TERMINAL OF THE ELECTRICAL SYSTEM*	ORM-C		NONE	173.120 173.250 173.257 173.306 175.305 176.905		NONE
	MURIATIC ACID. *SEE* HYDROCHLORIC ACID						
	NAPHTHA	COMBUSTIBLE LIQUID	UN2553	NONE	173.118a	NONE	COMBUSTIBLE[1]
	NAPHTHA	FLAMMABLE LIQUID	UN2553	FLAMMABLE LIQUID	173.118	173.119	FLAMMABLE
AW	NAPHTHALENE *OR* NAPHTHALIN	ORM-A	UN1334	NONE	173.505	173.655	NONE
	NAPHTHALENE DIOZONIDE	FORBIDDEN					
	NAPHTHA PETROLEUM. SEE PETROLEUM NAPHTHA						
	NAPHTHYL AMINEPERCHLORATE	FORBIDDEN					
	NATURAL GASOLINE. SEE GASOLINE						

(1) +/ A/ W/	(2) Hazardous materials descriptions and proper shipping names	(3) Hazard Class	(3A) Identification number	(4) Label(s) required (if not excepted)	(5) Packaging (a) Exceptions	(5) Packaging (b) Specific requirements	Placards required (*denotes placard for any quantity)
	NATURAL GAS, REFRIGERATED LIQUID (*WITH A HIGH METHANE CONTENT) (CRYOGENIC LIQUID*)	FLAMMABLE GAS	UN1972	FLAMMABLE GAS	NONE	173.318	FLAMMABLE GAS
	NEOHEXANE	FLAMMABLE LIQUID	UN1208	FLAMMABLE LIQUID	173.118	173.119	FLAMMABLE
	NEON *OR* NEON, COMPRESSED	NONFLAMMABLE GAS	UN1065	NON FLAMMABLE GAS	173.306	173.302	NONFLAMMABLE GAS
	NEON, REFRIGERATED LIQUID (*CRYOGENIC LIQUID*)	NONFLAMMABLE GAS	UN1913	NONFLAMMABLE GAS	173.320	173.316	NONFLAMMABLE GAS
	NEW EXPLOSIVE OR EXPLOSIVE DEVICE. SEE 173.51 AND 173.86						
	NICKEL CARBONYL	FLAMMABLE LIQUID	UN1259	FLAMMABLE LIQUID AND POISON	NONE	173.126	FLAMMABLE[2]
+	NICKEL CYANIDE, SOLID	POISON B	UN1653	POISON	173.370		POISON
	NICKEL NITRATE	OXIDIZER	UN2725	OXIDIZER	173.153	173.182	OXIDIZER
	NICKEL PICRATE	FORBIDDEN					
	NICOTINE HYDROCHLORIDE	POISON B	UN1656	POISON	173.345	173.346	POISON
	NICOTINE, LIQUID	POISON B	UN1654	POISON	NONE	173.346	POISON
	NICOTINE SALICYLATE	POISON B	UN1657	POISON	173.364	173.365	POISON
+	NICOTINE SULFATE, LIQUID	POISON B	UN1658	POISON	173.345	173.346	POISON
+	NICOTINE SULFATE, SOLID	POISON B	UN1658	POISON	173.364	173.365	POISON
	NICOTINE TARTRATE	POISON B	UN1659	POISON	173.364	173.365	POISON
	NITRATED PAPER (UNSTABLE)	FORBIDDEN					
	NITRATE, N.O.S.	OXIDIZER	NA1477	OXIDIZER	173.153	173.182	OXIDIZER
	NITRATE OF AMMONIA EXPLOSIVES. SEE HIGH EXPLOSIVE						
	NITRATES OF DIAZONIUM COMPOUNDS	FORBIDDEN					
	NITRATING ACID, MIXTURE (WITH MORE THAN 50% NITRIC ACID)	OXIDIZER	UN1796	OXIDIZER AND CORROSIVE	NONE	173.267	OXIDIZER
	NITRATING ACID, MIXTURE (WITH NOT MORE THAN 50% NITRIC ACID)	CORROSIVE MATERIAL	UN1796	CORROSIVE	NONE	173.267	CORROSIVE
	NITRATING ACID, SPENT	CORROSIVE MATERIAL	NA1826	CORROSIVE	NONE	173.248	CORROSIVE
	NITRIC ACID (*OVER 40%*)	OXIDIZER	UN2031	OXIDIZER AND CORROSIVE	NONE	173.268	OXIDIZER
	NITRIC ACID, 40% OR LESS	CORROSIVE MATERIAL	NA1760	CORROSIVE	NONE	173.268	CORROSIVE
	NITRIC ACID, FUMING	OXIDIZER	UN2032	OXIDIZER AND POISON	NONE	173.268	OXIDIZER[2]
	NITRIC ETHER. SEE ETHYL NITRATE						
	NITRIC OXIDE	POISON A	UN1660	POISON GAS	NONE	173.337	POISON GAS*
	2-NITRO-2-METHYLPROPANOL NITRATE	FORBIDDEN					
	6-NITRO-4-DIAZOTOLUENE-3 SULFONIC ACID (DRY)	*FORBIDDEN*					
	P-NITROANILINE. *SEE* NITROANILINE						
	N-NITROANILINE	FORBIDDEN					

2-SEE PLACARDING NOTES, PAGE 5

§172.101 HAZARDOUS MATERIALS TABLE

(1) +/A/W	(2) Hazardous materials descriptions and proper shipping names	(3) Hazard Class	(3A) Identification number	(4) Label(s) required (if not excepted)	(5) Packaging (a) Exceptions	(5) Packaging (b) Specific requirements	Placards required (*denotes placard for any quantity)
	M-NITROBENZENE DIAZONIUM PERCHLORATE	FORBIDDEN					
+	NITROANILINE	POISON B	UN1661	POISON	173.364	173.373	POISON
	NITROBENZENE, LIQUID *OR* NITROBENZOL, LIQUID (*OIL OF MIRBANE*)	POISON B	UN1662	POISON	173.345	173.346	POISON
	NITRO CARBONITRATE. SEE BLASTING AGENT, N.O.S						
	NITROCELLULOSE, COLLOIDED, GRANULAR *OR* FLAKE, WET WITH NOT LESS THAN 20% ALCOHOL *OR* SOLVENT, *OR* BLOCK, WET WITH NOT LESS THAN 25% ALCOHOL	FLAMMABLE LIQUID	NA2059	FLAMMABLE LIQUID	173.118	173.127	FLAMMABLE
	NITROCELLULOSE, COLLOIDED, GRANULAR *OR* FLAKE, WET WITH NOT LESS THAN 20% WATER	FLAMMABLE SOLID	NA2555	FLAMMABLE SOLID	173.153	173.184	FLAMMABLE SOLID[5]
	NITROCELLULOSE, DRY. SEE HIGH EXPLOSIVE						
	NITROCELLULOSE, WET WITH NOT LESS THAN 30% ALCOHOL *OR* SOLVENT	FLAMMABLE LIQUID	NA2556	FLAMMABLE LIQUID	173.118	173.127	FLAMMABLE
	NITROCELLULOSE, WET WITH NOT LESS THAN 20% WATER	FLAMMABLE SOLID	NA2555	FLAMMABLE SOLID	173.153	173.184	FLAMMABLE SOLID[5]
+	NITROCHLOROBENZENE, META *OR* PARA, SOLID	POISON B	UN1578	POISON	173.364	173.374	POISON
+	NITROCHLOROBENZENE, ORTHO, LIQUID	POISON B	UN1578	POISON	173.345	173.346	POISON
	NITROETHYLENE POLYMER	FORBIDDEN					
	NITROETHYL NITRATE	FORBIDDEN					
	NITROETHANE	FLAMMABLE LIQUID	UN2842	FLAMMABLE LIQUID	173.118	173.119	FLAMMABLE
	NITROGEN *OR* NITROGEN, COMPRESSED	NONFLAMMABLE GAS	UN1066	NONFLAMMABLE GAS	173.306	173.302 173.314	NONFLAMMABLE GAS
	NITROGEN, REFRIGERATED LIQUID (*CRYOGENIC LIQUID*)	NONFLAMMABLE GAS	UN1977	NONFLAMMABLE GAS	173.320	173.316 173.318	NONFLAMMABLE GAS
	NITROGEN DIOXIDE, LIQUID	POISON A	UN1067	POISON GAS AND OXIDIZER	NONE	173.336	POISON GAS*
	NITROGEN PEROXIDE, LIQUID	POISON A	NA1067	POISON GAS AND OXIDIZER	NONE	173.336	POISON GAS*
	NITROGEN TETROXIDE, LIQUID	POISON A	NA1067	POISON GAS AND OXIDIZER	NONE	173.336	POISON GAS*
	NITROGEN TRICHLORIDE	FORBIDDEN					
	NITROGEN TRIFLUORIDE	NONFLAMMABLE GAS	UN2451	NONFLAMMABLE GAS	NONE	173.302	NONFLAMMABLE GAS
	NITROGEN TRIIODIDE	FORBIDDEN					
	NITROGEN TRIIODIDE MONOAMINE	FORBIDDEN					
	NITROGLYCERIN, LIQUID, DESENSITIZED. SEE HIGH EXPLOSIVE, LIQUID						
	NITROGLYCERIN, LIQUID, NOT DESENSITIZED. SEE 173.51	FORBIDDEN					
	NITROGLYCERIN, SPIRITS OF. SEE SPIRITS OF NITROGLYCERIN						
	NITROGUANIDINE, DRY. SEE HIGH EXPLOSIVE						
	NITROGUANIDINE NITRATE	FORBIDDEN					

5-SEE PLACARDING NOTES, PAGE 5

§172.101 HAZARDOUS MATERIALS TABLE

(1) +/ A/ W/	(2) Hazardous materials descriptions and proper shipping names	(3) Hazard Class	(3A) Identi-fication number	(4) Label(s) required (if not excepted)	(5) Packaging (a) Exceptions	 (b) Specific require-ments	Placards required (*denotes placard for any quantity)
	NITROGUANIDINE, WET WITH NOT LESS THAN 20% WATER	FLAMMABLE SOLID	UN1336	FLAMMABLE SOLID	173.153	173.184	FLAMMABLE SOLID[5]
	1-NITRO HYDANTOIN	FORBIDDEN					
	NITROHYDROCHLORIC ACID	CORROSIVE MATERIAL	UN1798	CORROSIVE	NONE	173.278	CORROSIVE
	NITROHYDROCHLORIC ACID, DILUTED	CORROSIVE MATERIAL	UN1798	CORROSIVE	NONE	173.278	CORROSIVE
	NITRO ISOBUTANE TRIOL TRINITRATE	FORBIDDEN					
	NITROMANNITE. SEE HIGH EXPLOSIVE						
	NITROMANNITE (DRY)	FORBIDDEN					
	NITROMETHANE	FLAMMABLE LIQUID	UN1261	FLAMMABLE LIQUID	173.118	173.149a	FLAMMABLE
	NITROMURIATIC ACID. *SEE* NITROHYDROCHLORIC ACID						
	N-NITRO-N-METHYLGLYCOLAMIDE NITRATE	FORBIDDEN					
	NITROPHENOL PESTICIDE, SUBSTITUTED, LIQUID *OR* SOLID, N.O.S. (*COMPOUNDS AND PREPARATIONS*). *SEE* SUBSTITUTED NITROPHENOL PESTICIDE, LIQUID *OR* SOLID, N.O.S. (*COMPOUNDS AND PREPARATIONS*)						
	NITROSOGUANIDINE. *SEE* INITIATING EXPLOSIVE						
	M-NITROPHENYLDINITRO METHANE	*FORBIDDEN*					
	NITROPROPANE-FLAMMABLE LIQUID	FLAMMABLE LIQUID	UN2608	173.118	173.119	FLAMMABLE	
	NITROSTARCH, DRY. *SEE* HIGH EXPLOSIVE						
	NITROSTARCH, WET WITH NOT LESS THAN 30% ALCOHOL *OR* SOLVENT	FLAMMABLE LIQUID	UN1337	FLAMMABLE LIQUID	173.118	173.127	FLAMMABLE
	NITROSTARCH, WET WITH NOT LESS THAN 20% WATER	FLAMMABLE SOLID	UN1337	FLAMMABLE SOLID	173.153	173.184	FLAMMABLE SOLID[5]
	NITROSUGARS (DRY)	FORBIDDEN					
	NITROSYL CHLORIDE	NONFLAMM-ABLE GAS	UN1069	NONFLAMM-ABLE GAS	173.306	173.304 173.314	NONFLAMM-ABLE GAS
	NITROUREA. SEE HIGH EXPLOSIVE						
	NITROUS OXIDE *OR* NITROUS OXIDE, COMPRESSED	NONFLAMM-ABLE GAS	UN1070	NONFLAMM-ABLE GAS	173.306	173.304	NONFLAMM-ABLE GAS
	NITROUS OXIDE, REFRIGERATED LIQUID	NONFLAMM-ABLE GAS	UN2201	NONFLAMM-ABLE GAS	173.306	173.315	NONFLAMM-ABLE GAS
+	NITROXYLOL	POISON B	NA1665	POISON	173.345	173.346	POISON
	NONFLAMMABLE GAS, N.O.S. *SEE* COMPRESSED GAS, N.O.S.						
+	NONLIQUEFIED HYDROCARBON GAS. *SEE* HYDROCARBON GAS, NONLIQUEFIED						
	NONYLTRICHLOROSILANE	CORROSIVE MATERIAL	UN1799	CORROSIVE	NONE	173.280	CORROSIVE
	NORDHAUSEN ACID. *SEE* SULFURIC ACID						
	OCTADECYLTRICHLOROSILANE	CORROSIVE MATERIAL	UN1800	CORROSIVE	NONE	173.280	CORROSIVE
	1,7-OCTADIENE-3,5-DIYNE-1,8-DIMETHOXY-9-OCTADECYNOIC ACID	FORBIDDEN					

5-SEE PLACARDING NOTES, PAGE 5

§172.101 HAZARDOUS MATERIALS TABLE

(1) +/ A/ W	(2) Hazardous materials descriptions and proper shipping names	(3) Hazard Class	(3A) Identi-fication number	(4) Label(s) required (if not excepted)	(5) Packaging (a) Exceptions	(5) Packaging (b) Specific require-ments	Placards required (*denotes placard for any quantity)
	OCTANE	FLAMMABLE LIQUID	UN1262	FLAMMABLE LIQUID	173.118	173.119	FLAMMABLE
	N-OCTANOYL PEROXIDE, *TECHNICALLY PURE. SEE* ORGANIC PEROXIDE, LIQUID *OR* SOLUTION, N.O.S.		UN2129				
	OCTYLTRICHLOROSILANE	CORROSIVE MATERIAL	UN1801	CORROSIVE	NONE	173.280	CORROSIVE
	OIL, *DESCRIBED AS OIL*, OIL, N.O.S., PETROLEUM OIL, *OR* PETROLEUM OIL, N.O.S.	COMBUSTIBLE LIQUID	NA1270	NONE	173.118a	NONE	COMBUSTIBLE[1]
	OIL, *DESCRIBED AS OIL*, OIL, N.O.S., PETROLEUM OIL, *OR* PETROLEUM OIL, N.O.S.	FLAMMABLE LIQUID	NA1270	FLAMMABLE LIQUID	173.118	173.119	FLAMMABLE
	OIL OF MIRBANE. SEE NITROBENZENE, LIQUID						
	OIL OF VITRIOL. SEE SULFURIC ACID						
	OIL WELL CARTRIDGE	CLASS C EXPLOSIVE		EXPLOSIVES C	NONE	173.112	DANGEROUS
	OLEUM (*FUMING SULFURIC ACID*)	CORROSIVE MATERIAL	NA1831	CORROSIVE	NONE	173.272	CORROSIVE
	ORGANIC PEROXIDE, LIQUID *OR* SOLUTION, N.O.S.	FLAMMABLE LIQUID	NA1993	FLAMMABLE LIQUID AND ORGANIC PEROXIDE	NONE	173.119	FLAMMABLE
	ORGANIC PEROXIDE, LIQUID *OR* SOLUTION, N.O.S.	ORGANIC PEROXIDE	NA9183	ORGANIC PEROXIDE	173.153	173.221	ORGANIC PEROXIDE
	ORGANIC PEROXIDE, MIXTURE. *SEE* ORGANIC PEROXIDE, SOLID, N.O.S. *OR* ORGANIC PEROXIDE, LIQUID *OR* SOLUTION, N.O.S., *AS APPROPRIATE.*		UN2756				
	ORGANIC PEROXIDE, SAMPLE, N.O.S. *SEE* ORGANIC PEROXIDE, SOLID, N.O.S. *OR* ORGANIC PEROXIDE, LIQUID *OR* SOLUTION, N.O.S., *AS APPROPRIATE.*		UN2255				
	ORGANIC PEROXIDE, SOLID, N.O.S.	ORGANIC PEROXIDE	NA9187	ORGANIC PEROXIDE	173.153	173.154	ORGANIC PEROXIDE
	ORGANIC PEROXIDE, TRIAL QUANTITY, N.O.S. *SEE* ORGANIC PEROXIDE, SOLID, N.O.S. *OR* ORGANIC PEROXIDE, LIQUID *OR* SOLUTION, N.O.S., *AS APPROPRIATE.*		UN2899				
	ORGANIC PHOSPHATE MIXTURE, ORGANIC PHOSPHATE COMPOUND MIXTURE, *OR* ORGANIC PHOSPHORUS COMPOUND MIXTURE; LIQUID	POISON B	NA2783	POISON	173.359	173.359	POISON
	ORGANIC PHOSPHATE MIXTURE, ORGANIC PHOSPHATE COMPOUND MIXTURE, *OR* ORGANIC PHOSPHORUS COMPOUND MIXTURE; SOLID *OR* DRY	POISON B	NA2783	POISON	173.377	173.377	POISON
	ORGANIC PHOSPHATE, ORGANIC PHOSPHATE COMPOUND, *OR* ORGANIC PHOSPHORUS COMPOUND; MIXED WITH COMPRESSED GAS	POISON A	NA1955	POISON GAS	NONE	173.334	POISON GAS*
	ORGANIC PHOSPHATE, ORGANIC PHOSPHATE COMPOUND, *OR* ORGANIC PHOSPHORUS COMPOUND; LIQUID	POISON B	NA2783	POISON	NONE	173.358	POISON
	ORGANIC PHOSPHATE, ORGANIC PHOSPHATE COMPOUND, *OR* ORGANIC PHOSPHORUS COMPOUND; SOLID *OR* DRY	POISON B	NA2783	POISON	NONE	173.377	POISON
	ORGANOCHLORINE PESTICIDE, LIQUID, N.O.S. (*COMPOUNDS AND PREPARATIONS*)	FLAMMABLE LIQUID	UN2762	FLAMMABLE LIQUID	173.118	173.119	FLAMMABLE
	ORGANOCHLORINE PESTICIDE, LIQUID, N.O.S. (*COMPOUNDS AND PREPARATIONS*)	POISON B	UN2761	POISON	173.345	173.346	POISON
	ORGANOCHLORINE PESTICIDE, SOLID, N.O.S. (*COMPOUNDS AND PREPARATIONS*)	POISON B	UN2761	POISON	173.364	173.365	POISON

1-SEE PLACARDING NOTES, PAGE 5

§172.101 HAZARDOUS MATERIALS TABLE

(1) +/ A/ W/	(2) Hazardous materials descriptions and proper shipping names	(3) Hazard Class	(3A) Identification number	(4) Label(s) required (if not excepted)	(5) Packaging (a) Exceptions	(5) Packaging (b) Specific requirements	Placards required (*denotes placard for any quantity)
	ORGANOPHOSPHORUS PESTICIDE, LIQUID, N.O.S. (*COMPOUNDS AND PREPARATIONS*)	FLAMMABLE LIQUID	UN2784	FLAMMABLE LIQUID	NONE	173.119	FLAMMABLE
	ORGANOPHOSPHORUS PESTICIDE, LIQUID, N.O.S. (*COMPOUNDS AND PREPARATIONS*)	POISON B	UN2783	POISON	173.359	173.359	POISON
	ORGANOPHOSPHORUS PESTICIDE, SOLID, N.O.S. (*COMPOUNDS AND PREPARATIONS*)	POISON B	UN2783	POISON	173.377	173.377	POISON
	ORGANOTIN PESTICIDE, LIQUID, N.O.S. (*COMPOUNDS AND PREPARATIONS*)	FLAMMABLE LIQUID	UN2787	FLAMMABLE LIQUID	173.118	173.119	FLAMMABLE
	ORGANOTIN PESTICIDE, LIQUID, N.O.S. (*COMPOUNDS AND PREPARATIONS*)	POISON B	UN2786	POISON	173.345	173.346	POISON
	ORGANOTIN PESTICIDE, SOLID, N.O.S. (*COMPOUNDS AND PREPARATIONS*)	POISON B	UN2786	POISON	173.364	173.365	POISON
A	ORM-A, N.O.S.	ORM-A	NA1693	NONE	173.505	173.510	NONE
A	ORM-B, N.O.S.	ORM-B	NA1760	NONE	173.505	173.510	NONE
	ORM-C. *SEE 173.500 AND 176.900*						
	ORM-E LIQUID OR SOLID, N.O.S., *SEE* HAZARDOUS SUBSTANCE, LIQUID OR SOLID, N.O.S.						
	ORTHONITROANILINE. *SEE* NITROANILINE						
	OXIDIZER, CORROSIVE, LIQUID, N.O.S.	OXIDIZER	NA9193	OXIDIZER AND CORROSIVE	NONE	173.245	OXIDIZER
	OXIDIZER, CORROSIVE, SOLID, N.O.S.	OXIDIZER	NA9194	OXIDIZER AND CORROSIVE	173.153	173.154	OXIDIZER
	OXIDIZER MATERIAL PACKED WITH OTHER ARTICLES. SEE 173.152						
	OXIDIZER, N.O.S. *OR* OXIDIZING MATERIAL, N.O.S.	OXIDIZER	UN1479	OXIDIZER	173.153	173.154	OXIDIZER
	OXIDIZER, POISONOUS, LIQUID, N.O.S.	OXIDIZER	NA9199	OXIDIZER AND POISON	NONE	173.154	OXIDIZER
	OXIDIZER, POISONOUS, SOLID, N.O.S.	OXIDIZER	NA9200	OXIDIZER AND POISON	173.153	173.154	OXIDIZER
	OXYGEN *OR* OXYGEN, COMPRESSED	NONFLAMMABLE GAS	UN1072	OXIDIZER	173.306	173.302 173.314	NONFLAMMABLE GAS
	OXYGEN, REFRIGERATED LIQUID (*CRYOGENIC LIQUID*)	NONFLAMMABLE GAS	UN1073	OXIDIZER	173.320	173.316 173.318	OXYGEN
	PAINT	COMBUSTIBLE LIQUID	UN1263	NONE	173.118a	NONE	COMBUSTIBLE[1]
	PAINT	FLAMMABLE LIQUID	UN1263	FLAMMABLE LIQUID	173.118	173.128	FLAMMABLE
	PAINT *OR* PAINT RELATED MATERIAL	CORROSIVE MATERIAL	NA1760	CORROSIVE	173.244	173.245	CORROSIVE
	PAINT RELATED MATERIAL	COMBUSTIBLE LIQUID	NA1263	NONE	173.118a	NONE	COMBUSTIBLE[1]
	PAINT RELATED MATERIAL	FLAMMABLE LIQUID	UN1263	FLAMMABLE LIQUID	173.118 173.128	173.128	FLAMMABLE
	PAPER CAPS. SEE TOY CAPS						
AW	PARAFORMALDEHYDE	ORM-A	UN2213	NONE	173.505	173.510	NONE
	PARALDEHYDE	FLAMMABLE LIQUID	UN1264	FLAMMABLE LIQUID	173.118	173.119	FLAMMABLE
	PARAMENTHANE HYDROPEROXIDE	ORGANIC PEROXIDE	UN2125	ORGANIC PEROXIDE	173.153	173.224	ORGANIC PEROXIDE
	PARANITROANILINE, SOLID. *SEE* NITROANILINE						
	PARATHION AND COMPRESSED GAS MIXTURE	POISON A	NA1967	POISON GAS	NONE	173.334	POISON GAS*

1-SEE PLACARDING NOTES, PAGE 5

(1) +/ A/ W/	(2) Hazardous materials descriptions and proper shipping names	(3) Hazard Class	(3A) Identi-fication number	(4) Label(s) required (if not excepted)	(5) Packaging (a) Exceptions	(5) Packaging (b) Specific require-ments	Placards required (*denotes placard for any quantity)
	PARATHION, LIQUID	POISON B	NA2783	POISON	NONE	173.358	POISON
	PARATHION MIXTURE, DRY	POISON B	NA2783	POISON	173.377	173.377	POISON
	PARATHION MIXTURE, LIQUID	POISON B	NA2783	POISON	NONE	173.359	POISON
	PARIS GREEN, SOLID. *SEE* COPPER ACETOARSENITE, SOLID						
	PELARGONYL PEROXIDE, *TECHNICALLY PURE. SEE* ORGANIC PEROXIDE, SOLID, N.O.S.		UN2130				
	PENTABORANE	FLAMMABLE LIQUID	UN1380	FLAMMABLE LIQUID AND POISON	NONE	173.138	FLAMMABLE[2]
	PENTAERYTHRITE TETRANITRATE. SEE INITIATING EXPLOSIVE						
	PENTAERYTHRITE TETRANITRATE, DESENSITIZED, WET. SEE HIGH EXPLOSIVE						
	PENTAERYTHRITE TETRANITRATE (DRY)	FORBIDDEN					
	PENTANE	FLAMMABLE LIQUID	UN1265	FLAMMABLE LIQUID	173.118	173.119	FLAMMABLE
	PENTANITROANILINE (DRY)	FORBIDDEN					
	PENTOLITE, DRY. SEE HIGH EXPLOSIVE						
	PERACETIC ACID SOLUTION, *NOT OVER 43% PERACETIC ACID AND NOT OVER 6% HYDROGEN PEROXIDE*	ORGANIC PEROXIDE	NA2131	ORGANIC PEROXIDE	173.223	173.223	ORGANIC PEROXIDE
	PERCHLORATE, N.O.S.	OXIDIZER	NA1481	OXIDIZER	173.153	173.154	OXIDIZER
	PERCHLORIC ACID, *EXCEEDING 50% BUT* NOT EXCEEDING 72% STRENGTH	OXIDIZER	UN1873	OXIDIZER	NONE	173.269	OXIDIZER
	PERCHLORIC ACID EXCEEDING 72% STRENGTH	FORBIDDEN					
	PERCHLORIC ACID, NOT OVER 50% ACID	OXIDIZER	UN1802	OXIDIZER	173.244	173.269	OXIDIZER
	PERCHLOROETHYLENE *SEE* TETRACHLOROETHYLENE		NA1897				
+	PERCHLOROMETHYL MERCAPTAN	POISON B	UN1670	POISON	173.345	173.360	POISON
	PERCUSSION CAP	CLASS C EXPLOSIVE		NONE	NONE	173.107	NONE
	PERCUSSION FUZE	CLASS C EXPLOSIVE		EXPLOSIVE C	NONE	173.105	DANGEROUS
A	PERFLUORO-2-BUTENE	ORM-A	NA2422	NONE	173.505	173.605	NONE
	PERMANGANATE, N.O.S.	OXIDIZER	NA1482	OXIDIZER	173.153	173.154	OXIDIZER
	PERMANGANATE OF POTASH. SEE POTASSIUM PERMANGANATE						
	PEROXIDE, ORGANIC. SEE ORGANIC PEROXIDE						
	PEROXYACETIC ACID, *NOT MORE THAN 43% AND WITH NOT MORE THAN 6% HYDROGEN PEROXIDE. SEE* PERACETIC ACID SOLUTION, *NOT OVER 43% PERACETIC ACID AND NOT OVER 6% HYDROGEN PEROXIDE*		UN2131				
	PEROXYACETIC ACID, MORE THAN 43% AND WITH MORE THAN 6% HYDROGEN PEROXIDE	FORBIDDEN					

2-SEE PLACARDING NOTES, PAGE 5

(1) +/ A/ W	(2) Hazardous materials descriptions and proper shipping names	(3) Hazard Class	(3A) Identi- fication number	(4) Label(s) required (if not excepted)	(5) Packaging (a) Exceptions	(5) Packaging (b) Specific require- ments	Placards required (*denotes placard for any quantity)
W	PESTICIDE, WATER REACTIVE, *INCLUDING BUT NOT LIMITED TO FUNGICIDES, AND HERBICIDES, ETC., WHICH CONTAIN MANGANESE ETHYLENEBISDI-THIOCARBAMATE*	ORM-C	NA2210	NONE	173.505	173.1040	NONE
W	PETROLEUM COKE (*UNCALCINED*)	ORM-C		NONE	173.505	173.1045	NONE
	PETROLEUM DISTILLATE	COMBUSTIBLE LIQUID	UN1268	NONE	173.118a	NONE	COMBUSTIBLE[1]
	PETROLEUM DISTILLATE	FLAMMABLE LIQUID	UN1268	FLAMMABLE LIQUID	173.118	173.119	FLAMMABLE
	PETROLEUM ETHER	FLAMMABLE LIQUID	UN1271	FLAMMABLE LIQUID	173.118	173.119	FLAMMABLE
	PETROLEUM GAS, LIQUEFIED. *SEE* LIQUEFIED PETROLEUM GAS						
	PETROLEUM NAPHTHA	COMBUSTIBLE LIQUID	UN1255	NONE	173.118a	NONE	COMBUSTIBLE[1]
	PETROLEUM NAPHTHA	FLAMMABLE LIQUID	UN1255	FLAMMABLE LIQUID	173.118	173.119	FLAMMABLE
	PETROLEUM OIL, N.O.S. *SEE* OIL						
A	PHENCAPTON	ORM-A	NA2783	NONE	173.505	173.510	NONE
+	PHENOL	POISON B	UN1671	POISON	173.364	173.369	POISON
	PHENOL, LIQUID *OR* SOLUTION (*LIQUID TAR ACID CONTAINING OVER 50% PHENOL*)	POISON B	NA2821	POISON	173.345	173.349	POISON
	PHENOXY PESTICIDE, LIQUID, N.O.S. (*COMPOUNDS AND PREPARATIONS*)	FLAMMABLE LIQUID	UN2766	FLAMMABLE LIQUID	173.118	173.119	FLAMMABLE
	PHENOXY PESTICIDE, LIQUID, N.O.S. (*COMPOUNDS AND PREPARATIONS*)	POISON B	UN2765	POISON	173.345	173.346	POISON
	PHENOXY PESTICIDE, SOLID, N.O.S. (*COMPOUNDS AND PREPARATIONS*)	POISON B	UN2765	POISON	173.364	173.365	POISON
A	PHENYLENEDIAMINE, META *OR* PARA, SOLID	ORM-A	UN1673	NONE	173.505	173.510	NONE
	PHENYLDICHLOROARSINE	POISON B	NA1556	POISON	NONE	173.355	POISON
	M-PHENYLENE DIAMINEDIPERCHLORATE (DRY)	FORBIDDEN					
	PHENYL MERCAPTAN	POISON B	UN2337	POISON	173.345	173.346	POISON
	PHENYLTRICHLOROSILANE	CORROSIVE MATERIAL	UN1804	CORROSIVE	NONE	173.280	CORROSIVE
	PHENYLUREA PESTICIDE, LIQUID, N.O.S. (*COMPOUNDS AND PREPARATIONS*)	FLAMMABLE LIQUID	UN2768	FLAMMABLE LIQUID	173.118	173.119	FLAMMABLE
	PHENYLUREA PESTICIDE, LIQUID, N.O.S. (*COMPOUNDS AND PREPARATIONS*)	POISON B	UN2767	POISON	173.345	173.346	POISON
	PHENYLUREA PESTICIDE, SOLID, N.O.S. (*COMPOUNDS AND PREPARATIONS*)	POISON B	UN2767	POISON	173.364	173.365	POISON
	PHOSGENE (*DIPHOSGENE*)	POISON A	UN1076	POISON GAS	NONE	173.333	POISON GAS*
	PHOSPHINE	POISON A	UN2199	POISON GAS AND FLAMMABLE GAS	NONE	173.328	POISON GAS*
	PHOSPHORIC ACID	CORROSIVE MATERIAL	UN1805	CORROSIVE	173.244	173.245	CORROSIVE
	PHOSPHORIC ACID TRIETHYLENEIMINE. *SEE* TRIS-(1-AZIRIDIYL) PHOSPHINE OXIDE						

1-SEE PLACARDING NOTES, PAGE 5

§172.101 HAZARDOUS MATERIALS TABLE

(1) +/ A/ W	(2) Hazardous materials descriptions and proper shipping names	(3) Hazard Class	(3A) Identification number	(4) Label(s) required (if not excepted)	(5) Packaging (a) Exceptions	(5) Packaging (b) Specific requirements	Placards required (*denotes placard for any quantity)
	PHOSPHORIC ANHYDRIDE (*PHOSPHORUS PENTOXIDE*)	CORROSIVE MATERIAL	NA1807	CORROSIVE	NONE	173.188	CORROSIVE
	PHOSPHORUS, AMORPHOUS, RED	FLAMMABLE SOLID	UN1338	FLAMMABLE SOLID	NONE	173.189	FLAMMABLE SOLID[5]
	PHOSPHORUS BROMIDE. *SEE* PHOSPHORUS TRIBROMIDE.						
	PHOSPHORUS CHLORIDE. *SEE* PHOSPHORUS TRICHLORIDE						
	PHOSPHORUS HEPTASULFIDE	FLAMMABLE SOLID	UN1339	FLAMMABLE SOLID	NONE	173.225	FLAMMABLE SOLID[5]
	PHOSPHORUS OXYBROMIDE	CORROSIVE MATERIAL	UN1939	CORROSIVE	NONE	173.271	CORROSIVE
	PHOSPHORUS OXYCHLORIDE	CORROSIVE MATERIAL	UN1810	CORROSIVE & POISON	NONE	173.271	CORROSIVE[2]
	PHOSPHORUS PENTACHLORIDE, SOLID	CORROSIVE MATERIAL	UN1806	CORROSIVE	NONE	173.191	CORROSIVE
	PHOSPHORUS PENTASULFIDE	FLAMMABLE SOLID	UN1340	FLAMMABLE SOLID AND DANGEROUS WHEN WET	NONE	173.225	FLAMMABLE SOLID -W-*
	PHOSPHORUS SESQUISULFIDE	FLAMMABLE SOLID	UN1341	FLAMMABLE SOLID AND DANGEROUS WHEN WET	NONE	173.225	FLAMMABLE SOLID -W-*
	PHOSPHORUS TRIBROMIDE	CORROSIVE MATERIAL	UN1808	CORROSIVE	NONE	173.270	CORROSIVE
	PHOSPHORUS TRICHLORIDE	CORROSIVE MATERIAL	UN1809	CORROSIVE & POISON	NONE	173.271	CORROSIVE[2]
	PHOSPHORUS TRISULFIDE	FLAMMABLE SOLID	UN1343	FLAMMABLE SOLID	NONE	173.225	FLAMMABLE SOLID[5]
	PHOSPHORUS, WHITE *OR* YELLOW, DRY	FLAMMABLE SOLID	UN1381	FLAMMABLE SOLID AND POISON	NONE	173.190	FLAMMABLE SOLID[5]
	PHOSPHORUS, WHITE *OR* YELLOW, IN WATER	FLAMMABLE SOLID	UN1381	FLAMMABLE SOLID AND POISON	NONE	173.190	FLAMMABLE SOLID[5]
	PHOSPHORUS (WHITE OR RED) AND A CHLORATE, MIXTURES OF	FORBIDDEN					
	PHOSPHORYL CHLORIDE. *SEE* PHOSPHORUS OXYCHLORIDE						
	PHOTOGRAPHIC FILM. SEE FILM						
	PHOTOGRAPHIC FLASH POWDER. SEE FIREWORKS, SPECIAL *OR* LOW EXPLOSIVE						
	PHTHALIMIDE DERIVATIVE PESTICIDE, LIQUID, N.O.S. (*COMPOUNDS AND PREPARATIONS*)	FLAMMABLE LIQUID	UN2774	FLAMMABLE LIQUID	173.118	173.119	FLAMMABLE
	PHTHALIMIDE DERIVATIVE PESTICIDE, LIQUID, N.O.S. (*COMPOUNDS AND PREPARATIONS*)	POISON B	UN2773	POISON	173.345	173.346	POISON
	PHTHALIMIDE DERIVATIVE PESTICIDE, SOLID, N.O.S. (*COMPOUNDS AND PREPARATIONS*)	POISON B	UN2773	POISON	173.364	173.365	POISON
	PICRATE, DRY. SEE HIGH EXPLOSIVE						
	PICRATE OF AMMONIA. SEE HIGH EXPLOSIVE						
	PICRIC ACID, DRY. SEE HIGH EXPLOSIVE						
	PICRIC ACID, WET, WITH NOT LESS THAN 10% WATER	FLAMMABLE SOLID	NA1344	FLAMMABLE SOLID	173.192	173.193	FLAMMABLE SOLID[5]
	PICRIC ACID, WET WITH NOT LESS THAN 10% WATER, OVER 25 POUNDS. SEE HIGH EXPLOSIVE						

2-SEE PLACARDING NOTES, PAGE 5

§172.101 HAZARDOUS MATERIALS TABLE

(1) +/ A/ W	(2) Hazardous materials descriptions and proper shipping names	(3) Hazard Class	(3A) Identi-fication number	(4) Label(s) required (if not excepted)	(5) Packaging (a) Exceptions	(5) Packaging (b) Specific require-ments	Placards required (*denotes placard for any quantity)
	PINANE HYDROPEROXIDE, *TECHNICALLY PURE. SEE* ORGANIC PEROXIDE, LIQUID *OR* SOLUTION, N.O.S.		UN2162				
	PINANE HYDROPEROXIDE SOLUTION *NOT OVER 45% PEROXIDE.*	ORGANIC PEROXIDE	UN2162	ORGANIC PEROXIDE	173.153	173.224	ORGANIC PEROXIDE
	PINENE	FLAMMABLE LIQUID	UN2368	FLAMMABLE LIQUID	173.118	173.119	FLAMMABLE
	PINE OIL	COMBUSTIBLE LIQUID	UN1272	NONE	173.118a	NONE	COMBUSTIBLE[1]
	PINWHEELS. *SEE* FIREWORKS, COMMON						
	PIVALOYL CHLORIDE. *SEE* TRIMETHYLACETYL CHLORIDE						
	POISONOUS LIQUID *OR* GAS, FLAMMABLE, N.O.S.	POISON A	NA1953	POISON GAS AND FLAMMABLE GAS	NONE	173.328	POISON GAS*
	POISONOUS LIQUID *OR* GAS, N.O.S.	POISON A	NA1955	POISON GAS	NONE	173.328	POISON GAS*
	POISONOUS LIQUID, N.O.S. *OR* POISON B, LIQUID, N.O.S.	POISON B	UN2810	POISON	173.345	173.346 173.3a	POISON[2]
	POISONOUS SOLID, CORROSIVE, N.O.S.	POISON B	UN2928	POISON AND CORROSIVE	173.364	173.365	POISON
	POISONOUS SOLID, N.O.S. *OR* POISON B, SOLID, N.O.S.	POISON B	UN2811	POISON	173.364	173.365	POISON
	POLYMERIZABLE MATERIAL. SEE 173.21						
	POTASSIUM ARSENATE, SOLID	POISON B	UN1677	POISON	173.364	173.365	POISON
	POTASSIUM ARSENITE, SOLID	POISON B	UN1678	POISON	173.364	173.365	POISON
	POTASSIUM BIFLUORIDE SOLUTION. *SEE* POTASSIUM HYDROGEN FLUORIDE SOLUTION						
	POTASSIUM BROMATE	OXIDIZER	UN1484	OXIDIZER	173.153	173.154	OXIDIZER
	POTASSIUM CARBONYL	FORBIDDEN					
	POTASSIUM CHLORATE (*POTASH CHLORATE*)	OXIDIZER	UN1485	OXIDIZER	173.153	173.163	OXIDIZER
	POTASSIUM CYANIDE, SOLID	POISON B	UN1680	POISON	173.370	173.370	POISON
	POTASSIUM CYANIDE SOLUTION	POISON B	UN1680	POISON	173.345	173.352	POISON
	POTASSIUM DICHLORO ISOCYANURATE. SEE POTASSIUM DICHLORO-S-TRIAZINETRIONE						
	POTASSIUM DICHLORO-S-TRIAZINETRIONE, *DRY (CONTAINING MORE THAN 39% AVAILABLE CHLORINE)*	OXIDIZER	NA2465	OXIDIZER	173.153	173.217	OXIDIZER
A	POTASSIUM DICHROMATE	ORM-A	NA1479	NONE	173.505	173.510	NONE
A	POTASSIUM FLUORIDE	ORM-B	UN1812	NONE	173.505	173.510	NONE
	POTASSIUM FLUORIDE SOLUTION	CORROSIVE MATERIAL	UN1812	CORROSIVE	173.244	173.249	CORROSIVE
	POTASSIUM HYDRATE. *SEE* POTASSIUM HYDROXIDE						
	POTASSIUM HYDROGEN FLUORIDE SOLUTION	CORROSIVE MATERIAL	NA1811	CORROSIVE	173.244	173.249	CORROSIVE
A	POTASSIUM HYDROGEN SULFATE, SOLID	ORM-B	UN2509	NONE	173.505	173.800	NONE
	POTASSIUM HYDROXIDE, DRY SOLID, FLAKE, BEAD, *OR* GRANULAR	CORROSIVE MATERIAL	UN1813	CORROSIVE	173.244	173.245b	CORROSIVE
	POTASSIUM HYDROXIDE, LIQUID *OR* SOLUTION	CORROSIVE MATERIAL	UN1814	CORROSIVE	173.244	173.249	CORROSIVE

1-SEE PLACARDING NOTES, PAGE 5

(1) +/ A/ W/	(2) Hazardous materials descriptions and proper shipping names	(3) Hazard Class	(3A) Identi-fication number	(4) Label(s) required (if not excepted)	(5) Packaging (a) Exceptions	(5) Packaging (b) Specific require-ments	Placards required (*denotes placard for any quantity)
	POTASSIUM HYPOCHLORITE SOLUTION. *SEE* HYPOCHLORITE SOLUTIONS CONTAINING MORE THAN 7% AVAILABLE CHLORINE BY WEIGHT						
A	POTASSIUM METABISULFITE	ORM-B	NA2693	NONE	173.505	173.510	NONE
	POTASSIUM, METAL *OR* METALLIC	FLAMMABLE SOLID	UN2257	FLAMMABLE SOLID AND DANGEROUS WHEN WET	NONE	173.206	FLAMMABLE SOLID -W-*
	POTASSIUM, METAL LIQUID ALLOY	FLAMMABLE SOLID	UN1420	FLAMMABLE SOLID AND DANGEROUS WHEN WET	NONE	173.202	FLAMMABLE SOLID -W-*
	POTASSIUM NITRATE	OXIDIZER	UN1486	OXIDIZER	173.153	173.182	OXIDIZER
	POTASSIUM NITRATE MIXED (*FUSED*) WITH SODIUM NITRITE. *SEE* SODIUM NITRITE MIXED (*FUSED*) WITH POTASSIUM NITRATE						
	POTASSIUM NITRITE	OXIDIZER	UN1488	OXIDIZER	173.153	173.154	OXIDIZER
	POTASSIUM PERCHLORATE	OXIDIZER	UN1489	OXIDIZER	173.153	173.219	OXIDIZER
	POTASSIUM PERMANGANATE	OXIDIZER	UN1490	OXIDIZER	173.153	173.154 173.194	OXIDIZER
	POTASSIUM PEROXIDE	OXIDIZER	UN1491	OXIDIZER	NONE	173.187	OXIDIZER
	POTASSIUM PERSULFATE	OXIDIZER	UN1492	OXIDIZER	173.153	173.154	OXIDIZER
	POTASSIUM SULFIDE	FLAMMABLE SOLID	UN1382	FLAMMABLE SOLID	173.153	173.207	FLAMMABLE SOLID[5]
	POTASSIUM SUPEROXIDE	OXIDIZER	UN2466	OXIDIZER	NONE	173.187	OXIDIZER
	PRESSURIZED PRODUCT. SEE COMPRESSED GAS, N.O.S.						
	PRIMER. SEE CANNON PRIMER, COMBINATION PRIMER, *OR* SMALL-ARMS PRIMER						
	PRIMERS, DETONATING. SEE DETONATING PRIMERS, CLASS A *OR* CLASS C EXPLOSIVES						
	PROJECTILE, EXPLOSIVE. SEE EXPLOSIVE PROJECTILE						
	PROJECTILE, GAS, NONEXPLOSIVE. SEE CHEMICAL AMMUNITION, NONEXPLOSIVE (*CONTAINING A POISON A, POISON B OR IRRITATING MATERIAL, AS APPROPRIATE*)						
	PROJECTILE, GAS, SMOKE, OR INCENDIARY, WITH BURSTER OR BOOSTER WITH OR WITHOUT DETONATING FUZE. SEE EXPLOSIVE PROJECTILE						
	PROJECTILE, ILLUMINATING, INCENDIARY OR SMOKE, WITH EXPELLING CHARGE BUT WITHOUT BURSTING CHARGE. SEE FIREWORKS, SPECIAL						
	PROJECTILE, SAND-LOADED, EMPTY OR SOLID. SEE 173.55						
	PROPANE *OR* LIQUEFIED PETROLEUM GAS. *SEE* LIQUEFIED PETROLEUM GAS						
	PROPARGYL ALCOHOL	FLAMMABLE LIQUID	NA1986	FLAMMABLE LIQUID AND POISON	NONE	173.119	FLAMMABLE
	PROPELLANT EXPLOSIVE	CLASS A EXPLOSIVE		EXPLOSIVE A	NONE	173.64	EXPLOSIVES A*
	PROPELLANT EXPLOSIVE IN WATER (*SMOKELESS POWDER*)	CLASS B EXPLOSIVE		EXPLOSIVE B	NONE	173.93	EXPLOSIVES B*

5-SEE PLACARDING NOTES, PAGE 5

§172.101 HAZARDOUS MATERIALS TABLE

(1) +/ A/ W/	(2) Hazardous materials descriptions and proper shipping names	(3) Hazard Class	(3A) Identi-fication number	(4) Label(s) required (if not excepted)	(5) Packaging (a) Exceptions	(5) Packaging (b) Specific require-ments	Placards required (*denotes placard for any quantity)
	PROPELLANT EXPLOSIVE IN WATER, UNSTABLE, CONDEMNED, *OR* DETERIORATED (*SMOKELESS POWDER*)	CLASS B EXPLOSIVE		EXPLOSIVE B	NONE	173.93	EXPLOSIVES B*
	PROPELLANT EXPLOSIVE, LIQUID	CLASS B EXPLOSIVE		EXPLOSIVE B	NONE	173.93	EXPLOSIVES B*
	PROPELLANT EXPLOSIVE, SOLID	CLASS B EXPLOSIVE		EXPLOSIVE B	NONE	173.93	EXPLOSIVES B*
	PROPIONALDEHYDE	FLAMMABLE LIQUID	UN1275	FLAMMABLE LIQUID	173.118	173.119	FLAMMABLE
	PROPIONIC ACID	CORROSIVE MATERIAL	UN1848	CORROSIVE	173.244	173.245	CORROSIVE
	PROPIONIC ACID, SOLUTION	CORROSIVE MATERIAL	UN1848	CORROSIVE	173.244	173.245	CORROSIVE
	PROPIONIC ANHYDRIDE	CORROSIVE MATERIAL	UN2496	CORROSIVE	173.244	173.245	CORROSIVE
	PROPIONYL PEROXIDE, *NOT MORE THAN 28% IN SOLUTION. SEE* ORGANIC PEROXIDE, LIQUID *OR* SOLUTION, N.O.S.		UN2132				
	PROPIONYL PEROXIDE, MORE THAN 28% IN SOLUTION	FORBIDDEN					
	PROPYL ACETATE	FLAMMABLE LIQUID	UN1276	FLAMMABLE LIQUID	173.118	173.119	FLAMMABLE
	PROPYL ALCOHOL	FLAMMABLE LIQUID	UN1274	FLAMMABLE LIQUID	173.118	173.125	FLAMMABLE
	PROPYLAMINE	FLAMMABLE LIQUID	UN1277	FLAMMABLE LIQUID	NONE	173.119	FLAMMABLE
	PROPYL CHLORIDE	FLAMMABLE LIQUID	UN1278	FLAMMABLE LIQUID	NONE	173.119	FLAMMABLE
	PROPYLENE *OR* LIQUEFIED PETROLEUM GAS. *SEE* LIQUEFIED PETROLEUM GAS						
	PROPYLENEDIAMINE	FLAMMABLE LIQUID	UN2258	FLAMMABLE LIQUID	173.118	173.119	FLAMMABLE
	PROPYLENE DICHLORIDE	FLAMMABLE LIQUID	UN1279	FLAMMABLE LIQUID	173.118	173.119	FLAMMABLE
	PROPYLENEIMINE, INHIBITED	FLAMMABLE LIQUID	UN1921	FLAMMABLE LIQUID	NONE	173.139	FLAMMABLE
	PROPYLENE OXIDE	FLAMMABLE LIQUID	UN1280	FLAMMABLE LIQUID	173.118	173.119	FLAMMABLE
	PROPYL FORMATE	FLAMMABLE LIQUID	UN1281	FLAMMABLE LIQUID	173.118	173.119	FLAMMABLE
	PROPYL MERCAPTAN	FLAMMABLE LIQUID	NA2402	FLAMMABLE LIQUID	NONE	173.141	FLAMMABLE
	PROPYL TRICHLOROSILANE	CORROSIVE MATERIAL	UN1816	CORROSIVE	NONE	173.280	CORROSIVE
	PRUSSIC ACID. SEE HYDROCYANIC ACID, AS APPROPRIATE						
	PYRIDINE	FLAMMABLE LIQUID	UN1282	FLAMMABLE LIQUID	173.118	173.119	FLAMMABLE
	PYRIDINE PERCHLORATE	FORBIDDEN					
	PYROPHORIC LIQUID, N.O.S. *OR* PYROFORIC LIQUID, N.O.S.	FLAMMABLE LIQUID	UN2845	FLAMMABLE LIQUID	NONE	173.134	FLAMMABLE
	PYROSULFURYL CHLORIDE	CORROSIVE MATERIAL	UN1817	CORROSIVE	173.244	173.247	CORROSIVE
	PYROXYLIN PLASTIC, RODS, SHEETS, ROLLS, *OR* TUBES	FLAMMABLE SOLID	NA1325	FLAMMABLE SOLID	173.197	173.197	FLAMMABLE SOLID[5]
	PYROXYLIN PLASTIC, SCRAP	FLAMMABLE SOLID	NA1325	FLAMMABLE SOLID	NONE	173.195	FLAMMABLE SOLID[5]
	PYRROLIDINE	FLAMMABLE LIQUID	UN1922	FLAMMABLE LIQUID	173.118	173.119	FLAMMABLE
	QUEBRACHITOL PENTANITRATE	FORBIDDEN					
	QUICKLIME. *SEE* CALCIUM OXIDE						

5-SEE PLACARDING NOTES, PAGE 5

§172.101 HAZARDOUS MATERIALS TABLE

(1) +/ A/ W	(2) Hazardous materials descriptions and proper shipping names	(3) Hazard Class	(3A) Identification number	(4) Label(s) required (if not excepted)	(5) Packaging (a) Exceptions	(5) Packaging (b) Specific requirements	Placards required (*denotes placard for any quantity)
	RADIOACTIVE MATERIAL, ARTICLES, MANUFACTURED FROM NATURAL *OR* DEPLETED URANIUM *OR* NATURAL THORIUM	RADIOACTIVE MATERIAL	UN2909	NONE	173.421-1 173.424	173.421-1 173.424	RADIOACTIVE[3]
	RADIOACTIVE MATERIAL, EMPTY PACKAGES	RADIOACTIVE MATERIAL	UN2908	EMPTY	173.427		RADIOACTIVE[3]
	RADIOACTIVE MATERIAL, FISSILE, N.O.S.	RADIOACTIVE MATERIAL	UN2918	RADIOACTIVE	173.453	173.417	RADIOACTIVE[3]
	RADIOACTIVE MATERIAL, INSTRUMENTS *AND* ARTICLES.	RADIOACTIVE MATERIAL	UN2911	NONE	173.421-1 173.422	173.421-1 173.422	
	RADIOACTIVE MATERIAL, LIMITED QUANTITY, N.O.S.	RADIOACTIVE MATERIAL	UN2910	NONE	173.421 173.421-1	173.421 173.421-1	
	RADIOACTIVE MATERIAL, LOW SPECIFIC ACTIVITY *OR* LSA, N.O.S.	RADIOACTIVE MATERIAL	UN2912	RADIOACTIVE	173.421 173.422 173.424	173.425	RADIOACTIVE[3]
	RADIOACTIVE MATERIAL, N.O.S.	RADIOACTIVE MATERIAL	UN2982	RADIOACTIVE	173.421 173.422 173.424	173.415 173.416	RADIOACTIVE[3]
	RADIOACTIVE MATERIAL, SPECIAL FORM, N.O.S.	RADIOACTIVE MATERIAL	UN2974	RADIOACTIVE	173.421 173.422	173.415 173.416	RADIOACTIVE[3]
	RAILWAY FUSEE. SEE FUSEE						
	RAILWAY TORPEDO. *SEE* TORPEDO, RAILWAY						
	REFRIGERANT GAS, N.O.S. *OR* DISPERSANT GAS, N.O.S.	NONFLAMMABLE GAS	UN1078	NONFLAMMABLE GAS	173.306	173.304 173.314 173.315	NONFLAMMABLE GAS
	REFRIGERANT GAS, N.O.S. *OR* DISPERSANT GAS, N.O.S.	FLAMMABLE GAS	NA1954	FLAMMABLE GAS	173.306	173.304 173.314 173.315	FLAMMABLE GAS
	REFRIGERATING MACHINE	NONFLAMMABLE GAS	UN2857	NONFLAMMABLE GAS	173.306 173.307		NONFLAMMABLE GAS
	REFRIGERATING MACHINE	FLAMMABLE GAS	NA1954	FLAMMABLE GAS	173.306		FLAMMABLE GAS
	REFRIGERATING MACHINE	FLAMMABLE LIQUID	NA1993	FLAMMABLE LIQUID	173.130 173.306		FLAMMABLE
	RESIN SOLUTION (*RESIN COMPOUND, LIQUID*)	FLAMMABLE LIQUID	UN1866	FLAMMABLE LIQUID	173.118	173.119	FLAMMABLE
	RIFLE GRENADE. SEE GRENADE, HAND *OR* RIFLE, EXPLOSIVE						
	RIFLE POWDER. SEE PROPELLANT EXPLOSIVE *OR* BLACK POWDER						
	ROAD ASPHALT *OR* TAR, LIQUID *SEE* ASPHALT, CUT BACK						
	ROAD ASPHALT *OR* TAR (*WHEN HEATED TO OR ABOVE ITS FLASH POINT*). *SEE* ASPHALT						
	ROCKET AMMUNITION WITH EMPTY, INERT, *OR* SOLID LOADED PROJECTILE	CLASS A EXPLOSIVE		EXPLOSIVE A	NONE	173.57	EXPLOSIVES A*
	ROCKET AMMUNITION WITH EMPTY PROJECTILE	CLASS B EXPLOSIVE		EXPLOSIVE B	NONE	173.90	EXPLOSIVES B*
	ROCKET AMMUNITION WITH EXPLOSIVE PROJECTILE	CLASS A EXPLOSIVE		EXPLOSIVE A	NONE	173.57	EXPLOSIVES A*
	ROCKET AMMUNITION WITH GAS PROJECTILE	CLASS A EXPLOSIVE		EXPLOSIVE A	NONE	173.57	EXPLOSIVES A*
	ROCKET AMMUNITION WITH ILLUMINATING PROJECTILE	CLASS A EXPLOSIVE		EXPLOSIVE A	NONE	173.57	EXPLOSIVES A*

3-SEE PLACARDING NOTES, PAGE 5

(1) +/ A/ W	(2) Hazardous materials descriptions and proper shipping names	(3) Hazard Class	(3A) Identi-fication number	(4) Label(s) required (if not excepted)	(5) Packaging (a) Exceptions	(5) Packaging (b) Specific require-ments	Placards required (*denotes placard for any quantity)
	ROCKET AMMUNITION WITH INCENDIARY PROJECTILE	CLASS A EXPLOSIVE		EXPLOSIVE A	NONE	173.57	EXPLOSIVES A*
	ROCKET AMMUNITION WITH INERT LOADED PROJECTILE	CLASS B EXPLOSIVE		EXPLOSIVE B	NONE	173.90	EXPLOSIVES B*
	ROCKET AMMUNITION WITH SMOKE PROJECTILE	CLASS A EXPLOSIVE		EXPLOSIVE A	NONE	173.57	EXPLOSIVES A*
	ROCKET AMMUNITION WITH SOLID PROJECTILE	CLASS B EXPLOSIVE		EXPLOSIVE B	NONE	173.90	EXPLOSIVES B*
	ROCKET BODY, WITH ELECTRIC PRIMER OR ELECTRIC SQUIB. SEE 173.55						
	ROCKET ENGINE, LIQUID	CLASS B EXPLOSIVE		EXPLOSIVE B	NONE	173.95	EXPLOSIVES B*
	ROCKET FIREWORKS. SEE FIREWORKS, COMMON						
	ROCKET HEAD. *SEE* EXPLOSIVE PROJECTILE						
	ROCKET MOTOR	CLASS A EXPLOSIVE		EXPLOSIVE A	NONE	173.79	EXPLOSIVES A*
	ROCKET MOTOR	CLASS B EXPLOSIVE		EXPLOSIVE B	NONE	173.92	EXPLOSIVES B*
	ROMAN CANDLE. SEE FIREWORKS, COMMON						
	RUBIDIUM METAL	FLAMMABLE SOLID	UN1423	FLAMMABLE SOLID AND DANGEROUS WHEN WET	NONE	173.206	FAMMABLE SOLID -W-*
	RUBIDIUM METAL, IN CARTRIDGES	FLAMMABLE SOLID	UN1423	FLAMMABLE SOLID AND DANGEROUS WHEN WET	173.206		FLAMMABLE SOLID -W-*
	SAFETY FUSE. *SEE* FUSE, SAFETY						
	SAFETY SQUIB	CLASS C EXPLOSIVE		EXPLOSIVE C	NONE	173.106	DANGEROUS
	SALUTE. SEE FIREWORKS COMMON *OR* SPECIAL						
	SAMPLES. SEE 172.101(C)(12)						
	SAND ACID. *SEE* HYDROFLUOROSILICIC ACID						
	SELENIC ACID, LIQUID	CORROSIVE MATERIAL	UN1905	CORROSIVE	NONE	173.245	CORROSIVE
	SELENIUM NITRIDE	FORBIDDEN					
	SELENIUM OXIDE	POISON B	NA2811	POISON	173.364	173.365	POISON
	SELF-LIGHTING CIGARETTE	FLAMMABLE SOLID	UN1867	FLAMMABLE SOLID	173.21		FLAMMABLE SOLID[5]
	SELF PROPELLED VEHICLE. *SEE* MOTOR VEHICLE						
	SHAPED CHARGE, COMMERCIAL. SEE HIGH EXPLOSIVE *(173.65(H))*						
	SHAPED CHARGES (COMMERCIAL) CONTAINING MORE THAN 8 OUNCES OF EXPLOSIVES	FORBIDDEN					
	SHELL, FIREWORKS. *SEE* FIREWORKS, COMMON *OR* SPECIAL						
	SHIP, DISTRESS SIGNAL. SEE FIREWORKS, SPECIAL						
	SIGNAL FLARE	CLASS C EXPLOSIVE		EXPLOSIVE C	NONE	173.108	DANGEROUS
	SILICOFLUORIC ACID. *SEE* HYDROFLUOROSILICIC ACID						

5-SEE PLACARDING NOTES, PAGE 5

§172.101 HAZARDOUS MATERIALS TABLE

(1) +/ A/ W	(2) Hazardous materials descriptions and proper shipping names	(3) Hazard Class	(3A) Identi- fication number	(4) Label(s) required (if not excepted)	(5) Packaging (a) Exceptions	(5) Packaging (b) Specific require- ments	Placards required (*denotes placard for any quantity)
	SILICON CHLORIDE *OR* SILICON TETRACHLORIDE	CORROSIVE MATERIAL	UN1818	CORROSIVE	173.244	173.247	CORROSIVE
	SILICON CHROME, EXOTHERMIC. *SEE* FERROCHROME, EXOTHERMIC						
+	SILICON TETRAFLUORIDE	NONFLAMM- ABLE GAS	UN1859	NONFLAMM- ABLE GAS	173.306	173.302	NONFLAMM- ABLE GAS
	SILVER ACETYLIDE (DRY)	FORBIDDEN					
	SILVER AZIDE (DRY)	FORBIDDEN					
	SILVER CHLORITE (DRY)	FORBIDDEN					
+	SILVER CYANIDE	POISON B	UN1684	POISON	173.370	173.370	POISON
	SILVER FULMINATE (DRY)	FORBIDDEN					
	SILVER NITRATE	OXIDIZER	UN1493	OXIDIZER	173.153	173.182	OXIDIZER
	SILVER OXALATE (DRY)	FORBIDDEN					
	SILVER PICRATE (DRY)	FORBIDDEN					
	SMALL ARMS AMMUNITION	ORM-D		NONE	173.101	173.1201	NONE
	SMALL ARMS AMMUNITION	CLASS C EXPLOSIVE		NONE	173.101		NONE
	SMALL ARMS AMMUNITION, IRRITATING (*TEAR GAS*) CARTRIDGE	CLASS C EXPLOSIVE		IRRITANT	NONE	173.101	DANGEROUS
	SMALL ARMS PRIMER	CLASS C EXPLOSIVE		NONE	NONE	173.107	NONE
	SMOKE CANDLE	CLASS C EXPLOSIVE		EXPLOSIVE C	NONE	173.108	DANGEROUS
	SMOKE GENERATOR. SEE CHEMICAL AMMUNITION, NONEXPLOSIVE (*CONTAINING A POISON A, POISON B, OR IRRITATING MATERIAL, AS APPROPRIATE*)						
	SMOKE GRENADE	CLASS C EXPLOSIVE		EXPLOSIVE C	NONE	173.108	DANGEROUS
	SMOKELESS POWDER FOR CANNON OR SMALL ARMS. SEE PROPELLANT EXPLOSIVE, *CLASS A OR B, AS APPROPRIATE*						
	SMOKELESS POWDER FOR SMALL ARMS (*100 POUNDS OR LESS*)	FLAMMABLE SOLID	NA1325	FLAMMABLE SOLID	173.88	173.197a	FLAMMABLE SOLID[5]
	SMOKE POT	CLASS C EXPLOSIVE		EXPLOSIVE C	NONE	173.108	DANGEROUS
	SMOKE PROJECTILE WITH BURSTING CHARGE. SEE EXPLOSIVE PROJECTILE						
	SMOKE PROJECTILE WITH EXPELLING CHARGE BUT WITHOUT BURSTING CHARGE. SEE FIREWORKS, SPECIAL						
	SMOKE SIGNAL	CLASS C EXPLOSIVE		EXPLOSIVE C	NONE	173.108	DANGEROUS
	SODA AMATOL. SEE HIGH EXPLOSIVE						
	SODA LIME, SOLID	CORROSIVE MATERIAL	UN1907	CORROSIVE	173.244	173.245b	CORROSIVE
	SODIUM ACID SULFATE, SOLID *OR* SOLUTION. *SEE APPROPRIATE* SODIUM HYDROGEN SULFATE *ENTRY*						
A	SODIUM ALUMINATE, SOLID	ORM-B	UN2812	NONE	173.505	173.800	NONE

5-SEE PLACARDING NOTES, PAGE 5

(1) +/A/W	(2) Hazardous materials descriptions and proper shipping names	(3) Hazard Class	(3A) Identi-fication number	(4) Label(s) required (if not excepted)	(5) Packaging (a) Exceptions	(5) Packaging (b) Specific require-ments	Placards required (*denotes placard for any quantity)
	SODIUM ALUMINATE SOLUTION	CORROSIVE MATERIAL	UN1819	CORROSIVE	173.244	173.249	CORROSIVE
	SODIUM ALUMINUM HYDRIDE	FLAMMABLE SOLID	UN2835	FLAMMABLE SOLID AND DANGEROUS WHEN WET	NONE	173.206	FLAMMABLE SOLID -W-*
	SODIUM AMIDE	FLAMMABLE SOLID	UN1425	FLAMMABLE SOLID AND DANGEROUS WHEN WET	NONE	173.206	FLAMMABLE SOLID -W-*
	SODIUM ARSENATE	POISON B	UN1685	POISON	173.364	173.365 173.368	POISON
	SODIUM ARSENITE, LIQUID (*SOLUTION*)	POISON B	UN1686	POISON	173.345	173.346	POISON
+	SODIUM AZIDE	POISON B	UN1687	POISON	173.364	173.375	POISON
	SODIUM BIFLUORIDE, SOLID	CORROSIVE MATERIAL	UN2439	CORROSIVE	173.244	173.245b	CORROSIVE
	SODIUM BIFLUORIDE, SOLUTION	CORROSIVE MATERIAL	UN2439	CORROSIVE	173.244	173.245	CORROSIVE
	SODIUM BISULFATE, SOLID *OR* SOLUTION. *SEE* SODIUM HYDROGEN SULFATE, SOLID *OR* SOLUTION						
	SODIUM BISULFITE, SOLID *OR* SOLUTION. *SEE* SODIUM HYDROGEN SULFITE, SOLID *OR* SOLUTION						
	SODIUM BROMATE	OXIDIZER	UN1494	OXIDIZER	173.153	173.154	OXIDIZER
	SODIUM CHLORATE (*SODA CHLORATE*)	OXIDIZER	UN1495	OXIDIZER	173.153	173.163	OXIDIZER
	SODIUM CHLORITE	OXIDIZER	UN1496	OXIDIZER	NONE	173.160	OXIDIZER
	SODIUM CHLORITE SOLUTION (*NOT EXCEEDING 42% SODIUM CHLORITE*)	CORROSIVE MATERIAL	UN1908	CORROSIVE	173.244	173.263	CORROSIVE
	SODIUM CYANIDE, SOLID	POISON B	UN1689	POISON	173.370	173.370	POISON
	SODIUM CYANIDE SOLUTION	POISON B	UN1689	POISON	173.345	173.352	POISON
	SODIUM DICHLOROISOCYANURATE. *SEE* SODIUM DICHLORO-S-TRIAZINETRIONE						
	SODIUM DICHLORO-S-TRIAZINETRIONE (*DRY, CONTAINING MORE THAN 39% AVAILABLE CHLORINE*)	OXIDIZER	UN2465	OXIDIZER	173.153	173.217	OXIDIZER
A	SODIUM DICHROMATE	ORM-A	NA1479	NONE	173.505	173.510	NONE
A	SODIUM FLUORIDE, SOLID	ORM-B	UN1690	NONE	173.505	173.510	NONE
	SODIUM FLUORIDE, SOLUTION	CORROSIVE MATERIAL	UN1690	CORROSIVE	173.244	173.245	CORROSIVE
	SODIUM HYDRATE. *SEE* SODIUM HYDROXIDE						
	SODIUM HYDRIDE	FLAMMABLE SOLID	UN1427	FLAMMABLE SOLID AND DANGEROUS WHEN WET	NONE	173.198	FLAMMABLE SOLID -W-*
A	SODIUM HYDROGEN SULFATE, SOLID	ORM-B	UN1821	NONE	173.505	173.800	NONE
	SODIUM HYDROGEN SULFATE SOLUTION	CORROSIVE MATERIAL	UN2837	CORROSIVE	173.244	173.245	CORROSIVE
A	SODIUM HYDROGEN SULFITE, SOLID	ORM-B	NA2693	NONE	173.505	173.800	NONE
	SODIUM HYDROGEN SULFITE, SOLUTION	CORROSIVE MATERIAL	NA2693	CORROSIVE	173.244	173.245	CORROSIVE
	SODIUM HYDROSULFIDE, SOLID (*WITH LESS THAN 25% WATER OF CRYSTALLIZATION*)	FLAMMABLE SOLID	UN2318	FLAMMABLE SOLID	173.153	173.154	FLAMMABLE SOLID[5]

§172.101 HAZARDOUS MATERIALS TABLE

(1) +/ A/ W/	(2) Hazardous materials descriptions and proper shipping names	(3) Hazard Class	(3A) Identi-fication number	(4) Label(s) required (if not excepted)	(5) Packaging (a) Exceptions	(5) Packaging (b) Specific require-ments	Placards required (*denotes placard for any quantity)
	SODIUM HYDROSULFIDE, SOLID (*WITH NOT LESS THAN 25% WATER OF CRYSTALLIZATION*)	CORROSIVE MATERIAL	NA2923	CORROSIVE	173.244	173.245b	CORROSIVE
	SODIUM HYDROSULFIDE, SOLUTION	CORROSIVE MATERIAL	NA2922	CORROSIVE	173.244	173.245	CORROSIVE
	SODIUM HYDROSULFITE (*SODIUM DITHIONITE*)	FLAMMABLE SOLID	UN1384	FLAMMABLE SOLID	173.153	173.204	FLAMMABLE SOLID[5]
	SODIUM HYDROXIDE, DRY SOLID, FLAKE, BEAD, *OR* GRANULAR	CORROSIVE MATERIAL	UN1823	CORROSIVE	173.244	173.245b	CORROSIVE
	SODIUM HYDROXIDE, LIQUID *OR* SOLUTION	CORROSIVE MATERIAL	UN1824	CORROSIVE	173.244	173.249	CORROSIVE
	SODIUM HYPOCHLORITE . SEE HYPOCHLORITE SOLUTION *OR* HYPOCHLORITE SOLUTION CONTAINING NOT MORE THAN 7% AVAILABLE CHLORINE						
A	SODIUM METABISULFITE	ORM-B	NA2693	NONE	173.505	173.510	NONE
	SODIUM, METAL *OR* METALLIC	FLAMMABLE SOLID	UN1428	FLAMMABLE SOLID AND DANGEROUS WHEN WET	NONE	173.206	FLAMMALE SOLID -W-*
	SODIUM, METAL DISPERSION IN ORGANIC SOLVENT	FLAMMABLE SOLID	UN1429	FLAMMABLE SOLID AND DANGEROUS WHEN WET	NONE	173.230	FLAMMABLE SOLID -W-*
	SODIUM, METAL LIQUID ALLOY	FLAMMABLE SOLID	NA1421	FLAMMABLE SOLID AND DANGEROUS WHEN WET	NONE	173.202	FLAMMABLE SOLID -W-*
	SODIUM METHYLATE, ALCOHOL MIXTURE	COMBUSTIBLE LIQUID	NA1289	NONE	173.118a	NONE	COMBUSTIBLE[1]
	SODIUM METHYLATE, ALCOHOL MIXTURE	FLAMMABLE LIQUID	NA1289	FLAMMABLE LIQUID	173.118	173.119	FLAMMABLE
	SODIUM METHYLATE, ALCOHOL MIXTURE	CORROSIVE MATERIAL	NA1289	CORROSIVE	173.244	173.245	CORROSIVE
	SODIUM METHYLATE, DRY	FLAMMABLE SOLID	UN1431	FLAMMABLE SOLID AND DANGEROUS WHEN WET	173.153	173.154	FLAMMABLE SOLID -W-*
	SODIUM MONOXIDE, SOLID	CORROSIVE MATERIAL	UN1825	CORROSIVE	173.244	173.245b	CORROSIVE
	SODIUM NITRATE	OXIDIZER	UN1498	OXIDIZER	173.153	173.182	OXIDIZER
	SODIUM NITRITE	OXIDIZER	UN1500	OXIDIZER	173.153	173.234	OXIDZER
	SODIUM NITRITE MIXED (*FUSED*) WITH POTASSIUM NITRATE	OXIDIZER	UN1487	OXIDIZER	173.153	173.183	OXIDIZER
	SODIUM NITRITE MIXTURE (*SODIUM NITRATE, SODIUM NITRITE, AND POTASSIUM NITRATE*)	OXIDIZER	NA1487	OXIDIZER	173.153	173.234	OXIDIZER
A	SODIUM PENTACHLOROPHENATE	ORM-A	UN2567	NONE	173.505	173.510	NONE
	SODIUM PERCHLORATE	OXIDIZER	UN1502	OXIDIZER	173.153	173.154	OXIDIZER
	SODIUM PERMANGANATE	OXIDIZER	UN1503	OXIDIZER	173.153	173.154	OXIDIZER
	SODIUM PEROXIDE	OXIDIZER	UN1504	OXIDIZER	NONE	173.187	OXIDIZER
	SODIUM PERSULFATE	OXIDIZER	UN1505	OXIDIZER	173.153	173.154	OXIDIZER
	SODIUM PHENOLATE, SOLID	CORROSIVE MATERIAL	UN2497	CORROSIVE	173.244	173.245b	CORROSIVE

5 - SEE PLACARDING NOTES, PAGE 5

§172.101 HAZARDOUS MATERIALS TABLE

(1) +/ A/ W	(2) Hazardous materials descriptions and proper shipping names	(3) Hazard Class	(3A) Identi-fication number	(4) Label(s) required (if not excepted)	(5) Packaging (a) Exceptions	(5) Packaging (b) Specific require-ments	Placards required (*denotes placard for any quantity)
	SODIUM PHOSPHIDE	FLAMMABLE SOLID	UN1432	FLAMMABLE SOLID AND DANGEROUS WHEN WET	NONE	173.154	FLAMMABLE SOLID -W-*
	SODIUM PICRAMATE, WET (*WITH AT LEAST 20% WATER*)	FLAMMABLE SOLID	UN1349	FLAMMABLE SOLID	NONE	173.205	FLAMMABLE SOLID[5]
	SODIUM PICRYL PEROXIDE	FORBIDDEN					
	SODIUM POTASSIUM ALLOY, LIQUID	FLAMMABLE SOLID	UN1422	FLAMMABLE SOLID AND DANGEROUS WHEN WET	NONE	173.202	FLAMMABLE SOLID -W-* PLACARD
	SODIUM POTASSIUM ALLOY, SOLID	FLAMMABLE SOLID	UN1422	FLAMMABLE SOLID AND DANGEROUS WHEN WET	NONE	173.206	FLAMMABLE SOLID -W-*
	SODIUM SELENITE	POISON B	UN2630	POISON	173.364	173.365	POISON
	SODIUM SULFIDE, ANHYDROUS OR SODIUM SUFIDE WITH LESS THAN 30% WATER OF CRYSTALLIZATION	FLAMMABLE SOLID	UN1385	FLAMMABLE SOLID	173.153	173.207	FLAMMABLE SOLID[5]
	SODIUM SULFIDE, HYDRATED WITH NOT LESS THAN 30% WATER	CORROSIVE MATERIAL	UN1849	CORROSIVE	173.244	173.245B	CORROSIVE
	SODIUM SUPEROXIDE	OXIDIZER	UN2547	OXIDIZER	NONE	173.187	OXIDIZER
	SODIUM TETRANITRIDE	FORBIDDEN					
	SPARKLERS. SEE FIREWORKS, COMMON						
	SPENT IRON MASS. SEE IRON MASS, SPENT						
	SPENT IRON SPONGE. SEE IRON SPONGE, SPENT						
	SPENT MIXED ACID. SEE NITRATING ACID, SPENT						
	SPENT SULFURIC ACID. SEE SULFURIC ACID, SPENT						
	SPIRITS OF NITROGLYCERIN, (*1 TO 10%*)	FLAMMABLE LIQUID	NA1204	FLAMMABLE LIQUID	NONE	173.133	FLAMMABLE
	SPIRITS OF NITROGLYCERIN, NOT EXCEEDING 1% NITROGLYCERIN BY WEIGHT	FLAMMABLE LIQUID	NA1204	FLAMMABLE LIQUID	173.118	173.133	FLAMMABLE
	SPORTING POWDER. SEE BLACK POWDER *OR* PROPELLANT EXPLOSIVE, SOLID						
	SPRAY STARTING FLUID. SEE ENGINE STARTING FLUID						
	SPREADER CARTRIDGE. SEE FIREWORKS, SPECIAL						
	SQUIB, ELECTRIC OR SAFETY. SEE ELECTRIC SQUIB *OR* SAFETY SQUIB						
	STANNIC PHOSPHIDE	FLAMMABLE SOLID	UN1433	FLAMMABLE SOLID AND DANGEROUS WHEN WET	NONE	173.154	FLAMMABLE SOLID -W-*
A	STANNOUS CHLORIDE, SOLID	ORM-B	NA1759	NONE	173.505	173.510	NONE
	STARTER CARTRIDGE	CLASS B EXPLOSIVE		EXPLOSIVE B	NONE	173.92	EXPLOSIVES B*
	STARTER CARTRIDGE	CLASS C EXPLOSIVE		EXPLOSIVE C	NONE	173.102	DANGEROUS
	STORAGE BATTERY, WET. SEE BATTERY, ELECTRIC STORAGE, WET						
	STRONTIUM ARSENITE, SOLID	POISON B	UN1691	POISON	173.364	173.365	POISON

5 - SEE PLACARDING NOTES, PAGE 5

§172.101 HAZARDOUS MATERIALS TABLE

(1) +/ A/ W/	(2) Hazardous materials descriptions and proper shipping names	(3) Hazard Class	(3A) Identi-fication number	(4) Label(s) required (if not excepted)	(5) Packaging (a) Exceptions	(5) Packaging (b) Specific require-ments	Placards required (*denotes placard for any quantity)
	STRONTIUM CHLORATE	OXIDIZER	UN1506	OXIDIZER	173.153	173.163	OXIDIZER
	STRONTIUM CHLORATE, WET	OXIDIZER	UN1506	OXIDIZER	173.153	173.163	OXIDIZER
	STRONTIUM NITRATE	OXIDIZER	UN1507	OXIDIZER	173.153	173.182	OXIDIZER
	STRONTIUM PEROXIDE	OXIDIZER	UN1509	OXIDIZER	173.153	173.154	OXIDIZER
	STRYCHNINE SALT, SOLID	POISON B	UN1692	POISON	173.364	173.365	POISON
	STRYCHNINE, SOLID	POISON B	UN1692	POISON	173.364	173.365	POISON
	STYPHNATE OF LEAD. SEE INITIATING EXPLOSIVE						
	STYRENE MONOMER, INHIBITED	FLAMMABLE LIQUID	UN2055	FLAMMABLE LIQUID	173.118	173.119	FLAMMABLE
	SUBSTITUTED NITROPHENOL PESTICIDE, LIQUID, N.O.S. (*COMPOUNDS AND PREPARATIONS*)	FLAMMABLE LIQUID	UN2780	FLAMMABLE LIQUID	173.118	173.119	FLAMMABLE
	SUBSTITUTED NITROPHENOL PESTICIDE, LIQUID, N.O.S. (*COMPOUNDS AND PREPARATIONS*)	POISON B	UN2779	POISON	173.345	173.346	POISON
	SUBSTITUTED NITROPHENOL PESTICIDE, SOLID, N.O.S. (*COMPOUNDS AND PREPARATIONS*)	POISON B	UN2779	POISON	173.364	173.365	POISON
	SUCCINIC ACID PEROXIDE	ORGANIC PEROXIDE	UN2135	ORGANIC PEROXIDE	173.153	173.157 173.158	ORGANIC PEROXIDE
	SUCCINIC ACID PEROXIDE, *TECHNICALLY PURE. SEE* SUCCINIC ACID PEROXIDE.		UN2135				
	SUCROSE OCTANITRATE (DRY)	FORBIDDEN					
	SULFUR AND CHLORATE, LOOSE MIXTURES OF	FORBIDDEN					
	SULFUR CHLORIDE (*DI*)	CORROSIVE MATERIAL	UN1828	CORROSIVE	NONE	173.247	CORROSIVE
	SULFUR CHLORIDE (*MONO*)	CORROSIVE MATERIAL	UN1828	CORROSIVE	NONE	173.247	CORROSIVE
	SULFUR DIOXIDE	NONFLAMM-ABLE GAS	UN1079	NONFLAMM-ABLE GAS	173.306	173.304 173.314 173.315	NONFLAMM-ABLE GAS
	SULFUR FLOWER. *SEE* SULFUR, SOLID						
	SULFUR HEXAFLUORIDE	NONFLAMM-ABLE GAS	UN1080	NONFLAMM-ABLE GAS	173.306	173.304	NONFLAMM-ABLE GAS
	SULFURIC ACID (*FOR FUMING SULFURIC ACID, SEE* OLEUM)	CORROSIVE MATERIAL	UN1830	CORROSIVE	173.244	173.272	CORROSIVE
	SULFURIC ACID, SPENT	CORROSIVE MATERIAL	UN1832	CORROSIVE	NONE	173.248	CORROSIVE
	SULFURIC ANHYDRIDE. *SEE* SULFUR TRIOXIDE						
	SULFUROUS ACID	CORROSIVE MATERIAL	UN1833	CORROSIVE	173.244	173.245	CORROSIVE
W	SULFUR, SOLID	ORM-C	UN1350	NONE	173.505	173.1080	NONE
	SULFUR TRIOXIDE	CORROSIVE MATERIAL	UN1829	CORROSIVE	173.244	173.273	CORROSIVE
	SULFURYL CHLORIDE	CORROSIVE MATERIAL	UN1834	CORROSIVE	173.244	173.247	CORROSIVE
	SULFURYL FLUORIDE	NONFLAMM-ABLE GAS	UN2191	NONFLAMM-ABLE GAS	173.306	173.304 173.314	NONFLAMM-ABLE GAS
	SULPHUR. *SEE* SULFUR, SOLID						
	SULFUR, MOLTEN OR SULPHUR, MOLTEN	ORM-C	UN2448	NONE	173.505	173.1080	NONE
	SUPPLEMENTARY CHARGE (EXPLOSIVE)	CLASS A EXPLOSIVE		EXPLOSIVE A	NONE	173.69	EXPLOSIVES A*

§172.101 HAZARDOUS MATERIALS TABLE

(1) +/ A/ W/	(2) Hazardous materials descriptions and proper shipping names	(3) Hazard Class	(3A) Identification number	(4) Label(s) required (if not excepted)	(5) Packaging (a) Exceptions	(5) Packaging (b) Specific requirements	Placards required (*denotes placard for any quantity)
A	2,4,5-T. *SEE* 2,4,5-TRICHLOROPHENOXYACETIC ACID.						
	TANK CAR, CONTAINING RESIDUAL PHOSPHORUS AND FILLED WITH WATER OR INERT GAS, SEE 173.190						
	TANK CAR, EMPTY (PREVIOUSLY USED FOR A HAZARDOUS MATERIAL). SEE 173.29						
	TANK CAR, EMPTY (PREVIOUSLY USED FOR A POISON A MATERIAL). SEE 172.510 AND 173.29						
	TANK, PORTABLE, EMPTY (PREVIOUSLY USED FOR A HAZARDOUS MATERIAL). SEE 172.510, 172.514 AND 173.29						
	TANK TRUCK, EMPTY. SEE 172.510, 172.514 AND 173.29						
A	TDE (*1,1-DICHLORO-2,2-BIS(P-CHLOROPHENYL) ETHANE*)	ORM-A	NA2761	NONE	173.505	173.510	NONE
	TEAR GAS AMMUNITION. SEE CHEMICAL AMMUNITION, NONEXPLOSIVE *(CONTAINING AN IRRITANT MATERIAL)*						
	TEAR GAS CANDLE	IRRITATING MATERIAL	UN1700	IRRITANT	NONE	173.385	DANGEROUS
	TEAR GAS CARTRIDGE. SEE SMALL ARMS AMMUNITION, IRRITATING (*TEAR GAS*) CARTRIDGE						
	TEAR GAS DEVICE	IRRITATING MATERIAL	NA1693	IRRITANT	NONE	173.385	DANGEROUS
	TEAR GAS GRENADE. SEE GRENADE, TEAR GAS						
	TERTIARY ALCOHOL. SEE ALCOHOL, N.O.S.						
	TETRAAZIDO BENZENE QUINONE	FORBIDDEN					
AW	TETRACHLOROETHANE	ORM-A	UN1702	NONE	173.505	173.620	NONE
A	TETRACHLOROETHYLENE *OR* PERCHLOROETHYLENE	ORM-A	UN1897	NONE	173.505	173.605	NONE
	TETRAETHYLAMMONIUM PERCHLORATE (DRY)	FLAMMABLE SOLID	UN1325	FLAMMABLE SOLID	173.153	173.154	FLAMMABLE SOLID
	TETRAETHYL DITHIOPYROPHOSPHATE AND COMPRESSED GAS MIXTURE	POISON A	UN1703	POISON GAS	NONE	173.334	POISON GAS*
	TETRAETHYL DITHIOPYROPHOSPHATE, LIQUID	POISON B	UN1704	POISON	NONE	173.358	POISON
	TETRAETHYL DITHIOPYROPHOSPHATE MIXTURE, DRY	POISON B	UN1704	POISON	NONE	173.377	POISON
	TETRAETHYL DITHIOPYROPHOSPHATE MIXTURE, LIQUID	POISON B	UN1704	POISON	NONE	173.359	POISON
	TETRAETHYL LEAD, LIQUID (*INCLUDING FLASH POINT FOR EXPORT SHIPMENT BY WATER*)	POISON B	NA1649	POISON	NONE	173.354	POISON
	TETRAETHYL PYROPHOSPHATE AND COMPRESSED GAS MIXTURE	POISON A	UN1705	POISON GAS	NONE	173.334	POISON GAS*
	TETRAETHYL PYROPHOSPHATE, LIQUID	POISON B	NA2783	POISON	NONE	173.358	POISON
	TETRAETHYL PYROPHOSPHATE MIXTURE, DRY	POISON B	NA2783	POISON	NONE	173.377	POISON
	TETRAETHYL PYROPHOSPHATE MIXTURE, LIQUID	POISON B	NA2783	POISON	NONE	173.359	POISON
+	TETRAFLUOROETHYLENE, INHIBITED	FLAMMABLE GAS	UN1081	FLAMMABLE GAS	173.306	173.304	FLAMMABLE GAS

§172.101 HAZARDOUS MATERIALS TABLE

(1) +/ A/ W/	(2) Hazardous materials descriptions and proper shipping names	(3) Hazard Class	(3A) Identi-fication number	(4) Label(s) required (if not excepted)	(5) Packaging (a) Exceptions	(5) Packaging (b) Specific require-ments	Placards required (*denotes placard for any quantity)
	1,2,3,6-TETRAHYDROBENZALDEHYDE	CORROSIVE MATERIAL	UN2498	CORROSIVE	173.244	173.245	CORROSIVE
	TETRAHYDROFURAN	FLAMMABLE LIQUID	UN2056	FLAMMABLE LIQUID	NONE	173.119	FLAMMABLE
	TETRALIN HYDROPEROXIDE, *TECHNICALLY PURE. SEE* ORGANIC PEROXIDE, SOLID, N.O.S.		UN2136				
	TETRAMETHYLAMMONIUM HYDROXIDE, LIQUID	CORROSIVE MATERIAL	UN1835	CORROSIVE	173.244	173.245	CORROSIVE
	1,1,3,3-TETRAMETHYLBUTYL HYDROPEROXIDE, *TECHNICALLY PURE.* , *SEE* ORGANIC PEROXIDE, LIQUID *OR* SOLUTION, N.O.S.		UN2160				
	1,1,3,3-TETRAMETHYLBUTYL PEROXY-2-ETHYLHEXANOATE, *TECHNICALLY PURE. SEE* ORGANIC PEROXIDE, LIQUID *OR* SOLUTION, N.O.S.		UN2161				
	TETRAMETHYLENE DIPEROXIDE DICARBAMIDE	FORBIDDEN					
A	TETRAMETHYLMETHYLENEDIAMINE	ORM-A	NA9069	NONE	173.505	173.510	NONE
	TETRANITRO DIGLYCERIN	FORBIDDEN					
	TETRANITROMETHANE	OXIDIZER	UN1510	OXIDIZER	NONE	173.203	OXIDIZER[2]
	2,3,4,6-TETRANITROPHENOL	FORBIDDEN					
	2,3,4,6-TETRANITROPHENYL METHYL NITRAMINE	FORBIDDEN					
	2,3,4,6-TETRANITROPHENYLNITRAMINE	FORBIDDEN					
	TETRANITRORESORCINOL (DRY)	FORBIDDEN					
	2,3,5,6-TETRANITROSO-1,4-DINITROBENZENE	FORBIDDEN					
	2,3,5,6-TETRANITROSO NITROBENZENE (DRY)	FORBIDDEN					
	TETRAZENE (GUANYL NITROSAMINO GUANYLTETRAZENE). SEE INITIATING EXPLOSIVE						
	TETRAZINE (DRY)	FORBIDDEN					
	TETRAZOLYL AZIDE (DRY)	FORBIDDEN					
	TETRYL. SEE HIGH EXPLOSIVE						
	TEXTILE TREATING COMPOUND *OR* MIXTURE, LIQUID	CORROSIVE MATERIAL	NA1760	CORROSIVE	173.244	173.249a	CORROSIVE
	THALLIUM SALT, SOLID, N.O.S.	POISON B	NA1707	POISON	173.364	173.365	POISON
	THALLIUM SULFATE, SOLID	POISON B	NA1707	POISON	173.364	173.365	POISON
	THINNER FOR RUST PREVENTIVE COATING. *SEE* PAINT RELATED MATERIAL						
	THIOCARBONYLCHLORIDE. SEE THIOPHOSGENE						
	THIOGLYCOLIC ACID	CORROSIVE MATERIAL	UN1940	CORROSIVE	173.244	173.245	CORROSIVE
	THIONYL CHLORIDE	CORROSIVE MATERIAL	UN1836	CORROSIVE	NONE	173.247	CORROSIVE
	THIOPHENOL. SEE PHENYL MERCAPTAN						
+	THIOPHOSGENE	POISON B	UN2474	POISON	NONE	173.356	POISON
	THIOPHOSPHORYL CHLORIDE	CORROSIVE MATERIAL	UN1837	CORROSIVE	NONE	173.271	CORROSIVE

2-SEE PLACARDING NOTES, PAGE 5

§172.101 HAZARDOUS MATERIALS TABLE

(1) +/ A/ W/	(2) Hazardous materials descriptions and proper shipping names	(3) Hazard Class	(3A) Identi-fication number	(4) Label(s) required (if not excepted)	(5) Packaging (a) Exceptions	(5) Packaging (b) Specific require-ments	Placards required (*denotes placard for any quantity)
A	THIRAM	ORM-A	NA2771	NONE	173.505	173.510	NONE
	THORIUM METAL, PYROPHORIC	RADIOACTIVE MATERIAL	UN2975	RADIOACTIVE AND FLAMM-ABLE SOLID	NONE	173.418	RADIOACTIVE[3]
	THORIUM NITRATE	RADIOACTIVE MATERIAL	UN2976	RADIOACTIVE AND OXIDIZER	NONE	173.419	RADIOACTIVE[3]
	TIME FUZE. SEE FUZE, TIME						
	TIN CHLORIDE, FUMING. *SEE* TIN TETRACHLORIDE, ANHYDROUS						
	TIN PERCHLORIDE. *SEE* TIN TETRACHLORIDE, ANHY-DROUS						
	TIN TETRACHLORIDE, ANHYDROUS	CORROSIVE MATERIAL	UN1827	CORROSIVE	173.244	173.247	CORROSIVE
	TITANIUM METAL POWDER, DRY *OR* WET *WITH LESS THAN 20% WATER*	FLAMMABLE SOLID	UN2546	FLAMMABLE SOLID	NONE	173.208	FLAMMABLE SOLID[5]
	TITANIUM METAL POWDER, WET WITH 20% OR MORE WATER	FLAMMABLE SOLID	UN1352	FLAMMABLE SOLID	NONE	173.208	FLAMMABLE SOLID[5]
	TITANIUM SULFATE SOLUTION *CONTAINING NOT MORE THAN 45% SULFURIC ACID*	CORROSIVE MATERIAL	NA1760	CORROSIVE	173.244	173.297	CORROSIVE
	TITANIUM TETRACHLORIDE	CORROSIVE MATERIAL	UN1838	CORROSIVE POISON	173.244	173.247	CORROSIVE[2]
	TOLUENE (*TOLUOL*)	FLAMMABLE LIQUID	UN1294	FLAMMABLE LIQUID	173.118	173.119	FLAMMABLE
A	TOLUENEDIAMINE	ORM-A	NA1709	NONE	173.505	173.510	NONE
+	TOLUENE DIISOCYANATE	POISON B	UN2078	POISON	173.345	173.346	POISON
	TOLUENE SULFONIC ACID, LIQUID	CORROSIVE MATERIAL	UN2584	CORROSIVE	173.244	173.245	CORROSIVE
	TORCH. SEE FIREWORKS, COMMON						
	TORPEDO, RAILWAY	CLASS B EXPLOSIVE		EXPLOSIVE B	NONE	173.91	EXPLOSIVES B*
A	TOXAPHENE	ORM-A	NA2761	NONE	173.505	173.510	NONE
	TOY CAPS	CLASS C EXPLOSIVE		EXPLOSIVE C	NONE	173.100 173.109	DANGEROUS
	TOY PROPELLANT DEVICE	CLASS C EXPLOSIVE		EXPLOSIVE C	NONE	173.111	DANGEROUS
	TOY SMOKE DEVICE	CLASS C EXPLOSIVE		EXPLOSIVE C	NONE	173.111	DANGEROUS
	TOY TORPEDO. SEE FIREWORKS, SPECIAL						
	2,4,5-TP. *SEE* 2,4,5-TRICHLOROPHENOXYPROPIONIC ACID						
	TRACER	CLASS C EXPLOSIVE		EXPLOSIVE C	NONE	173.105	DANGEROUS
	TRACER FUZE	CLASS C EXPLOSIVE		EXPLOSIVE C	NONE	173.105	DANGEROUS
	TRACTOR. *SEE* MOTOR VEHICLE						
	TRAILER *OR* TRUCK BODY WITH REFRIGERATION *OR* HEATING EQUIPMENT. *SEE* MOTOR VEHICLE						
	TRIAZINE PESTICIDE, LIQUID, N.O.S. (*COMPOUNDS AND PREPARATIONS*)	FLAMMABLE LIQUID	UN2764	FLAMMABLE LIQUID	173.118	173.119	FLAMMABLE

2,3,5-SEE PLACARDING NOTES, PAGE 5

§172.101 HAZARDOUS MATERIALS TABLE

(1) +/ A/ W/	(2) Hazardous materials descriptions and proper shipping names	(3) Hazard Class	(3A) Identi-fication number	(4) Label(s) required (if not excepted)	(5) Packaging (a) Exceptions	(5) Packaging (b) Specific require-ments	Placards required (*denotes placard for any quantity)
	TRIAZINE PESTICIDE, LIQUID, N.O.S. (*COMPOUNDS AND PREPARATIONS*)	POISON B	UN2763	POISON	173.345	173.346	POISON
	TRIAZINE PESTICIDE, SOLID, N.O.S. (*COMPOUNDS AND PREPARATIONS*)	POISON B	UN2763	POISON	173.364	173.365	POISON
	TRI-(B-NITROXYETHYL)AMMONIUM NITRATE	FORBIDDEN					
A	TRICHLORFON	ORM-A	NA2783	NONE	173.505	173.510	NONE
	TRICHLOROACETIC ACID, SOLID	CORROSIVE MATERIAL	UN1839	CORROSIVE	173.244	173.245b	CORROSIVE
	TRICHLOROACETIC ACID SOLUTION	CORROSIVE MATERIAL	UN2564	CORROSIVE	173.244	173.245	CORROSIVE
A	1,1,1-TRICHLOROETHANE	ORM-A	UN2831	NONE	173.505	173.605	NONE
A	TRICHLOROETHYLENE	ORM-A	UN1710	NONE	173.505	173.605	NONE
	TRICHLOROISOCYANURIC ACID, DRY	OXIDIZER	UN2468	OXIDIZER	173.153	173.217	OXIDIZER
	TRICHLOROMETHYL PERCHLORATE	FORBIDDEN					
A	TRICHLOROPHENOL	ORM-A	NA2020	NONE	173.505	173.510	NONE
A	2,4,5-TRICHLOROPHENOXYACETIC ACID	ORM-A	NA2765	NONE	173.505	173.510	NONE
A	2,4,5-TRICHLOROPHENOXYPROPIONIC ACID	ORM-A	NA2765	NONE	173.505	173.510	NONE
	TRICHLOROSILANE	FLAMMABLE LIQUID	UN1295	FLAMMABLE LIQUID	NONE	173.136	FLAMMABLE
	TRICHLORO-S-TRIAZINETRIONE *DRY, CONTAINING OVER 39% AVAILABLE CHLORINE*	OXIDIZER	NA2468	OXIDIZER	173.153	173.217	OXIDIZER
	MONO-(TRICHLORO) TETRA-(MONOPOTASSIUM DI-CHLORO)-PENTA-S-TRIAZINETRIONE, DRY *(CONTAINING OVER 39% AVAILABLE CHLORINE)*	OXIDIZER	NA2468	OXIDIZER	173.153	173.217	OXIDIZER
	TRICK MATCHES	CLASS C EXPLOSIVE		EXPLOSIVE C	NONE	173.111	DANGEROUS
	TRICK NOISE MAKER, EXPLOSIVE	CLASS C EXPLOSIVE		EXPLOSIVE C	NONE	173.111	DANGEROUS
	TRIETHYLAMINE	FLAMMABLE LIQUID	UN1296	FLAMMABLE LIQUID	173.118	173.119	FLAMMABLE
+	TRIFLUOROCHLOROETHYLENE	FLAMMABLE GAS	UN1082	FLAMMABLE GAS	173.306	173.304 173.314	FLAMMABLE GAS
	TRIFLUOROMETHANE AND CHLOROTRIFLUORO-METHANE MIXTURE (*CONSTANT BOILING MIXTURE*) (*R-503*). *SEE* REFRIGERANT GAS, N.O.S.						
	TRIFORMOXIME TRINITRATE	FORBIDDEN					
	1,3,5-TRIMETHYL-2,4,6-TRINITROBENZENE	FORBIDDEN					
	TRIMETHYLACETYL CHLORIDE	CORROSIVE MATERIAL	UN2438	CORROSIVE	173.244	173.247	CORROSIVE
	TRIMETHYLAMINE, ANHYDROUS	FLAMMABLE GAS	UN1083	FLAMMABLE GAS	173.306	173.304 173.314 173.315	FLAMMABLE GAS
	TRIMETHYLAMINE, AQUEOUS SOLUTION	FLAMMABLE LIQUID	UN1297	FLAMMABLE LIQUID	173.118	173.119	FLAMMABLE
	TRIMETHYLCHLOROSILANE	FLAMMABLE LIQUID	UN1298	FLAMMABLE LIQUID	NONE	173.135	FLAMMABLE
	TRIMETHYLENE GLYCOL DIPERCHLORATE	FORBIDDEN					
	TRIMETHYLOL NITROMETHANE TRINITRATE	FORBIDDEN					
	2,4,6-TRINITRO-1,3,5-TRIAZIDO BENZENE (DRY)	FORBIDDEN					

§172.101 HAZARDOUS MATERIALS TABLE

(1) +/ A/ W	(2) Hazardous materials descriptions and proper shipping names	(3) Hazard Class	(3A) Identi-fication number	(4) Label(s) required (if not excepted)	(5) Packaging (a) Exceptions	(5) Packaging (b) Specific require-ments	Placards required (*denotes placard for any quantity)
	2,4,6-TRINITRO-1,3-DIAZOBENZENE	FORBIDDEN					
	TRINITROACETIC ACID	FORBIDDEN					
	TRINITROACETONITRILE	FORBIDDEN					
	TRINITROAMINE COBALT	FORBIDDEN					
	TRINITROBENZENE, DRY. SEE HIGH EXPLOSIVE						
	TRINITROBENZENE, WET *CONTAINING AT LEAST 10% WATER*	FLAMMABLE SOLID	UN1354	FLAMMABLE SOLID	173.212		FLAMMABLE SOLID[5]
	TRINITROBENZENE, WET, CONTAINING AT LEAST 10% WATER, OVER 16 OUNCES IN ONE OUTSIDE PACKAGING. SEE HIGH EXPLOSIVE						
	TRINITROBENZOIC ACID, DRY. SEE HIGH EXPLOSIVE						
	TRINITROBENZOIC ACID, WET, *CONTAINING AT LEAST 10% WATER*	FLAMMABLE SOLID	UN1355	FLAMMABLE SOLID	173.192	173.193	FLAMMABLE SOLID[5]
	TRINITROBENZOIC ACID, WET, CONTAINING AT LEAST 10% WATER, OVER 25 POUNDS IN ONE OUTSIDE PACKAGING. SEE HIGH EXPLOSIVE						
	TRINITROETHANOL	FORBIDDEN					
	TRINITROETHYLNITRATE	FORBIDDEN					
	TRINITROMETHANE	FORBIDDEN					
	1,3,5-TRINITRONAPHTHALENE	FORBIDDEN					
	2,4,6-TRINITROPHENYL GUANIDINE (DRY)	FORBIDDEN					
	2,4,6-TRINITROPHENYL NITRAMINE	FORBIDDEN					
	2,4,6-TRINITROPHENYL TRIMETHYLOL METHYL NITRAMINE TRINITRATE (DRY)	FORBIDDEN					
	TRINITRORESORCINOL. SEE HIGH EXPLOSIVE						
	2,4,6-TRINITROSO-3-METHYL NITRAMINOANISOLE	FORBIDDEN					
	TRINITROTETRAMINE COBALT NITRATE	FORBIDDEN					
	TRINITROTOLUENE, DRY. SEE HIGH EXPLOSIVE						
	TRINITROTOLUENE, WET *CONTAINING AT LEAST 10% WATER*	FLAMMABLE SOLID	UN1356	FLAMMABLE SOLID	173.212		FLAMMABLE SOLID[5]
	TRINITROTOLUENE, WET, CONTAINING AT LEAST 10% WATER, OVER 16 OUNCES IN ONE OUTSIDE PACKAGING. SEE HIGH EXPLOSIVE						
	TRIS-(1-AZIRIDINYL) PHOSPHINE OXIDE	CORROSIVE MATERIAL	UN2501	CORROSIVE	173.244	173.299A	CORROSIVE
	TRIS, BIS-BIFLUOROAMINO DIETHOXY PROPANE (TVOPA)	FORBIDDEN					
	TUNGSTEN HEXAFLUORIDE	CORROSIVE MATERIAL	UN2196	CORROSIVE	NONE	173.284	CORROSIVE
	TURPENTINE	COMBUSTIBLE LIQUID	UN1299	NONE	173.118A	NONE	COMBUSTIBLE[1]
	TURPENTINE	FLAMMABLE LIQUID	UN1299	FLAMMABLE LIQUID	173.118	173.119	FLAMMABLE

1,5-SEE PLACARDING NOTES, PAGE 5

(1) +/ A/ W/	(2) Hazardous materials descriptions and proper shipping names	(3) Hazard Class	(3A) Identification number	(4) Label(s) required (if not excepted)	(5) Packaging (a) Exceptions	(5) Packaging (b) Specific requirements	Placards required (*denotes placard for any quantity)
	URANIUM HEXAFLUORIDE, FISSILE *(CONTAINING MORE THAN 1% U-235)*	RADIOACTIVE MATERIAL	UN2977	RADIOACTIVE AND CORROSIVE	173.453	173.417	RADIOACTIVE[7] CORROSIVE
	URANIUM HEXAFLUORIDE, LOW SPECIFIC ACTIVITY	RADIOACTIVE MATERIAL	UN2978	RADIOACTIVE AND CORROSIVE	173.421	173.425	RADIOACTIVE[7] CORROSIVE
	URANIUM METAL, PYROPHORIC	RADIOACTIVE MATERIAL	UN2979	RADIOACTIVE AND FLAMMABLE SOLID	NONE	173.418	RADIOACTIVE[3]
	URANYL ACETATE	RADIOACTIVE MATERIAL	NA9180	RADIOACTIVE	173.421 173.425	173.415 173.416	RADIOACTIVE[3]
	URANYL NITRATE HEXAHYDRATE SOLUTION	RADIOACTIVE MATERIAL	UN2980	RADIOACTIVE AND CORROSIVE	173.421 173.425	173.415 173.416 173.417	RADIOACTIVE[3]
	URANYL NITRATE, SOLID	RADIOACTIVE MATERIAL	UN2981	RADIOACTIVE AND OXIDIZER	NONE	173.419	RADIOACTIVE[3]
	UREA NITRATE, DRY. SEE HIGH EXPLOSIVE						
	UREA NITRATE, WET *WITH 10% OR MORE WATER*	FLAMMABLE SOLID	UN1357	FLAMMABLE SOLID	173.192	173.193	FLAMMABLE SOLID[5]
	UREA NITRATE, WET WITH 10% OR MORE WATER, OVER 25 POUNDS IN ONE OUTSIDE PACKAGING. SEE HIGH EXPLOSIVE						
	UREA PEROXIDE	ORGANIC PEROXIDE	NA1511	ORGANIC PEROXIDE	173.153	173.227	ORGANIC PEROXIDE
	VALERIC ACID	CORROSIVE MATERIAL	NA1760	CORROSIVE	173.244	173.245	CORROSIVE
	VALERYL CHLORIDE	CORROSIVE MATERIAL	UN2502	CORROSIVE	173.244	173.245	CORROSIVE
	VANADIUM OXYTRICHLORIDE	CORROSIVE MATERIAL	UN2443	CORROSIVE	173.244	173.247A	CORROSIVE
	VANADIUM OXYTRICHLORIDE AND TITANIUM TETRACHLORIDE MIXTURE	CORROSIVE MATERIAL	NA2443	CORROSIVE	NONE	173.245 173.245A	CORROSIVE
	VANADIUM TETRACHLORIDE	CORROSIVE MATERIAL	UN2444	CORROSIVE	173.244	173.247A	CORROSIVE
	VERY SIGNAL CARTRIDGE	CLASS C EXPLOSIVE		EXPLOSIVE C	NONE	173.108	DANGEROUS
	VINYL ACETATE	FLAMMABLE LIQUID	UN1301	FLAMMABLE LIQUID	173.118	173.119	FLAMMABLE
+	VINYL CHLORIDE	FLAMMABLE GAS	UN1086	FLAMMABLE GAS	173.306	173.304 173.314 173.315	FLAMMABLE GAS
	VINYL ETHYL ETHER, INHIBITED	FLAMMABLE LIQUID	UN1302	FLAMMABLE LIQUID	NONE	173.119	FLAMMABLE
	VINYL FLUORIDE, INHIBITED	FLAMMABLE GAS	UN1860	FLAMMABLE GAS	173.306	173.304 173.314 173.315	FLAMMABLE GAS
	VINYLIDENE CHLORIDE, INHIBITED	FLAMMABLE LIQUID	UN1303	FLAMMABLE LIQUID	173.118	173.119	FLAMMABLE
	VINYL ISOBUTYL ETHER	FLAMMABLE LIQUID	UN1304	FLAMMABLE LIQUID	173.118	173.119	FLAMMABLE
+	VINYL METHYL ETHER	FLAMMABLE GAS	UN1087	FLAMMABLE GAS	173.306	173.304 173.314 173.315	FLAMMABLE GAS
	VINYL NITRATE POLYMER	FORBIDDEN					
	VINYL TRICHLOROSILANE	FLAMMABLE LIQUID	UN1305	FLAMMABLE LIQUID	NONE	173.135	FLAMMABLE
	VITRIOL, OIL OF. SEE SULFURIC ACID						
	WAR HEAD. SEE EXPLOSIVE PROJECTILE						
	WASTE, HAZARDOUS. SEE HAZARDOUS WASTE, LIQUID *OR* SOLID, N.O.S						

3,7-SEE PLACARDING NOTES, PAGE 5

(1) +/ A/ W/	(2) Hazardous materials descriptions and proper shipping names	(3) Hazard Class	(3A) Identi-fication number	(4) Label(s) required (if not excepted)	(5) Packaging (a) Exceptions	(5) Packaging (b) Specific require-ments	Placards required (*denotes placard for any quantity)
	WATER PUMP SYSTEM *TANK CHARGED WITH COMPRESSED AIR OR NITROGEN*	NONFLAMMABLE GAS	NA1956	NONE	173.306		NONE
	WATER REACTIVE SOLID, N.O.S.	FLAMMABLE SOLID	UN2813	FLAMMABLE SOLID AND DANGEROUS WHEN WET	173.153	173.154	FLAMMABLE SOLID -W-*
	WHEELCHAIR, BATTERY-EQUIPED. SEE BATTERY, *ELECTRIC STORAGE,* WET, WITH WHEELCHAIR						
	WHITE ACID (*AMMONIUM BIFLUORIDE AND HYDROFLUORIC ACID MIXTURE*)	CORROSIVE MATERIAL	NA1760	CORROSIVE	173.244	173.264	CORROSIVE
	XENON	NONFLAMMABLE GAS	UN2036	NONFLAMMABLE GAS	173. 306	173.302	NONFLAMMABLE GAS
	X-RAY FILM. *SEE* FILM						
	XYLENE (*XYLOL*)	FLAMMABLE LIQUID	UN1307	FLAMMABLE LIQUID	173.118	173.119	FLAMMABLE
A	XYLENOL	ORM-A	UN2261	NONE	173.505	173.510	NONE
	XYLYL BROMIDE	IRRITATING MATERIAL	UN1701	IRRITANT	NONE	173.382	DANGEROUS
	P-XYLYL DIAZINE	FORBIDDEN					
	ZINC AMMONIUM NITRITE	OXIDIZER	UN1512	OXIDIZER	NONE	173.228	OXIDIZER
	ZINC ARSENATE	POISON B	UN1712	POISON	173.364	173.365	POISON
	ZINC ARSENITE, SOLID	POISON B	UN1712	POISON	173.364	173.365	POISON
	ZINC CHLORATE	OXIDIZER	UN1513	OXIDIZER	173.153	173.163	OXIDIZER
	ZINC CHLORIDE SOLUTION	CORROSIVE MATERIAL	UN1840	CORROSIVE	173.244	173.245	CORROSIVE
	ZINC CYANIDE	POISON B	UN1713	POISON	173.370	173.370	POISON
	ZINC ETHYL. SEE PYROPHORIC LIQUID, N.O.S.						
A	ZINC HYDROSULFITE	ORM-A	UN1931	NONE	173.505	173.510	NONE
	ZINC MURIATE SOLUTION. *SEE* ZINC CHLORIDE SOLUTION						
	ZINC NITRATE	OXIDIZER	UN1514	OXIDIZER	173.153	173.182	OXIDIZER
	ZINC PERMANGANATE	OXIDIZER	UN1515	OXIDIZER	173.153	173.154	OXIDIZER
	ZINC PEROXIDE	OXIDIZER	UN1516	OXIDIZER	173.153	173.154	OXIDIZER
	ZINC PHOSPHIDE	POISON B	UN1714	POISON	173.364	173.365	POISON
	ZIRCONIUM HYDRIDE	FLAMMABLE SOLID	UN1437	FLAMMABLE SOLID AND DANGEROUS WHEN WET	NONE	173.206	FLAMMABLE SOLID -W-*
	ZIRCONIUM METAL, DRY, *CHEMICALLY PRODUCED, FINER THAN 20 MESH PARTICLE SIZE*	FLAMMABLE SOLID	UN2008	FLAMMABLE SOLID	NONE	173.214	FLAMMABLE SOLID[5]
	ZIRCONIUM METAL, DRY, *MECHANICALLY PRODUCED, FINER THAN 270 MESH PARTICLE SIZE*	FLAMMABLE SOLID	UN2008	FLAMMABLE SOLID	NONE	173.214	FLAMMABLE SOLID[5]
	ZIRCONIUM, METAL, LIQUID, SUSPENSIONS	FLAMMABLE LIQUID	UN1308	FLAMMABLE LIQUID	NONE	173.140	FLAMMABLE[5]

5-SEE PLACARDING NOTES, PAGE 5

(1) +/ A/ W	(2) Hazardous materials descriptions and proper shipping names	(3) Hazard Class	(3A) Identi-fication number	(4) Label(s) required (if not excepted)	(5) Packaging (a) Exceptions	(5) Packaging (b) Specific require-ments	Placards required (*denotes placard for any quantity)
	ZIRCONIUM METAL, WET, *CHEMICALLY PRODUCED, FINER THAN 20 MESH PARTICLE SIZE*	FLAMMABLE SOLID	UN1358	FLAMMABLE SOLID	NONE	173.214	FLAMMABLE SOLID[5]
	ZIRCONIUM METAL, WET, *MECHANICALLY PRODUCED, FINER THAN 270 MESH PARTICLE SIZE*	FLAMMABLE SOLID	UN1358	FLAMMABLE SOLID	NONE	173.214	FLAMMABLE SOLID[5]
	ZIRCONIUM NITRATE	OXIDIZER	UN2728	OXIDIZER	173.153	173.182	OXIDIZER
	ZIRCONIUM PICRAMATE, WET *WITH AT LEAST 20% OF WATER*	FLAMMABLE SOLID	UN1517	FLAMMABLE SOLID	NONE	173.216	FLAMMABLE SOLID[5]
	ZIRCONIUM SCRAP (*BORINGS, CLIPPINGS, SHAVINGS, SHEETS, OR TURNINGS*)	FLAMMABLE SOLID	UN1932	FLAMMABLE SOLID	173.153	173.220	FLAMMABLE SOLID[5]
A	ZIRCONIUM SULFATE	ORM-B	NA9163	NONE	NONE	173.510	NONE
	ZIRCONIUM TETRACHLORIDE, SOLID	CORROSIVE MATERIAL	UN2503	CORROSIVE	173.244	173.245B	CORROSIVE

5-SEE PLACARDING NOTES, PAGE 5

APPENDIX TO §172.101 - LIST OF HAZARDOUS SUBSTANCES AND REPORTABLE QUANTITIES

This appendix lists materials and their corresponding reportable quantities (RQs) which are listed or designated as "hazardous substances" under section 101(14) of the Comprehensive Environmental Response, Compensation and Liability Act (CERCLA; Pub. L. 96-510). This Appendix is divided into 2 TABLES which are entitled "TABLE 1 - HAZARDOUS SUBSTANCES OTHER THAN RADIONUCLIDES" and "TABLE 2 - RADIONUCLIDES". A material listed in this Appendix is regulated as a hazardous material and a hazardous substance under this subchapter if it meets the definition of a hazardous substance in §171.8 of this subchapter.

The procedure for selecting a proper shipping name for a hazardous substance is set forth in §172.101(c)(9).

Column 1 of TABLE 1, entitled *"Hazardous substance"*, contains the names of those elements and compounds which are hazardous substances. Following the listing of elements and compounds is a listing of waste streams. These waste streams appear on the list in numerical sequence and are referenced by the appropriate "F" or "K" numbers. Column 2 of TABLE 1, entitled *"Synonyms"*, contains the names of synonyms for certain elements and compounds listed in Column 1. No synonyms are listed for waste streams. Synonyms are useful in identifying hazardous substances and in identifying proper shipping names. Column 3 of TABLE 1, entitled *"Reportable quantity (RQ)"*, contains the reportable quantity (RQ), in pounds and kilograms, for each hazardous substance listed in Column 1 of TABLE 1.

A series of notes are used throughout TABLE 1 and TABLE 2 to provide additional information concerning certain hazardous substances. These notes are explained at the end of each TABLE.

TABLE 2 lists radionuclides which are hazardous substances and their corresponding RQs. The RQs in Table 2 for radionuclides are expressed in units of curies and terabecquerels, whereas those in Table 1 are expressed in units of pounds. If a material is listed in both Table 1 and Table 2, the lowest RQ shall apply. Radionuclides are listed in alphabetical order. The RQs for radionuclides are given in the radiological unit of measure of curie, abbreviated "Ci", followed, in parentheses, by an equivalent unit measured in terabecquerels, abbreviated "TBq".

For mixtures of radionuclides, the following determinations shall be used in determining if a package contains an RQ of a hazardous substance: (i) If the identity and quantity (in curies or terabecquerels) of each radionuclide in a mixture or solution is known, the ratio between the quantity per package (in curies or terabecquerels) and the RQ for the radionuclide must be determined for each radionuclide. A package contains an RQ of a hazardous substance when the sum of the ratios for the radionuclides in the mixture or solution is equal to or greater than one; (ii) if the identity of each radionuclide in a mixture or solution is known but the quantity per package (in curies or terabecquerels) of one or more of the radionuclides is unknown, an RQ of a hazardous substance is present in a package when the total quantity (in curies or terabecquerels) of the mixture or solution is equal to or greater than the lowest RQ of any individual radionuclide in the mixture or solution; and (iii) if the identity of one or more radionuclides in a mixture or solution is unknown (or if the identity of a radionuclide by itself is unknown), an RQ of a hazardous substance is present when the total quantity (in curies or terabecquerels) in a package is equal to or greater than either one curie or the lowest RQ of any known individual radionuclide in the mixture or solution, whichever is lower.

TABLE 1 - LIST OF HAZARDOUS SUBSTANCES OTHER THAN RADIONUCLIDES

Hazardous Substance	Synonyms	Reportable Quantity (RQ) Pounds (Kilograms)
Acenaphthene		100 (45.4)
Acenaphthylene		5000 (2270)
Acetaldehyde *	Ethanal	1000 (454)
Acetaldehyde, chloro-	Chloroacetaldehyde	1000 (454)
Acetaldehyde, trichloro-	Chloral	5000 (2270)
Acetamide, N-(aminothioxomethyl)-	1-Acetyl-2-thiourea	1000 (454)
Acetamide, N-(4-ethoxyphenyl)-	Phenacetin	100 (45.4)
Acetamide, N-fluoren-2-yl-	2-Acetylaminofluorene	1 (0.454)
Acetamide, 2-fluoro-	Fluoroacetamide	100 (45.4)
Acetic acid *		5000 (2270)
Acetic acid, ethyl ester	Ethyl acetate	5000 (2270)
Acetic acid, fluoro-, sodium salt	Fluoroacetic acid, sodium salt	10 (4.54)
Acetic acid, lead salt	Lead acetate	5000 (2270)
Acetic acid, thallium(I) salt	Thallium(I) acetate	100 (45.4)
Acetic anhydride *		5000 (2270)
Acetimidic acid, N-[(methylcarbamoyl)oxy]thio-methyl ester	Methomyl	100 (45.4)
Acetone *	2-Propanone	5000 (2270)
Acetone cyanohydrin *	Propanenitrile, 2-hydroxy-2-methyl- 2-Methyllactonitrile	10 (4.54)
Acetonitrile *	Ethanenitrile	5000 (2270)
3-(alpha-Acetonylbenzyl)-4-hydroxycoumarin and salts	Warfarin	100 (45.4)
Acetophenone	Ethanone, 1-phenyl-	5000 (2270)
2-Acetylaminofluorene	Acetamide, N-fluoren-2-yl-	1 (0.454)
Acetyl bromide *		5000 (2270)
Acetyl chloride *	Ethanoyl chloride	5000 (2270)
1-Acetyl-2-thiourea	Acetamide, N-(aminothioxomethyl)-	1000 (454)
Acrolein *	2-Propenal	1 (0.454)
Acrylamide	2-Propenamide	5000 (2270)
Acrylic acid *	2-Propenoic acid	5000 (2270)
Acrylonitrile *	2-Propenenitrile	100 (45.4)
Adipic acid		5000 (2270)
Alanine, 3-[p-bis(2-chloroethyl)amino]phenyl-, L-	Melphalan	1 (0.454)
Aldicarb	Propanal, 2-methyl-2-(methylthio)-, O-[(methylamino)carbonyl]oxime	1 (0.454)
Aldrin *	1,2,3,4,10-10-Hexachloro-1,4,4a,5,8,8a-hexahydro- 1,4:5,8-endo,exo-dimethanonaphthalene	1 (0.454)
Allyl alcohol *	2-Propen-1-ol	100 (45.4)
Allyl chloride *		1000 (454)
Aluminum phosphide *		100 (45.4)
Aluminum sulfate *		5000 (2270)
2-Amino-1-methyl benzene	o-Toluidine	100 (45.4)
4-Amino-1-methyl benzene	p-Toluidine	100 (45.4)
5-(Aminomethyl)-3-isoxazolol	3(2H)-Isoxazolone, 5-(aminomethyl)-	1000 (454)
4-Aminopyridine	4-Pyridinamine	1000 (454)
Amitrole	1H-1,2,4-Triazol-3-amine	10 (4.54)
Ammonia *		100 (45.4)
Ammonium acetate		5000 (2270)
Ammonium benzoate		5000 (2270)
Ammonium bicarbonate		5000 (2270)
Ammonium bichromate	Ammonium dichromate @	10 (4.54)
Ammonium bifluoride *		100 (45.4)
Ammonium bisulfite *		5000 (2270)
Ammonium carbamate *		5000 (2270)
Ammonium carbonate *		5000 (2270)
Ammonium chloride		5000 (2270)
Ammonium chromate		10 (4.54)
Ammonium citrate, dibasic		5000 (2270)
Ammonium dichromate @	Ammonium bichromate	10 (4.54)
Ammonium fluoborate *		5000 (2270)
Ammonium fluoride *		100 (45.4)
Ammonium hydroxide *		1000 (454)
Ammonium oxalate *		5000 (2270)
Ammonium picrate *	Phenol, 2,4,6-trinitro-, ammonium salt	10 (4.54)
Ammonium silicofluoride *		1000 (454)
Ammonium sulfamate		5000 (2270)
Ammonium sulfide *		100 (45.4)
Ammonium sulfite		5000 (2270)

TABLE 1 - LIST OF HAZARDOUS SUBSTANCES OTHER THAN RADIONUCLIDES - Continued

Hazardous Substance	Synonyms	Reportable Quantity (RQ) Pounds (Kilograms)
Ammonium tartrate		5000 (2270)
Ammonium thiocyanate		5000 (2270)
Ammonium vanadate	Vanadic acid, ammonium salt	1000 (454)
Amyl acetate *		5000 (2270)
iso-Amyl acetate		
sec-Amyl acetate		
tert-Amyl acetate		
Aniline *	Benzenamine	5000 (2270)
Anthracene		5000 (2270)
Antimony ¢		5000 (2270)
Antimony pentachloride *		1000 (454)
Antimony potassium tartrate *		100 (45.4)
Antimony tribromide *		1000 (454)
Antimony trichloride *		1000 (454)
Antimony trifluoride *		1000 (454)
Antimony trioxide		1000 (454)
Aroclor 1016	POLYCHLORINATED BIPHENYLS (PCBs)	1 (0.454)
Aroclor 1221	POLYCHLORINATED BIPHENYLS (PCBs)	1 (0.454)
Aroclor 1232	POLYCHLORINATED BIPHENYLS (PCBs)	1 (0.454)
Aroclor 1242	POLYCHLORINATED BIPHENYLS (PCBs)	1 (0.454)
Aroclor 1248	POLYCHLORINATED BIPHENYLS (PCBs)	1 (0.454)
Aroclor 1254	POLYCHLORINATED BIPHENYLS (PCBs)	1 (0.454)
Aroclor 1260	POLYCHLORINATED BIPHENYLS (PCBs)	1 (0.454)
Arsenic * ¢		1 (0.454)
Arsenic acid *		1 (0.454)
Arsenic disulfide *		1 (0.454)
Arsenic(III) oxide	Arsenic trioxide *	1 (0.454)
Arsenic(V) oxide	Arsenic pentoxide *	1 (0.454)
Arsenic pentoxide *	Arsenic(V) oxide	1 (0.454)
Arsenic trichloride *		1 (0.454)
Arsenic trioxide *	Arsenic(III) oxide	1 (0.454)
Arsenic trisulfide *		1 (0.454)
Arsine, diethyl-	Diethylarsine	1 (0.454)
Asbestos * ¢¢		1 (0.454)
Auramine	Benzenamine, 4,4'-carbonimidoylbis (N,N-dimethyl-	100 (45.4)
Azaserine	L-Serine, diazoacetate (ester)	1 (0.454)
Azinphos methyl @	Guthion *	1 (0.454)
Aziridine	Ethylenimine	1 (0.454)
Azirino(2',3':3,4)pyrrolo(1,2-a)indole-4,7-dione,6- amino-8-[((aminocarbonyl)oxy) methyl]-1,1a,2,8,8a, 8b-hexahydro-8a-methoxy-5-methyl-.	Mitomycin C	10 (4.54)
Barium cyanide *		10 (4.54)
Benz[j]aceanthrylene, 1,2-dihydro-3-methyl-	3-Methylcholanthrene	10 (4.54)
Benz[c]acridine	3,4-Benzacridine	100 (45.4)
3,4-Benzacridine	Benz[c]acridine	100 (45.4)
Benzal chloride	Benzene, dichloromethyl-	5000 (2270)
Benz[a]anthracene	Benzo[a]anthracene 1,2-Benzanthracene	10 (4.54)
1,2-Benzanthracene	Benz[a]anthracene Benzo[a]anthracene	10 (4.54)
1,2-Benzanthracene, 7,12-dimethyl-	7,12-Dimethylbenz[a]anthracene	1 (0.454)
Benzenamine	Aniline *	5000 (2270)
Benzenamine, 4,4'-carbonimidoylbis (N,N-dimethyl-	Auramine	100 (45.4)
Benzenamine, 4-chloro-	p-Chloroaniline	1000 (454)
Benzenamine, 4-chloro-2-methyl-, hydrochloride	4-Chloro-o-toluidine, hydrochloride	100 (45.4)
Benzenamine, N,N-dimethyl-4-phenylazo-	Dimethylaminoazobenzene	10 (4.54)
Benzenamine, 4,4'-methylenebis(2-chloro-	4,4'-Methylenebis(2-chloroaniline)	10 (4.54)
Benzenamine, 2-methyl-, hydrochloride	o-Toluidine hydrochloride	100 (45.4)
Benzenamine, 2-methyl-5-nitro-	5-Nitro-o-toluidine	100 (45.4)
Benzenamine, 4-nitro-	p-Nitroaniline	5000 (2270)
Benzene *		10 (4.54)
Benzene, 1-bromo-4-phenoxy-	4-Bromophenyl phenyl ether	100 (45.4)
Benzene, chloro-	Chlorobenzene *	100 (45.4)
Benzene, chloromethyl-	Benzyl chloride *	100 (45.4)
Benzene, 1,2-dichloro-	o-Dichlorobenzene * 1,2-Dichlorobenzene	100 (45.4)
Benzene, 1,3-dichloro-	m-Dichlorobenzene 1,3-Dichlorobenzene	100 (45.4)
Benzene, 1,4-dichloro-	p-Dichlorobenzene * 1,4-Dichlorobenzene	100 (45.4)
Benzene, dichloromethyl-	Benzal chloride	5000 (2270)
Benzene, 2,4-diisocyanatomethyl-	Toluene diisocyanate *	100 (45.4)
Benzene, dimethyl	Xylene * (mixed)	1000 (454)
m-	m-	
o-	o-	
p-	p-	
Benzene, hexachloro-	Hexachlorobenzene	10 (4.54)

TABLE 1 - LIST OF HAZARDOUS SUBSTANCES OTHER THAN RADIONUCLIDES - Continued

Hazardous Substance	Synonyms	Reportable Quantity (RQ) Pounds (Kilograms)
Benzene, hexahydro-	Cyclohexane *	1000 (454)
Benzene, hydroxy-	Phenol *	1000 (454)
Benzene, methyl-	Toluene *	1000 (454)
Benzene, 1-methyl-2,4-dinitro-	2,4-Dinitrotoluene	10 (4.54)
Benzene, 1-methyl-2,6-dinitro-	2,6-Dinitrotoluene	100 (45.4)
Benzene, 1,2-methylenedioxy-4-allyl-	Safrole	100 (45.4)
Benzene, 1,2-methylenedioxy-4-propenyl-	Isosafrole	100 (45.4)
Benzene, 1,2-methylenedioxy-4-propyl-	Dihydrosafrole	10 (4.54)
Benzene, 1-methylethyl-	Cumene	5000 (2270)
Benzene, nitro-	Nitrobenzene *	1000 (454)
Benzene, pentachloro-	Pentachlorobenzene	10 (4.54)
Benzene, pentachloronitro-	Pentachloronitrobenzene	100 (45.4)
Benzene, 1,2,4,5-tetrachloro-	1,2,4,5-Tetrachlorobenzene	5000 (2270)
Benzene, trichloromethyl-	Benzotrichloride	10 (4.54)
Benzene, 1,3,5-trinitro-	sym-Trinitrobenzene *	10 (4.54)
Benzeneacetic acid, 4-chloro-alpha-(4-chlorophenyl)-alpha-hydroxy-, ethyl ester.	Ethyl 4,4'-dichlorobenzilate	10 (4.54)
1,2-Benzenedicarboxylic acid anhydride	Phthalic anhydride	5000 (2270)
1,2-Benzenedicarboxylic acid, [bis(2-ethylhexyl)] ester	Bis(2-ethylhexyl)phthalate	100 (45.4)
1,2-Benzenedicarboxylic acid, dibutyl ester	Di-n-butyl phthalate	10 (4.54)
	Dibutyl phthalate	
	n-Butyl phthalate	
1,2-Benzenedicarboxylic acid, diethyl ester	Diethyl phthalate	1000(454)
1,2-Benzenedicarboxylic acid, dimethyl ester	Dimethyl phthalate	5000 (2270)
1,2-Benzenedicarboxylic acid, di-n-octyl ester	Di-n-octyl phthalate	5000 (2270)
1,3-Benzenediol	Resorcinol	5000 (2270)
1,2-Benzenediol,4-[1-hydroxy-2-(methylamino)ethyl]-	Epinephrine	1000 (454)
Benzenesulfonic acid chloride	Benzenesulfonyl chloride	100 (45.4)
Benzenesulfonyl chloride	Benzenesulfonic acid chloride	100 (45.4)
Benzenethiol	Phenyl mercaptan @	100 (45.4)
	Thiophenol	
Benzidine *	(1,1'-Biphenyl)-4,4'diamine	1 (0.454)
1,2-Benzisothiazolin-3-one,1,1-dioxide, and salts	Saccharin and salts	100 (45.4)
Benzo[a]anthracene	Benz[a]anthracene	10 (4.54)
	1,2-Benzanthracene	
Benzo[b]fluoranthene		1 (0.454)
Benzo[k]fluoranthene		5000 (2270)
Benzo[j,k]fluorene	Fluoranthene	100 (45.4)
Benzoic acid		5000 (2270)
Benzonitrile *		5000 (2270)
Benzo[g,h,i]perylene		5000 (2270)
Benzo[a]pyrene	3,4-Benzopyrene	1 (0.454)
3,4-Benzopyrene	Benzo[a]pyrene	1 (0.454)
p-Benzoquinone	1,4-Cyclohexadienedione	10 (4.54)
Benzotrichloride	Benzene, trichloromethyl-	10 (4.54)
Benzoyl chloride *		1000 (454)
1,2-Benzphenanthrene	Chrysene	100 (45.4)
Benzyl chloride *	Benzene, chloromethyl-	100 (45.4)
Beryllium ¢	Beryllium dust ¢	10 (4.54)
Beryllium chloride *		1 (0.454)
Beryllium dust ¢	Beryllium ¢	10 (4.54)
Beryllium fluoride *		1 (0.454)
Beryllium nitrate *		1 (0.454)
alpha - BHC		10 (4.54)
beta - BHC		1 (0.454)
delta - BHC		1 (0.454)
gamma - BHC	Hexachlorocyclohexane (gamma isomer)	1 (0.454)
	Lindane *	
2,2'-Bioxirane	1,2:3,4-Diepoxybutane	10 (4.54)
(1,1'-Biphenyl)-4,4'-diamine	Benzidine *	1 (0.454)
(1,1'-Biphenyl)-4,4'-diamine,3,3'-dichloro-	3,3'-Dichlorobenzidine	1 (0.454)
(1,1'-Biphenyl)-4,4'-diamine,3,3'-dimethoxy-	3,3'-Dimethoxybenzidine	10 (4.54)
(1,1'-Biphenyl)-4,4'-diamine,3,3'-dimethyl-	3,3'-Dimethylbenzidine	10 (4.54)
Bis(2-chloroethoxy) methane	Ethane, 1,1'-[methylenebis(oxy)]bis(2-chloro-	1000 (454)
Bis(2-chloroethyl) ether	Dichloroethyl ether	10 (4.54)
	Ethane, 1,1'-oxybis(2-chloro-	
Bis(2-chloroisopropyl) ether	Propane, 2,2'-oxybis(2-chloro-	1000 (454)
Bis(chloromethyl) ether	Methane, oxybis(chloro-	1 (0.454)
Bis(dimethylthiocarbamoyl) disulfide	Thiram	10 (4.54)
Bis(2-ethylhexyl)phthalate	1,2-Benzenedicarboxylic acid, [bis(2-ethylhexyl)]ester	100 (45.4)
Bromine cyanide	Cyanogen bromide *	1000 (454)
Bromoacetone *	2-Propanone, 1-bromo-	1000 (454)
Bromoform	Methane, tribromo-	100 (45.4)
4-Bromophenyl phenyl ether	Benzene, 1-bromo-4-phenoxy-	100 (45.4)
Brucine	Strychnidin-10-one, 2,3-dimethoxy-	100 (45.4)
1,3-Butadiene, 1,1,2,3,4,4-hexachloro-	Hexachlorobutadiene *	1 (0.454)
1-Butanamine, N-butyl-N-nitroso-	N-Nitrosodi-n-butylamine	10 (4.54)

TABLE 1 - LIST OF HAZARDOUS SUBSTANCES OTHER THAN RADIONUCLIDES - Continued

Hazardous Substance	Synonyms	Reportable Quantity (RQ) Pounds (Kilograms)
Butanoic acid, 4-[bis(2-chloroethyl)amino]benzene-	Chlorambucil	10 (4.54)
1-Butanol	n-Butyl alcohol *	5000 (2270)
2-Butanone	Ethyl methyl ketone @ Methyl ethyl ketone *	5000 (2270)
2-Butanone peroxide	Methyl ethyl ketone peroxide *	10 (4.54)
2-Butenal	Crotonaldehyde *	100 (45.4)
2-Butene, 1,4-dichloro-	1,4-Dichloro-2-butene	1 (0.454)
Butyl acetate *		5000 (2270)
iso-Butyl acetate		
sec-Butyl acetate		
tert-Butyl acetate		
n-Butyl alcohol *	1-Butanol	5000 (2270)
Butylamine *		1000 (454)
iso-Butylamine		
sec-Butylamine		
tert-Butylamine		
Butyl benzyl phthalate		100 (45.4)
n-Butyl phthalate	Di-n-butyl phthalate Dibutyl phthalate 1,2-Benzenedicarboxylic acid, dibutyl ester	10 (4.54)
Butyric acid *		5000 (2270)
iso-Butyric acid		
Cacodylic acid	Hydroxydimethylarsine oxide	1 (0.454)
Cadmium ¢		10 (4.54)
Cadmium acetate		10 (4.54)
Cadmium bromide		10 (4.54)
Cadmium chloride		10 (4.54)
Calcium arsenate *		1 (0.454)
Calcium arsenite *		1 (0.454)
Calcium carbide *		10 (4.54)
Calcium chromate	Chromic acid, calcium salt	10 (4.54)
Calcium cyanide *		10 (4.54)
Calcium dodecylbenzene sulfonate		1000 (454)
Calcium hypochlorite *		10 (4.54)
Camphene, octachloro-	Toxaphene *	1 (0.454)
Captan		10 (4.54)
Carbamic acid, ethyl ester	Ethyl carbamate (Urethan)	100 (45.4)
Carbamic acid, methylnitroso-, ethyl ester	N-Nitroso-N-methylurethane	1 (0.454)
Carbamide, N-ethyl-N-nitroso-	N-Nitroso-N-ethylurea	1 (0.454)
Carbamide, N-methyl-N-nitroso-	N-Nitroso-N-methylurea	1 (0.454)
Carbamide, thio-	Thiourea	10 (4.54)
Carbamimidoselenoic acid	Selenourea	1000 (454)
Carbamoyl chloride, dimethyl-	Dimethylcarbamoyl chloride	1 (0.454)
Carbaryl *		100 (45.4)
Carbofuran *		10 (4.54)
Carbon bisulfide *	Carbon disulfide *	100 (45.4)
Carbon disulfide *	Carbon bisulfide *	100 (45.4)
Carbonic acid, dithallium (I) salt	Thallium(I) carbonate	100 (45.4)
Carbonochloridic acid, methyl ester	Methyl chlorocarbonate * Methyl chloroformate @	1000 (454)
Carbon oxyfluoride	Carbonyl fluoride	1000 (454)
Carbon tetrachloride *	Methane, tetrachloro-	10 (4.54)
Carbonyl chloride	Phosgene *	10 (4.54)
Carbonyl fluoride	Carbon oxyfluoride	1000 (454)
Chloral	Acetaldehyde, trichloro-	5000(2270)
Chlorambucil	Butanoic acid, 4-[bis(2-chloroethyl)amino]benzene-	10 (4.54)
Chlordane *	Chlordane, technical * 4,7-Methanoindan, 1,2,4,5,6,7,8,8-octachloro-3a,4,7,7a-tetrahydro-	1 (0.454)
Chlordane, technical *	Chlordane * 4,7-Methanoindan, 1,2,4,5,6,7,8,8-octachloro-3a,4,7,7a-tetrahydro-	1 (0.454)
Chlorine *		10 (4.54)
Chlorine cyanide	Cyanogen chloride *	10 (4.54)
Chlornaphazine	2-Naphthylamine, N,N-bis(2-chloroethyl)-	100 (45.4)
Chloroacetaldehyde	Acetaldehyde, chloro-	1000 (454)
p-Chloroaniline	Benzenamine, 4-chloro-	1000 (454)
Chlorobenzene *	Benzene, chloro-	100 (45.4)
4-Chloro-m-cresol	p-Chloro-m-cresol Phenol, 4-chloro-3-methyl-	5000 (2270)
p-Chloro-m-cresol	Phenol, 4-chloro-3-methyl- 4-Chloro-m-cresol	5000 (2270)
Chlorodibromomethane		100 (45.4)
1-Chloro-2,3-epoxypropane	Epichlorohydrin * Oxirane, 2-(chloromethyl)-	100 (45.4)
Chloroethane	Ethyl chloride @	100 (45.4)
2-Chloroethyl vinyl ether	Ethene, 2-chloroethoxy-	1000 (454)
Chloroform *	Methane, trichloro-	10 (4.54)

TABLE 1 - LIST OF HAZARDOUS SUBSTANCES OTHER THAN RADIONUCLIDES - Continued

Hazardous Substance	Synonyms	Reportable Quantity (RQ) Pounds (Kilograms)
Chloromethane	Methane, chloro-	1 (0.454)
	Methyl chloride *	
Chloromethyl methyl ether	Methane, chloromethoxy-	1 (0.454)
	Methylchloromethyl ether @	
beta-Chloronaphthalene	Naphthalene, 2-chloro-	5000 (2270)
	2-Chloronaphthalene	
2-Chloronaphthalene	beta-Chloronaphthalene	5000 (2270)
	Naphthalene, 2-chloro-	
2-Chlorophenol	o-Chlorophenol	100 (45.4)
	Phenol, 2-chloro-	
o-Chlorophenol	Phenol, 2-chloro-	100 (45.4)
	2-Chlorophenol	
4-Chlorophenyl phenyl ether		5000 (2270)
1-(o-Chlorophenyl)thiourea	Thiourea, (2-chlorophenyl)-	100 (45.4)
3-Chloropropionitrile	Propanenitrile, 3-chloro-	1000 (454)
Chlorosulfonic acid *		1000 (454)
4-Chloro-o-toluidine, hydrochloride	Benzenamine, 4-chloro-2-methyl-, hydrochloride	100 (45.4)
Chlorpyrifos *		1 (0.454)
Chromic acetate		1000 (454)
Chromic acid *		10 (4.54)
Chromic acid, calcium salt	Calcium chromate	10 (4.54)
Chromic sulfate		1000 (454)
Chromium ¢		5000 (2270)
Chromous chloride		1000 (454)
Chrysene	1,2-Benzphenanthrene	100 (45.4)
Cobaltous bromide		1000 (454)
Cobaltous formate		1000 (454)
Cobaltous sulfamate		1000 (454)
Coke Oven Emissions		1 (0.454)
Copper ¢		5000 (2270)
Copper cyanide *		10 (4.54)
Coumaphos *		10 (4.54)
Creosote		1 (0.454)
Cresols *	Cresylic acid	1000 (454)
m-Cresols	m-Cresylic acid	
o-Cresols	o-Cresylic acid	
p-Cresols	p-Cresylic acid	
Cresylic acid	Cresols *	1000 (454)
m-Cresols	m-Cresylic acid	
o-Cresols	o-Cresylic acid	
p-Cresols	p-Cresylic acid	
Crotonaldehyde *	2-Butenal	100 (45.4)
Cumene	Benzene, 1-methylethyl-	5000 (2270)
Cumene hydroperoxide @	alpha,alpha-Dimethylbenzylhydroperoxide	10 (4.54)
	Hydroperoxide, 1-methyl-1-phenylethyl-	
Cupric acetate		100 (45.4)
Cupric acetoarsenite *		1 (0.454)
Cupric chloride *		10 (4.54)
Cupric nitrate *		100 (45.4)
Cupric oxalate		100 (45.4)
Cupric sulfate		10 (4.54)
Cupric sulfate ammoniated		100 (45.4)
Cupric tartrate		100 (45.4)
Cyanides (soluble cyanide salts), not elsewhere specified *		10 (4.54)
Cyanogen *		100 (45.4)
Cyanogen bromide *	Bromine cyanide	1000 (454)
Cyanogen chloride *	Chlorine cyanide	10 (4.54)
1,4-Cyclohexadienedione	p-Benzoquinone	10 (4.54)
Cyclohexane *	Benzene, hexahydro-	1000 (454)
Cyclohexanone		5000 (2270)
1,3-Cyclopentadiene, 1,2,3,4,5,5-hexachloro-	Hexachlorocyclopentadiene *	10 (4.54)
Cyclophosphamide	2H-1,3,2-Oxazaphosphorine,2-[bis(2-chloroethyl)amino] tetrahydro-2-oxide.	10 (4.54)
2,4-D Acid	2,4-D *, salts and esters	100 (45.4)
	2,4-Dichlorophenoxyacetic acid *, salts and esters	
2,4-D Esters		100 (45.4)
2,4-D *, salts and esters	2,4-D Acid	100 (45.4)
	2,4-Dichlorophenoxyacetic acid *, salts and esters	
Daunomycin	5,12-Naphthacenedione, (8S-cis)-8-acetyl-10-[3-amino-2,3,6-trideoxy-alpha-L-lyxo-hexopyranosyl)oxy]-7,8,9,10-tetrahydro-6,8,11-trihydroxy-1-methoxy-.	10 (4.54)
DDD	Dichlorodiphenyl dichloroethane	1 (0.454)
	TDE *	
	4,4'-DDD	
4,4'-DDD	DDD	1 (0.454)
	Dichlorodiphenyl dichloroethane	
	TDE *	

Hazardous Substance	Synonyms	Reportable Quantity (RQ) Pounds (Kilograms)
DDE	4,4'-DDE	1 (0.454)
4,4'-DDE	DDE	1 (0.454)
DDT *	Dichlorodiphenyl trichloroethane *	1 (0.454)
	4,4'-DDT	
4,4'-DDT	DDT *	1 (0.454)
	Dichlorodiphenyl trichloroethane *	
Decachlorooctahydro-1,3,4-metheno-2H-cyclobuta[c,d]-pentalen-2-one	Kepone *	1 (0.454)
Diallate	S-(2,3-Dichloroallyl) diisopropylthiocarbamate	100 (45.4)
Diamine	Hydrazine *	1 (0.454)
Diaminotoluene	Toluenediamine *	10 (4.54)
Diazinon *		1 (0.454)
Dibenz[a,h]anthracene	Dibenzo[a,h]anthracene	1 (0.454)
	1,2:5,6-Dibenzanthracene	
1,2:5,6-Dibenzanthracene	Dibenz[a,h]anthracene	1 (0.454)
	Dibenzo[a,h]anthracene	
Dibenzo[a,h]anthracene	Dibenz[a,h]anthracene	1 (0.454)
	1,2:5,6-Dibenzanthracene	
1,2:7,8-Dibenzopyrene	Dibenz[a,i]pyrene	10 (4.54)
Dibenz[a,i]pyrene	1,2:7,8-Dibenzopyrene	10 (4.54)
1,2-Dibromo-3-chloropropane	Propane, 1,2-dibromo-3-chloro-	1 (0.454)
Dibutyl phthalate	Di-n-butyl phthalate	10 (4.54)
	n-Butyl phthalate *	
	1,2-Benzenedicarboxylic acid, dibutyl ester	
Di-n-butyl phthalate	Dibutyl phthalate	10 (4.54)
	n-Butyl phthalate *	
	1,2-Benzenedicarboxylic acid, dibutyl ester	
Dicamba		1000 (454)
Dichlobenil		100 (45.4)
Dichlone		1 (0.454)
S-(2,3-Dichloroallyl) diisopropylthiocarbamate	Diallate	100 (45.4)
3,5-Dichloro-N-(1,1-dimethyl-2-propynyl)benzamide	Pronamide	5000 (2270)
Dichlorobenzene (mixed)		100 (45.4)
1,2-Dichlorobenzene	Benzene, 1,2-dichloro-	100 (45.4)
	o-Dichlorobenzene *	
1,3-Dichlorobenzene	Benzene, 1,3-dichloro-	100 (45.4)
	m-Dichlorobenzene	
1,4-Dichlorobenzene	Benzene, 1,4-dichloro-	100 (45.4)
	p-Dichlorobenzene *	
m-Dichlorobenzene	Benzene, 1,3-dichloro-	100 (45.4)
	1,3-Dichlorobenzene	
o-Dichlorobenzene *	Benzene, 1,2-dichloro-	100 (45.4)
	1,2-Dichlorobenzene	
p-Dichlorobenzene *	Benzene, 1,4-dichloro-	100 (45.4)
	1,4-Dichlorobenzene	
3,3'-Dichlorobenzidine	(1,1'-Biphenyl)-4,4'-diamine,3,3'-dichloro-	1 (0.454)
Dichlorobromomethane		5000 (2270)
1,4-Dichloro-2-butene	2-Butene, 1,4-dichloro-	1 (0.454)
Dichlorodifluoromethane *	Methane, dichlorodifluoro-	5000 (2270)
Dichlorodiphenyl dichloroethane	DDD	1 (0.454)
	TDE *	
	4,4'-DDD	
Dichlorodiphenyl trichloroethane *	DDT *	1 (0.454)
	4,4'-DDT	
1,1-Dichloroethane	Ethane, 1,1-dichloro-	1000 (454)
	Ethylidene dichloride	
1,2-Dichloroethane	Ethane, 1,2-dichloro-	100 (45.4)
	Ethylene dichloride *	
1,1-Dichloroethylene	Ethene, 1,1-dichloro-	100 (45.4)
	Vinylidene chloride *	
1,2-trans-Dichloroethylene	Ethene, trans-1,2-dichloro-	1000 (454)
Dichloroethyl ether	Bis (2-chloroethyl) ether	10 (4.54)
	Ethane, 1,1'-oxybis(2-chloro-	
2,4-Dichlorophenol	Phenol, 2,4-dichloro-	100 (45.4)
2,6-Dichlorophenol	Phenol, 2,6-dichloro-	100 (45.4)
2,4-Dichlorophenoxyacetic acid *, salts and esters	2,4-D Acid	100 (45.4)
	2,4-D *, salts and esters	
Dichlorophenylarsine	Phenyl dichloroarsine *	1 (0.454)
Dichloropropane *		1000 (454)
1,1-Dichloropropane		
1,3-Dichloropropane		
1,2-Dichloropropane	Propylene dichloride *	1000 (454)
Dichloropropane - Dichloropropene (mixture)		100 (45.4)
Dichloropropene(s) *		100 (45.4)
2,3-Dichloropropene (isomer)		
1,3-Dichloropropene	Propene, 1,3-dichloro-	100 (45.4)
2,2-Dichloropropionic acid *		5000 (2270)
Dichlorvos *		10 (4.54)

TABLE 1 - LIST OF HAZARDOUS SUBSTANCES OTHER THAN RADIONUCLIDES - Continued

Hazardous Substance	Synonyms	Reportable Quantity (RQ) Pounds (Kilograms)
Dicofol		10 (4.54)
Dieldrin *	1,2,3,4,10,10-Hexachloro-6,7-epoxy-1,4,4a,5,6,7,8,8a-octahydro-endo,exo-1,4:5,8-dimethanonaphthalene.	1 (0.454)
1,2:3,4-Diepoxybutane	2,2'-Bioxirane	10 (4.54)
Diethylamine *		100 (45.4)
Diethylarsine	Arsine, diethyl-	1 (0.454)
1,4-Diethylene dioxide	1,4-Dioxane	100 (45.4)
O,O-Diethyl S-[2-(ethylthio)ethyl] phosphorodithioate	Disulfoton *	1 (0.454)
N,N'-Diethylhydrazine	Hydrazine, 1,2-diethyl-	10 (4.54)
O,O-Diethyl S-methyl dithiophosphate	Phosphorodithioic acid, O,O-diethylS-methyl ester	5000 (2270)
Diethyl-p-nitrophenyl phosphate	Phosphoric acid, diethyl p-nitrophenyl ester	100 (45.4)
Diethyl phthalate	1,2-Benzenedicarboxylic acid, diethyl ester	1000(454)
O,O-Diethyl O-pyrazinyl phosphorothioate	Phosphorothioic acid, O,O-diethyl O-pyrazinyl ester	100 (45.4)
Diethylstilbestrol	4,4'-Stilbenediol, alpha,alpha'-diethyl-	1 (0.454)
1,2-Dihydro-3,6-pyridazinedione	Maleic hydrazide	5000 (2270)
Dihydrosafrole	Benzene, 1,2-methylenedioxy-4-propyl-	10 (4.54)
Diisopropyl fluorophosphate	Phosphorofluoridic acid, bis(1-methylethyl) ester	100 (45.4)
Dimethoate	Phosphorodithioic acid, O,O-dimethyl S-[2(methylamino)-2-oxoethyl] ester.	10 (4.54)
3,3'-Dimethoxybenzidine	(1,1'-Biphenyl)-4,4'-diamine,3,3'-dimethoxy-	10 (4.54)
Dimethylamine *	Methanamine, N-methyl-	1000 (454)
Dimethylaminoazobenzene	Benzenamine, N,N-dimethyl-4-phenylazo-	10 (4.54)
7,12-Dimethylbenz[a]anthracene	1,2-Benzanthracene, 7,12-dimethyl-	1 (0.454)
3,3'-Dimethylbenzidine	(1,1'-Biphenyl)-4,4'-diamine,3,3'-dimethyl-	10 (4.54)
alpha,alpha-Dimethylbenzylhydroperoxide	Hydroperoxide, 1-methyl-1-phenylethyl-	10 (4.54)
	Cumene hydroperoxide @	
3,3-Dimethyl-1-(methylthio)-2-butanone, O-[(methylamino)carbonyl] oxime.	Thiofanox	100 (45.4)
Dimethylcarbamoyl chloride	Carbamoyl chloride, dimethyl-	1 (0.454)
1,1-Dimethylhydrazine	Dimethylhydrazine, unsymmetrical @	10 (4.54)
	Hydrazine, 1,1-dimethyl-	
1,2-Dimethylhydrazine	Hydrazine, 1,2-dimethyl-	1 (0.454)
Dimethylhydrazine, unsymmetrical @	1,1-Dimethylhydrazine	10 (4.54)
	Hydrazine, 1,1-dimethyl-	
O,O-Dimethyl O-p-nitrophenyl phosphorothioate	Methyl parathion *	100 (45.4)
Dimethylnitrosamine	(-Nitrosodimethylamine	10 (4.54)
alpha,alpha-Dimethylphenethylamine	Ethanamine, 1,1-dimethyl-2-phenyl-	5000 (2270)
2,4-Dimethylphenol	Phenol, 2,4-dimethyl-	100 (45.4)
Dimethyl phthalate	1,2-Benzenedicarboxylic acid, dimethyl ester	5000 (2270)
Dimethyl sulfate *	Sulfuric acid, dimethyl ester	100(45.4)
Dinitrobenzene * (mixed)		100 (45.4)
m-Dinitrobenzene		
o-Dinitrobenzene		
p-Dinitrobenzene		
4,6-Dinitro-o-cresol and salts	Phenol, 2,4-dinitro-6-methyl-, and salts	10 (4.54)
4,6-Dinitro-o-cyclohexylphenol	Phenol, 2-cyclohexyl-4,6-dinitro-	100 (45.4)
Dinitrophenol		10 (4.54)
2,5-Dinitrophenol		
2,6-Dinitrophenol		
2,4-Dinitrophenol	Phenol, 2,4-dinitro-	10 (4.54)
Dinitrotoluene		10 (4.54)
3,4-Dinitrotoluene		
2,4-Dinitrotoluene	Benzene, 1-methyl-2,4-dinitro-	10 (4.54)
2,6-Dinitrotoluene	Benzene, 1-methyl-2,6-dinitro-	100 (45.4)
Dinoseb	Phenol, 2,4-dinitro-6-(1-methylpropyl)-	1000 (454)
Di-n-octyl phthalate	1,2-Benzenedicarboxylic acid, di-n-octyl ester	5000 (2270)
1,4-Dioxane	1,4-Diethylene dioxide	100 (45.4)
1,2-Diphenylhydrazine	Hydrazine, 1,2-diphenyl-	10 (4.54)
Diphosphoramide, octamethyl-	Octamethylpyrophosphoramide	100 (45.4)
Dipropylamine	1-Propanamine, N-propyl-	5000 (2270)
Di-n-propylnitrosamine	N-Nitrosodi-n-propylamine	10 (4.54)
Diquat		1000 (454)
Disulfoton *	O,O-Diethyl S-[2-(ethylthio)ethyl]phosphorodithioate	1 (0.454)
2,4-Dithiobiuret	Thioimidodicarbonic diamide	100 (45.4)
Dithiopyrophosphoric acid, tetraethyl ester	Tetraethyldithiopyrophosphate	100 (45.4)
Diuron		100 (45.4)
Dodecylbenzenesulfonic acid *		1000 (454)
Endosulfan *	5-Norbornene-2,3-dimethanol,1,4,5,6,7,7-hexachloro,cyclic sulfite	1 (0.454)
alpha - Endosulfan		1 (0.454)
beta - Endosulfan		1 (0.454)
Endosulfan sulfate		1 (0.454)
Endothall	7-Oxabicyclo[2.2.1]heptane-2,3-dicarboxylic acid	1000 (454)
Endrin *	1,2,3,4,10,10-Hexachloro-6,7-epoxy-1,4,4a,5,6,7,8,8a-octahydro-endo,endo-1,4:5,8-dimethanonaphthalene.	1 (0.454)
Endrin aldehyde		1 (0.454)
Epichlorohydrin *	1-Chloro-2,3-epoxypropane	100 (45.4)
	Oxirane, 2-(chloromethyl)-	

TABLE 1 - LIST OF HAZARDOUS SUBSTANCES OTHER THAN RADIONUCLIDES - Continued

Hazardous Substance	Synonyms	Reportable Quantity (RQ) Pounds (Kilograms)
Epinephrine	1,2-Benzenediol,4-[1-hydroxy-2-(methylamino)ethyl]	1000 (454)
Ethanal	Acetaldehyde *	1000 (454)
Ethanamine, 1,1-dimethyl-2-phenyl-	alpha,alpha-Dimethylphenethylamine	5000 (2270)
Ethanamine, N-ethyl-N-nitroso-	N-Nitrosodiethylamine	1 (0.454)
Ethane, 1,2-dibromo-	Ethylene dibromide *	1 (0.454)
Ethane, 1,1-dichloro-	Ethylidene dichloride	1000 (454)
	1,1-Dichloroethane	
Ethane, 1,2-dichloro-	Ethylene dichloride *	100 (45.4)
	1,2-Dichloroethane	
Ethane, 1,1,1,2,2,2-hexachloro-	Hexachloroethane *	100 (45.4)
Ethane, 1,1'-[methylenebis(oxy)]bis(2-chloro-	Bis(2-chloroethoxy)methane	1000 (454)
Ethane, 1,1'-oxybis-	Ethyl ether *	100 (45.4)
Ethane, 1,1'-oxybis(2-chloro-	Bis (2-chloroethyl) ether	10 (4.54)
	Dichloroethyl ether	
Ethane, pentachloro-	Pentachloroethane	10 (4.54)
Ethane, 1,1,1,2-tetrachloro-	1,1,1,2-Tetrachloroethane	100 (45.4)
Ethane, 1,1,2,2-tetrachloro-	1,1,2,2-Tetrachloroethane	100 (45.4)
Ethane, 1,1,2-trichloro-	1,1,2-Trichloroethane	100 (45.4)
Ethane, 1,1,1-trichloro-2,2-bis(p-methoxyphenyl)-	Methoxychlor	1 (0.454)
1,2-Ethanediylbiscarbamodithioic acid	Ethylenebis(dithiocarbamic acid)	5000 (2270)
Ethanenitrile	Acetonitrile *	5000 (2270)
Ethanethioamide	Thioacetamide	10 (4.54)
Ethanol, 2,2'-(nitrosoimino)bis-	N-Nitrosodiethanolamine	1 (0.454)
Ethanone, 1-phenyl-	Acetophenone	5000 (2270)
Ethanoyl chloride	Acetyl chloride *	5000 (2270)
Ethenamine, N-methyl-N-nitroso-	N-Nitrosomethylvinylamine	10 (4.54)
Ethene, chloro-	Vinyl chloride *	1 (0.454)
Ethene, 2-chloroethoxy-	2-Chloroethyl vinyl ether	1000 (454)
Ethene, 1,1-dichloro-	Vinylidene chloride *	100 (45.4)
	1,1-Dichloroethylene	
Ethene, 1,1,2,2-tetrachloro-	Perchloroethylene	100 (45.4)
	Tetrachloroethene	
	Tetrachloroethylene	
Ethene, trans-1,2-dichloro-	1,2-trans-Dichloroethylene	1000 (454)
Ethion *		10 (4.54)
2-Ethoxyethanol	Ethylene glycol monoethyl ether *	1000 (454)
Ethyl acetate	Acetic acid, ethyl ester	5000 (2270)
Ethyl acrylate *	2-Propenoic acid, ethyl ester	1000 (454)
Ethylbenzene *		1000 (454)
Ethyl carbamate (Urethan)	Carbamic acid, ethyl ester	100 (45.4)
Ethyl chloride @	Chloroethane	100 (45.4)
Ethyl cyanide	Propanenitrile	10 (4.54)
Ethyl 4,4'-dichlorobenzilate	Benzeneacetic acid, 4-chloro-alpha-(4-chlorophenyl)-alpha-hydroxy-, ethyl ester.	10 (4.54)
Ethylene dibromide *	Ethane, 1,2-dibromo-	1 (0.454)
Ethylene dichloride *	1,2-Dichloroethane	100 (45.4)
	Ethane, 1,2-dichloro-	
Ethylene glycol monoethyl ether *	2-Ethoxyethanol	1000 (454)
Ethylene oxide *	Oxirane	10 (4.54)
Ethylenebis(dithiocarbamic acid)	1,2-Ethanediylbiscarbamodithioicacid	5000 (2270)
Ethylenediamine *		5000 (2270)
Ethylenediamine tetraacetic acid (EDTA)		5000 (2270)
Ethylenethiourea	2-Imidazolidinethione	10 (4.54)
Ethylenimine	Aziridine	1 (0.454)
Ethyl ether *	Ethane, 1,1'-oxybis-	100 (45.4)
Ethylidene dichloride	Ethane, 1,1-dichloro-	1000 (454)
	1,1-Dichloroethane	
Ethyl methacrylate	2-Propenoic acid, 2-methyl-, ethyl ester	1000 (454)
Ethyl methanesulfonate	Methanesulfonic acid, ethyl ester	1 (0.454)
Ethyl methyl ketone @	2-Butanone	5000 (2270)
	Methyl ethyl ketone *	
Famphur	Phosphorothioic acid, O,O-dimethyl O-[p-[(dimethylamino)-sulfonyl] phenyl] ester.	1000 (454)
Ferric ammonium citrate		1000 (454)
Ferric ammonium oxalate		1000 (454)
Ferric chloride		1000 (454)
Ferric fluoride		100 (45.4)
Ferric nitrate *		1000 (454)
Ferric sulfate		1000 (454)
Ferrous ammonium sulfate		1000 (454)
Ferrous chloride *		100 (45.4)
Ferrous sulfate		1000 (454)
Fluoranthene	Benzo[j,k]fluorene	100 (45.4)
Fluorene		5000 (2270)
Fluorine *		10 (4.54)
Fluoroacetamide	Acetamide, 2-fluoro-	100 (45.4)
Fluoroacetic acid, sodium salt	Acetic acid, fluoro-, sodium salt	10 (4.54)

TABLE 1 - LIST OF HAZARDOUS SUBSTANCES OTHER THAN RADIONUCLIDES - Continued

Hazardous Substance	Synonyms	Reportable Quantity (RQ) Pounds (Kilograms)
Formaldehyde *	Methylene oxide	100(45.4)
Formic acid *	Methanoic acid	5000 (2270)
Fulminic acid, mercury(II)salt	Mercury fulminate	10 (4.54)
Fumaric acid		5000 (2270)
Furan *	Furfuran	100 (45.4)
Furan, tetrahydro-	Tetrahydrofuran *	1000 (454)
2-Furancarboxaldehyde	Furfural *	5000 (2270)
2,5-Furandione	Maleic anhydride *	5000 (2270)
Furfural *	2-Furancarboxaldehyde	5000 (2270)
Furfuran	Furan *	100 (45.4)
D-Glucopyranose, 2-deoxy-2-(3-methyl-3-nitrosoureido)-	Streptozotocin	1 (0.454)
Glycidylaldehyde	1-Propanal, 2,3-epoxy-	10 (4.54)
Guanidine, N-nitroso-N-methyl-N'-nitro-	N-Methyl-N'-nitro-N-nitrosoguanidine	10 (4.54)
Guthion *	Azinphos methyl @	1 (0.454)
Heptachlor	4,7-Methano-1H-indene, 1,4,5,6,7,8,8-heptachloro-3a,4,7,7a-tetrahydro-	1 (0.454)
Heptachlor epoxide		1 (0.454)
Hexachlorobenzene	Benzene, hexachloro-	10 (4.54)
Hexachlorobutadiene *	1,3-Butadiene, 1,1,2,3,4,4-hexachloro-	1 (0.454)
Hexachlorocyclohexane (gamma isomer)	gamma - BHC	1 (0.454)
	Lindane *	
Hexachlorocyclopentadiene *	1,3-Cyclopentadiene, 1,2,3,4,5,5-hexachloro-	10 (4.54)
1,2,3,4,10,10-Hexachloro-6,7-epoxy-1,4,4a,5,6,7,8,8a-octahydro-endo,endo-1,4:5,8-dimethanonaphthalene.	Endrin *	1 (0.454)
1,2,3,4,10,10-Hexachloro-6,7-epoxy-1,4,4a,5,6,7,8,8a-octahydro-endo,exo-1,4:5,8-dimethanonaphthalene.	Dieldrin *	1 (0.454)
Hexachloroethane *	Ethane, 1,1,1,2,2,2-hexachloro-	100·(45.4)
Hexachlorohexahydro-endo,endo-dimethanonaphthalene	1,2,3,4,10,10-Hexachloro-1,4,4a,5,8,8a-hexahydro-1,4,5,8-endo,endo-dimethanonaphthalene.	1 (0.454)
1,2,3,4,10,10-Hexachloro-1,4,4a,5,8,8a-hexahydro-1,4,5,8-endo,endo-dimethanonaphthalene.	Hexachlorohexahydro-endo,endo-dimethanonaphthalene	1 (0.454)
1,2,3,4,10-10-Hexachloro-1,4,4a,5,8,8a-hexahydro-1,4:5,8-endo,exo-dimethanonaphthalene.	Aldrin *	1 (0.454)
Hexachlorophene	2,2'-Methylenebis(3,4,6-trichlorophenol)	100 (45.4)
Hexachloropropene	1-Propene, 1,1,2,3,3,3-hexachloro-	1000 (454)
Hexaethyl tetraphosphate *	Tetraphosphoric acid, hexaethyl ester	100 (45.4)
Hydrazine *	Diamine	1 (0.454)
Hydrazine, 1,2-diethyl-	N,N'-Diethylhydrazine	10 (4.54)
Hydrazine, 1,1-dimethyl-	1,1-Dimethylhydrazine	10 (4.54)
	Dimethylhydrazine, unsymmetrical @	
Hydrazine, 1,2-dimethyl-	1,2-Dimethylhydrazine	1 (0.454)
Hydrazine, 1,2-diphenyl-	1,2-Diphenylhydrazine	10 (4.54)
Hydrazine, methyl-	Methyl hydrazine *	10 (4.54)
Hydrazinecarbothioamide	Thiosemicarbazide	100 (45.4)
Hydrochloric acid *		5000 (2270)
Hydrocyanic acid *	Hydrogen cyanide	10 (4.54)
Hydrofluoric acid *	Hydrogen fluoride *	100 (45.4)
Hydrogen cyanide	Hydrocyanic acid *	10 (4.54)
Hydrogen fluoride *	Hydrofluoric acid *	100 (45.4)
Hydrogen phosphide	Phosphine *	100 (45.4)
Hydrogen sulfide *	Hydrosulfuric acid	100 (45.4)
	Sulfur hydride	
Hydroperoxide, 1-methyl-1-phenylethyl-	alpha,alpha-Dimethylbenzylhydroperoxide	10 (4.54)
	Cumene hydroperoxide @	
Hydrosulfuric acid	Hydrogen sulfide *	100 (45.4)
	Sulfur hydride	
Hydroxydimethylarsine oxide	Cacodylic acid	1 (0.454)
2-Imidazolidinethione	Ethylenethiourea	10 (4.54)
Indeno(1,2,3-cd)pyrene	1,10-(1,2-Phenylene)pyrene	100 (45.4)
Isobutyl alcohol	1-Propanol, 2-methyl-	5000 (2270)
Isocyanic acid, methyl ester	Methyl isocyanate *	1 (0.454)
Isophorone		5000 (2270)
Isoprene *		100 (45.4)
Isopropanolamine dodecylbenzene sulfonate		1000 (454)
Isosafrole	Benzene, 1,2-methylenedioxy-4-propenyl-	100 (45.4)
3(2H)-Isoxazolone, 5-(aminomethyl)-	5-(Aminomethyl)-3-isoxazolol	1000 (454)
Kepone	Decachlorooctahydro-1,3,4-metheno-2H-cyclobuta[c,d]-pentalen-2-one	1 (0.454)
Lasiocarpine		10 (4.54)
Lead ¢		1 (0.454)
Lead acetate	Acetic acid, lead salt	5000 (2270)
Lead arsenate *		1 (0.454)
Lead chloride *		100 (45.4)
Lead fluoborate *		100 (45.4)
Lead fluoride *		100 (45.4)
Lead iodide		100 (45.4)
Lead nitrate *		100 (45.4)
Lead phosphate	Phosphoric acid, lead salt	1 (0.454)
Lead stearate		5000 (2270)

Hazardous Substance	Synonyms	Reportable Quantity (RQ) Pounds (Kilograms)
Lead subacetate		100 (45.4)
Lead sulfate *		100 (45.4)
Lead sulfide		5000 (2270)
Lead thiocyanate		100 (45.4)
Lindane *	gamma - BHC	1 (0.454)
	Hexachlorocyclohexane (gamma isomer)	
Lithium chromate		10 (4.54)
Malathion *		100 (45.4)
Maleic acid *		5000 (2270)
Maleic anhydride *	2,5-Furandione	5000 (2270)
Maleic hydrazide	1,2-Dihydro-3,6-pyridazinedione	5000 (2270)
Malononitrile	Propanedinitrile	1000 (454)
Melphalan	Alanine, 3-[p-bis(2-chloroethyl)amino]phenyl-, L-	1 (0.454)
Mercaptodimethur		10 (4.54)
Mercuric cyanide *		1 (0.454)
Mercuric nitrate *		10 (4.54)
Mercuric sulfate *		10 (4.54)
Mercuric thiocyanate		10 (4.54)
Mercurous nitrate *		10 (4.54)
Mercury *		1 (0.454)
Mercury, (acetato-O)phenyl-	Phenylmercuric acetate	100 (45.4)
Mercury fulminate	Fulminic acid, mercury(II)salt	10 (4.54)
Methacrylonitrile	2-Propenenitrile, 2-methyl-	1000 (454)
Methanamine, N-methyl-	Dimethylamine *	1000 (454)
Methane, bromo-	Methyl bromide *	1000 (454)
Methane, chloro-	Chloromethane	100 (45.4)
	Methyl chloride *	
Methane, chloromethoxy-	Chloromethyl methyl ether	1 (0.454)
	Methylchloromethyl ether @	
Methane, dibromo-	Methylene bromide	1000 (454)
Methane, dichloro-	Methylene chloride *	1000 (454)
Methane, dichlorodifluoro-	Dichlorodifluoromethane *	5000 (2270)
Methane, iodo-	Methyl iodide	100 (45.4)
Methane, oxybis(chloro-	Bis(chloromethyl) ether	1 (0.454)
Methane, tetrachloro-	Carbon tetrachloride *	10 (4.54)
Methane, tetranitro-	Tetranitromethane *	10 (4.54)
Methane, tribromo-	Bromoform	100 (45.4)
Methane, trichloro-	Chloroform *	10 (4.54)
Methane, trichlorofluoro-	Trichloromonofluoromethane	5000 (2270)
Methanesulfenyl chloride, trichloro-	Perchloromethyl mercaptan @	100 (45.4)
	Trichloromethanesulfenyl chloride	
Methanesulfonic acid, ethyl ester	Ethyl methanesulfonate	1 (0.454)
Methanethiol	Methyl mercaptan *	100 (45.4)
	Thiomethanol	
Methanoic acid	Formic acid *	5000 (2270)
4,7-Methanoindan, 1,2,4,5,6,7,8,8-octachloro-3a,4,7,7a-tetrahydro-	Chlordane *	1 (0.454)
	Chlordane, technical *	
4,7-Methano-1H-indene, 1,4,5,6,7,8,8-heptachloro-a,4,7,7a-tetrahydro-	Heptachlor	1 (0.454)
Methanol *	Methyl alcohol *	5000 (2270)
Methapyrilene	Pyridine, 2-[(2-(dimethylamino)ethyl)-2-thenylamino]-	5000 (2270)
Methomyl	Acetimidic acid, N-[(methylcarbamoyl)oxy]thio-, methyl ester	100 (45.4)
Methoxychlor	Ethane, 1,1,1-trichloro-2,2-bis(p-methoxyphenyl)-	1 (0.454)
Methyl alcohol *	Methanol *	5000 (2270)
Methylamine @	Monomethylamine	100 (45.4)
2-Methylaziridine	1,2-Propylenimine	1 (0.454)
Methyl bromide *	Methane, bromo-	1000 (454)
1-Methylbutadiene	1,3-Pentadiene	100 (45.4)
Methyl chloride *	Chloromethane	100 (45.4)
	Methane, chloro-	
Methyl chlorocarbonate *	Carbonochloridic acid, methyl ester	1000 (454)
	Methyl chloroformate @	
Methyl chloroform *	1,1,1-Trichloroethane *	1000 (454)
Methyl chloroformate @	Carbonochloridic acid, methyl ester	1000 (454)
	Methyl chlorocarbonate *	
Methylchloromethyl ether @	Chloromethyl methyl ether	1 (0.454)
	Methane, chloromethoxy-	
3-Methylcholanthrene	Benz[j]aceanthrylene, 1,2-dihydro-3-methyl-	10 (4.54)
4,4'-Methylenebis(2-chloroaniline)	Benzenamine, 4,4'-methylenebis(2-chloro-	10 (4.54)
2,2'-Methylenebis(3,4,6-trichlorophenol)	Hexachlorophene	100 (45.4)
Methylene bromide	Methane, dibromo-	1000 (454)
Methylene chloride *	Methane, dichloro-	1000 (454)
Methylene oxide	Formaldehyde *	10 (4.54)
Methyl ethyl ketone *	2-Butanone	5000 (2270)
	Ethyl methyl ketone @	
Methyl ethyl ketone peroxide *	2-Butanone peroxide	10 (4.54)
Methyl hydrazine *	Hydrazine, methyl-	10 (4.54)
Methyl iodide	Methane, iodo-	100 (45.4)

TABLE 1 - LIST OF HAZARDOUS SUBSTANCES OTHER THAN RADIONUCLIDES - Continued

Hazardous Substance	Synonyms	Reportable Quantity (RQ) Pounds (Kilograms)
Methyl isobutyl ketone	4-Methyl-2-pentanone	5000 (2270)
Methyl isocyanate *	Isocyanic acid, methyl ester	1 (0.454)
2-Methyllactonitrile	Acetone cyanohydrin *	10 (4.54)
	Propanenitrile, 2-hydroxy-2-methyl-	
Methyl mercaptan *	Methanethiol	100 (45.4)
	Thiomethanol	
Methyl methacrylate *	2-Propenoic acid, 2-methyl-, methyl ester	1000 (454)
N-Methyl-N'-nitro-N-nitrosoguanidine	Guanidine, N-nitroso-N-methyl-N'-nitro-	10 (4.54)
Methyl parathion *	O,O-Dimethyl O-p-nitrophenyl phosphorothioate	100 (45.4)
4-Methyl-2-pentanone	Methyl isobutyl ketone	5000 (2270)
Methylthiouracil	4(1H)-Pyrimidinone, 2,3-dihydro-6-methyl-2-thioxo-	10 (4.54)
Mevinphos *		10 (4.54)
Mexacarbate *		1000 (454)
Mitomycin C	Azirino(2',3':3,4)pyrrolo(1,2-a)indole-4,7-dione,6-amino-8-[((aminocarbonyl)oxy) methyl]-1,1a,2,8,8a,8b-hexahydro-8a-methoxy-5-methyl-.	10 (4.54)
Monoethylamine *		100 (45.4)
Monomethylamine	Methylamine @	100 (45.4)
Naled		10 (4.54)
5,12-Naphthacenedione, (8S-cis)-8-acetyl-10-[3-amino-2,3,6-trideoxy-alpha-L-lyxo-hexopyranosyl) oxy]-7,8,9,10-tetrahydro-6,8,11-trihydroxy-1-methoxy-.	Daunomycin	10 (4.54)
Naphthalene *		100 (45.4)
Naphthalene, 2-chloro-	beta-Chloronaphthalene	5000 (2270)
	2-Chloronaphthalene	
1,4-Naphthalenedione	1,4-Naphthoquinone	5000 (2270)
2,7-Naphthalenedisulfonic acid, 3,3'-[(3,3'- dimethyl-(!,1'-biphenyl)-4,4'-diyl)-bis(azo)]bis(5-amino-4-hydroxy)-tetrasodium salt.	Trypan blue	10 (4.54)
Naphthenic acid		100 (45.4)
1,4-Naphthoquinone	1,4-Naphthalenedione	5000 (2270)
alpha-Naphthylamine	1-Naphthylamine	100 (45.4)
beta-Naphthylamine	2-Naphthylamine	1 (0.454)
1-Naphthylamine	alpha-Naphthylamine	100 (45.4)
2-Naphthylamine	beta-Naphthylamine	1 (0.454)
2-Naphthylamine, N,N-bis(2-chloroethyl)-	Chlornaphazine	100 (45.4)
alpha-Naphthylthiourea	Thiourea, 1-naphthalenyl-	100 (45.4)
Nickel ¢		100 (45.4)
Nickel ammonium sulfate		100 (45.4)
Nickel carbonyl *	Nickel tetracarbonyl	10 (4.54)
Nickel chloride		100 (45.4)
Nickel cyanide *	Nickel(II) cyanide	10 (4.54)
Nickel(II) cyanide	Nickel cyanide *	10 (4.54)
Nickel hydroxide		10 (4.54)
Nickel nitrate *		100 (45.4)
Nickel sulfate		100 (45.4)
Nickel tetracarbonyl	Nickel carbonyl *	10 (4.54)
Nicotine * and salts *	Pyridine, (S)-3-(1-methyl-2-pyrrolidinyl)-, and salts	100 (45.4)
Nitric acid *		1000 (454)
Nitric oxide *	Nitrogen(II) oxide	10 (4.54)
p-Nitroaniline *	Benzenamine, 4-nitro-	5000 (2270)
Nitrobenzene *	Benzene, nitro-	1000 (454)
Nitrogen dioxide *	Nitrogen(IV) oxide	10 (4.54)
	Nitrogen peroxide @	
	Nitrogen tetroxide @	
Nitrogen(II) oxide	Nitric oxide *	10 (4.54)
Nitrogen(IV) oxide	Nitrogen dioxide *	10 (4.54)
	Nitrogen peroxide @	
	Nitrogen tetroxide @	
Nitrogen peroxide @	Nitrogen dioxide *	10 (4.54)
	Nitrogen(IV) oxide	
	Nitrogen tetroxide @	
Nitrogen tetroxide @	Nitrogen dioxide *	10 (4.54)
	Nitrogen(IV) oxide	
	Nitrogen peroxide	
Nitroglycerine *	1,2,3-Propanetriol, trinitrate-	10 (4.54)
Nitrophenol (mixed)		100 (45.4)
m-		
o-	2-Nitrophenol	
p-	4-Nitrophenol	
	Phenol, 4-nitro	
o-Nitrophenol	2-Nitrophenol	100 (45.4)
p-Nitrophenol	Phenol, 4-nitro-	100 (45.4)
	4-Nitrophenol	
2-Nitrophenol	o-Nitrophenol	100 (45.4)
4-Nitrophenol	p-Nitrophenol	100 (45.4)
	Phenol, 4-nitro-	
2-Nitropropane	Propane, 2-nitro-	10 (4.54)

TABLE 1 - LIST OF HAZARDOUS SUBSTANCES OTHER THAN RADIONUCLIDES - Continued

Hazardous Substance	Synonyms	Reportable Quantity (RQ) Pounds (Kilograms)
N-Nitrosodi-n-butylamine	1-Butanamine, N-butyl-N-nitroso-	10 (4.54)
N-Nitrosodiethanolamine	Ethanol, 2,2'-(nitrosoimino)bis-	1 (0.454)
N-Nitrosodiethylamine	Ethanamine, N-ethyl-N-nitroso-	1 (0.454)
N-Nitrosodimethylamine	Dimethylnitrosamine	10 (4.54)
N-Nitrosodiphenylamine		100 (45.4)
N-Nitrosodi-n-propylamine	Di-n-propylnitrosamine	10 (4.54)
N-Nitroso-N-ethylurea	Carbamide, N-ethyl-N-nitroso-	1 (0.454)
N-Nitroso-N-methylurea	Carbamide, N-methyl-N-nitroso-	1 (0.454)
N-Nitroso-N-methylurethane	Carbamic acid, methylnitroso-, ethyl ester	1 (0.454)
N-Nitrosomethylvinylamine	Ethenamine, N-methyl-N-nitroso-	10 (4.54)
N-Nitrosopiperidine	Pyridine, hexahydro-N-nitroso-	10 (4.54)
N-Nitrosopyrrolidine	Pyrrole, tetrahydro-N-nitroso-	1 (0.454)
Nitrotoluene		1000 (454)
m-Nitrotoluene		
o-Nitrotoluene		
p-Nitrotoluene		
5-Nitro-o-toluidine	Benzenamine, 2-methyl-5-nitro-	100 (45.4)
5-Norbornene-2,3-dimethanol,1,4,5,6,7,7-hexachloro,cyclic sulfite	Endosulfan *	1 (0.454)
Octamethylpyrophosphoramide	Diphosphoramide, octamethyl-	100 (45.4)
Osmium oxide	Osmium tetroxide	1000 (454)
Osmium tetroxide	Osmium oxide	1000 (454)
7-Oxabicyclo[2.2.1]heptane-2,3-dicarboxylic acid	Endothall	1000 (454)
1,2-Oxathiolane, 2,2-dioxide	1,3-Propane sultone	10 (4.54)
2H-1,3,2-Oxazaphosphorine,2-[bis(2-chloroethyl) amino] tetrahydro-2-oxide.	Cyclophosphamide	10 (4.54)
Oxirane	Ethylene oxide *	10 (4.54)
Oxirane, 2-(chloromethyl)-	Epichlorohydrin * 1-Chloro-2,3-epoxypropane	100 (45.4)
Paraformaldehyde *		1000 (454)
Paraldehyde *	1,3,5-Trioxane, 2,4,6-trimethyl-	1000 (454)
Parathion *	Phosphorothioic acid, O,O-diethyl O-(p-nitrophenyl)ester	10 (4.54)
Pentachlorobenzene	Benzene, pentachloro-	10 (4.54)
Pentachloroethane	Ethane, pentachloro-	10 (4.54)
Pentachloronitrobenzene	Benzene, pentachloronitro-	100 (45.4)
Pentachlorophenol	Phenol, pentachloro-	10 (4.54)
1,3-Pentadiene	1-Methylbutadiene	100 (45.4)
Perchloroethylene *	Ethene, 1,1,2,2-tetrachloro Tetrachloroethene Tetrachloroethylene *	100 (45.4)
Perchloromethyl mercaptan @	Methanesulfenyl chloride, trichloro- Trichloromethanesulfenyl chloride	100 (45.4)
Phenacetin	Acetamide, N-(4-ethoxyphenyl)-	100 (45.4)
Phenanthrene		5000 (2270)
Phenol *	Benzene, hydroxy-	1000 (454)
Phenol, 2-chloro-	o-Chlorophenol 2-Chlorophenol	100 (45.4)
Phenol, 4-chloro-3-methyl-	p-Chloro-m-cresol 4-Chloro-m-cresol	5000 (2270)
Phenol, 2-cyclohexyl-4,6-dinitro-	4,6-Dinitro-o-cyclohexylphenol	100 (45.4)
Phenol, 2,4-dichloro-	2,4-Dichlorophenol	100 (45.4)
Phenol, 2,6-dichloro-	2,6-Dichlorophenol	100 (45.4)
Phenol, 2,4-dimethyl-	2,4-Dimethylphenol	100 (45.4)
Phenol, 2,4-dinitro-	2,4-Dinitrophenol	10 (4.54)
Phenol, 2,4-dinitro-6-(1-methylpropyl)-	Dinoseb	1000 (454)
Phenol, 2,4-dinitro-6-methyl-, and salts	4,6-Dinitro-o-cresol and salts	10 (4.54)
Phenol, 4-nitro-	p-Nitrophenol * 4-Nitrophenol *	100 (45.4)
Phenol, pentachloro-	Pentachlorophenol	10 (4.54)
Phenol, 2,3,4,6-tetrachloro-	2,3,4,6-Tetrachlorophenol	10 (4.54)
Phenol, 2,4,5-trichloro-	2,4,5-Trichlorophenol	10 (4.54)
Phenol, 2,4,6-trichloro-	2,4,6-Trichlorophenol	10 (4.54)
Phenol, 2,4,6-trinitro-, ammonium salt	Ammonium picrate *	10 (4.54)
Phenyl dichloroarsine *	Dichlorophenylarsine	1 (0.454)
1,10-(1,2-Phenylene)pyrene	Indeno(1,2,3-cd)pyrene	100 (45.4)
Phenyl mercaptan @	Benzenethiol Thiophenol *	100 (45.4)
Phenylmercuric acetate	Mercury, (acetato-O)phenyl-	100 (45.4)
N-Phenylthiourea	Thiourea, phenyl-	100 (45.4)
Phorate	Phosphorodithioic acid, O,O-diethyl S-(ethylthio), methylester	10 (4.54)
Phosgene *	Carbonyl chloride	10 (4.54)
Phosphine *	Hydrogen phosphide	100 (45.4)
Phosphoric acid *		5000 (2270)
Phosphoric acid, diethyl p-nitrophenyl ester	Diethyl-p-nitrophenyl phosphate	100 (45.4)
Phosphoric acid, lead salt	Lead phosphate	1 (0.454)
Phosphorodithioic acid, O,O-diethyl S-(ethylthio), methyl ester	Phorate	10 (4.54)
Phosphorodithioic acid, O,O-diethyl S-methyl ester	O,O-Diethyl S-methyl dithiophosphate	5000 (2270)

TABLE 1 - LIST OF HAZARDOUS SUBSTANCES OTHER THAN RADIONUCLIDES - Continued

Hazardous Substance	Synonyms	Reportable Quantity (RQ) Pounds (Kilograms)
Phosphorodithioic acid, O,O-dimethyl S-[2 (methylamino)-2-oxoethyl] ester.	Dimethoate	10 (4.54)
Phosphorofluoridic acid, bis(1-methylethyl) ester	Diisopropyl fluorophosphate	100 (45.4)
Phosphorothioic acid, O,O-diethyl O-(p-nitrophenyl) ester	Parathion *	10 (4.54)
Phosphorothioic acid, O,O-diethyl O-pyrazinyl ester	O,O-Diethyl O-pyrazinyl phosphorothioate	100 (45.4)
Phosphorothioic acid, O,O-dimethyl O-[p-[(dimethylamino)-sulfonyl] phenyl] ester.	Famphur	1000 (454)
Phosphorus *		1 (0.454)
Phosphorus oxychloride *		1000 (454)
Phosphorus pentasulfide *	Phosphorus sulfide	100 (45.4)
	Sulfur phosphide	
Phosphorus sulfide	Phosphorus pentasulfide *	100 (45.4)
	Sulfur phosphide	
Phosphorus trichloride *		1000 (454)
Phthalic anhydride	1,2-Benzenedicarboxylic acid anhydride	5000 (2270)
2-Picoline	Pyridine, 2-methyl-	5000 (2270)
Plumbane, tetraethyl-	Tetraethyl lead *	10 (4.54)
POLYCHLORINATED BIPHENYLS (PCBs)	Aroclor 1016	1 (0.454)
	Aroclor 1221	
	Aroclor 1232	
	Aroclor 1242	
	Aroclor 1248	
	Aroclor 1254	
	Aroclor 1260	
Potassium arsenate *		1 (0.454)
Potassium arsenite *		1 (0.454)
Potassium bichromate	Potassium dichromate @	10 (4.54)
Potassium chromate		10 (4.54)
Potassium cyanide *		10 (4.54)
Potassium dichromate @	Potassium bichromate	10 (4.54)
Potassium hydroxide *		1000 (454)
Potassium permanganate *		100 (45.4)
Potassium silver cyanide		1 (0.454)
Pronamide	3,5-Dichloro-N-(1,1-dimethyl-2-propynyl)benzamide	5000 (2270)
1-Propanal, 2,3-epoxy-	Glycidylaldehyde	10 (4.54)
Propanal, 2-methyl-2-(methylthio)-,O-[(methylamino)carbonyl]oxime	Aldicarb	1 (0.454)
1-Propanamine	n-Propylamine *	5000 (2270)
1-Propanamine, N-propyl-	Dipropylamine	5000 (2270)
Propane, 1,2-dibromo-3-chloro-	1,2-Dibromo-3-chloropropane	1 (0.454)
Propane, 2-nitro-	2-Nitropropane	10 (4.54)
Propane, 2,2'-oxybis(2-chloro-	Bis(2-chloroisopropyl) ether	1000 (454)
1,3-Propane sultone	1,2-Oxathiolane, 2,2-dioxide	10 (4.54)
Propanedinitrile	Malononitrile	1000 (454)
Propanenitrile	Ethyl cyanide	10 (4.54)
Propanenitrile, 3-chloro-	3-Chloropropionitrile	1000 (454)
Propanenitrile, 2-hydroxy-2-methyl-	Acetone cyanohydrin *	10 (4.54)
	2-Methyllactonitrile	
1,2,3-Propanetriol, trinitrate-	Nitroglycerine *	10 (4.54)
1-Propanol, 2,3-dibromo-, phosphate (3:1)	Tris(2,3-dibromopropyl)phosphate	10 (4.54)
1-Propanol, 2-methyl-	Isobutyl alcohol	5000 (2270)
2-Propanone	Acetone *	5000 (2270)
2-Propanone, 1-bromo-	Bromoacetone *	1000 (454)
Propargite		10 (4.54)
Propargyl alcohol *	2-Propyn-1-ol	1000 (454)
2-Propenal	Acrolein *	1 (0.454)
2-Propenamide	Acrylamide	5000 (2270)
Propene, 1,3-dichloro-	1,3-Dichloropropene	100 (45.4)
1-Propene, 1,1,2,3,3,3-hexachloro-	Hexachloropropene	1000 (454)
2-Propenenitrile	Acrylonitrile *	100 (45.4)
2-Propenenitrile, 2-methyl-	Methacrylonitrile	1000 (454)
2-Propenoic acid	Acrylic acid *	5000 (2270)
2-Propenoic acid, ethyl ester	Ethyl acrylate *	1000 (454)
2-Propenoic acid, 2-methyl-, ethyl ester	Ethyl methacrylate	1000 (454)
2-Propenoic acid, 2-methyl-, methyl ester	Methyl methacrylate *	1000 (454)
2-Propen-1-ol	Allyl alcohol *	100 (45.4)
Propionic acid *		5000 (2270)
Propionic acid, 2-(2,4,5-trichlorophenoxy)-	Silvex	100 (45.4)
	2,4,5-TP @	
	2,4,5-TP acid	
Propionic anhydride		5000 (2270)
n-Propylamine *	1-Propanamine	5000 (2270)
Propylene dichloride *	1,2-Dichloropropane	1000 (454)
Propylene oxide *		100 (45.4)
1,2-Propylenimine *	2-Methylaziridine	1 (0.454)
2-Propyn-1-ol	Propargyl alcohol *	1000 (454)
Pyrene		5000 (2270)
Pyrethrins		1 (0.454)

TABLE 1 - LIST OF HAZARDOUS SUBSTANCES OTHER THAN RADIONUCLIDES - Continued

Hazardous Substance	Synonyms	Reportable Quantity (RQ) Pounds (Kilograms)
4-Pyridinamine	4-Aminopyridine	1000 (454)
Pyridine *		1000 (454)
Pyridine, 2-[(2-(dimethylamino)ethyl)-2- thenylamino]-	Methapyrilene	5000 (2270)
Pyridine, hexahydro-N-nitroso-	N-Nitrosopiperidine	10 (4.54)
Pyridine, 2-methyl-	2-Picoline	5000 (2270)
Pyridine, (S)-3-(1-methyl-2-pyrrolidinyl)-, and salts	Nicotine * and salts *	100 (45.4)
4(1H)-Pyrimidinone, 2,3-dihydro-6-methyl-2-thioxo-	Methylthiouracil	10 (4.54)
Pyrophosphoric acid, tetraethyl ester	Tetraethyl pyrophosphate *	10 (4.54)
Pyrrole, tetrahydro-N-nitroso-	N-Nitrosopyrrolidine	1 (0.454)
Quinoline		5000 (2270)
RADIONUCLIDES		1 (0.454)
Reserpine	Yohimban-16-carboxylic acid,11,17-dimethoxy-18-[(3,4,5-trimethoxybenzoyl)oxy]-, methyl ester.	5000 (2270)
Resorcinol	1,3-Benzenediol	5000 (2270)
Saccharin and salts	1,2-Benzisothiazolin-3-one,1,1-dioxide, and salts	100 (45.4)
Safrole	Benzene, 1,2-methylenedioxy-4-allyl-	100 (45.4)
Selenious acid		10 (4.54)
Selenium ¢		100 (45.4)
Selenium dioxide	Selenium oxide *	10 (4.54)
Selenium disulfide	Sulfur selenide	10 (4.54)
Selenium oxide *	Selenium dioxide	10 (4.54)
Selenourea	Carbamimidoselenoic acid	1000 (454)
L-Serine, diazoacetate (ester)	Azaserine	1 (0.454)
Silver ¢		1000 (454)
Silver cyanide *		1 (0.454)
Silver nitrate *		1 (0.454)
Silvex	Propionic acid, 2-(2,4,5-trichlorophenoxy)- 2,4,5-TP @ 2,4,5-TP acid	100 (45.4)
Sodium *		10 (4.54)
Sodium arsenate *		1 (0.454)
Sodium arsenite *		1 (0.454)
Sodium azide *		1000 (454)
Sodium bichromate	Sodium dichromate @	10 (4.54)
Sodium bifluoride *		100 (45.4)
Sodium bisulfite *		5000 (2270)
Sodium chromate		10 (4.54)
Sodium cyanide *		10 (4.54)
Sodium dichromate @	Sodium bichromate	10 (4.54)
Sodium dodecylbenzene sulfonate		1000 (454)
Sodium fluoride *		1000 (454)
Sodium hydrosulfide *		5000 (2270)
Sodium hydroxide *		1000 (454)
Sodium hypochlorite *		100 (45.4)
Sodium methylate *		1000 (454)
Sodium nitrite *		100 (45.4)
Sodium phosphate, dibasic		5000 (2270)
Sodium phosphate, tribasic		5000 (2270)
Sodium selenite *		100 (45.4)
4,4'-Stilbenediol, alpha,alpha'-diethyl-	Diethylstilbestrol	1 (0.454)
Streptozotocin	D-Glucopyranose, 2-deoxy-2-(3-methyl-3-nitrosoureido)-	1 (0.454)
Strontium chromate		10 (4.54)
Strychnidin-10-one, 2,3-dimethoxy-	Brucine	100 (45.4)
Strychnidin-10-one, and salts	Strychnine * and salts *	10 (4.54)
Strychnine * and salts *	Strychnidin-10-one, and salts	10 (4.54)
Styrene		1000 (454)
Sulfur hydride	Hydrogen sulfide * Hydrosulfuric acid	100 (45.4)
Sulfur monochloride		1000 (454)
Sulfur phosphide	Phosphorus pentasulfide * Phosphorus sulfide	100 (45.4)
Sulfur selenide	Selenium disulfide	10 (4.54)
Sulfuric acid *		1000 (454)
Sulfuric acid, dimethyl ester	Dimethyl sulfate *	10 (4.54)
Sulfuric acid, thallium(I) salt	Thallium(I) sulfate *	100 (45.4)
2,4,5-T *	2,4,5-T acid 2,4,5-Trichlorophenoxyacetic acid *	1000 (454)
2,4,5-T acid	2,4,5-T * 2,4,5-Trichlorophenoxyacetic acid *	1000 (454)
2,4,5-T amines		5000 (2270)
2,4,5-T esters		1000 (454)
2,4,5-T salts		1000 (454)
TDE *	DDD Dichlorodiphenyl dichloroethane 4,4'-DDD	1 (0.454)
1,2,4,5-Tetrachlorobenzene	Benzene, 1,2,4,5-tetrachloro-	5000 (2270)
2,3,7,8-Tetrachlorodibenzo-p-dioxin (TCDD)		1 (0.454)

TABLE 1 - LIST OF HAZARDOUS SUBSTANCES OTHER THAN RADIONUCLIDES - Continued

Hazardous Substance	Synonyms	Reportable Quantity (RQ) Pounds (Kilograms)
1,1,1,2-Tetrachloroethane	Ethane, 1,1,1,2-tetrachloro-	100 (45.4)
1,1,2,2-Tetrachloroethane	Ethane, 1,1,2,2-tetrachloro-	100 (45.4)
Tetrachloroethene	Ethene, 1,1,2,2-tetrachloro-	100 (45.4)
	Perchloroethylene *	
	Tetrachloroethylene *	
Tetrachloroethylene *	Ethene, 1,1,2,2-tetrachloro-	100 (45.4)
	Perchloroethylene *	
	Tetrachloroethene	
2,3,4,6-Tetrachlorophenol	Phenol, 2,3,4,6-tetrachloro-	10 (4.54)
Tetraethyl lead *	Plumbane, tetraethyl-	10 (4.54)
Tetraethyl pyrophosphate *	Pyrophosphoric acid, tetraethyl ester	10 (4.54)
Tetraethyldithiopyrophosphate	Dithiopyrophosphoric acid, tetraethyl ester	100 (45.4)
Tetrahydrofuran *	Furan, tetrahydro-	1000 (454)
Tetranitromethane *	Methane, tetranitro-	10 (4.54)
Tetraphosphoric acid, hexaethyl ester	Hexaethyl tetraphosphate *	100 (45.4)
Thallic oxide	Thallium(III) oxide	100 (45.4)
Thallium ¢		1000 (454)
Thallium(I) acetate	Acetic acid, thallium(I) salt	100 (45.4)
Thallium(I) carbonate	Carbonic acid, dithallium (I) salt	100 (45.4)
Thallium(I) chloride		100 (45.4)
Thallium(I) nitrate		100 (45.4)
Thallium(III) oxide	Thallic oxide	100 (45.4)
Thallium(I) selenide		1000 (454)
Thallium(I) sulfate *	Sulfuric acid, thallium(I) salt	100 (45.4)
Thioacetamide	Ethanethioamide	10 (4.54)
Thiofanox	3,3-Dimethyl-1-(methylthio)-2-butanone,O-[(methylamino)carbonyl] oxime.	100 (45.4)
Thioimidodicarbonic diamide	2,4-Dithiobiuret	100 (45.4)
Thiomethanol	Methanethiol	100 (45.4)
	Methyl mercaptan *	
Thiophenol *	Benzenethiol	100 (45.4)
	Phenyl mercaptan @	
Thiosemicarbazide	Hydrazinecarbothioamide	100 (45.4)
Thiourea	Carbamide, thio-	10 (4.54)
Thiourea, (2-chlorophenyl)-	1-(o-Chlorophenyl)thiourea	100 (45.4)
Thiourea, 1-naphthalenyl-	alpha-Naphthylthiourea	100 (45.4)
Thiourea, phenyl-	N-Phenylthiourea	100 (45.4)
Thiram	Bis(dimethylthiocarbamoyl) disulfide	10 (4.54)
Toluene *	Benzene, methyl-	1000 (454)
Toluenediamine *	Diaminotoluene	10 (4.54)
Toluene diisocyanate *	Benzene, 2,4-diisocyanatomethyl-	100 (45.4)
o-Toluidine	2-Amino-1-methyl benzene	100 (45.4)
p-Toluidine	4-Amino-1-methyl benzene	100 (45.4)
o-Toluidine hydrochloride	Benzenamine, 2-methyl-, hydrochloride	100 (45.4)
Toxaphene *	Camphene, octachloro-	1 (0.454)
2,4,5-TP @	Propionic acid, 2-(2,4,5-trichlorophenoxy)-	100 (45.4)
	Silvex	
	2,4,5-TP acid	
2,4,5-TP acid	Propionic acid, 2-(2,4,5-trichlorophenoxy)-	100 (45.4)
	Silvex	
	2,4,5-TP @	
2,4,5-TP acid esters		100 (45.4)
1H-1,2,4-Triazol-3-amine	Amitrole	10 (4.54)
Trichlorfon		100 (45.4)
1,2,4-Trichlorobenzene		100 (45.4)
1,1,1-Trichloroethane *	Methyl chloroform *	1000 (454)
1,1,2-Trichloroethane	Ethane, 1,1,2-trichloro-	100 (45.4)
Trichloroethene	Trichloroethylene *	100 (45.4)
Trichloroethylene *	Trichloroethene	100 (45.4)
Trichloromethanesulfenyl chloride	Methanesulfenyl chloride, trichloro-	100 (45.4)
	Perchloromethyl mercaptan @	
Trichloromonofluoromethane	Methane, trichlorofluoro-	5000 (2270)
Trichlorophenol *		10 (4.54)
2,3,4-Trichlorophenol		
2,3,5-Trichlorophenol		
2,3,6-Trichlorophenol		
2,4,5-Trichlorophenol	Phenol, 2,4,5-trichloro-	
2,4,6-Trichlorophenol	Phenol, 2,4,6-trichloro-	
3,4,5-Trichlorophenol		
2,4,5-Trichlorophenol	Phenol, 2,4,5-trichloro-	10 (4.54)
2,4,6-Trichlorophenol	Phenol, 2,4,6-trichloro-	10 (4.54)
2,4,5-Trichlorophenoxyacetic acid *	2,4,5-T *	1000 (454)
	2,4,5-T acid	
Triethanolamine dodecylbenzene sulfonate		1000 (454)
Triethylamine		5000 (2270)
Trimethylamine *		100 (45.4)
sym-Trinitrobenzene *	Benzene, 1,3,5-trinitro-	10 (4.54)

TABLE 1 - LIST OF HAZARDOUS SUBSTANCES OTHER THAN RADIONUCLIDES - Continued

Hazardous Substance	Synonyms	Reportable Quantity (RQ) Pounds (Kilograms)
1,3,5-Trioxane, 2,4,6-trimethyl-	Paraldehyde	1000 (454)
Tris(2,3-dibromopropyl) phosphate	1-Propanol, 2,3-dibromo-, phosphate (3:1)	10 (4.54)
Trypan blue	2,7-Naphthalenedisulfonic acid, 3,3'-[(3,3'-dimethyl-(l,1'-biphenyl)-4,4'-diyl)-bis(azo)]bis(5-amino-4-hydroxy)-tetrasodium salt.	10 (4.54)
Unlisted Hazardous Wastes Characteristic of Corrosivity D002		100 (45.4)
Unlisted Hazardous Wastes Characteristic of EP Toxicity		
Arsenic D004		1 (0.454)
Barium D005		1000 (454)
Cadmium D006		10 (4.54)
Chromium D007		10 (4.54)
Lead D008		1 (0.454)
Mercury D009		1 (0.454)
Selenium D010		10 (4.54)
Silver D011		1 (0.454)
Endrin D012		1 (0.454)
Lindane D013		1 (0.454)
Methoxychlor D014		1 (0.454)
Toxaphene D015		1 (0.454)
2,4-D D016		100 (45.4)
2,4,5-TP D017		100 (45.4)
Unlisted Hazardous Wastes Characteristic of Ignitability D001		100 (45.4)
Unlisted Hazardous Wastes Characteristic of Reactivity D003		100 (45.4)
Uracil, 5-[bis(2-chloroethyl)amino]-	Uracil mustard	10 (4.54)
Uracil mustard	Uracil, 5-[bis(2-chloroethyl)amino]-	10 (4.54)
Uranyl acetate *		100 (45.4)
Uranyl nitrate *		100 (45.4)
Vanadic acid, ammonium salt	Ammonium vanadate	1000 (454)
Vanadium(V) oxide	Vanadium pentoxide	1000 (454)
Vanadium pentoxide	Vanadium(V) oxide	1000 (454)
Vanadyl sulfate		1000 (454)
Vinyl acetate *		5000 (2270)
Vinyl chloride *	Ethene, chloro-	1 (0.454)
Vinylidene chloride *	Ethene, 1,1-dichloro- 1,1-Dichloroethylene	100 (45.4)
Warfarin	3-(alpha-Acetonylbenzyl)-4-hydroxycoumarin and salts	100 (45.4)
Xylene * (mixed)	Benzene, dimethyl	1000 (454)
m-	m-	
o-	o-	
p-	p-	
Xylenol *		1000 (454)
Yohimban-16-carboxylic acid,11,17-dimethoxy-18-[(3,4,5-trimethoxybenzoyl)oxy]-, methyl ester.	Reserpine	5000 (2270)
Zinc ¢		1000 (454)
Zinc acetate		1000 (454)
Zinc ammonium chloride		1000 (454)
Zinc borate		1000 (454)
Zinc bromide		1000 (454)
Zinc carbonate		1000 (454)
Zinc chloride		1000 (454)
Zinc cyanide *		10 (4.54)
Zinc fluoride		1000 (454)
Zinc formate		1000 (454)
Zinc hydrosulfite *		1000 (454)
Zinc nitrate *		1000 (454)
Zinc phenolsulfonate		5000 (2270)
Zinc phosphide *		100 (45.4)
Zinc silicofluoride		5000 (2270)
Zinc sulfate		1000 (454)
Zirconium nitrate *		5000 (2270)
Zirconium potassium fluoride		1000 (454)
Zirconium sulfate *		5000 (2270)
Zirconium tetrachloride *		5000 (2270)
F001		10 (4.54)
The following spent halogenated solvents used in degreasing; all spent solvent mixtures/blends used in degreasing containing, before use, a total of ten percent or more (by volume) of one or more of the above halogenated solvents or those solvents listed in F002, F004, and F005; and still bottoms from the recovery of these spent solvents and spent solvent mixtures..		
(a) Tetrachloroethylene		100 (45.4)
(b) Trichloroethylene		100 (45.4)
(c) Methylene chloride		1000 (454)
(d) 1,1,1-Trichloroethane		1000 (454)
(e) Carbon tetrachloride		10 (4.54)
(f) Chlorinated fluorocarbons		5000 (2270)

TABLE 1 - LIST OF HAZARDOUS SUBSTANCES OTHER THAN RADIONUCLIDES - Continued

Hazardous Substance	Synonyms	Reportable Quantity (RQ) Pounds (Kilograms)
F002		10 (4.54)
The following spent halogenated solvents; all spent solvent mixtures/blends containing, before use, a total of ten percent or more (by volume) of one or more of the above halogenated solvents or those listed in F001, F004, F005; and still bottoms from the recovery of these spent solvents and spent solvent mixtures..		
(a) Tetrachloroethylene		100 (45.4)
(b) Methylene chloride		1000 (454)
(c) Trichloroethylene		100 (45.4)
(d) 1,1,1-Trichloroethane		1000 (454)
(e) Chlorobenzene		100 (45.4)
(f) 1,1,2-Trichloro-1,2,2-trifluoroethane		5000 (2270)
(g) o-Dichlorobenzene		100 (45.4)
(h) Trichlorofluoromethane		5000 (2270)
(i) 1,1,2 Trichloroethane		100 (45.4)
F003		100 (45.4)
The following spent non-halogenated solvents and solvents:		
(a) Xylene		1000 (454)
(b) Acetone		5000 (2270)
(c) Ethyl acetate		5000 (2270)
(d) Ethylbenzene		1000 (454)
(e) Ethyl ether		100 (45.4)
(f) Methyl isobutyl ketone		5000 (2270)
(g) n-Butyl alcohol		5000 (2270)
(h) Cyclohexanone		5000 (2270)
(i) Methanol		5000 (2270)
F004		1000 (454)
The following spent non-halogenated solvents and the stillbottoms from the recovery of these solvents:.		
(a) Cresols/Cresylic acid		1000 (454)
(b) Nitrobenzene		1000 (454)
F005		100 (45.4)
The following spent non-halogenated solvents and the stillbottoms from the recovery of these solvents:.		
(a) Toluene		1000 (454)
(b) Methyl ethyl ketone		5000 (2270)
(c) Carbon disulfide		100 (45.4)
(d) Isobutanol		5000 (2270)
(e) Pyridine		1000 (454)
F006		10 (4.54)
Wastewater treatment sludges from electroplating operations except from the following processes: (1) sulfuric acid anodizing of aluminum,(2) tin plating on carbon steel, (3) zinc plating (segregated basis) on carbonsteel, (4) aluminum or zinc-aluminum plating on carbon steel, (5) cleaning/stripping associated with tin, zinc and aluminum plating on carbon steel, and (6) chemical etchingand milling of aluminum.		
F007		10 (4.54)
Spent cyanide plating bath solutions from electroplating operations		
F008		10 (4.54)
Plating bath sludges from the bottom of plating baths fromelectroplatingoperations where cyanides are used in the process (except for precious metals electroplating plating bath sludges).		
F009		10 (4.54)
Spent stripping and cleaning bath solutions from electroplating operations where cyanides are used in the process (except for precious metals electroplating spent stripping and cleaning bath solutions).		
F010		10 (4.54)
Quenching bath sludge from oil baths from metal heat treating operationswhere cyanides are used in the process (except for precious metals heat-treating quenching bath sludges).		
F011		10 (4.54)
Spent cyanide solutions from salt bath pot cleaning from metal heat treating operations (except for precious metals heat treating spent cyanide solutions from salt bath pot cleaning).		
F012		10 (4.54)
Quenching wastewater treatment sludges from metal heat treating operations where cyanides are used in the process (except for precious metals heat treating quenching wastewater treatment sludges).		
F019		10 (4.54)
Wastewater treatment sludges from the chemical conversion coating of aluminum.		

TABLE 1 - LIST OF HAZARDOUS SUBSTANCES OTHER THAN RADIONUCLIDES - Continued

Hazardous Substance	Synonyms	Reportable Quantity (RQ) Pounds (Kilograms)
F020 Wastes (except wastewater and spent carbon from hydrogen chloride purification) from the production or manufacturing use (as a reactant, chemical intermediate, or component in a formulating process) of tri- or tetrachlorophenol, or of intermediates used to produce their pesticide derivatives. (This listing does not include wastes from the production of hexachlorophene from highly purified 2,4,5-trichlorophenol.).		1 (0.454)
F021 Wastes (except wastewater and spent carbon from hydrogen chloride purification) from the production or manufacturing use (as a reactant, chemical intermediate, or component in a formulating process) of pentachlorophenol, or of intermediates used to produce its derivatives..		1 (0.454)
F022 Wastes (except wastewater and spent carbon from hydrogen chloride purification) from the manufacturing use (as a reactant, chemical intermediate, or component in a formulating process) of tetra-, penta- or hexachlorobenzenes under alkaline conditions..		1 (0.454)
F023 Wastes (except wastewater and spent carbon from hydrogen chloride purification) from the production of materials on equipment previously used for the production or manufacturing use (as a reactant, chemical intermediate, or component in a formulating process) of tri- and tetrachlorophenols. (This listing does not include wastes from equipment used only for the productionor use of hexachlorophene from highly purified 2,4,5-trichlorophenol.).		1 (0.454)
F024 Wastes, including but not limited to distillation residues, heavy ends, tars, and reactor cleanout wastes, from the production of chlorinated aliphatichydrocarbons, having carbon content from one to five, utilizing free radical catalyzed processes. (This listing does not include light ends, spent filters and filter aids, spent dessicants(sic), wastewater, wastewater treatment sludges,spent catalysts, and wastes listed in 40 CFR 261.32.).		1 (0.454)
F026 Wastes (except wastewater and spent carbon from hydrogen chloride purification) from the production of materials on equipment previously used for the manufacturing use (as a reactant, chemical intermediate, or component in a formulating process) of tetra-, penta-, or hexachlorobenzene under alkaline conditions..		1 (0.454)
F027 Discarded unused formulations containing tri-, tetra-, or pentachlorophenol or discarded unused formulations containing compounds derivedfrom these chlorophenols. (This listing does not include formulations containing hexachlorophene synthesized from prepurified 2,4,5-trichlorophenol as the sole component.).		1 (0.454)
F028 Residues resulting from the incineration or thermal treatment of soil contaminated with EPA Hazardous Waste Nos. F020, F021, F022, F023, F026, and F027..		1 (0.454)
K001 Bottom sediment sludge from the treatment of wastewaters from wood preserving processes that use creosote and/or pentachlorophenol.		1 (0.454)
K002 Wastewater treatment sludge from the production of chrome yellow and orange pigments.		1 (0.454)
K003 Wastewater treatment sludge from the production of molybdate orange pigments.		1 (0.454)
K004 Wastewater treatment sludge from the production of zinc yellow pigments.		10 (4.54)
K005 Wastewater treatment sludge from the production of chrome green pigments.		1 (0.454)
K006 Wastewater treatment sludge from the production of chrome oxide green pigments (anhydrous and hydrated).		10 (4.54)
K007 Wastewater treatment sludge from the production of iron blue pigments..		10 (4.54)
K008 Oven residue from the production of chrome oxide green pigments		10 (4.54)
K009 Distillation bottoms from the production of acetaldehyde from ethylene		10 (4.54)
K010 Distillation side cuts from the production of acetaldehydefrom ethylene		10 (4.54)

TABLE 1 - LIST OF HAZARDOUS SUBSTANCES OTHER THAN RADIONUCLIDES - Continued

Hazardous Substance	Synonyms	Reportable Quantity (RQ) Pounds (Kilograms)
K011 Bottom stream from the wastewater stripper in the production of acrylonitrile.		10 (4.54)
K013 Bottom stream from the acetonitrile column in the production of acrylonitrile.		10 (4.54)
K014 Bottoms from the acetonitrile purification column in the production of acrylonitrile.		5000 (2270)
K015 Still bottoms from the distillation of benzyl chloride		10 (4.54)
K016 Heavy ends or distillation residues from the production of carbon tetrachloride.		1 (0.454)
K017 Heavy ends (still bottoms) from the purification column in the production of epichlorohydrin.		1 (0.454)
K018 Heavy ends from the fractionation column in ethyl chloride production		1 (0.454)
K019 Heavy ends from the distillation of ethylene dichloride in ethylene dichloride production..		1 (0.454)
K020 Heavy ends from the distillation of vinyl chloride in vinyl chloride monomer production.		1 (0.454)
K021 Aqueous spent antimony catalyst waste from fluoromethanes production.		10 (4.54)
K022 Distillation bottom tars from the production of phenol/acetone from cumene.		1 (0.454)
K023 Distillation light ends from the production of phthalic anhydride from naphthalene.		5000 (2270)
K024 Distillation bottoms from the production of phthalic anhydride from naphthalene.		5000 (2270)
K025 Distillation bottoms from the production of nitrobenzene by the nitration of benzene.		10 (4.54)
K026 Stripping still tails from the production of methyl ethyl pyridines		1000 (454)
K027 Centrifuge and distillation residues from toluene diisocyanate production.		10 (4.54)
K028 Spent catalyst from the hydrochlorinator reactor in the production of 1,1,1-trichloroethane.		1 (0.454)
K029 Waste from the product steam stripper in the production of 1,1,1-trichloroethane.		1 (0.454)
K030 Column bottoms or heavy ends from the combined production of trichloroethylene and perchloroethylene.		1 (0.454)
K031 By-product salts generated in the production of MSMA and cacodylic acid.		1 (0.454)
K032 Wastewater treatment sludge from the production of chlordane		10 (4.54)
K033 Wastewater and scrub water from the chlorination of cyclopentadiene in the production of chlordane.		10 (4.54)
K034 Filter solids from the filtration of hexachlorocyclopentadiene in the production of chlordane.		10 (4.54)
K035 Wastewater treatment sludges generated in the production of creosote		1 (0.454)
K036 Still bottoms from toluene reclamation distillation in the production ofdisulfoton.		1 (0.454)
K037 Wastewater treatment sludges from the production of disulfoton		1 (0.454)
K038 Wastewater from the washing and stripping of phorate production		10 (4.54)
K039 Filter cake from the filtration of diethylphosphorodithioic acid in the production of phorate.		10 (4.54)

TABLE 1 - LIST OF HAZARDOUS SUBSTANCES OTHER THAN RADIONUCLIDES - Continued

Hazardous Substance	Synonyms	Reportable Quantity (RQ) Pounds (Kilograms)
K040 Wastewater treatment sludge from the production of phorate		10 (4.54)
K041 Wastewater treatment sludge from the production of toxaphene		1 (0.454)
K042 Heavy ends or distillation residues from the distillation of tetrachlorobenzene in the production of 2,4,5-T.		10 (4.54)
K043 2,6-dichlorophenol waste from the production of 2,4-D		10 (4.54)
K044 Wastewater treatment sludges from the manufacturing and processing of explosives.		10 (4.54)
K045 Spent carbon from the treatment of wastewater containing explosives		10 (4.54)
K046 Wastewater treatment sludges from the manufacturing, formulation and loading of lead-based initiating compounds.		100 (45.4)
K047 Pink/red water from TNT operations		10 (4.54)
K048 Dissolved air flotation (DAF) float from the petroleum refining industry		1 (0.454)
K049 Slop oil emulsion solids from the petroleum refining industry		1 (0.454)
K050 Heat exchanger bundle cleaning sludge from the petroleum refining industry.		10 (4.54)
K051 API separator sludge from the petroleum refining industry		1 (0.454)
K052 Tank bottoms (leaded) from the petroleum refining industry		10 (4.54)
K060 Ammonia still lime sludge from coking operations		1 (0.454)
K061 Emission control dust/sludge from the primary production of steel in electric furnaces.		1 (0.454)
K062 Spent pickle liquor from steel finishing operations		1 (0.454)
K064 Acid plant blowdown slurry/sludge resulting from thickening of blowdoen slurry from primary copper production..		1 (0.454)
K065 Surface impoundment solids contained in and dredged from surface impoundments at primary lead smelting facilities..		1 (0.454)
K066 Sludge from treatment of process wastewater and /or acid plant blowdown from primary zinc production..		1 (0.454)
K069 Emission control dust/sludge from secondary lead smelting		1 (0.454)
K071 Brine purification muds from the mercury cell process in chlorine production, where separately prepurified brine is not used.		1 (0.454)
K073 Chlorinated hydrocarbon waste from the purification step of the diaphragm cell process using graphite anodes in chlorine production..		10 (4.54)
K083 Distillation bottoms from aniline extraction		100 (45.4)
K084 Wastewater treatment sludges generated during the production of veterinary pharmaceuticals from arsenic or organo-arsenic compounds.		1 (0.454)
K085 Distillation or fractionation column bottoms from the production of chlorobenzenes.		10 (4.54)
K086 Solvent washes and sludges, caustic washes and sludges, orwater washes and sludges from cleaning tubs and equipment used in the formulation of ink frompigments, driers, soaps, and stabilizers containing chromium and lead.		1 (0.454)
K087 Decanter tank tar sludge from coking operations		100 (45.4)
K088 Spent potliners from primary aluminum reduction		1 (0.454)
K090 Emission control dust or sludge from ferrochromiumsilicon production		1 (0.454)
K091 Emission control dust or sludge from ferrochromium production.		1 (0.454)

TABLE 1 - LIST OF HAZARDOUS SUBSTANCES OTHER THAN RADIONUCLIDES - Continued

Hazardous Substance	Synonyms	Reportable Quantity (RQ) Pounds (Kilograms)
K093 Distillation light ends from the production of phthalic anhydride from ortho-xylene.		5000 (2270)
K094 Distillation bottoms from the production of phthalic anhydride from ortho-xylene.		5000 (2270)
K095 Distillation bottoms from the production of 1,1,1-trichloroethane.		100 (45.4)
K096 Heavy ends from the heavy ends column from the production of 1,1,1-trichloroethane..		100 (45.4)
K097 Vacuum stripper discharge from the chlordane chlorinator in the production of chlordane.		1 (0.454)
K098 Untreated process wastewater from the production of toxaphene		1 (0.454)
K099 Untreated wastewater from the production of 2,4-D		10 (4.54)
K100 Waste leaching solution from acid leaching of emission control dust/sludge from secondary lead smelting.		1 (0.454)
K101 Distillation tar residues from the distillation of aniline-based compounds in the production of veterinary pharmaceuticals from arsenic or organo-arsenic compounds.		1 (0.454)
K102 Residue from the use of activated carbon for decolorization in the production of veterinary pharmaceuticals from arsenic or organo-arsenic compounds.		1 (0.454)
K103 Process residues from aniline extraction from the production of aniline		100 (45.4)
K104 Combined wastewater streams generated from nitrobenzene/aniline chlorobenzenes.		10 (4.54)
K105 Separated aqueous stream from the reactor product washing step in the production of chlorobenzenes.		10 (4.54)
K106 Wastewater treatment sludge from the mercury cell process in chlorine production.		1 (0.454)
K111 Product washwaters from the production of dinitrotoluene via nitration of toluene..		10 (4.54)
K112 Reaction by-product water from the drying column in the production of toluenediamine via hydrogenation of dinitrotoluene..		10 (4.54)
K113 Condensed liquid light ends from the purification of toluenediamine in the production of toluenediamine via hydrogenation of dinitrotoluene..		10 (4.54)
K114 Vicinals from the purification of toluenediamine in the production of toluenediamine via hydrogenation of dinitrotoluene..		10 (4.54)
K115 Heavy ends from the purification of toluenediamine in the production of toluenediamine via hydrogenation of dinitrotoluene..		10 (4.54)
K116 Organic condensate from the solvent recovery column in theproduction of toluene diisocyanate via phosgenation of toluenediamine..		10 (4.54)
K117 Wastewater from the reaction vent gas scrubber in the production of ethylene bromide via bromination of ethene..		1 (0.454)
K118 Spent absorbent solids from purification of ethylene dibromide in the production of ethylene dibromide..		1 (0.454)
K123 Process wastewater (including supernates, filtrates, and washwaters) from the production of ethylenebisdithiocarbamic acid and its salts..		10 (4.54)
K124 Reactor vent scrubber water from the production of ethylenebisdithiocarbamic acid and its salts..		10 (4.54)
K125 Filtration, evaporation, and centrifugation solids from the production of ethylenebisdithiocarbamic acid and its salts..		10 (4.54)
K126 Baghouse dust and floor sweepings in milling and packaging operations from the production or formulation of ethylenebisdithiocarbamic acid and its salts..		10 (4.54)

TABLE 1 - LIST OF HAZARDOUS SUBSTANCES OTHER THAN RADIONUCLIDES - Continued

Hazardous Substance	Synonyms	Reportable Quantity (RQ) Pounds (Kilograms)
K136 Still bottoms from the purification of ethylene dibromide in the production of ethylene dibromide via bromination of ethene..		1 (0.454)

Footnotes:
¢ - the RQ for these hazardous substances is limited to those pieces of the metal having a diameter smaller than 100 micrometers (0.004 inches)
¢¢ - the RQ for asbestos is limited to friable forms only
* - indicates that this material appears by name in the Hazardous Materials Table
@ - indicates that the name was added by RSPA because (1) the name is a synonym for a specific hazardous substance and (2) the name appears in the Hazardous Materials Tables as a proper shipping name.

Issued in Washington, DC on August 14, 1989, under authority delegated in 49 CFR part 1.

Travis P. Dungan,

Administrator, Research and Special Programs Administration.

[FR Doc. 89-19419 Filed 8-18-89; 8:45 am]

BILLING CODE 4910-60-M

List of Hazardous Substances and Reportable Quantities

TABLE 2.—RADIONUCLIDES

(1)—Radionuclide	(2)—Atomic number	(3)—Reportable quantity (RQ) Ci (TBq)
Actinium-224	89	100 (3.7)
Actinium-225	89	1 (.037)
Actinium-226	89	10 (.37)
Actinium-227	89	0.001 (.000037)
Actinium-228	89	10 (.37)
Aluminum-26	13	10 (.37)
Americium-237	95	1000 (37)
Americium-238	95	100 (3.7)
Americium-239	95	100 (3.7)
Americium-240	95	10 (.37)
Americium-241	95	0.01 (.00037)
Americium-242	95	100 (3.7)
Americium-242m	95	0.01 (.00037)
Americium-243	95	0.01 (.00037)
Americium-244	95	10 (.37)
Americium-244m	95	1000 (37)
Americium-245	95	1000 (37)
Americium-246	95	1000 (37)
Americium-246m	95	1000 (37)
Antimony-115	51	1000 (37)
Antimony-116	51	1000 (37)
Antimony-116m	51	100 (3.7)
Antimony-117	51	1000 (37)
Antimony-118m	51	10 (.37)
Antimony-119	51	1000 (37)
Antimony-120 (16 min)	51	1000 (37)
Antimony-120 (5.76 day)	51	10 (.37)
Antimony-122	51	10 (.37)
Antimony-124	51	10 (.37)
Antimony-124m	51	1000 (37)
Antimony-125	51	10 (.37)
Antimony-126	51	10 (.37)
Antimony-126m	51	1000 (37)
Antimony-127	51	10 (.37)
Antimony-128 (10.4 min)	51	1000 (37)
Antimony-128 (9.01 hr)	51	10 (.37)
Antimony-129	51	100 (3.7)
Antimony-130	51	100 (3.7)
Antimony-131	51	1000 (37)
Argon-39	18	1000 (37)
Argon-41	18	10 (.37)
Arsenic-69	33	1000 (37)
Arsenic-70	33	100 (3.7)
Arsenic-71	33	100 (3.7)
Arsenic-72	33	10 (.37)
Arsenic-73	33	100 (3.7)
Arsenic-74	33	10 (.37)
Arsenic-76	33	100 (3.7)
Arsenic-77	33	1000 (37)
Arsenic-78	33	100 (3.7)
Astatine-207	85	100 (3.7)
Astatine-211	85	100 (3.7)
Barium-126	56	1000 (37)
Barium-128	56	10 (.37)
Barium-131	56	10 (.37)
Barium-131m	56	1000 (37)
Barium-133	56	10 (.37)
Barium-133m	56	100 (3.7)
Barium-135m	56	1000 (37)
Barium-139	56	1000 (37)
Barium-140	56	10 (.37)
Barium-141	56	1000 (37)
Barium-142	56	1000 (37)
Berkelium-245	97	100 (3.7)
Berkelium-246	97	10 (.37)
Berkelium-247	97	0.01 (.00037)
Berkelium-249	97	1 (.037)
Berkelium-250	97	100 (3.7)
Beryllium-10	4	1 (.037)
Beryllium-7	4	100 (3.7)
Bismuth-200	83	100 (3.7)
Bismuth-201	83	100 (3.7)
Bismuth-202	83	1000 (37)
Bismuth-203	83	10 (.37)
Bismuth-205	83	10 (.37)
Bismuth-206	83	10 (.37)
Bismuth-207	83	10 (.37)
Bismuth-210	83	10 (.37)
Bismuth-210m	83	0.1 (.0037)
Bismuth-212	83	100 (3.7)
Bismuth-213	83	100 (3.7)
Bismuth-214	83	100 (3.7)
Bromine-74	35	100 (3.7)
Bromine-74m	35	100 (3.7)
Bromine-75	35	100 (3.7)
Bromine-76	35	10 (.37)
Bromine-77	35	100 (3.7)
Bromine-80	35	1000 (37)
Bromine-80m	35	1000 (37)
Bromine-82	35	10 (.37)
Bromine-83	35	1000 (37)
Bromine-84	35	100 (3.7)
Cadmium-104	48	1000 (37)
Cadmium-107	48	1000 (37)
Cadmium-109	48	1 (.037)
Cadmium-113	48	0.1 (.0037)
Cadmium-113m	48	0.1 (.0037)
Cadmium-115	48	100 (3.7)
Cadmium-115m	48	10 (.37)
Cadmium-117	48	100 (3.7)
Cadmium-117m	48	10 (.37)
Calcium-41	20	10 (.37)
Calcium-45	20	10 (.37)
Calcium-47	20	10 (.37)
Californium-244	98	1000 (37)
Californium-246	98	10 (.37)
Californium-248	98	0.1 (.0037)
Californium-249	98	0.01 (.00037)
Californium-250	98	0.01 (.00037)
Californium-251	98	0.01 (.00037)
Californium-252	98	0.1 (.0037)
Californium-253	98	10 (.37)
Californium-254	98	0.1 (.0037)
Carbon-11	6	1000 (37)
Carbon-14	6	10 (.37)
Cerium-134	58	10 (.37)
Cerium-135	58	10 (.37)
Cerium-137	58	1000 (37)
Cerium-137m	58	100 (3.7)
Cerium-139	58	100 (3.7)
Cerium-141	58	10 (.37)
Cerium-143	58	100 (3.7)
Cerium-144	58	1 (.037)
Cesium-125	55	1000 (37)
Cesium-127	55	100 (3.7)
Cesium-129	55	100 (3.7)
Cesium-130	55	1000 (37)
Cesium-131	55	1000 (37)
Cesium-132	55	10 (.37)
Cesium-134	55	1 (.037)
Cesium-134m	55	1000 (37)
Cesium-135	55	10 (.37)
Cesium-135m	55	100 (3.7)
Cesium-136	55	10 (.37)
Cesium-137	55	1 (.037)
Cesium-138	55	100 (3.7)
Chlorine-36	17	10 (.37)
Chlorine-38	17	100 (3.7)
Chlorine-39	17	100 (3.7)
Chromium-48	24	100 (3.7)
Chromium-49	24	1000 (37)
Chromium-51	24	1000 (37)
Cobalt-55	27	10 (.37)
Cobalt-56	27	10 (.37)
Cobalt-57	27	100 (3.7)
Cobalt-58	27	10 (.37)
Cobalt-58m	27	1000 (37)
Cobalt-60	27	10 (.37)
Cobalt-60m	27	1000 (37)
Cobalt-61	27	1000 (37)
Cobalt-62m	27	1000 (37)
Copper-60	29	100 (3.7)
Copper-61	29	100 (3.7)
Copper-64	29	1000 (37)
Copper-67	29	100 (3.7)
Curium-238	96	1000 (37)
Curium-240	96	1 (.037)
Curium-241	96	10 (.37)
Curium-242	96	1 (.037)
Curium-243	96	0.01 (.00037)
Curium-244	96	0.01 (.00037)
Curium-245	96	0.01 (.00037)
Curium-246	96	0.01 (.00037)
Curium-247	96	0.01 (.00037)
Curium-248	96	0.001 (.000037)
Curium-249	96	1000 (37)
Dysprosium-155	66	100 (3.7)
Dysprosium-157	66	100 (3.7)
Dysprosium-159	66	100 (3.7)
Dysprosium-165	66	1000 (37)
Dysprosium-166	66	10 (.37)
Einsteinium-250	99	10 (.37)
Einsteinium-251	99	1000 (37)
Einsteinium-253	99	10 (.37)
Einsteinium-254	99	0.1 (.0037)
Einsteinium-254m	99	1 (.037)
Erbium-161	68	100 (3.7)
Erbium-165	68	1000 (37)
Erbium-169	68	100 (3.7)
Erbium-171	68	100 (3.7)
Erbium-172	68	10 (.37)
Europium-145	63	10 (.37)
Europium-146	63	10 (.37)
Europium-147	63	10 (.37)
Europium-148	63	10 (.37)
Europium-149	63	100 (3.7)
Europium-150 (12.6 hr)	63	1000 (37)
Europium-150 (34.2 yr)	63	10 (.37)
Europium-152	63	10 (.37)
Europium-152m	63	100 (3.7)
Europium-154	63	10 (.37)
Europium-155	63	10 (.37)
Europium-156	63	10 (.37)
Europium-157	63	10 (.37)
Europium-158	63	1000 (37)
Fermium-252	100	10 (.37)
Fermium-253	100	10 (.37)
Fermium-254	100	100 (3.7)
Fermium-255	100	100 (3.7)
Fermium-257	100	1 (.037)
Fluorine-18	9	1000 (37)
Francium-222	87	100 (3.7)
Francium-223	87	100 (3.7)
Gadolinium-145	64	100 (3.7)
Gadolinium-146	64	10 (.37)
Gadolinium-147	64	10 (.37)
Gadolinium-148	64	0.001 (.000037)
Gadolinium-149	64	100 (3.7)
Gadolinium-151	64	100 (3.7)
Gadolinium-152	64	0.001 (.000037)
Gadolinium-153	64	10 (.37)
Gadolinium-159	64	1000 (37)
Gallium-65	31	1000 (37)
Gallium-66	31	10 (.37)
Gallium-67	31	100 (3.7)
Gallium-68	31	1000 (37)
Gallium-70	31	1000 (37)
Gallium-72	31	10 (.37)
Gallium-73	31	100 (3.7)
Germanium-66	32	100 (3.7)
Germanium-67	32	1000 (37)
Germanium-68	32	10 (.37)

TABLE 2.—RADIONUCLIDES—Continued

(1)—Radionuclide	(2)—Atomic number	(3)—Reportable quantity (RQ) Ci (TBq)
Germanium-69	32	10 (.37)
Germanium-71	32	1000 (37)
Germanium-75	32	1000 (37)
Germanium-77	32	10 (.37)
Germanium-78	32	1000 (37)
Gold-193	79	100 (3.7)
Gold-194	79	10 (.37)
Gold-195	79	100 (3.7)
Gold-198	79	100 (3.7)
Gold-198m	79	10 (.37)
Gold-199	79	100 (3.7)
Gold-200	79	1000 (37)
Gold-200m	79	10 (.37)
Gold-201	79	1000 (37)
Hafnium-170	72	100 (3.7)
Hafnium-172	72	1 (.037)
Hafnium-173	72	100 (3.7)
Hafnium-175	72	100 (3.7)
Hafnium-177m	72	1000 (37)
Hafnium-178m	72	0.1 (.0037)
Hafnium-179m	72	100 (3.7)
Hafnium-180m	72	100 (3.7)
Hafnium-181	72	10 (.37)
Hafnium-182	72	0.1 (.0037)
Hafnium-182m	72	100 (3.7)
Hafnium-183	72	100 (3.7)
Hafnium-184	72	100 (3.7)
Holmium-155	67	1000 (37)
Holmium-157	67	1000 (37)
Holmium-159	67	1000 (37)
Holmium-161	67	1000 (37)
Holmium-162	67	1000 (37)
Holmium-162m	67	1000 (37)
Holmium-164	67	1000 (37)
Holmium-164m	67	1000 (37)
Holmium-166	67	100 (3.7)
Holmium-166m	67	1 (.037)
Holmium-167	67	100 (3.7)
Hydrogen-3	1	100 (3.7)
Indium-109	49	100 (3.7)
Indium-110 (4.9 hr)	49	10 (.37)
Indium-110 (69.1 min)	49	100 (3.7)
Indium-111	49	100 (3.7)
Indium-112	49	1000 (37)
Indium-113m	49	1000 (37)
Indium-114m	49	10 (.37)
Indium-115	49	0.1 (.0037)
Indium-115m	49	100 (3.7)
Indium-116m	49	100 (3.7)
Indium-117	49	1000 (37)
Indium-117m	49	100 (3.7)
Indium-119m	49	1000 (37)
Iodine-120	53	10 (.37)
Iodine-120m	53	100 (3.7)
Iodine-121	53	100 (3.7)
Iodine-123	53	10 (.37)
Iodine-124	53	0.1 (.0037)
Iodine-125	53	0.01 (.00037)
Iodine-126	53	0.01 (.00037)
Iodine-128	53	1000 (37)
Iodine-129	53	0.001 (.000037)
Iodine-130	53	1 (.037)
Iodine-131	53	0.01 (.00037)
Iodine-132	53	10 (.37)
Iodine-132m	53	10 (.37)
Iodine-133	53	0.1 (.0037)
Iodine-134	53	100 (3.7)
Iodine-135	53	10 (.37)
Iridium-182	77	1000 (37)
Iridium-184	77	100 (3.7)
Iridium-185	77	100 (3.7)
Iridium-186	77	10 (.37)
Iridium-187	77	100 (3.7)
Iridium-188	77	10 (.37)
Iridium-189	77	100 (3.7)
Iridium-190	77	10 (.37)
Iridium-190m	77	1000 (37)
Iridium-192	77	10 (.37)
Iridium-192m	77	100 (3.7)
Iridium-194	77	100 (3.7)
Iridium-194m	77	10 (.37)
Iridium-195	77	1000 (37)
Iridium-195m	77	100 (3.7)
Iron-52	26	100 (3.7)
Iron-55	26	100 (3.7)
Iron-59	26	10 (.37)
Iron-60	26	0.1 (.0037)
Krypton-74	36	10 (.37)
Krypton-76	36	10 (.37)
Krypton-77	36	10 (.37)
Krypton-79	36	100 (3.7)
Krypton-81	36	1000 (37)
Krypton-83m	36	1000 (37)
Krypton-85	36	1000 (37)
Krypton-85m	36	100 (3.7)
Krypton-87	36	10 (.37)
Krypton-88	36	10 (.37)
Lanthanum-131	57	1000 (37)
Lanthanum-132	57	100 (3.7)
Lanthanum-135	57	1000 (37)
Lanthanum-137	57	10 (.37)
Lanthanum-138	57	1 (.037)
Lanthanum-140	57	10 (.37)
Lanthanum-141	57	1000 (37)
Lanthanum-142	57	100 (3.7)
Lanthanum-143	57	1000 (37)
Lead-195m	82	1000 (37)
Lead-198	82	100 (3.7)
Lead-199	82	100 (3.7)
Lead-200	82	100 (3.7)
Lead-201	82	100 (3.7)
Lead-202	82	1 (.037)
Lead-202m	82	10 (.37)
Lead-203	82	100 (3.7)
Lead-205	82	100 (3.7)
Lead-209	82	1000 (37)
Lead-210	82	0.01 (.00037)
Lead-211	82	100 (3.7)
Lead-212	82	10 (.37)
Lead-214	82	100 (3.7)
Lutetium-169	71	10 (.37)
Lutetium-170	71	10 (.37)
Lutetium-171	71	10 (.37)
Lutetium-172	71	10 (.37)
Lutetium-173	71	100 (3.7)
Lutetium-174	71	10 (.37)
Lutetium-174m	71	10 (.37)
Lutetium-176	71	1 (.037)
Lutetium-176m	71	1000 (37)
Lutetium-177	71	100 (3.7)
Lutetium-177m	71	10 (.37)
Lutetium-178	71	1000 (37)
Lutetium-178m	71	1000 (37)
Lutetium-179	71	1000 (37)
Magnesium-28	12	10 (.37)
Manganese-51	25	1000 (37)
Manganese-52	25	10 (.37)
Manganese-52m	25	1000 (37)
Manganese-53	25	1000 (37)
Manganese-54	25	10 (.37)
Manganese-56	25	100 (3.7)
Mendelevium-257	101	100 (3.7)
Mendelevium-258	101	1 (.037)
Mercury-193	80	100 (3.7)
Mercury-193m	80	10 (.37)
Mercury-194	80	0.1 (.0037)
Mercury-195	80	100 (3.7)
Mercury-195m	80	100 (3.7)
Mercury-197	80	1000 (37)
Mercury-197m	80	1000 (37)
Mercury-199m	80	1000 (37)
Mercury-203	80	10 (.37)
Molybdenum-101	42	1000 (37)
Molybdenum-90	42	100 (3.7)
Molybdenum-93	42	100 (3.7)
Molybdenum-93m	42	10 (.37)
Molybdenum-99	42	100 (3.7)
Neodymium-136	60	1000 (37)
Neodymium-138	60	1000 (37)
Neodymium-139	60	1000 (37)
Neodymium-139m	60	100 (3.7)
Neodymium-141	60	1000 (37)
Neodymium-147	60	10 (.37)
Neodymium-149	60	100 (3.7)
Neodymium-151	60	1000 (37)
Neptunium-232	93	1000 (37)
Neptunium-233	93	1000 (37)
Neptunium-234	93	10 (.37)
Neptunium-235	93	1000 (37)
Neptunium-236 (1.2 E 5 yr)	93	0.1 (.0037)
Neptunium-236 (22.5 hr)	93	100 (3.7)
Neptunium-237	93	0.01 (.00037)
Neptunium-238	93	10 (.37)
Neptunium-239	93	100 (3.7)
Neptunium-240	93	100 (3.7)
Nickel-56	28	10 (.37)
Nickel-57	28	10 (.37)
Nickel-59	28	100 (3.7)
Nickel-63	28	100 (3.7)
Nickel-65	28	100 (3.7)
Nickel-66	28	10 (.37)
Niobium-88	41	100 (3.7)
Niobium-89 (122 min)	41	100 (3.7)
Niobium-89 (66 min)	41	100 (3.7)
Niobium-90	41	10 (.37)
Niobium-93m	41	100 (3.7)
Niobium-94	41	10 (.37)
Niobium-95	41	10 (.37)
Niobium-95m	41	100 (3.7)
Niobium-96	41	10 (.37)
Niobium-97	41	100 (3.7)
Niobium-98	41	1000 (37)
Osmium-180	76	1000 (37)
Osmium-181	76	100 (3.7)
Osmium-182	76	100 (3.7)
Osmium-185	76	10 (.37)
Osmium-189m	76	1000 (37)
Osmium-191	76	100 (3.7)
Osmium-191m	76	1000 (37)
Osmium-193	76	100 (3.7)
Osmium-194	76	1 (.037)
Palladium-100	46	100 (3.7)
Palladium-101	46	100 (3.7)
Palladium-103	46	100 (3.7)
Palladium-107	46	100 (3.7)
Palladium-109	46	1000 (37)
Phosphorus-32	15	0.1 (.0037)
Phosphorus-33	15	1 (.037)
Platinum-186	78	100 (3.7)
Platinum-188	78	100 (3.7)
Platinum-189	78	100 (3.7)
Platinum-191	78	100 (3.7)
Platinum-193	78	1000 (37)
Platinum-193m	78	100 (3.7)
Platinum-195m	78	100 (3.7)
Platinum-197	78	1000 (37)
Platinum-197m	78	1000 (37)
Platinum-199	78	1000 (37)
Platinum-200	78	100 (3.7)
Plutonium-234	94	1000 (37)
Plutonium-235	94	1000 (37)
Plutonium-236	94	0.1 (.0037)
Plutonium-237	94	1000 (37)
Plutonium-238	94	0.01 (.00037)
Plutonium-239	94	0.01 (.00037)
Plutonium-240	94	0.01 (.00037)
Plutonium-241	94	1 (.037)
Plutonium-242	94	0.01 (.00037)

TABLE 2.—RADIONUCLIDES—Continued

(1)—Radionuclide	(2)—Atomic number	(3)—Reportable quantity (RQ) Ci (TBq)
Plutonium-243	94	1000 (37)
Plutonium-244	94	0.01 (.00037)
Plutonium-245	94	100 (3.7)
Polonium-203	84	100 (3.7)
Polonium-205	84	100 (3.7)
Polonium-207	84	10 (.37)
Polonium-210	84	0.01 (.00037)
Potassium-40	19	1 (.037)
Potassium-42	19	100 (3.7)
Potassium-43	19	10 (.37)
Potassium-44	19	100 (3.7)
Potassium-45	19	1000 (37)
Praseodymium-136	59	1000 (37)
Praseodymium-137	59	1000 (37)
Praseodymium-138m	59	100 (3.7)
Praseodymium-139	59	1000 (37)
Praseodymium-142	59	100 (3.7)
Praseodymium-142m	59	1000 (37)
Praseodymium-143	59	10 (.37)
Praseodymium-144	59	1000 (37)
Praseodymium-145	59	1000 (37)
Praseodymium-147	59	1000 (37)
Promethium-141	61	1000 (37)
Promethium-143	61	100 (3.7)
Promethium-144	61	10 (.37)
Promethium-145	61	100 (3.7)
Promethium-146	61	10 (.37)
Promethium-147	61	10 (.37)
Promethium-148	61	10 (.37)
Promethium-148m	61	10 (.37)
Promethium-149	61	100 (3.7)
Promethium-150	61	100 (3.7)
Promethium-151	61	100 (3.7)
Protactinium-227	91	100 (3.7)
Protactinium-228	91	10 (.37)
Protactinium-230	91	10 (.37)
Protactinium-231	91	0.01 (.00037)
Protactinium-232	91	10 (.37)
Protactinium-233	91	100 (3.7)
Protactinium-234	91	10 (.37)
RADIONUCLIDES § †		1 (.037)
Radium-223	88	1 (.037)
Radium-224	88	10 (.37)
Radium-225	88	1 (.037)
Radium-226 **	88	0.1 (.0037)
Radium-227	88	1000 (37)
Radium-228	88	0.1 (.0037)
Radon-220	86	0.1 (.0037)
Radon-222	86	0.1 (.0037)
Rhenium-177	75	1000 (37)
Rhenium-178	75	1000 (37)
Rhenium-181	75	100 (3.7)
Rhenium-182 (12.7 hr)	75	10 (.37)
Rhenium-182 (64.0 hr)	75	10 (.37)
Rhenium-184	75	10 (.37)
Rhenium-184m	75	10 (.37)
Rhenium-186	75	100 (3.7)
Rhenium-186m	75	10 (.37)
Rhenium-187	75	1000 (37)
Rhenium-188	75	1000 (37)
Rhenium-188m	75	1000 (37)
Rhenium-189	75	1000 (37)
Rhodium-100	45	10 (.37)
Rhodium-101	45	10 (.37)
Rhodium-101m	45	100 (3.7)
Rhodium-102	45	10 (.37)
Rhodium-102m	45	10 (.37)
Rhodium-103m	45	1000 (37)
Rhodium-105	45	100 (3.7)
Rhodium-106m	45	10 (.37)
Rhodium-107	45	1000 (37)
Rhodium-99	45	10 (.37)
Rhodium-99m	45	100 (3.7)
Rubidium-79	37	1000 (37)

TABLE 2.—RADIONUCLIDES—Continued

(1)—Radionuclide	(2)—Atomic number	(3)—Reportable quantity (RQ) Ci (TBq)
Rubidium-81	37	100 (3.7)
Rubidium-81m	37	1000 (37)
Rubidium-82m	37	10 (.37)
Rubidium-83	37	10 (.37)
Rubidium-84	37	10 (.37)
Rubidium-86	37	10 (.37)
Rubidium-87	37	10 (.37)
Rubidium-88	37	1000 (37)
Rubidium-89	37	1000 (37)
Ruthenium-103	44	10 (.37)
Ruthenium-105	44	100 (3.7)
Ruthenium-106	44	1 (.037)
Ruthenium-94	44	1000 (37)
Ruthenium-97	44	100 (3.7)
Samarium-141	62	1000 (37)
Samarium-141m	62	1000 (37)
Samarium-142	62	1000 (37)
Samarium-145	62	100 (3.7)
Samarium-146	62	0.01 (.00037)
Samarium-147	62	0.01 (.00037)
Samarium-151	62	10 (.37)
Samarium-153	62	100 (3.7)
Samarium-155	62	1000 (37)
Samarium-156	62	100 (3.7)
Scandium-43	21	1000 (37)
Scandium-44	21	100 (3.7)
Scandium-44m	21	10 (.37)
Scandium-46	21	10 (.37)
Scandium-47	21	100 (3.7)
Scandium-48	21	10 (.37)
Scandium-49	21	1000 (37)
Selenium-70	34	1000 (37)
Selenium-73	34	10 (.37)
Selenium-73m	34	100 (3.7)
Selenium-75	34	10 (.37)
Selenium-79	34	10 (.37)
Selenium-81	34	1000 (37)
Selenium-81m	34	1000 (37)
Selenium-83	34	1000 (37)
Silicon-31	14	1000 (37)
Silicon-32	14	1 (.037)
Silver-102	47	100 (3.7)
Silver-103	47	1000 (37)
Silver-104	47	1000 (37)
Silver-104m	47	1000 (37)
Silver-105	47	10 (.37)
Silver-106	47	1000 (37)
Silver-106m	47	10 (.37)
Silver-108m	47	10 (.37)
Silver-110m	47	10 (.37)
Silver-111	47	10 (.37)
Silver-112	47	100 (3.7)
Silver-115	47	1000 (37)
Sodium-22	11	10 (.37)
Sodium-24	11	10 (.37)
Strontium-80	38	100 (3.7)
Strontium-81	38	1000 (37)
Strontium-83	38	100 (3.7)
Strontium-85	38	10 (.37)
Strontium-85m	38	1000 (37)
Strontium-87m	38	100 (3.7)
Strontium-89	38	10 (.37)
Strontium-90	38	0.1 (.0037)
Strontium-91	38	10 (.37)
Strontium-92	38	100 (3.7)
Sulfur-35	16	1 (.037)
Tantalum-172	73	100 (3.7)
Tantalum-173	73	100 (3.7)
Tantalum-174	73	100 (3.7)
Tantalum-175	73	100 (3.7)
Tantalum-176	73	10 (.37)
Tantalum-177	73	1000 (37)
Tantalum-178	73	1000 (37)
Tantalum-179	73	1000 (37)
Tantalum-180	73	100 (3.7)
Tantalum-180m	73	1000 (37)
Tantalum-182	73	10 (.37)

TABLE 2.—RADIONUCLIDES—Continued

(1)—Radionuclide	(2)—Atomic number	(3)—Reportable quantity (RQ) Ci (TBq)
Tantalum-182m	73	1000 (37)
Tantalum-183	73	100 (3.7)
Tantalum-184	73	10 (.37)
Tantalum-185	73	1000 (37)
Tantalum-186	73	1000 (37)
Technetium-101	43	1000 (37)
Technetium-104	43	1000 (37)
Technetium-93	43	100 (3.7)
Technetium-93m	43	1000 (37)
Technetium-94	43	10 (.37)
Technetium-94m	43	100 (3.7)
Technetium-96	43	10 (.37)
Technetium-96m	43	1000 (37)
Technetium-97	43	100 (3.7)
Technetium-97m	43	100 (3.7)
Technetium-98	43	10 (.37)
Technetium-99	43	10 (.37)
Technetium-99m	43	100 (3.7)
Tellurium-116	52	1000 (37)
Tellurium-121	52	10 (.37)
Tellurium-121m	52	10 (.37)
Tellurium-123	52	10 (.37)
Tellurium-123m	52	10 (.37)
Tellurium-125m	52	10 (.37)
Tellurium-127	52	1000 (37)
Tellurium-127m	52	10 (.37)
Tellurium-129	52	1000 (37)
Tellurium-129m	52	10 (.37)
Tellurium-131	52	1000 (37)
Tellurium-131m	52	10 (.37)
Tellurium-132	52	10 (.37)
Tellurium-133	52	1000 (37)
Tellurium-133m	52	1000 (37)
Tellurium-134	52	1000 (37)
Terbium-147	65	100 (3.7)
Terbium-149	65	100 (3.7)
Terbium-150	65	100 (3.7)
Terbium-151	65	10 (.37)
Terbium-153	65	100 (3.7)
Terbium-154	65	10 (.37)
Terbium-155	65	100 (3.7)
Terbium-156	65	10 (.37)
Terbium-156m (24.4 hr)	65	1000 (37)
Terbium-156m (5.0 hr)	65	1000 (37)
Terbium-157	65	100 (3.7)
Terbium-158	65	10 (.37)
Terbium-160	65	10 (.37)
Terbium-161	65	100 (3.7)
Thallium-194	81	1000 (37)
Thallium-194m	81	100 (3.7)
Thallium-195	81	100 (3.7)
Thallium-197	81	100 (3.7)
Thallium-198	81	10 (.37)
Thallium-198m	81	100 (3.7)
Thallium-199	81	100 (3.7)
Thallium-200	81	10 (.37)
Thallium-201	81	1000 (37)
Thallium-202	81	10 (.37)
Thallium-204	81	10 (.37)
Thorium (Irradiated)	90	***
Thorium (Natural)	90	**
Thorium-226	90	100 (3.7)
Thorium-227	90	1 (.037)
Thorium-228	90	0.01 (.00037)
Thorium-229	90	0.001 (.000037)
Thorium-230	90	0.01 (.00037)
Thorium-231	90	100 (3.7)
Thorium-232 **	90	0.001 (.000037)
Thorium-234	90	100 (3.7)
Thulium-162	69	1000 (37)
Thulium-166	69	10 (.37)
Thulium-167	69	100 (3.7)
Thulium-170	69	10 (.37)
Thulium-171	69	100 (3.7)
Thulium-172	69	100 (3.7)

TABLE 2.—RADIONUCLIDES—Continued

(1)—Radionuclide	(2)—Atomic number	(3)—Reportable quantity (RQ) Ci (TBq)
Thulium-173	69	100 (3.7)
Thulium-175	69	1000 (37)
Tin-110	50	100 (3.7)
Tin-111	50	1000 (37)
Tin-113	50	10 (.37)
Tin-117m	50	100 (3.7)
Tin-119m	50	10 (.37)
Tin-121	50	1000 (37)
Tin-121m	50	10 (.37)
Tin-123	50	10 (.37)
Tin-123m	50	1000 (37)
Tin-125	50	10 (.37)
Tin-126	50	1 (.037)
Tin-127	50	100 (3.7)
Tin-128	50	1000 (37)
Titanium-44	22	1 (.037)
Titanium-45	22	1000 (37)
Tungsten-176	74	1000 (37)
Tungsten-177	74	100 (3.7)
Tungsten-178	74	100 (3.7)
Tungsten-179	74	1000 (37)
Tungsten-181	74	100 (3.7)
Tungsten-185	74	10 (.37)
Tungsten-187	74	100 (3.7)
Tungsten-188	74	10 (.37)
Uranium (Depleted)	92	***
Uranium (Irradiated)	92	***
Uranium (Natural)	92	**
Uranium Enriched 20% or greater	92	***
Uranium Enriched less than 20%	92	***
Uranium-230	92	1 (.037)
Uranium-231	92	1000 (37)
Uranium-232	92	0.01 (.00037)
Uranium-233	92	0.1 (.0037)
Uranium-234 **	92	0.1 (.0037)
Uranium-235 **	92	0.1 (.0037)
Uranium-236	92	0.1 (.0037)
Uranium-237	92	100 (3.7)
Uranium-238 **	92	0.1 (.0037)
Uranium-239	92	1000 (37)
Uranium-240	92	1000 (37)
Vanadium-47	23	1000 (37)
Vanadium-48	23	10 (.37)
Vanadium-49	23	1000 (37)
Xenon-120	54	100 (3.7)
Xenon-121	54	10 (.37)
Xenon-122	54	100 (3.7)
Xenon-123	54	10 (.37)
Xenon-125	54	100 (3.7)
Xenon-127	54	100 (3.7)
Xenon-129m	54	1000 (37)
Xenon-131m	54	1000 (37)
Xenon-133	54	1000 (37)
Xenon-133m	54	1000 (37)
Xenon-135	54	100 (3.7)
Xenon-135m	54	10 (.37)
Xenon-138	54	10 (.37)
Ytterbium-162	70	1000 (37)
Ytterbium-166	70	10 (.37)
Ytterbium-167	70	1000 (37)
Ytterbium-169	70	10 (.37)
Ytterbium-175	70	100 (3.7)
Ytterbium-177	70	1000 (37)
Ytterbium-178	70	1000 (37)
Yttrium-86	39	10 (.37)
Yttrium-86m	39	1000 (37)
Yttrium-87	39	10 (.37)
Yttrium-88	39	10 (.37)
Yttrium-90	39	10 (.37)
Yttrium-90m	39	100 (3.7)
Yttrium-91	39	10 (.37)
Yttrium-91m	39	1000 (37)
Yttrium-92	39	100 (3.7)
Yttrium-93	39	100 (3.7)
Yttrium-94	39	1000 (37)
Yttrium-95	39	1000 (37)
Zinc-62	30	100 (3.7)
Zinc-63	30	1000 (37)
Zinc-65	30	10 (.37)
Zinc-69	30	1000 (37)
Zinc-69m	30	100 (3.7)
Zinc-71m	30	100 (3.7)
Zinc-72	30	100 (3.7)
Zirconium-86	40	100 (3.7)
Zirconium-88	40	10 (.37)
Zirconium-89	40	100 (3.7)
Zirconium-93	40	1 (.037)
Zirconium-95	40	10 (.37)
Zirconium-97	40	10 (.37)

§ The RQs for all radionuclides apply to chemical compounds containing the radionuclides and elemental forms regardless of the diameter of pieces of solid material.

† The RQ of one curie applies to all radionuclides not otherwise listed. Whenever the RQs in TABLE 1—HAZARDOUS SUBSTANCES OTHER THAN RADIONUCLIDES and this table conflict, the lowest RQ shall apply. For example, uranyl acetate and uranyl nitrate have RQs shown in TABLE 1 of 100 pounds, equivalent to about one-tenth the RQ level for uranium-238 in this table.

** The method to determine the RQs for mixtures or solutions of radionuclides can be found in paragraph 6 of the note preceding TABLE 1 of this Appendix. RQs for the following four common radionuclide mixtures are provided: radium-226 in secular equilibrium with its' daughters (0.053 curie); natural uranium (0.1 curie); natural uranium in secular equilibrium with its' daughters (0.052 curie); and natural thorium in secular equilibrium with its' daughters (0.011 curie).

*** Indicates that the name was added by RSPA because it appears in the list of radionuclides in 49 CFR 173.435. The reportable quantity (RQ), if not specifically listed elsewhere in this Appendix, shall be determined in accordance with the procedures in Paragraph 6 of this Appendix.

Optional Hazardous Materials Table

International shipments of hazardous materials which move at any point by water are subject to the requirements of the International Maritime Organization (IMO) *(49 CFR, 172.102)*. The full text of these regulations is in the International Maritime Dangerous Goods (IMDG) Code.

To facilitate the movement of international shipments within the United States, the Department of Transportation has developed the Optional Hazardous Materials Table (Section 172.102 of the regulations) to relate the IMO descriptions in the U. S. Regulations. Use of the Optional Table will, in many cases, provide relief from dual markings and dual descriptions for international shipments of hazardous materials being transported within the United States.

The "Optional Table" is helpful to the motor carrier in verifying that an international shipment is properly described and classified for transportation within this country. The following cautions must be observed in using the "Optional Table."

1. Materials are not listed unless they are subject to the IMDG Code. Therefore, a material that is defined as a hazardous material by DOT, but which is not subject to IMDG will not be shown.

2. Materials that are subject to the IMDG Code are listed in the Optional Table even though they are not subject to DOT regulations.

3. IMDG class descriptions are by number and may, or may not, bear the optional notation "IMDG Class" preceding or following the class number.

4. When a material is transported by a motor vehicle and the IMDG class or division is used in the description, the shipping paper must also include the hazard class name from Part 173 that most closely corresponds to the IMDG class assigned to that material; e.g. "Ethylene oxide, IMDG class 2.1, Flammable Gas, UN1040." *NOTE: The IMDG classes and the corresponding DOT hazard classes are provided on page 104.*

5. When the IMDG description is used and the material is in a different hazard class under IMDG regulations, the appropriate IMDG class designation must be used; e.g. ethylene oxide is a Flammable Liquid under DOT regulations and is classified as a Flammable Gas under the IMDG Code. A shipment being made under the IMDG must show the IMDG class following the proper shipping name.

6. If a hazardous substance is transported under an acceptable shipping name from the Optional Table, 172.102, that does not contain the name of the hazardous substance, the additional description of the hazardous substance from the Appendix to the Hazardous Materials Table (172.101) must be shown on shipping papers and packages in parentheses in association with the proper shipping name. For waste streams, the EPA identification number, or the notation, "EPA" followed by the word "ignitability," "corrosivity," "reactivity," or "EP toxicity," as appropriate, must be shown in a similar manner. *(49 CFR 172.102)*

7. A material described, classed, and labeled as authorized by 172.102 must be in conformance with all additional requirements of the IMDG code.

The columns of the Optional Hazardous Materials Table (172.102) provide information as follows:

Column 1—If the letter "N" appears in this column, the IMO description cannot be used. In this case, the description given in the Hazardous Materials Table, 172.101, must be used.

Column 2—Gives the hazardous materials description and proper shipping name of the material as shown in the IMO Code. Only those names not in italics may be used.

Column 3—Shows the IMO class (see explanation of the relationship between IMO class and DOT class on the following page).

Column 4—Shows the identification number of the material. UN numbers only will be found in this column of the Optional Table.

Column 5—Shows the required label(s) for packagings. IMDG rules provide for two labels not required by DOT rules.

a) **St. Andrews Cross** - a special label required on packaging of material which cannot be loaded next to foodstuffs.

b) **Spontaneously Combustible** - the presence of this label has been previously authorized for international shipments when required by the destination country, and it will be found on the specified material for overseas shipments by water.

The following tables appear in the IMDG Code and in 172.102, however they are not included in this manual because of their limited use by motor carriers:

Column 6—Shows the package group specified for the material in the IMDG Code.

Column 7— Shows loading and storage restrictions for transportation by water.

Comparison Of IMO And DOT Hazardous Materials Classes

IMO CLASS, DIVISION AND GENERAL DESCRIPTION	DOT HAZARDOUS MATERIAL CLASS
Class 1 - Explosives	
Division 1.1 - mass explosion hazard	Explosive A
Division 1.2 - projection hazard but no mass explosion hazard	Explosive A or B
Division 1.3 - fire hazard with minor projection or blast hazard but no mass explosion hazard	Explosive B
Division 1.4 - no significant hazard	Explosive C
Division 1.5 - very insensitive substance	Blasting Agent
Class 2 - Gases (Compressed, liquified or dissolved)	
Division 2.1 - Flammable gases	Flammable Gas
Division 2.2 - Nonflammable gases	Nonflammable Gas
Division 2.3 - Poison gases	Poison A
Class 3 - Flammable liquids	
Division 3.1 - Low flash point group (liquids with flashpoint below 0° F.).	Flammable Liquid
Division 3.2 - Intermediate flash point group (liquids with flash points 0° F. or above but less than 73° F.).	Flammable Liquid
Division 3.3 - Liquids high flash point group (liquids with flash points of 73° F. or above up to and including 141° F.).	Flammable Liquid or Combustible Liquid
Class 4 - Flammable solids or substances	
Division 4.1 - Flammable solids	Flammable Solid
Division 4.2 - Substances liable to spontaneous combustion	Flammable Solid (solid Flammable Liquid (pyroforic liquid)
Division 4.3 - Substances which emit flammable gases when wet	Flammable Solid (with Dangerous When Wet label)
Class 5 - Oxidizing Substances	
Division 5.1 - Oxidizing substances or agents	Oxidizer
Division 5.2 - Organic peroxides	Organic Peroxide
Class 6 - Poisonous or infectious substances	
Division 6.1 - Poisonous substances	Poison B
Division 6.2 - Infectious substances	Etiologic Agent
Class 7 - Radioactive materials	Radioactive Material
Class 8 - Corrosives	Corrosive Materials
Class 9 - Miscellaneous dangerous substances	Other Regulated Materials (ORM)

NOTE: *A letter following one of the above designations refers to loading and storage restrictions on shipboard. Observe DOT loading and storage restrictions for highway transportation within the United States.*

§172.102 OPTIONAL HAZARDOUS MATERIALS TABLE

(1) Notes and Symbols	(2) Hazardous Materials and Description and Proper Shipping Names	(3) IMO Class	(4) Identification Number	(5) Label(s) required
	ACETAL	3.1	UN1088	FLAMMABLE LIQUID
	ACETALDEHYDE	3.1	UN1089	FLAMMABLE LIQUID
	ACETALDEHYDE AMMONIA	9	UN1841	NONE
	ACETALDEHYDE OXIME	3.3	UN2332	FLAMMABLE LIQUID
	ACETIC ACID, GLACIAL *OR* ACETIC ACID SOLUTION, *MORE THAN 80% ACID, BY WEIGHT*	8	UN2789	CORROSIVE, FLAMMABLE LIQUID
	ACETIC ACID SOLUTION, *MORE THAN 25% BUT NOT MORE THAN 80% ACID, BY WEIGHT*	8	UN2790	CORROSIVE
	ACETIC ANHYDRIDE	8	UN1715	CORROSIVE, FLAMMABLE LIQUID
	ACETONE	3.1	UN1090	FLAMMABLE LIQUID
	ACETONE CYANOHYDRIN, *STABLILIZED*	6.1	UN1541	POISON
	ACETONE OILS	3.2	UN1091	FLAMMABLE LIQUID
	ACETONITRILE. *SEE* METHYL CYANIDE			
	ACETYL ACETONE PEROXIDE, *MAXIMUM CONCENTRATION 40% IN SOLUTION*	5.2	UN2080	ORGANIC PEROXIDE
	ACETYL BENZOYL PEROXIDE; *MAXIMUM CONCENTRATION 45% IN SOLUTION*	5.2	UN2081	ORGANIC PEROXIDE
	ACETYL BROMIDE	8	UN1716	CORROSIVE
	ACETYL CHLORIDE	3.2	UN1717	FLAMMABLE LIQUID, CORROSIVE
	ACETYL CYCLOHEXANE SULPHONYL PEROXIDE, *MAXIMUM CONCENTRATION 82%, WETTED WITH MINIMUM 12% WATER*	5.2	UN2082	ORGANIC PEROXIDE
	ACETYL CYCLOHEXANE SULPHONYL PEROXIDE, *MAXIMUM CONCENTRATION 32% IN SOLUTION*	5.2	UN2083	ORGANIC PEROXIDE
	ACETYLENE, *DISSOLVED*	2.1	UN1001	FLAMMABLE GAS
	ACETYL IODIDE	8	UN1898	CORROSIVE
	ACETYL METHYL CARBINOL	3.3	UN2621	FLAMMABLE LIQUID
	ACETYL PEROXIDE, *MAXIMUM CONCENTRATION 27% IN SOLUTION*	5.2	UN2084	ORGANIC PEROXIDE
	ACID MIXTURES, SPENT, NITRATING	8	UN1826	CORROSIVE
	ACIDS, LIQUID, N.O.S. *SEE* CORROSIVE LIQUIDS, N.O.S.			
	ACRIDINE	4.1	UN2713	FLAMMABLE SOLID
	ACROLEIN DIMER, *STABILIZED*	3.3	UN2607	FLAMMABLE LIQUID
	ACROLEIN, INHIBITED	3.1	UN1092	FLAMMABLE LIQUID, POISON
	ACRYLAMIDE	6.1	UN2074	ST. ANDREWS CROSS
	ACRYLIC ACID, INHIBITED	8	UN2218	CORROSIVE, FLAMMABLE LIQUID
	ACRYLONITRILE, INHIBITED	3.2	UN1093	FLAMMABLE LIQUID, POISON
	ACTIVATED CARBON. *SEE* CARBON, ACTIVATED			
	ACTIVATED CHARCOAL. *SEE* CARBON, ACTIVATED			
	ADHESIVES, *CONTAINING A FLAMMABLE LIQUID*	3.1 3.2 3.3	UN1133 UN1133 UN1133	FLAMMABLE LIQUID FLAMMABLE LIQUID FLAMMABLE LIQUID
	ADIPONITRILE	6.1	UN2205	ST. ANDREWS CROSS
	AEROSOL DISPENSERS, *WITH A CAPACITY BELOW 1400 CUBIC CM.* :			
	(1) MORE THAN 10% BY WEIGHT OF TOTAL CONTENTS CONSISTING OF FLAMMABLE GAS	2.1	UN1950	FLAMMABLE GAS
	(2) INTERNAL PRESSURE GREATER THAN 160 PSIG AT 130 DEG F.	2.2	UN1950	NONFLAMMABLE GAS
	(3) MORE THAN 45% BY WEIGHT OF TOTAL CONTENTS CONSISTING OF FLAMMABLE LIQUID. THIS LIMIT IS REDUCED TO 35% IF THERE IS ANY FLAMMABLE GAS PRESENT.	3.1 3.2 3.3	UN1950 UN1950 UN1950	FLAMMABLE LIQUID FLAMMABLE LIQUID FLAMMABLE LIQUID
	(4) MORE THAN 10% BY WEIGHT OF TOXIC SUBSTANCES IN THE LIQUID CONCENTRATE	6.1 6.1	UN1950	POISON ST. ANDREWS CROSS
	(5) MORE THAN 5% BY WEIGHT OF CORROSIVE SUBSTANCES IN THE LIQUID CONCENTRATE	8	UN1950	CORROSIVE
	(6) AS SPECIFIED UNDER GROUP 2 ON PAGE 9011 OF IMCO CODE	9	UN1950	NONE
N	AEROSOL DISPENSERS, *WITH A CAPACITY OF 1400 CUBIC CM. OR MORE*	2	UN1950	
	AGENTS, BLASTING, TYPE B. *SEE* EXPLOSIVES, BLASTING, TYPE B			
	AGENTS, BLASTING, TYPE E. *SEE* EXPLOSIVES, BLASTING, TYPE E			
	AIR, *COMPRESSED*	2.2	UN1002	NONFLAMMABLE GAS
	AIRCRAFT THRUST DEVICE *FOR ASSISTED TAKE-OFF*	4.1	UN2791	FLAMMABLE SOLID
	AIR, *REFRIGERATED LIQUID*	2.2	UN1003	NONFLAMMABLE GAS, OXIDIZER
N	ALARM DEVICES, EXPLOSIVE *AUTOMATIC*	1.4 S	UN0001	NONE. PACKAGE TO BE MARKED '1.4 S'
	ALCOHOLS, N.O.S.	3.1 3.2 3.3	UN1987 UN1987 UN1987	FLAMMABLE LIQUID FLAMMABLE LIQUID FLAMMABLE LIQUID

§172.102 OPTIONAL HAZARDOUS MATERIALS TABLE

(1) Notes and Symbols	(2) Hazardous Materials and Description and Proper Shipping Names	(3) IMO Class	(4) Identification Number	(5) Label(s) required
	ALCOHOLS, TOXIC, N.O.S.	3.1	UN1986	FLAMMABLE LIQUID, POISON
		3.2	UN1986	FLAMMABLE LIQUID, POISON
		3.3	UN1986	FLAMMABLE LIQUID, POISON
	ALDEHYDES, N.O.S.	3.1	UN1989	FLAMMABLE LIQUID
		3.2	UN1989	FLAMMABLE LIQUID
		3.3	UN1989	FLAMMABLE LIQUID
	ALDEHYDES, TOXIC, N.O.S.	3.1	UN1988	FLAMMABLE LIQUID, POISON
		3.2	UN1988	FLAMMABLE LIQUID, POISON
		3.3	UN1988	FLAMMABLE LIQUID, POISON
	ALDOL	6.1	UN2839	POISON
	ALKALI METAL ALLOYS, LIQUID	4.3	UN1421	DANGEROUS WHEN WET
	ALKALI METAL AMALGAMS, N.O.S.	4.3	UN1389	DANGEROUS WHEN WET
	ALKALI METAL AMIDES, N.O.S.	4.3	UN1390	DANGEROUS WHEN WET
	ALKALI METAL DISPERSIONS, N.O.S. *OR* ALKALI EARTH METAL DISPERSIONS, N.O.S.	4.3	UN1391	DANGEROUS WHEN WET
	ALKALINE CAUSTIC LIQUIDS, N.O.S. *SEE* CAUSTIC ALKALI LIQUIDS, , N.O.S.			
	ALKALINE CORROSIVE LIQUIDS, N.O.S. *SEE* CORROSIVE LIQUIDS, N.O.S.			
	ALKALINE EARTH METAL ALLOYS, N.O.S.	4.3	UN1393	DANGEROUS WHEN WET
	ALKALINE EARTH METAL AMALGAMS, N.O.S.	4.3	UN1392	DANGEROUS WHEN WET
	ALKALOIDS, N.O.S. *OR* ALKALOID SALTS, N.O.S., *POISONOUS*	6.1	UN1544	POISON
		6.1	UN1544	ST. ANDREWS CROSS
	ALKYLAMINES AND POLYAMINES, *FLASHPOINT BELOW 23 DEG C AND BOILING POINT ABOVE 35 DEG C BUT NOT MORE THAN 200 DEG C*, N.O.S.	3.2	UN2733	FLAMMABLE LIQUID, CORROSIVE
	ALKYLAMINES AND POLYAMINES, *FLASHPOINT 23 DEG C OR ABOVE, BOILING POINT ABOVE 35 DEG C BUT NOT MORE THAN 200 DEG C*, N.O.S.	8	UN2734	CORROSIVE, FLAMMABLE LIQUID (ONLY IF FLASHPOINT 61 DEG C OR BELOW)
	ALKYLAMINES AND POLYAMINES, *FLASHPOINT OF 23 DEG C OR ABOVE AND BOILING POINT ABOVE 200 DEG C*, N.O.S.	8	UN2735	CORROSIVE, FLAMMABLE LIQUID (ONLY IF FLASHPOINT 61 DEG C OR BELOW)
	ALKYL, ARYL *OR* TOLUENE SULPHONIC ACID, LIQUID, *CONTAINING MORE THAN 5% FREE SULPHURIC ACID*	8	UN2584	CORROSIVE
	ALKYL, ARYL *OR* TOLUENE SULPHONIC ACID, LIQUID, *CONTAINING NOT MORE THAN 5% FREE SULPHURIC ACID*	8	UN2586	CORROSIVE
	ALKYL, ARYL *OR* TOLUENE SULPHONIC ACID, SOLID, *CONTAINING MORE THAN 5% FREE SULPHURIC ACID*	8	UN2583	CORROSIVE
	ALKYL, ARYL *OR* TOLUENE SULPHONIC ACID, SOLID, *CONTAINING NOT MORE THAN 5% FREE SULPHURIC ACID*	8	UN2585	CORROSIVE
	ALKYL PHENOLS, N.O.S.	6.1	UN2430	ST. ANDREWS CROSS
	ALLYL ACETATE	3.2	UN2333	FLAMMABLE LIQUID, POISON
	ALLYL ALCOHOL	3.2	UN1098	FLAMMABLE LIQUID, POISON
	ALLYLAMINE	3.1	UN2334	FLAMMABLE LIQUID, POISON
	ALLYL BROMIDE	3.2	UN1099	FLAMMABLE LIQUID, POISON
	ALLYL CHLORIDE	3.1	UN1100	FLAMMABLE LIQUID, POISON
	ALLYL CHLOROCARBONATE. *SEE* ALLYL CHLOROFORMATE			
	ALLYL CHLOROFORMATE	8	UN1722	CORROSIVE, FLAMMABLE LIQUID
	ALLYL ETHYL ETHER	3.2	UN2335	FLAMMABLE LIQUID, POISON
	ALLYL FORMATE	3.2	UN2336	FLAMMABLE LIQUID, POISON
	ALLYL GLYCIDYL ETHER	3.3	UN2219	FLAMMABLE LIQUID
	ALLYL IODIDE	3.2	UN1723	FLAMMABLE LIQUID, CORROSIVE
	ALLYL ISOTHIOCYANATE, *STABILIZED*	6.1	UN1545	POISON, FLAMMABLE LIQUID
	ALLYL TRICHLOROSILANE, *STABILIZED*	8	UN1724	CORROSIVE, FLAMMABLE LIQUID
	ALUMINIUM ALKYL HALIDES, *IN* SOLUTION	4.2	UN2220	SPONTANEOUSLY COMBUSTIBLE
	ALUMINIUM ALKYL HALIDES, *PURE*	4.2	UN2221	SPONTANEOUSLY COMBUSTIBLE
	ALUMINIUM ALKYLS. *SEE* METAL ALKYLS, N.O.S.			
	ALUMINIUM BOROHYDRIDE *OR* ALUMINIUM BOROHYDRIDE IN DEVICES	4.2	UN2870	SPONTANEOUSLY COMBUSTIBLE DANGEROUS WHEN WET
	ALUMINIUM BROMIDE, ANHYDROUS	8	UN1725	CORROSIVE
	ALUMINIUM BROMIDE SOLUTION	8	UN2580	CORROSIVE
	ALUMINIUM CARBIDE	4.3	UN1394	DANGEROUS WHEN WET
	ALUMINIUM CHLORIDE, ANHYDROUS	8	UN1726	CORROSIVE
	ALUMINIUM CHLORIDE SOLUTION	8	UN2581	CORROSIVE
	ALUMINIUM FERROSILICON, *POWDER*	4.3	UN1395	DANGEROUS WHEN WET
	ALUMINIUM HYDRIDE	4.2	UN2463	SPONTANEOUSLY COMBUSTIBLE

§172.102 OPTIONAL HAZARDOUS MATERIALS TABLE

(1) Notes and Symbols	(2) Hazardous Materials and Description and Proper Shipping Names	(3) IMO Class	(4) Identi-fication Number	(5) Label(s) required
	ALUMINIUM NITRATE	5.1	UN1438	OXIDIZER
	ALUMINIUM PHOSPHIDE	4.3	UN1397	DANGEROUS WHEN WET, POISON
	ALUMINIUM POWDER, COATED	4.1	UN1309	FLAMMABLE SOLID
	ALUMINIUM, POWDER, PYROPHORIC. *SEE* PYROPHORIC METALS			
	ALUMINIUM, POWDER, UNCOATED, *NON-PYROPHORIC*	4.3	UN1396	DANGEROUS WHEN WET
	ALUMINIUM RESINATE	4.1	UN2715	FLAMMABLE SOLID
	ALUMINIUM SILICON, *POWDER, UNCOATED*	4.3	UN1398	DANGEROUS WHEN WET
	2-AMINO-4-CHLOROPHENOL	6.1	UN2673	POISON
	2-AMINO-5-DIETHYLAMINOPENTANE	6.1	UN2946	ST. ANDREWS CROSS
	N-AMINOETHYLPIPERAZINE	8	UN2815	CORROSIVE
	AMINOPHENOLS (*O-, M-, P-*)	6.1	UN2512	ST. ANDREWS CROSS
	AMINOPYRIDINES (*O-, M-, P-*)	6.1	UN2671	POISON
	AMMONIA, *ANHYDROUS, LIQUEFIED, OR AMMONIA SOLUTIONS, DENSITY (SPECIFIC GRAVITY) LESS THAN 0.880 AT 15 DEGREES C, IN WATER, CONTAINING MORE THAN 50% AMMONIA.*	2.3	UN1005	POISON GAS
	AMMONIA SOLUTIONS *DENSITY (SPECIFIC GRAVITY) BETWEEN 0.880 AND 0.957 AT 15 DEG C, IN WATER, WITH MORE THAN 10% AND NOT MORE THAN 35% AMMONIA , BY WEIGHT*	8	UN2672	CORROSIVE
	AMMONIA SOLUTIONS, *DENSITY (SPECIFIC GRAVITY) LESS THAN 0.880 AT 15 DEGREES C, IN WATER, CONTAINING MORE THAN 35% AND NOT MORE THAN 50% AMMONIA.*	2.2	UN2073	NONFLAMMABLE GAS
	AMMONIUM ARSENATE	6.1	UN1546	POISON
	AMMONIUM BIFLUORIDE. *SEE* AMMONIUM HYDROGEN FLUORIDE			
	AMMONIUM DICHROMATE	5.1	UN1439	OXIDIZER
	AMMONIUM DINITRO-O-CRESOLATE	6.1	UN1843	POISON
	AMMONIUM FLUORIDE	6.1	UN2505	ST. ANDREWS CROSS
	AMMONIUM HYDROGEN FLUORIDE, SOLID	8	UN1727	CORROSIVE
	AMMONIUM HYDROGEN FLUORIDE, SOLUTION	8	UN2817	CORROSIVE, POISON
	AMMONIUM HYDROGEN SULPHATE	8	UN2506	CORROSIVE
	AMMONIUM METAVANADATE	6.1	UN2859	POISON
	AMMONIUM NITRATE FERTILIZERS, N.O.S.	5.1	UN2072	OXIDIZER
	AMMONIUM NITRATE FERTILIZERS, *OF THE SAME COMPOSITION AS DEFINED IN CLASS 5.1 ON PAGES 5015 AND 5016 OF THE IMCO CODE BUT CONTAINING GREATER AMOUNTS OF ORGANIC AND/OR COMBUSTIBLE MATERIAL THAN SPECIFIED IN THESE ENTRIES*	1.1 D	UN0223	EXPLOSIVE (1.1D)
	AMMONIUM NITRATE FERTILIZERS, *TYPE A*			
	(1) UNIFORM NON- SEGREGATING MIXTURES OF AMMONIUM NITRATE WITH ADDED MATTER WHICH IS INORGANIC AND CHEMICALLY INERT TOWARDS AMMONIUM NITRATE, CONTAINING NOT LESS THAN 90% OF AMMONIUM NITRATE AND NOT MORE THAN 0.2% OF COMBUSTIBLE MATERIAL (INCLUDING ORGANIC MATERIAL CALCULATED AS CARBON), OR CONTAINING LESS THAN 90% BUT MORE THAN 70% OF AMMONIUM NITRATE AND NOT MORE THAN 0.4% OF TOTAL COMBUSTIBLE MATERIAL	5.1	UN2067	OXIDIZER
	(2) UNIFORM NON-SEGRATING MIXTURES OF AMMONIUM NITRATE WITH CALCIUM CARBONATE AND/OR DOLOMITE, CONTAINING MORE THAN 80% BUT LESS THAN 90% OF AMMONIUM NITRATE AND NOT MORE THAN 0.4% OF TOTAL COMBUSTIBLE MATERIAL	5.1	UN2068	OXIDIZER
	(3) UNIFORM NON-SEGREGATING MIXTURES OF AMMONIUM NITRATE/AMMONIUM SULPHATE, CONTAINING MORE THAN 45% BUT NOT MORE THAN 70% OF AMMONIUM NITRATE AND CONTAINING NOT MORE THAN 0.4% OF TOTAL COMBUSTIBLE MATERIAL	5.1	UN2069	OXIDIZER
	(4) UNIFORM NON-SEGREGATING MIXTURES OF NITROGEN/PHOSPHATE OR NITROGEN/POTASH TYPES OR COMPLETE FERTILIZERS OF NITROGEN/PHOSPHATE/POTASH TYPE, CONTAINING MORE THAN 70% BUT LESS THAN 90% OF AMMONIUM NITRATE AND NOT MORE THAN 0.4% OF TOTAL COMBUSTIBLE MATERIAL	5.1	UN2070	OXIDIZER
	AMMONIUM NITRATE FERTILIZERS, *TYPE B. UNIFORM NON-SEGREGATING MIXTURES OF NITROGEN/PHOSPHATE OR NITROGEN/POTASH TYPES OR COMPLETE FERTILIZERS OF NITROGEN/PHOSPHATE/POTASH TYPE, CONTAINING NOT MORE THAN 70% OF AMMONIUM NITRATE AND NOT MORE THAN 0.4% OF TOTAL ADDED COMBUSTIBLE MATERIAL OR CONTAINING NOT MORE THAN 45% OF AMMONIUM NITRATE WITH UNRESTRICTED COMBUSTIBLE MATERIAL*	9	UN2071	NONE
	AMMONIUM NITRATE, *WITH MORE THAN 0.2% BY WEIGHT OF COMBUSTIBLE SUBSTANCES, INCLUDING ANY ORGANIC SUBSTANCE CALCULATED AS CARBON, TO THE EXCLUSION OF ANY OTHER ADDED SUBSTANCE*	1.1 D	UN0222	EXPLOSIVE (1.,1D)

§172.102 OPTIONAL HAZARDOUS MATERIALS TABLE

(1) Notes and Symbols	(2) Hazardous Materials and Description and Proper Shipping Names	(3) IMO Class	(4) Identification Number	(5) Label(s) required
	AMMONIUM NITRATE, *WITH NOT MORE THAN 0.2% OF COMBUSTIBLE SUBSTANCE INCLUDING ANY ORGANIC SUBSTANCE CALCULATED AS CARBON, TO THE EXCLUSION OF ANY OTHER ADDED SUBSTANCE*	5.1	UN1942	OXIDIZER
	AMMONIUM PERCHLORATE	5.1	UN1442	OXIDIZER
N	AMMONIUM PERCHLORATE, *AVERAGE PARTICLE SIZE LESS THAN 45 MICRONS*	1.1D	UN0402	EXPLOSIVE (1.1D)
	AMMONIUM PERSULPHATE	5.1	UN1444	OXIDIZER
N	AMMONIUM PICRATE, *DRY OR WETTED WITH LESS THAN 10% WATER, BY WEIGHT*	1.1D	UN0004	EXPLOSIVE (1.1D)
	AMMONIUM PICRATE, WETTED, *WITH NOT LESS THAN 10% WATER, BY WEIGHT*	4.1	UN1310	FLAMMABLE SOLID
	AMMONIUM PICRATE, WETTED, *WITH NOT LESS THAN 33 1/3 % WATER, BY WEIGHT*	4.1	UN1310	FLAMMABLE SOLID
	AMMONIUM POLYSULPHIDE, SOLUTION	8	UN2818	CORROSIVE POISON
	AMMONIUM POLYVANADATE	6.1	UN2861	POISON
	AMMONIUM SILICOFLUORIDE	6.1	UN2854	ST. ANDREWS CROSS
	AMMONIUM SULPHIDE, SOLUTION	8	UN2683	CORROSIVE, POISON, FLAMMABLE LIQUID (ONLY IF FLASHPOINT BELOW 61 DEG C)
	AMMUNITION, ILLUMINATING, *WITH OR WITHOUT BURSTER, EXPELLING CHARGE OR PROPELLING CHARGE*	1.4 G	UN0297	EXPLOSIVE (1.4G)
N	AMMUNITION, ILLUMINATING, *WITH OR WITHOUT BURSTER, EXPELLING CHARGE OR PROPELLING CHARGE*	1.2G	UN0171	EXPLOSIVE (1.2G)
N	AMMUNITION, ILLUMINATING, *WITH OR WITHOUT BURSTER, EXPELLING CHARGE OR PROPELLING CHARGE*	1.3G	UN0254	EXPLOSIVE (1.3G)
N	AMMUNITION, INCENDIARY, *LIQUID OR GEL, WITH BURSTER, EXPELLING CHARGE OR PROPELLING CHARGE*	1.3J	UN0247	EXPLOSIVE (1.3J)
	AMMUNITION, INCENDIARY *(OTHER THAN WATER-ACTIVATED AMMUNITION) WITHOUT WHITE PHOSPHORUS OR PHOSPHIDES, WITH OR WITHOUT BURSTER, EXPELLING CHARGE OR PROPELLING CHARGE*	1.4 G	UN0300	EXPLOSIVE *(1.4G)*
N	AMMUNITION, INCENDIARY *(OTHER THAN WATER-ACTIVATED AMMUNITION), WITHOUT WHITE PHOSPHORUS OR PHOSPHIDES, WITH OR WITHOUT BURSTER, EXPELLING CHARGE OR PROPELLING CHARGE*	1.2G	UN0009	EXPLOSIVE (1.2G)
N	AMMUNITION, INCENDIARY *(OTHER THAN WATER-ACTIVATED AMMUNITION), WITHOUT WHITE PHOSPHORUS OR PHOSPHIDES, WITH OR WITHOUT BURSTER, EXPELLING CHARGE OR PROPELLING CHARGE*	1.3G	UN0010	EXPLOSIVE (1.3G)
N	AMMUNITION, INCENDIARY, WHITE PHOSPHORUS, *WITH BURSTER, EXPELLING CHARGE OR PROPELLING CHARGE*	1.2H	UN0243	EXPLOSIVE (1.2H)
N	AMMUNITION, INCENDIARY, WHITE PHOSPHORUS, *WITH BURSTER, EXPELLING CHARGE OR PROPELLING CHARGE*	1.3H	UN0244	EXPLOSIVE (1.3H)
	AMMUNITION, PRACTICE	1.4G	UN0362	EXPLOSIVE (1.4G)
	AMMUNITION, PROOF	1.4G	UN0363	EXPLOSIVE (1.4G)
N	AMMUNITION, SMOKE *(OTHER THAN WATER-ACTIVATED AMMUNITION) WITHOUT WHITE PHOSPHORUS OR PHOSPHIDES, WITH OR WITHOUT BURSTER, EXPELLING CHARGE OR PROPELLING CHARGE*	1.2G	UN0015	EXPLOSIVE (1.2G), CORROSIVE
N	AMMUNITION, SMOKE *(OTHER THAN WATER-ACTIVATED AMMUNITION) WITHOUT WHITE PHOSPHORUS OR PHOSPHIDES, WITH OR WITHOUT BURSTER, EXPELLING CHARGE OR PROPELLING CHARGE*	1.3G	UN0016	EXPLOSIVE (1.3G), CORROSIVE
N	AMMUNITION, SMOKE *(OTHER THAN WATER-ACTIVATED AMMUNITION) WITHOUT WHITE PHOSPHORUS OR PHOSPHIDES, WITH OR WITHOUT BURSTER, EXPELLING CHARGE OR PROPELLING CHARGE*	1.4G	UN0303	EXPLOSIVE (1.4G), CORROSIVE
N	AMMUNITION, SMOKE, WHITE PHOSPHORUS *(OTHER THAN WATER-ACTIVATED AMMUNITION), WITH BURSTER, EXPELLING CHARGE OR PROPELLING CHARGE*	1.2H	UN0245	EXPLOSIVE (1.2H)
N	AMMUNITION, SMOKE, WHITE PHOSPHORUS *(OTHER THAN WATER-ACTIVATED AMMUNITION), WITH BURSTER, EXPELLING CHARGE OR PROPELLING CHARGE*	1.3H	UN0246	EXPLOSIVE (1.3H)
	AMMUNITION, TEAR PRODUCING, NON-EXPLOSIVE, *WITH NEITHER BURSTER NOR EXPELLING CHARGE, NON-FUZED*	6.1	UN2017	POISON, CORROSIVE
N	AMMUNITION, TEAR PRODUCING, *WITH BURSTER, EXPELLING CHARGE OR PROPELLING CHARGE*	1.2G	UN0018	EXPLOSIVE (1.2G), POISON, CORROSIVE
N	AMMUNITION, TEAR PRODUCING, *WITH BURSTER, EXPELLING CHARGE OR PROPELLING CHARGE*	1.3G	UN0019	EXPLOSIVE (1.3G), POISON, CORROSIVE
	AMMUNITION, TEAR-PRODUCING, *WITH BURSTER, EXPELLING CHARGE OR PROPELLING CHARGE*	1.4 G	UN0301	EXPLOSIVE (1.4G), POISON, CORROSIVE
	AMMUNITION, *TOXIC, NON-EXPLOSIVE, WITH NEITHER BURSTER NOR EXPELLING CHARGE, NON-FUZED*	6.1	UN2016	POISON

§172.102 OPTIONAL HAZARDOUS MATERIALS TABLE

(1) Notes and Symbols	(2) Hazardous Materials and Description and Proper Shipping Names	(3) IMO Class	(4) Identification Number	(5) Label(s) required
N	AMMUNITION, TOXIC *(OTHER THAN WATER-ACTIVATED AMMUNITION) WITH BURSTER, EXPELLING CHARGE OR PROPELLING CHARGE*	1.2K	UN0020	EXPLOSIVE (1.2K), POISON
N	AMMUNITION, TOXIC *(OTHER THAN WATER-ACTIVATED AMMUNITION) WITH BURSTER, EXPELLING CHARGE OR PROPELLING CHARGE*	1.3K	UN0021	EXPLOSIVE (1.3K), POISON
	AMYL ACETATES	3.2 3.3	UN1104 UN1104	FLAMMABLE LIQUID FLAMMABLE LIQUID
	AMYL ACID PHOSPHATE	8	UN2819	CORROSIVE
	AMYL ALCOHOLS	3.2 3.3	UN1105 UN1105	FLAMMABLE LIQUID FLAMMABLE LIQUID
	AMYLAMINE	3.2	UN1106	FLAMMABLE LIQUID
	AMYL BUTYRATES	3.3	UN2620	FLAMMABLE LIQUID
	AMYL CHLORIDE	3.2	UN1107	FLAMMABLE LIQUID
	N-AMYLENE	3.1	UN1108	FLAMMABLE LIQUID
	AMYL FORMATES	3.3	UN1109	FLAMMABLE LIQUID
	AMYL MERCAPTAN	3.2	UN1111	FLAMMABLE LIQUID
	AMYL METHYL KETONE	3.3	UN1110	FLAMMABLE LIQUID
	AMYL NITRATE	3.3	UN1112	FLAMMABLE LIQUID
	AMYL NITRITE	3.1	UN1113	FLAMMABLE LIQUID
	TERT-AMYL PEROXY-2-ETHYLHEXANOATE, *TECHNICAL PURE*	5.2	UN2898	ORGANIC PEROXIDE
	TERT-AMYL PEROXYNEODECANOATE *WITH AT LEAST 25% PHLEGMATIZER*	5.2	UN2891	ORGANIC PEROXIDE
	TERT-AMYL PEROXYPIVALATE *MAXIMUM CONCENTRATION 77% IN SOLUTION*	5.2	UN2957	ORGANIC PEROXIDE
	AMYL TRICHLOROSILANE	8	UN1728	CORROSIVE
	ANILINE	6.1	UN1547	POISON
	ANILINE HYDROCHLORIDE	6.1	UN1548	ST. ANDREWS CROSS
	ANILINE OIL. *SEE* ANILINE			
	O-ANISIDINE	6.1	UN2431	ST. ANDREWS CROSS
	ANISOLE	3.3	UN2222	FLAMMABLE LIQUID
	ANISOYL CHLORIDE	8	UN1729	CORROSIVE
	ANTI-FREEZE. *SEE* FLAMMABLE LIQUID PREPARATIONS, N.O.S.			
	ANTIMONY CHLORIDE. *SEE* ANTIMONY TRICHLORIDE, *LIQUID OR SOLID*			
	ANTIMONY COMPOUNDS, INORGANIC, N.O.S.	6.1 6.1	UN1549 UN1549	POISON ST. ANDREWS CROSS
	ANTIMONY LACTATE	6.1	UN1550	ST. ANDREWS CROSS
	ANTIMONY PENTACHLORIDE, *LIQUID*	8	UN1730	CORROSIVE
	ANTIMONY PENTACHLORIDE, *SOLUTIONS*	8	UN1731	CORROSIVE
	ANTIMONY PENTAFLUORIDE	8	UN1732	CORROSIVE POISON
	ANTIMONY POTASSIUM TARTRATE	6.1	UN1551	ST. ANDREWS CROSS
	ANTIMONY POWDER	6.1	UN2871	ST. ANDREWS CROSS
	ANTIMONY TRICHLORIDE, *LIQUID*	8	UN1733	CORROSIVE
	ANTIMONY TRICHLORIDE, *SOLID*	8	UN1733	CORROSIVE
	ARGON, *COMPRESSED*	2.2	UN1006	NONFLAMMABLE GAS
	ARGON, REFRIGERATED LIQUID	2.2	UN1951	NONFLAMMABLE GAS
	ARSENIC ACID, LIQUID	6.1	UN1553	POISON
	ARSENIC ACID, SOLID	6.1	UN1554	POISON
	ARSENICAL DUST	6.1	UN1562	POISON
	ARSENICAL FLUE DUST. *SEE* ARSENICAL DUST			
	ARSENICAL PESTICIDES, LIQUID, FLAMMABLE, TOXIC, N.O.S., *FLASH POINT LESS THAN 23 DEG C*	3.2	UN2760	FLAMMABLE LIQUID AND POISON OR ST. ANDREWS CROSS (ACCORDING TO TOXICITY)
	ARSENICAL PESTICIDES, LIQUID, TOXIC, FLAMMABLE, N.O.S. *FLASHPOINT BETWEEN 23 DEG C AND 61 DEG C*	6.1 6.1 6.1	UN2993 UN2993 UN2993	POISON, FLAMMABLE LIQUID POISON, FLAMMABLE LIQUID ST. ANDREWS CROSS FLAMMABLE LIQUID
	ARSENICAL PESTICIDES, LIQUID, TOXIC, N.O.S.	6.1 6.1 6.1	UN2994 UN2994 UN2994	POISON POISON ST. ANDREWS CROSS
	ARSENICAL PESTICIDES, SOLID, TOXIC, N.O.S.	6.1 6.1	UN2759 UN2759	POISON ST. ANDREWS CROSS
	ARSENIC BROMIDE	6.1	UN1555	POISON
	ARSENIC CHLORIDE. *SEE* ARSENIC TRICHLORIDE			

§172.102 OPTIONAL HAZARDOUS MATERIALS TABLE

(1) Notes and Symbols	(2) Hazardous Materials and Description and Proper Shipping Names	(3) IMO Class	(4) Identi-fication Number	(5) Label(s) required
	ARSENIC COMPOUNDS, LIQUID, N.O.S.	6.1 6.1	UN1556 UN1556	POISON, FLAMMABLE LIQUID (IF FLASHPOINT BETWEEN 23 DEG C AND 61 DEG C) ST. ANDREWS CROSS, FLAMMABLE LIQUID (IF FLASHPOINT BETWEEN 23 DEG C AND 61 DEG C)
	ARSENIC COMPOUNDS, SOLID, N.O.S.	6.1 6.1 6.1	UN1557 UN1557 UN1557	POISON POISON ST. ANDREWS CROSS
	ARSENIC, *METALLIC*	6.1	UN1558	POISON
	ARSENIC PENTOXIDE	6.1	UN1559	POISON
	ARSENIC SULPHIDES, SOLID, N.O.S. *SEE* ARSENIC COMPOUNDS, SOLID, N.O.S.			
	ARSENIC TRICHLORIDE	6.1	UN1560	POISON
	ARSENIC TRIOXIDE	6.1	UN1561	POISON
	ARSINE	2.3	UN2188	POISON GAS, FLAMMABLE GAS
	ARTICLES, EXPLOSIVE, N.O.S.	1.4B	UN0350	EXPLOSIVE (1.4B)
	ARTICLES, EXPLOSIVE, N.O.S.	1.4C	UN0351	EXPLOSIVE (1.4C)
	ARTICLES, EXPLOSIVE, N.O.S.	1.4D	UN0352	EXPLOSIVE (1.4D)
	ARTICLES, EXPLOSIVE, N.O.S.	1.4G	UN0353	EXPLOSIVE (1.4G)
	ARTICLES, EXPLOSIVE, N.O.S.	1.4S	UN0349	NONE. PACKAGE TO BE MARKED '1.4S'
N	ARTICLES, EXPLOSIVE, N.O.S.	1.1L	UN0354	EXPLOSIVE (1.1L)
N	ARTICLES, EXPLOSIVE, N.O.S.	1.2L	UN0355	EXPLOSIVE (1.2L)
N	ARTICLES, EXPLOSIVE, N.O.S.	1.3L	UN0356	EXPLOSIVE (1.3L)
N	ARTICLES, PYROPHORIC	1.2L	UN0380	EXPLOSIVE (1.2L)
	ASBESTOS, BLUE	9	UN2212	NONE
	ASBESTOS, WHITE	9	UN2590	NONE
	ASPHALT CUT-BACKS. *SEE* CUT-BACKS, *ASPHALT OR BITUMEN*			
	2,2'-AZODI-(2,4-DIMETHYL-4-METHOXYVALERONITRILE)	4.1	UN2955	FLAMMABLE SOLID
	2,2'-AZODI-(2,4-DIMETHYLVALERONITRILE)	4.1	UN2953	FLAMMABLE SOLID
	AZODI-(1,1'-HEXAHYDROBENZONITRILE)	4.1	UN2954	FLAMMABLE SOLID
	AZODIISOBUTYRONITRILE	4.1	UN2952	FLAMMABLE SOLID
	BAGS, *EMPTY AND UNWASHED, HAVING CONTAINED* POTASSIUM NITRATE *OR* SODIUM NITRATE	4.1	UN1359	FLAMMABLE SOLID
	BARIUM, *ALLOYS, NON-PYROPHORIC*	4.3	UN1399	DANGEROUS WHEN WET
	BARIUM ALLOYS, PYROPHORIC	4.2	UN1854	SPONTANEOUSLY COMBUSTIBLE
N	BARIUM AZIDE, *DRY OR CONTAINING, BY WEIGHT, LESS THAN 50% WATER*	1.1A	UN0224	EXPLOSIVE (1.1A), POISON
	BARIUM AZIDE, WETTED *WITH NOT LESS THAN 50% WATER, BY WEIGHT*	4.1	UN1571	FLAMMABLE SOLID, POISON
	BARIUM BROMATE	5.1	UN2719	OXIDIZER POISON
	BARIUM CHLORATE	5.1	UN1445	OXIDIZER POISON
	BARIUM COMPOUNDS, N.O.S.	6.1 6.1	UN1564 UN1564	POISON ST. ANDREWS CROSS
	BARIUM CYANIDE	6.1	UN1565	POISON
	BARIUM HYPOCHLORITE, *WITH MORE THAN 22% AVAILABLE CHLORINE*	5.1	UN2741	OXIDIZER POISON
	BARIUM, *METAL, NON-PYROPHORIC*	4.3	UN1400	DANGEROUS WHEN WET
	BARIUM NITRATE	5.1	UN1446	OXIDIZER POISON
	BARIUM OXIDE	6.1	UN1884	ST. ANDREWS CROSS
	BARIUM PERCHLORATE	5.1	UN1447	OXIDIZER, POISON
	BARIUM PERMANGANATE	5.1	UN1448	OXIDIZER, POISON
	BARIUM PEROXIDE	5.1	UN1449	OXIDIZER, POISON
	BATTERIES, WET, FILLED WITH ACID, *ELECTRIC, STORAGE*	8	UN2794	CORROSIVE
	BATTERIES, WET, FILLED WITH ALKALI, *ELECTRIC, STORAGE*	8	UN2795	CORROSIVE
	BATTERIES, WET, NON-SPILLABLE, *ELECTRIC, STORAGE*	8	UN2800	NONE. PACKAGE TO BE MARKED 'CLASS 8'
	BATTERY FLUID, ACID	8	UN2796	CORROSIVE
	BATTERY FLUID, ALKALI	8	UN2797	CORROSIVE
	BENZENE	3.2	UN1114	FLAMMABLE LIQUID
	BENZENE SULPHONYL CHLORIDE	8	UN2225	CORROSIVE
	BENZIDINE	6.1	UN1885	POISON
	BENZINE	3.1 3.2 3.3	UN1115 UN1115 UN1115	FLAMMABLE LIQUID FLAMMABLE LIQUID FLAMMABLE LIQUID

§172.102 OPTIONAL HAZARDOUS MATERIALS TABLE

(1) Notes and Symbols	(2) Hazardous Materials and Description and Proper Shipping Names	(3) IMO Class	(4) Identification Number	(5) Label(s) required
	BENZOIC DERIVATIVE PETICIDE, LIQUID, FLAMMABLE, TOXIC, N.O.S. *FLASHPOINT LESS THAN 23 DEG C*	3.2	UN2770	FLAMMABLE LIQUID AND POISON OR ST. ANDREWS CROSS (ACCORDING TO TOXICITY)
	BENZOIC DERIVATIVE PESTICIDES, LIQUID, TOXIC, FLAMMABLE, N.O.S. *FLASHPOINT BETWEEN 23 DEG C AND 61 DEG C*	6.1	UN3003	POISON, FLAMMABLE LIQUID
		6.1	UN3003	POISON, FLAMMABLE LIQUID
		6.1	UN3003	ST. ANDREWS CROSS, FLAMMABLE LIQUID
	BENZOIC DERIVATIVE PESTICIDES LIQUID, TOXIC, N.O.S.	6.1	UN3004	POISON
		6.1	UN3004	POISON
		6.1	UN3004	ST. ANDREWS CROSS
	BENZOIC DERIVATIVE PESTICIDES, SOLID, TOXIC, N.O.S.	6.1	UN2769	POISON
		6.1	UN2769	ST. ANDREWS CROSS
	BENZONITRILE	6.1	UN2224	POISON
	BENZOQUINONE	6.1	UN2587	POISON
	BENZOTRICHLORIDE	8	UN2226	CORROSIVE
	BENZOTRIFLUORIDE	3.2	UN2338	FLAMMABLE LIQUID
	BENZOYL CHLORIDE	8	UN1736	CORROSIVE
	BENZOYL PEROXIDE, *IN A CONCENTRATION OF MORE THAN 77% BUT LESS THAN 95% WITH WATER*	5.2	UN2088	ORGANIC PEROXIDE
	BENZOYL PEROXIDE, *IN A CONCENTRATION OF MORE THAN 72% BUT LESS THAN 95% AS A PASTE*	5.2	UN2086	ORGANIC PEROXIDE
	BENZOYL PEROXIDE, *IN A CONCENTRATION OF NOT MORE THAN 72% AS A PASTE*	5.2	UN2087	ORGANIC PEROXIDE
	BENZOYL PEROXIDE, *IN A CONCENTRATION OF NOT MORE THAN 77% WITH WATER*	5.2	UN2090	ORGANIC PEROXIDE
	BENZOYL PEROXIDE, *IN CONCENTRATIONS FROM 30% TO MAXIMUM 52% WITH INERT SOLID*	5.2	UN2089	ORGANIC PEROXIDE
	BENZOYL PEROXIDE, *TECHNICAL PURE OR IN A CONCENTRATION OF MORE THAN 52% WITH INERT SOLID*	5.2	UN2085	ORGANIC PEROXIDE
	BENZYL BROMIDE	8	UN1737	CORROSIVE
	BENZYL CHLORIDE	6.1	UN1738	POISON, CORROSIVE
	BENZYL CHLOROFORMATE	8	UN1739	CORROSIVE
	BENZYL DIMETHYLAMINE	8	UN2619	CORROSIVE, FLAMMABLE LIQUID
	BENZYLIDENE CHLORIDE	6.1	UN1886	POISON
	BENZYL IODIDE	6.1	UN2653	POISON
	BERYLLIUM COMPOUNDS	6.1	UN1566	POISON
	BERYLLIUM, *METAL POWDER*	6.1	UN1567	POISON, FLAMMABLE SOLID
	BERYLLIUM NITRATE	5.1	UN2464	OXIDIZER, POISON
	BHUSA	4.1	UN1327	NONE
	BIFLUORIDES, N.O.S.	8	UN1740	CORROSIVE
	BIPYRIDILIUM PESTICIDES, LIQUID, FLAMMABLE, TOXIC, N.O.S., *FLASHPOINT LESS THAN 23 DEG C*	3.2	UN2782	FLAMMABLE LIQUID AND POISON OR ST. ANDREWS CROSS (ACCORDING TO TOXICITY)
	BIPYRIDILIUM PESTICIDES, LIQUID, TOXIC, FLAMMABLE, N.O.S., *FLASHPOINT BETWEEN 23 DEG C AND 61 DEG C*	6.1	UN3015	POISON, FLAMMABLE LIQUID
		6.1	UN3015	POISON, FLAMMABLE LIQUID
		6.1	UN3015	ST. ANDREWS CROSS, FLAMMABLE LIQUID
	BIPYRIDILIUM PESTICIDES, LIQUID, TOXIC, N.O.S.	6.1	UN3016	POISON
		6.1	UN3016	POISON
		6.1	UN3016	ST. ANDREWS CROSS
	BIPYRIDILIUM PESTICIDES, SOLID, TOXIC, N.O.S.	6.1	UN2781	POISON
		6.1	UN2781	POISON
		6.1	UN2781	ST. ANDREWS CROSS
	BISULPHITES, INORGANIC, AQUEOUS SOLUTION, N.O.S.	8	UN2693	CORROSIVE
N	BLACK POWDER, COMPRESSED	1.1D	UN0028	EXPLOSIVE (1.1D)
N	BLACK POWDER, *GRANULAR OR AS MEAL*	1.1D	UN0027	EXPLOSIVE (1.1D)
	BLEACHING POWDER. *SEE* CALCIUM HYPOCHLORITE MIXTURES, DRY, *WITH MORE THAN 10% BUT NOT MORE THAN 39% AVAILABLE CHLORINE*			
	BLUE ASBESTOS. *SEE* ASBESTOS, BLUE			
N	BOMBS, CONTAINING FLAMMABLE LIQUID, *WITH BURSTING CHARGE*	1.1J	UN0399	EXPLOSIVE (1.1J)
N	BOMBS, CONTAINING FLAMMABLE LIQUID, *WITH BURSTING CHARGE*	1.2J	UN0400	EXPLOSIVE (1.2J)
N	BOMBS, PHOTO-FLASH	1.1D	UN0038	EXPLOSIVE (1.1D)
N	BOMBS, PHOTO-FLASH	1.1F	UN0037	EXPLOSIVE (1.1F)
N	BOMBS, PHOTO-FLASH	1.2G	UN0039	EXPLOSIVE (1.2G)

§172.102 OPTIONAL HAZARDOUS MATERIALS TABLE

(1) Notes and Symbols	(2) Hazardous Materials and Description and Proper Shipping Names	(3) IMO Class	(4) Identification Number	(5) Label(s) required
N	BOMBS, PHOTO-FLASH	1.3G	UN0299	EXPLOSIVE (1.3G)
	BOMBS, SMOKE, *CONTAINING A CORROSIVE LIQUID, NON-EXPLOSIVE, WITHOUT INITIATING DEVICE*	8	UN2028	CORROSIVE
N	BOMBS, *WITH BURSTING CHARGE*	1.1D	UN0034	EXPLOSIVE (1.1D)
N	BOMBS, *WITH BURSTING CHARGE*	1.2D	UN0035	EXPLOSIVE (1.2D)
N	BOMBS, *WITH BURSTING CHARGE*	1.1F	UN0033	EXPLOSIVE (1.1F)
N	BOMBS, *WITH BURSTING CHARGE*	1.2F	UN0291	EXPLOSIVE (1.2F)
N	BOOSTERS, WITH DETONATOR	1.1B	UN0225	EXPLOSIVE (1.1B)
N	BOOSTERS, WITH DETONATOR	1.2B	UN0268	EXPLOSIVE (1.2B)
N	BOOSTERS, *WITHOUT DETONATOR*	1.1D	UN0042	EXPLOSIVE (1.1D)
N	BOOSTERS, *WITHOUT DETONATOR*	1.2D	UN0283	EXPLOSIVE (1.2D)
	BORATE AND CHLORATE, MIXTURES. *SEE* CHLORATE AND BORATE *MIXTURES*			
	BORNEOL	4.1	UN1312	NONE. PACKAGE TO BE MARKED 'CLASS 4.1'
	BORON TRIBROMIDE	8	UN2692	CORROSIVE
	BORON TRICHLORIDE	2.2	UN1741	NONFLAMMABLE GAS, CORROSIVE
	BORON TRIFLUORIDE	2.3	UN1008	POISON GAS
	BORON TRIFLUORIDE ACETIC ACID COMPLEX	8	UN1742	CORROSIVE
	BORON TRIFLUORIDE DIETHYLETHERATE	4.3	UN2604	DANGEROUS WHEN WET, FLAMMABLE LIQUID, CORROSIVE
	BORON TRIFLUORIDE DIHYDRATE	8	UN2851	CORROSIVE
	BORON TRIFLUORIDE DIMETHYL ETHERATE	4.3	UN2965	DANGEROUS WHEN WET, FLAMMABLE LIQUID, CORROSIVE
	BORON TRIFLUORIDE PROPIONIC ACID COMPLEX	8	UN1743	CORROSIVE
	BOX TOE GUM. *SEE* NITROCELLULOSE			
	BRAKE FLUID, *HYDRAULIC*	3.2 3.3	UN1118 UN1118	FLAMMABLE LIQUID FLAMMABLE LIQUID
	BROMATES, INORGANIC, N.O.S.	5.1	UN1450	OXIDIZER
	BROMINE, *(AND SOLUTIONS)*	8	UN1744	CORROSIVE
	BROMINE CHLORIDE	2.3	UN2901	POISON GAS, OXIDIZER, CORROSIVE
	BROMINE PENTAFLUORIDE	5.1	UN1745	OXIDIZER, POISON, CORROSIVE
	BROMINE TRIFLUORIDE	5.1	UN1746	OXIDIZER, POISON, CORROSIVE
	BROMOACETIC ACID, *SOLID*	8	UN1938	CORROSIVE
	BROMOACETIC ACID, *SOLUTION*	8	UN1938	CORROSIVE
	BROMOACETONE	6.1	UN1569	POISON, FLAMMABLE LIQUID
	BROMOACETYL BROMIDE	8	UN2513	CORROSIVE
	BROMOBENZENE	3.3	UN2514	FLAMMABLE LIQUID
	BROMOBENZYL CYANIDES	6.1	UN1694	POISON
	2-BROMOBUTANE	3.2	UN2339	FLAMMABLE LIQUID
	BROMOCHLOROMETHANE	6.1	UN1887	ST. ANDREWS CROSS
	2-BROMOETHYL ETHYL ETHER	3.2	UN2340	FLAMMABLE LIQUID
	BROMOFORM	6.1	UN2515	ST. ANDREWS CROSS
	1-BROMO-3-METHYLBUTANE	3.2	UN2341	FLAMMABLE LIQUID
	BROMOMETHYLPROPANES	3.2	UN2342	FLAMMABLE LIQUID
	BROMOMETHYLPROPANES	3.3	UN2342	FLAMMABLE LIQUID
	2-BROMOPENTANE	3.2	UN2343	FLAMMABLE LIQUID
	BROMOPROPANES	3.2	UN2344	FLAMMABLE LIQUID
	BROMOPROPANES	3.3	UN2344	FLAMMABLE LIQUID
	3-BROMOPROPYNE	3.2	UN2345	FLAMMABLE LIQUID
	BROMOTRIFLUOROETHYLENE	2.1	UN2419	FLAMMABLE GAS
	BROMOTRIFLUOROMETHANE	2.2	UN1009	NONFLAMMABLE GAS
	BRUCINE	6.1	UN1570	POISON
N	BURSTERS, *EXPLOSIVE*	1.1D	UN0043	EXPLOSIVE (1.1D)
	BUTADIENE, *INHIBITED*	2.1	UN1010	FLAMMABLE GAS
	BUTANEDIONE	3.2	UN2346	FLAMMABLE LIQUID
	BUTANE *OR* BUTANE MIXTURES	2.1	UN1011	FLAMMABLE GAS
	SEC-BUTANOL	3.3	UN1120	FLAMMABLE LIQUID
	TERT-BUTANOL	3.2	UN1120	FLAMMABLE LIQUID
	BUTANOL	3.3	UN1120	FLAMMABLE LIQUID

§172.102 OPTIONAL HAZARDOUS MATERIALS TABLE

(1) Notes and Symbols	(2) Hazardous Materials and Description and Proper Shipping Names	(3) IMO Class	(4) Identification Number	(5) Label(s) required
	BUTENE. *SEE* BUTYLENE			
	BUTOXYL	3.3	UN2708	FLAMMABLE LIQUID
	BUTYL ACETATES	3.2	UN1123	FLAMMABLE LIQUID
	BUTYL ACID PHOSPHATE	8	UN1718	CORROSIVE
	BUTYLACRYLATE, INHIBITED	3.3	UN2348	FLAMMABLE LIQUID
	BUTYL ALCOHOL. *SEE* BUTANOL			
	SEC-BUTYL ALCOHOL. *SEE* SEC-BUTANOL			
	TERT-BUTYL ALCOHOL. *SEE* TERT-BUTANOL			
	N-BUTYLAMINE	3.2	UN1125	FLAMMABLE LIQUID
	N-N-BUTYLANILINE	6.1	UN2738	POISON
	BUTYL BENZENES	3.3	UN2709	FLAMMABLE LIQUID
	N-BUTYL BROMIDE	3.3	UN1126	FLAMMABLE LIQUID
	N-BUTYL CHLORIDE. *SEE* CHLOROBUTANES			
	N-BUTYLCHLOROFORMATE	6.1	UN2743	POISON, CORROSIVE, FLAMMABLE LIQUID
	TERT-BUTYL CUMENE PEROXIDE. *SEE* TERT-BUTYL CUMYL PEROXIDE			
	TERT-BUTYL CUMYL PEROXIDE, *TECHNICAL PURE*	5.2	UN2091	ORGANIC PEROXIDE
	TERT-BUTYLCYCLOHEXYLCHLOROFORMATE	6.1	UN2747	ST. ANDREWS CROSS
	N-BUTYL-4,4-DI-(TERT-BUTYL-PEROXY) VALERATE, *MAXIMUM CONCENTRATION 52% WITH INERT SOLID*	5.2	UN2141	ORGANIC PEROXIDE
	N-BUTYL-4,4-DI-(TERT-BUTYL-PEROXY) VALERATE, *TECHNICAL PURE*	5.2	UN2140	ORGANIC PEROXIDE
	BUTYLENE	2.1	UN1012	FLAMMABLE GAS
	BUTYL ETHER. *SEE* DIBUTYL ETHERS			
	N-BUTYL FORMATE	3.2	UN1128	FLAMMABLE LIQUID
	TERT-BUTYL HYDROPEROXIDE, *IN A CONCENTRATION OVER 72% TO MAXIMUM 90% WITH WATER*	5.2	UN2094	ORGANIC PEROXIDE
	TERT-BUTYL HYDROPEROXIDE, *MAXIMUM CONCENTRATION 80% IN DI-TERT-BUTYL PEROXIDE AND/OR SOLVENT*	5.2	UN2092	ORGANIC PEROXIDE, FLAMMABLE LIQUID (ONLY IF FLASHPOINT OF SOLVENT IS 61 DEG C. OR BELOW)
	TERT-BUTYL HYDROPEROXIDE, *MAXIMUM CONCENTRATION 72% WITH WATER*	5.2	UN2093	ORGANIC PEROXIDE
	N-N-BUTYL IMIDAZOLE	6.1	UN2690	POISON
	TERT-BUTYL ISOCYANATE	3.2	UN2484	FLAMMABLE LIQUID, POISON
	N-BUTYL ISOCYANATE	3.2	UN2485	FLAMMABLE LIQUID, POISON
	BUTYL MERCAPTAN	3.2	UN2347	FLAMMABLE LIQUID
	N-BUTYL METHACRYLATE	3.3	UN2227	FLAMMABLE LIQUID
	BUTYL METHYL ETHER	3.2	UN2350	FLAMMABLE LIQUID
	TERT-BUTYL MONOPEROXYMALEATE, *MAXIMUM CONCENTRATION 55% AS A PASTE*	5.2	UN2101	ORGANIC PEROXIDE
	TERT-BUTYL MONOPEROXYMALEATE, *MAXIMUM CONCENTRATION 55% IN SOLUTION*	5.2	UN2100	ORGANIC PEROXIDE
	TERT-BUTYL MONOPEROXYMALEATE, *TECHNICAL PURE*	5.2	UN2099	ORGANIC PEROXIDE
	TERT-BUTYL MONOPEROXYPHTHALATE, *TECHNICAL PURE*	5.2	UN2105	ORGANIC PEROXIDE
	BUTYL NITRITE	3.2	UN2351	FLAMMABLE LIQUID
	TERT-BUTYL PEROXIDE. *SEE* DI-TERT-BUTYL PEROXIDE			
	TERT-BUTYL PEROXYACETATE, *IN A CONCENTRATION OF MORE THAN 52% TO A MAXIMUM CONCENTRATION OF 76% IN SOLUTION*	5.2	UN2095	ORGANIC PEROXIDE
	TERT-BUTYL PEROXYACETATE, *MAXIMUM CONCENTRATION 52% IN SOLUTION*	5.2	UN2096	ORGANIC PEROXIDE
	TERT-BUTYL PEROXYBENZOATE, *MAXIMUM CONCENTRATION 75% IN SOLUTION*	5.2	UN2098	ORGANIC PEROXIDE
	TERT-BUTYL PEROXYBENZOATE, *TECHNICAL PURE OR IN A CONCENTRATION OF MORE THAN 75% IN SOLUTION*	5.2	UN2097	ORGANIC PEROXIDE
	TERT-BUTYL PEROXYBENZOATE *WITH AT LEAST 50% INERT INORGANIC SOLID*	5.2	UN2890	ORGANIC PEROXIDE
	TERT-BUTYL PEROXYCROTONATE, *MAXIMUM CONCENTRATION 76% IN SOLUTION*	5.2	UN2183	ORGANIC PEROXIDE
	N-BUTYL PEROXYDICARBONATE, *IN A CONCENTRATION OF MORE THAN 27% , TO A MAXIMUM CONCENTRATION OF 52% IN SOLUTION*	5.2	UN2169	ORGANIC PEROXIDE
	N-BUTYL PEROXYDICARBONATE, *MAXIMUM CONCENTRATION 27% IN SOLUTION*	5.2	UN2170	ORGANIC PEROXIDE
	TERT-BUTYL PEROXYDIETHYLACETATE, *(IN A MAXIMUM CONCENTRATION OF 33%)* , WITH TERT-BUTYL PEROXYBENZOATE, *(IN A MAXIMUM CONCENTRATION OF 33%), AND WITH SOLVENT*	5.2	UN2551	ORGANIC PEROXIDE

§172.102 OPTIONAL HAZARDOUS MATERIALS TABLE

(1) Notes and Symbols	(2) Hazardous Materials and Description and Proper Shipping Names	(3) IMO Class	(4) Identification Number	(5) Label(s) required
	TERT-BUTYL PEROXYDIETHYLACETATE, *TECHNICAL PURE*	5.2	UN2144	ORGANIC PEROXIDE
	TERT-BUTYL PEROXY-2-ETHYLHEXANOATE, *MAXIMUM CONCENTRATION 30%* , WITH 2,2-DI-(TERT-BUTYL PEROXY) BUTANE *MAXIMUM CONCENTRATION 35% AND WITH AT LEAST 35% PHLEGMATIZER*	5.2	UN2886	ORGANIC PEROXIDE
	TERT-BUTYL PEROXY-2-ETHYLHEXANOATE *MAXIMUM CONCENTRATION 12%* WITH 2,2-DI-(TERT-BUTYL PEROXY)BUTANE *MAXIMUM CONCENTRATION 14%, AND WITH AT LEAST 14% PHLEGMATIZER AND 60% INERT INORGANIC SOLID*	5.2	UN2887	ORGANIC PEROXIDE
	TERT-BUTYL PEROXY-2-ETHYL HEXANOATE, *TECHNICAL PURE*	5.2	UN2143	ORGANIC PEROXIDE
	TERT-BUTYL PEROXY-2-ETHYLHEXANOATE, *WITH AT LEAST 50% PHLEGMATIZER*	5.2	UN2888	ORGANIC PEROXIDE
	TERT-BUTYL PEROXYISOBUTYRATE, *IN A CONCENTRATION OF MORE THAN 52% TO A MAXIMUM CONCENTRATION OF 77% IN SOLUTION*	5.2	UN2142	ORGANIC PEROXIDE
	TERT-BUTYL PEROXYISOBUTYRATE, *MAXIMUM CONCENTRATION 52% IN SOLUTION*	5.2	UN2562	ORGANIC PEROXIDE
	TERT-BUTYL PEROXY ISOPROPYL CARBONATE, *TECHNICAL PURE*	5.2	UN2103	ORGANIC PEROXIDE
	TERT-BUTYL PEROXYNEODECANOATE, *MAXIMUM CONCENTRATION 77% IN SOLUTION*	5.2	UN2177	ORGANIC PEROXIDE
	TERT-BUTYL PEROXYNEODECANOATE, *TECHNICAL PURE*	5.2	UN2594	ORGANIC PEROXIDE
	3-TERT-BUTYLPEROXY-3-PHENYL PHTHALIDE, *TECHNICAL PURE*	5.2	UN2596	ORGANIC PEROXIDE
	TERT-BUTYL PEROXYPIVALATE, *MAXIMUM CONCENTRATION 77% IN SOLUTION*	5.2	UN2110	ORGANIC PEROXIDE
	TERT-BUTYL PEROXY-3,5,5-TRIMETHYL HEXANOATE, *TECHNICAL PURE*	5.2	UN2104	ORGANIC PEROXIDE
	BUTYLPHENOLS, LIQUID	6.1	UN2228	ST. ANDREWS CROSS
	BUTYLPHENOLS, SOLID	6.1	UN2229	ST. ANDREWS CROSS
	BUTYL PHOSPHORIC ACID. *SEE* BUTYL ACID PHOSPHATE			
	BUTYL PROPIONATE	3.3	UN1914	FLAMMABLE LIQUID
	BUTYL TOLUENES	6.1	UN2667	POISON, FLAMMABLE LIQUID (ONLY IF FLASHPOINT BELOW 61 DEG C)
	BUTYL TRICHLOROSILANE	8	UN1747	CORROSIVE, FLAMMABLE LIQUID
	BUTYL VINYL ETHER, *INHIBITED*	3.2	UN2352	FLAMMABLE LIQUID
	1,4-BUTYNEDIOL	4.1	UN2716	FLAMMABLE SOLID
	BUTYRALDEHYDE	3.2	UN1129	FLAMMABLE LIQUID
	BUTYRALDOXIME	3.3	UN2840	FLAMMABLE LIQUID
	N-BUTYRIC ACID	8	UN2820	CORROSIVE
	BUTYRIC ANHYDRIDE	8	UN2739	CORROSIVE
	BUTYRONITRILE	3.2	UN2411	FLAMMABLE LIQUID, POISON
	BUTYRYL CHLORIDE	3.2	UN2353	FLAMMABLE LIQUID, CORROSIVE
	CACODYLIC ACID	6.1	UN1572	POISON
	CADMIUM COMPOUNDS	6.1 6.1	UN2570 UN2570	POISON ST. ANDREWS CROSS
	CAESIUM HYDROXIDE, SOLID	8	UN2682	CORROSIVE
	CAESIUM HYDROXIDE, SOLUTION	8	UN2681	CORROSIVE
	CAESIUM, *METAL*	4.3	UN1407	DANGEROUS WHEN WET
	CAESIUM NITRATE	5.1	UN1451	OXIDIZER
	CAESIUM, *POWDERED. SEE* PYROPHORIC METALS			
	CALCIUM ARSENATE	6.1	UN1573	POISON
	CALCIUM ARSENATE AND ARSENITE, *SOLID MIXTURES*	6.1	UN1574	POISON
	CALCIUM BISULPHITE, *SOLUTION. SEE* CALCIUM HYDROGEN SULPHITE, *SOLUTION*			
	CALCIUM CARBIDE	4.3	UN1402	DANGEROUS WHEN WET
	CALCIUM CHLORATE	5.1	UN1452	OXIDIZER
	CALCIUM CHLORATE, *AQUEOUS* SOLUTION	5.1	UN2429	OXIDIZER
	CALCIUM CHLORITE	5.1	UN1453	OXIDIZER
	CALCIUM CYANAMIDE, *CONTAINING MORE THAN 0.1% AND NOT MORE THAN 0.5% OF CALCIUM CARBIDE*	4.3	UN1403	DANGEROUS WHEN WET
	CALCIUM CYANAMIDE, *CONTAINING MORE THAN 0.5% OF CALCIUM CARBIDE*	4.3	UN1403	DANGEROUS WHEN WET
	CALCIUM CYANIDE	6.1	UN1575	POISON
	CALCIUM HYDRIDE	4.3	UN1404	DANGEROUS WHEN WET
	CALCIUM HYDROSULPHITE	4.2	UN1923	SPONTANEOUSLY COMBUSTIBLE

§172.102 OPTIONAL HAZARDOUS MATERIALS TABLE

(1) Notes and Symbols	(2) Hazardous Materials and Description and Proper Shipping Names	(3) IMO Class	(4) Identification Number	(5) Label(s) required
	CALCIUM HYPOCHLORATE, HYDRATED *OR* CALCIUM HYPOCHLORITE, HYDRATED MIXTURES *WITH NOT LESS THAN 5.5% BUT NOT MORE THAN 10% WATER*	5.1	UN2880	OXIDIZER
	CALCIUM HYPOCHLORITE, DRY *OR* CALCIUM HYPOCHLORITE MIXTURES, *WITH MORE THAN 39% AVAILABLE CHLORINE (8.8% AVAILABLE OXYGEN)*	5.1	UN1748	OXIDIZER
	CALCIUM HYPOCHLORITE MIXTURES, DRY, *WITH MORE THAN 10%, BUT NOT MORE THAN 39%, AVAILABLE CHLORINE*	5.1	UN2208	OXIDIZER
	CALCIUM MANGANESE SILICON	4.3	UN2844	DANGEROUS WHEN WET
	CALCIUM, *METAL AND ALLOYS, NON-PYROPHORIC*	4.3	UN1401	DANGEROUS WHEN WET
	CALCIUM NITRATE	5.1	UN1454	OXIDIZER
	CALCIUM PERCHLORATE	5.1	UN1455	OXIDIZER
	CALCIUM PERMANGANATE	5.1	UN1456	OXIDIZER
	CALCIUM PEROXIDE	5.1	UN1457	OXIDIZER
	CALCIUM PHOSPHIDE	4.3	UN1360	DANGEROUS WHEN WET
	CALCIUM, PYROPHORIC *OR* CALCIUM ALLOYS, PYROPHORIC	4.2	UN1855	SPONTANEOUSLY COMBUSTIBLE
	CALCIUM RESINATE, FUSED	4.1	UN1314	FLAMMABLE SOLID
	CALCIUM RESINATE, *TECHNICAL PURE*	4.1	UN1313	FLAMMABLE SOLID
	CALCIUM SILICIDE	4.3	UN1405	DANGEROUS WHEN WET
	CALCIUM SILICON	4.3	UN1406	DANGEROUS WHEN WET
	CAMPHOR, *NATURAL OR SYNTHETIC*	4.1	UN2717	FLAMMABLE SOLID
	CAMPHOR OIL	3.3	UN1130	FLAMMABLE LIQUID
	CAPRYLOYL PEROXIDE. *SEE* N-OCTANOYL PEROXIDE			
	CAPS, BLASTING. *SEE* BLASTING CAPS			
	CARBAMATE PESTICIDES, LIQUID, FLAMMABLE, TOXIC, N.O.S., *FLASH POINT LESS THAN 23 DEG C*	3.2	UN2758	FLAMMABLE LIQUID AND POISON OR ST. ANDREWS CROSS (ACCORDING TO TOXICITY)
	CARBAMATE PESTICIDES, LIQUID, TOXIC, FLAMMABLE, N.O.S., *FLASH POINT BETWEEN 23 DEG C AND 61 DEG C*	6.1 6.1 6.1	UN2991 UN2991 UN2991	POISON, FLAMMABLE LIQUID POISON, FLAMMABLE LIQUID ST. ANDREWS CROSS, FLAMMABLE LIQUID
	CARBAMATE PESTICIDES, LIQUID, TOXIC, N.O.S.	6.1 6.1 6.1	UN2992 UN2992 UN2992	POISON POISON ST. ANDREWS CROSS
	CARBAMATE PESTICIDES, SOLID, TOXIC, N.O.S.	6.1 6.1	UN2757 UN2757	POISON ST. ANDREWS CROSS
	CARBOLIC ACID. *SEE* PHENOLS, PHENOL SOLUTIONS *OR* PHENOL, MOLTEN			
	CARBON, ACTIVATED	4.2	UN1362	SPONTANEOUSLY COMBUSTIBLE
	CARBON, *ANIMAL OR VEGETABLE ORIGIN*	4.2	UN1361	SPONTANEOUSLY COMBUSTIBLE
	CARBON BISULPHIDE. *SEE* CARBON DISULPHIDE			
	CARBON DIOXIDE	2.2	UN1013	NONFLAMMABLE GAS
	CARBON DIOXIDE AND ETHYLENE OXIDE MIXTURES *WITH MORE THAN 6% ETHYLENE OXIDE*	2.3	UN1041	POISON GAS, FLAMMABLE GAS
	CARBON DIOXIDE AND ETHYLENE OXIDE MIXTURES *WITH NOT MORE THAN 6% ETHYLENE OXIDE*	2.3	UN1952	POISON GAS
	CARBON DIOXIDE AND NITROUS OXIDE, *MIXTURES*	2.2	UN1015	NONFLAMMABLE GAS
	CARBON DIOXIDE AND OXYGEN, *MIXTURES*	2.2	UN1014	NONFLAMMABLE GAS
	CARBON DIOXIDE, REFRIGERATED LIQUID	2.2	UN2187	NON FLAMMABLE GAS
	CARBON DIOXIDE, SOLID	9	UN1845	NONE
	CARBON DISULPHIDE	3.1	UN1131	FLAMMABLE LIQUID, POISON
	CARBON MONOXIDE	2.1	UN1016	FLAMMABLE GAS, POISON GAS
	CARBON MONOXIDE AND HYDROGEN MIXTURE	2.3	UN2600	POISON GAS, FLAMMABLE GAS
	CARBON PAPER. SEE PAPER, *TREATED WITH UNSATURATED OILS, INCOMPLETELY DRIED*			
	CARBON REMOVER, *LIQUID*	3.2	UN1132	FLAMMABLE LIQUID
	CARBON SULPHIDE. *SEE* CARBON DISULPHIDE			
	CARBON TETRABROMIDE	6.1	UN2516	ST. ANDREWS CROSS
	CARBON TETRACHLORIDE	6.1	UN1846	POISON
	CARBONYL CHLORIDE. SEE PHOSGENE			
	CARBONYL FLUORIDE	2.3	UN2417	POISON GAS
	CARBONYL SULFIDE	2.3	UN2204	POISON GAS, FLAMMABLE GAS
	CARTRIDGE CASES. SEE CASES, CARTRIDGES			

§172.102 OPTIONAL HAZARDOUS MATERIALS TABLE

(1) Notes and Symbols	(2) Hazardous Materials and Description and Proper Shipping Names	(3) IMO Class	(4) Identification Number	(5) Label(s) required
N	CARTRIDGES, FLASH	1.1G	UN0049	EXPLOSIVE (1.1G)
N	CARTRIDGES, FLASH	1.3G	UN0050	EXPLOSIVE (1.3G)
	CARTRIDGES FOR WEAPONS, BLANK	1.4 C	UN0338	EXPLOSIVE (1.4C)
N	CARTRIDGES FOR WEAPONS, BLANK	1.2C	UN0413	EXPLOSIVE (1.2C)
N	CARTRIDGES FOR WEAPONS, BLANK	1.1C	UN0326	EXPLOSIVE (1.1C)
N	CARTRIDGES FOR WEAPONS, BLANK	1.3C	UN0327	EXPLOSIVE (1.3C)
	CARTRIDGES FOR WEAPONS, BLANK *OR* CARTRIDGES, SAFETY, BLANK	1.4 S	UN0014	NONE. PACKAGE TO BE MARKED '1.4 S'
	CARTRIDGES FOR WEAPONS, *OTHER THAN BLANK*	1.4 S	UN0012	NONE. PACKAGE TO BE MARKED '1.4 S'
N	CARTRIDGES FOR WEAPONS, *WITH BURSTING CHARGE*	1.1E	UN0006	EXPLOSIVE (1.1E)
N	CARTRIDGES FOR WEAPONS, *WITH BURSTING CHARGE*	1.2E	UN0321	EXPLOSIVE (1.2E)
N	CARTRIDGES FOR WEAPONS, *WITH BURSTING CHARGE*	1.1F	UN0005	EXPLOSIVE (1.1F)
N	CARTRIDGES FOR WEAPONS, *WITH BURSTING CHARGE*	1.2F	UN0007	EXPLOSIVE (1.2F)
	CARTRIDGES FOR WEAPONS, *WITH BURSTING CHARGE*	1.4E	UN0412	EXPLOSIVE (1.4E)
N	CARTRIDGES FOR WEAPONS, *WITH BURSTING CHARGE*	1.4F	UN0348	EXPLOSIVE (1.4F)
	CARTRIDGES FOR WEAPONS, WITH INERT PROJECTILE	1.4 C	UN0339	EXPLOSIVE (1.4C)
N	CARTRIDGES FOR WEAPONS, WITH INERT PROJECTILE	1.2C	UN0328	EXPLOSIVE (1.2C)
N	CARTRIDGES FOR WEAPONS, WITH INERT PROJECTILE	1.3C	UN0417	EXPLOSIVE (1.3C)
	CARTRIDGES, OIL WELL	1.4 C	UN0278	EXPLOSIVE (1.4C)
N	CARTRIDGES, OIL WELL	1.3C	UN0277	EXPLOSIVE (1.3C)
	CARTRIDGES, POWER DEVICE	1.4 C	UN0276	EXPLOSIVE (1.4C)
	CARTRIDGES, POWER DEVICE	1.4 S	UN0323	NONE. PACKAGE TO BE MARKED '1.4 S'
N	CARTRIDGES, POWER DEVICE	1.2C	UN0381	EXPLOSIVE (1.2C)
N	CARTRIDGES, POWER DEVICE	1.3C	UN0275	EXPLOSIVE (1.3C)
	CARTRIDGES, SAFETY. *SEE* CARTIDGES FOR WEAPONS, *OTHER THAN BLANK (UN0012) OR* CARTRIDGES FOR WEAPONS, BLANK *(UN0014)*			
	CARTRIDGES, SIGNAL	1.4 G	UN0312	EXPLOSIVE (1.4G)
N	CARTRIDGES, SIGNAL	1.3G	UN0054	EXPLOSIVE (1.3G)
	CARTRIDGES, SIGNAL	1.4S	UN0405	NONE. PACKAGE TO BE MARKED '1.4S'
	CASES, CARTRIDGE, EMPTY, WITH PRIMER	1.4C	UN0379	EXPLOSIVE (1.4C)
	CASES, CARTRIDGES, EMPTY, WITH PRIMER	1.4S	UN0055	NONE. PACKAGE TO BE MARKED '1.4 S'
	CASINGHEAD GASOLINE	3.1	UN1257	FLAMMABLE LIQUID
	CASTOR BEANS, CASTOR MEAL, CASTOR POMACE *OR* CASTOR FLAKE	9	UN2969	NONE
	CAUSTIC ALKALI LIQUIDS, N.O.S.	8	UN1719	CORROSIVE
	CAUSTIC POTASH. *SEE* POTASSIUM HYDROXIDE, *SOLUTION*			
	CAUSTIC POTASH, SOLID. *SEE* POTASSIUM HYDROXIDE, *SOLID*			
	CELLULOID, *IN BLOCKS, RODS, ROLLS, SHEETS, TUBES, ETC. (SCRAP EXCLUDED)*	4.1	UN2000	FLAMMABLE SOLID
	CELLULOID, SCRAP	4.2	UN2002	SPONTANEOUSLY COMBUSTIBLE
	CEMENT, *LIQUID. SEE* ADHESIVES, *CONTAINING A FLAMMABLE LIQUID*			
	CERIUM, CRUDE, *POWDER*	4.1	UN1333	FLAMMABLE SOLID
	CERIUM, CRUDE, *SLABS OR INGOTS*	4.1	UN1333	FLAMMABLE SOLID
	CHARCOAL, ACTIVATED. *SEE* CARBON, ACTIVATED			
	CHARCOAL, NON-ACTIVATED, OF ANIMAL OR VEGETABLE ORIGIN. SEE CARBON, *OF ANIMAL OR VEGETABLE ORIGIN*			
N	CHARGES, DEMOLITION	1.1D	UN0048	EXPLOSIVE (1.1D)
N	CHARGES, DEPTH	1.1D	UN0056	EXPLOSIVE (1.1D)
N	CHARGES, PROPELLING, FOR CANNON	1.1C	UN0279	EXPLOSIVE (1.1C)
N	CHARGES, PROPELLING, FOR CANNON	1.2C	UN0414	EXPLOSIVE (1.2C)
N	CHARGES, PROPELLING, FOR CANNON	1.3C	UN0242	EXPLOSIVE (1.3C)
N	CHARGES, PROPELLING, FOR ROCKET MOTORS	1.1C	UN0271	EXPLOSIVE (1.1C)
N	CHARGES, PROPELLING, FOR ROCKET MOTORS	1.2C	UN0415	EXPLOSIVE (1.2C)
N	CHARGES, PROPELLING, FOR ROCKET MOTORS	1.3C	UN0272	EXPLOSIVE (1.3C)
N	CHARGES, PROPELLING, FOR ROCKET MOTORS, *COMPOSITE MIXTURE*	1.1C	UN0273	EXPLOSIVE (1.1C)
N	CHARGES, PROPELLING, FOR ROCKET MOTORS, *COMPOSITE MIXTURE*	1.2C	UN0416	EXPLOSIVE (1.2C)
N	CHARGES, PROPELLING, FOR ROCKET MOTORS, *COMPOSITE MIXTURE*	1.3C	UN0274	EXPLOSIVE (1.3C)
N	CHARGES, SHAPED, COMMERCIAL, *WITHOUT DETONATOR*	1.1D	UN0059	EXPLOSIVE (1.1D)

§172.102 OPTIONAL HAZARDOUS MATERIALS TABLE

(1) Notes and Symbols	(2) Hazardous Materials and Description and Proper Shipping Names	(3) IMO Class	(4) Identification Number	(5) Label(s) required
	CHARGES, SHAPED, FLEXIBLE, LINEAR, *METAL CLAD*	1.4 D	UN0237	EXPLOSIVE (1.4D)
N	CHARGES, SHAPED, FLEXIBLE, LINEAR, *METAL CLAD*	1.1D	UN0288	EXPLOSIVE (1.1D)
N	CHARGES, SUPPLEMENTARY, EXPLOSIVE	1.1D	UN0060	EXPLOSIVE (1.1D)
	CHLORAL, *ANHYDROUS, INHIBITED*	6.1	UN2075	POISON
	CHLORATE AND BORATE, MIXTURES	5.1	UN1458	OXIDIZER
	CHLORATE AND MAGNESIUM CHLORIDE, MIXTURE	5.1	UN1459	OXIDIZER
	CHLORATES, INORGANIC, N.O.S.	5.1	UN1461	OXIDIZER
	CHLORIC ACID SOLUTION *WITH NOT MORE THAN 10% ACID*	5.1	UN2626	OXIDIZER
	CHLORINE	2.3	UN1017	POISON GAS
	CHLORINE PENTAFLUORIDE	2.3	UN2548	POISON GAS, OXIDIZER CORROSIVE
	CHLORINE TRIFLUORIDE	2.3	UN1749	POISON GAS, OXIDIZER CORROSIVE
	CHLORITES, INORGANIC, N.O.S.	5.1	UN1462	OXIDIZER
	CHLOROACETALDEHYDE	6.1	UN2232	POISON
	CHLOROACETIC ACID, *LIQUID*	8	UN1750	CORROSIVE
	CHLOROACETIC ACID, SOLID	8	UN1751	CORROSIVE
	CHLOROACETONE, *STABILIZED*	6.1	UN1695	POISON
	CHLOROACETONITRILE	6.1	UN2668	POISON, FLAMMABLE LIQUID
	CHLOROACETOPHENONE	6.1	UN1697	POISON
	CHLOROACETYL CHLORIDE	8	UN1752	CORROSIVE
	CHLOROANILINES, *LIQUID*	6.1	UN2019	POISON
	CHLOROANILINES, *SOLID*	6.1	UN2018	POISON
	P-CHLORO-O-ANISIDINE	6.1	UN2233	ST. ANDREWS CROSS
	CHLOROBENZENE	3.3	UN1134	FLAMMABLE LIQUID
	CHLOROBENZOTRIFLUORIDES	3.3	UN2234	FLAMMABLE LIQUID
	P-CHLOROBENZOYL PEROXIDE. *SEE* DI-(4-CHLOROBENZOYL) PEROXIDE			
	P-CHLOROBENZYL CHLORIDE	6.1	UN2235	ST. ANDREWS CROSS
	1-CHLORO-3-BROMOPROPANE	6.1	UN2688	ST. ANDREWS CROSS
	CHLOROBUTANES	3.2	UN1127	FLAMMABLE LIQUID
	CHLOROCRESOLS	6.1	UN2669	POISON
	CHLORODIFLUOROBROMOMETHANE	2.2	UN1974	NONFLAMMABLE GAS
	CHLORODIFLUOROETHANES	2.1	UN2517	FLAMMABLE GAS
	CHLORODIFLUOROMETHANE	2.2	UN1018	NONFLAMMABLE GAS
	CHLORODIFLUOROMETHANE AND CHLOROPENTAFLUOROETHANE, *MIXTURE WITH A FIXED BOILING POINT CONTAINING ABOUT 49% OF CHLORODIFLUOROMETHANE*	2.2	UN1973	NONFLAMMABLE GAS
	CHLORODINITROBENZENE	6.1	UN1577	POISON
	CHLOROFORM	6.1	UN1888	POISON
	CHLOROFORMATES, N.O.S., *WITH A FLASHPOINT NOT LESS THAN 23 DEG C*	6.1	UN2742	POISON, CORROSIVE FLAMMABLE LIQUID (ONLY IF FLASHPOINT BELOW 61 DEG C)
	CHLOROMETHYLCHLOROFORMATE	6.1	UN2745	POISON CORROSIVE
	CHLOROMETHYL ETHYL ETHER	3.2	UN2354	FLAMMABLE LIQUID, POISON
	3-CHLORO-4-METHYLPHENYL ISOCYANATE	6.1	UN2236	POISON
	CHLORONITROANILINES	6.1	UN2237	ST. ANDREWS CROSS
	CHLORONITROBENZENES *(O-, M-, P-)*	6.1	UN1578	POISON
	CHLORO-O-NITROTOLUENE	6.1	UN2433	ST. ANDREWS CROSS
	CHLOROPENTAFLUOROETHANE	2.2	UN1020	NONFLAMMABLE GAS
	3-CHLOROPEROXYBENZOIC ACID, *MAXIMUM CONCENTRATION 86%*	5.2	UN2755	ORGANIC PEROXIDE
	CHLOROPHENATES, LIQUID	8	UN2904	CORROSIVE
	CHLOROPHENATES, SOLID	8	UN2905	CORROSIVE
	CHLOROPHENOLS, LIQUID	6.1	UN2021	ST. ANDREWS CROSS
	CHLOROPHENOLS, SOLID	6.1	UN2020	ST. ANDREWS CROSS
	CHLOROPHENYL TRICHLOROSILANE	8	UN1753	CORROSIVE
	CHLOROPICRIN	6.1	UN1580	POISON
	CHLOROPICRIN AND METHYL BROMIDE, *MIXTURES*	2.3	UN1581	POISON GAS
	CHLOROPICRIN AND METHYL CHLORIDE, *MIXTURES*	2.3	UN1582	POISON GAS
	CHLOROPICRIN MIXTURES, N.O.S.	6.1 6.1	UN1583 UN1583	POISON ST. ANDREWS CROSS
	CHLOROPLATINIC ACID, SOLID	8	UN2507	CORROSIVE

§172.102 OPTIONAL HAZARDOUS MATERIALS TABLE

(1) Notes and Symbols	(2) Hazardous Materials and Description and Proper Shipping Names	(3) IMO Class	(4) Identi- fication Number	(5) Label(s) required
	CHLOROPRENE, INHIBITED	3.2	UN1991	FLAMMABLE LIQUID, POISON
	2-CHLOROPROPANE	3.1	UN2356	FLAMMABLE LIQUID
	3-CHLOROPROPANOL-1	6.1	UN2849	ST. ANDREWS CROSS
	2-CHLOROPROPENE	3.1	UN2456	FLAMMABLE LIQUID
	CHLOROPROPIONIC ACID	8	UN2511	CORROSIVE
	2-CHLOROPYRIDINE	6.1	UN2822	POISON
	CHLOROSULPHONIC ACID, *WITH OR WITHOUT SULPHUR TRIOXIDE*	8	UN1754	CORROSIVE
	CHLOROTETRAFLUOROETHANE	2.2	UN1021	NONFLAMMABLE GAS
	CHLOROTOLUENES *(O-, M-, P-)*	3.3	UN2238	FLAMMABLE LIQUID
	4-CHLORO-O-TOLUIDINE HYDROCHLORIDE	6.1	UN1579	ST. ANDREWS CROSS
	CHLOROTOLUIDINES	6.1	UN2239	ST. ANDREWS CROSS
	CHLOROTRIFLUOROETHANE	2.2	UN1983	NONFLAMMABLE GAS
	CHLOROTRIFLUOROETHYLENE. SEE TRIFLUOROCHLOROETHYLENE			
	CHLOROTRIFLUOROMETHANE	2.2	UN1022	NONFLAMMABLE GAS
	CHLOROTRIFLUOROMETHANE AND TRIFLUORMETHANE AZEOTROPIC MIXTURE *WITH APPROXIMATELY 60% CHLOROTRIFLUOROMETHANE*	2.2	UN2599	NONFLAMMABLE GAS
	CHROMIC ACID, SOLID. *SEE* CHROMIUM TRIOXIDE, ANHYDROUS			
	CHROMIC ACID, *SOLUTION*	8	UN1755	CORROSIVE
	CHROMIC ANHYDRIDE. *SEE* CHROMIUM TRIOXIDE, ANHYDROUS			
	CHROMIC FLUORIDE, SOLID	8	UN1756	CORROSIVE
	CHROMIC FLUORIDE, SOLUTION	8	UN1757	CORROSIVE
	CHROMIUM NITRATE	5.1	UN2720	OXIDIZER
	CHROMIUM OXYCHLORIDE	8	UN1758	CORROSIVE
	CHROMIUM TRIOXIDE, ANHYDROUS	5.1	UN1463	OXIDIZER CORROSIVE
	CHROMOSULPHURIC ACID	8	UN2240	CORROSIVE
	CHRYSOTILE. SEE ASBESTOS, WHITE			
	CIGARETTES, *SELF-LIGHTING*	4.1	UN1867	FLAMMABLE SOLID
	CLEANING COMPOUND. *SEE* FLAMMABLE LIQUID PREPARATIONS, N.O.S.			
	COAL GAS	2.3	UN1023	POISON GAS, FLAMMABLE GAS
	COAL TAR DISTILLATE	3.2 3.3	UN1136 UN1136	FLAMMABLE LIQUID FLAMMABLE LIQUID
	COAL TAR NAPHTHA	3.2	UN2553	FLAMMABLE LIQUID
	COAL TAR OIL. *SEE* COAL TAR DISTILLATE			
	COATING SOLUTION	3.2 3.3	UN1139 UN1139	FLAMMABLE LIQUID FLAMMABLE LIQUID
	COBALT NAPHTHENATES, *POWDER*	4.1	UN2001	FLAMMABLE SOLID
	COBALT RESINATE, *PRECIPITATED*	4.1	UN1318	FLAMMABLE SOLID
	COCCULUS, *SOLID*	6.1	UN1584	POISON
	COLOGNE SPIRITS. *SEE* ETHANOL			
	COLUMBIAN SPIRITS. *SEE* METHANOL			
	COMPONENTS, EXPLOSIVE TRAIN, N.O.S.	1.4B	UN0383	EXPLOSIVE (1.4B)
	COMPONENTS, EXPLOSIVE TRAIN, N.O.S.	1.4S	UN0384	NONE. PACKAGE TO BE MARKED '1.4S'
N	COMPONENTS, EXPLOSIVE TRAIN, N.O.S.	1.2B	UN0382	EXPLOSIVE (1.2B)
	COMPRESSED *OR* LIQUEFIED GASES, FLAMMABLE, *NON-TOXIC*, N.O.S.	2.1	UN1954	FLAMMABLE GAS
	COMPRESSED *OR* LIQUEFIED GASES, FLAMMABLE, TOXIC, N.O.S.	2.1	UN1953	FLAMMABLE GAS, POISON GAS
	COMPRESSED *OR* LIQUEFIED GASES, *NON-FLAMMABLE, NON-TOXIC*, N.O.S.	2.2	UN1956	NONFLAMMABLE GAS
	COMPRESSED *OR* LIQUEFIED GASES, *NON-FLAMMABLE*, TOXIC, N.O.S.	2.3	UN1955	POISON GAS
N	CONTRIVANCES, WATER-ACTIVATED, *WITH BURSTER, EXPELLING CHARGE OR PROPELLING CHARGE*	1.2L	UN0248	EXPLOSIVE (1.2L), DANGEROUS WHEN WET
N	CONTRIVANCES, WATER-ACTIVATED, *WITH BURSTER, EXPELLING CHARGE OR PROPELLING CHARGE*	1.3L	UN0249	EXPLOSIVE (1.3L), DANGEROUS WHEN WET
	COPPER ACETOARSENITE	6.1	UN1585	POISON
	COPPER ARSENITE	6.1	UN1586	POISON
	COPPER BASED PESTICIDES, LIQUID, FLAMMABLE, TOXIC, N.O.S. *FLASHPOINT LESS THAN 23 DEG C*	3.2	UN2776	FLAMMABLE LIQUID AND POISON OR ST. ANDREWS CROSS (ACCORDING TO TOXICITY)
	COPPER BASED PESTICIDES, LIQUID, TOXIC, FLAMMABLE, N.O.S., *FLASHPOINT BETWEEN 23 DEG C AND 61 DEG C*	6.1 6.1 6.1	UN3009 UN3009 UN3009	POISON, FLAMMABLE LIQUID POISON, FLAMMABLE LIQUID ST. ANDREWS CROSS, FLAMMABLE LIQUID

§172.102 OPTIONAL HAZARDOUS MATERIALS TABLE

(1) Notes and Symbols	(2) Hazardous Materials and Description and Proper Shipping Names	(3) IMO Class	(4) Identification Number	(5) Label(s) required
	COPPER BASED PESTICIDES, LIQUID, TOXIC, N.O.S.	6.1	UN3010	POISON
		6.1	UN3010	POISON
		6.1	UN3010	ST. ANDREWS CROSS
	COPPER BASED PESTICIDES, SOLID, TOXIC, N.O.S.	6.1	UN2775	POISON
		6.1	UN2775	ST. ANDREWS CROSS
	COPPER CHLORATE	5.1	UN2721	OXIDIZER
	COPPER CYANIDE	6.1	UN1587	POISON
	COPRA	4.2	UN1363	NONE. PACKAGE TO BE MARKED 'CLASS 4.2'
	CORD, DETONATING, *FLEXIBLE*	1.4 D	UN0289	EXPLOSIVE (1.4D)
N	CORD, DETONATING, *FLEXIBLE*	1.1D	UN0065	EXPLOSIVE (1.1D)
N	CORD, DETONATING, *METAL CLAD*	1.1D	UN0290	EXPLOSIVE (1.1D)
N	CORD, DETONATING, *METAL CLAD*	1.2D	UN0102	EXPLOSIVE (1.2D)
	CORD, DETONATING, MILD EFFECT, *METAL CLAD*	1.4 D	UN0104	EXPLOSIVE (1.4D)
	CORD, IGNITER	1.4 G	UN0066	EXPLOSIVE (1.4G)
	CORROSIVE LIQUIDS, FLAMMABLE, N.O.S.	8	UN2920	CORROSIVE, FLAMMABLE LIQUID
	CORROSIVE LIQUIDS, N.O.S.	8	UN1760	CORROSIVE
		8	UN1760	CORROSIVE
	CORROSIVE LIQUIDS, POISONOUS, N.O.S.	8	UN2922	CORROSIVE, POISON
	CORROSIVE SOLIDS, FLAMMABLE, N.O.S.	8	UN2921	CORROSIVE, FLAMMABLE SOLID
	CORROSIVE SOLIDS, N.O.S.	8	UN1759	CORROSIVE
		8	UN1759	CORROSIVE
	CORROSIVE SOLIDS, POISONOUS, N.O.S.	8	UN2923	CORROSIVE, POISON
	COSMETICS. *SEE* PERFUMERY PRODUCTS			
	COTTON, DRY. *SEE* FIBRES, *VEGETABLE, DRY*			
	COTTON WASTE, OILY *IMPREGNATED WITH LESS THAN 5% OF ANIMAL OR VEGETABLE OIL*	4.2	UN1364	SPONTANEOUSLY COMBUSTIBLE
	COTTON WASTE, OILY *IMPREGNATED WITH 5%, OR MORE, OF ANIMAL OR VEGETABLE OIL*	4.2	UN1364	SPONTANEOUSLY COMBUSTIBLE
	COTTON, *WET OR CONTAMINATED*	4.2	UN1365	SPONTANEOUSLY COMBUSTIBLE
	CRESOLS *(O-, M-, P-)*	6.1	UN2076	POISON
	CRESYLIC ACID	6.1	UN2022	POISON
	CROCIDOLITE. *SEE* ASBESTOS, BLUE			
	CROTONALDEHYDE, INHIBITED	3.2	UN1143	FLAMMABLE LIQUID
	CROTONIC ACID	8	UN2823	CORROSIVE
	CROTONYLENE	3.1	UN1144	FLAMMABLE LIQUID
	CUMENE HYDROPEROXIDE. *SEE* CUMYL HYDROPEROXIDE			
	CUMYL HYDROPEROXIDE, *TECHNICAL PURE*	5.2	UN2116	ORGANIC PEROXIDE
	CUMYL PEROXYNEODECANOATE, *MAXIMUM CONCENTRATION 77% IN SOLUTION*	5.2	UN2963	ORGANIC PEROXIDE
	CUMYL PEROXYPIVALATE, *MAXIMUM CONCENTRATION 77% IN SOLUTION*	5.2	UN2964	ORGANIC PEROXIDE
	CUPRIC CYANIDE. SEE COPPER CYANIDE			
	CUPRIETHYLENEDIAMINE, SOLUTION	8	UN1761	CORROSIVE, POISON
	CUT-BACKS, *ASPHALT OR BITUMEN*	3.2	UN1999	FLAMMABLE LIQUID
		3.3	UN1999	FLAMMABLE LIQUID
	CUTTERS, CABLE, EXPLOSIVE	1.4 S	UN0070	NONE. PACKAGE TO BE MARKED '1.4 S'
	CYANIDE MIXTURES. *SEE* CYANIDES, *(INORGANIC)*, N.O.S.			
	CYANIDES, INORGANIC, N.O.S.	6.1	UN1588	POISON
		6.1	UN1588	ST. ANDREWS CROSS
	CYANIDE, SOLUTIONS	6.1	UN1935	POISON
	CYANOGEN	2.3	UN1026	POISON GAS, FLAMMABLE GAS
	CYANOGEN BROMIDE	6.1	UN1889	POISON CORROSIVE
	CYANOGEN CHLORIDE, *INHIBITED*	2.3	UN1589	POISON GAS
	CYANURIC CHLORIDE	8	UN2670	CORROSIVE
	CYCLOBUTANE	2.1	UN2601	FLAMMABLE GAS
	CYCLOBUTYLCHLOROFORMATE	6.1	UN2744	POISON, CORROSIVE, FLAMMABLE LIQUID
	1,5,9-CYCLODODECATRIENE	8	UN2518	CORROSIVE
	CYCLOHEPTANE	3.2	UN2241	FLAMMABLE LIQUID
	CYCLOHEPTATRIENE	3.2	UN2603	FLAMMABLE LIQUID, POISON
	CYCLOHEPTENE	3.2	UN2242	FLAMMABLE LIQUID

§172.102 OPTIONAL HAZARDOUS MATERIALS TABLE

(1) Notes and Symbols	(2) Hazardous Materials and Description and Proper Shipping Names	(3) IMO Class	(4) Identi-fication Number	(5) Label(s) required
	CYCLOHEXANE	3.1	UN1145	FLAMMABLE LIQUID
	CYCLOHEXANONE	3.3	UN1915	FLAMMABLE LIQUID
	CYCLOHEXANONE PEROXIDES, *IN A CONCENTRATION OF MORE THAN 90% WITH LESS THAN 10% WATER*	5.2	UN2117	ORGANIC PEROXIDE
	CYCLOHEXANONE PEROXIDES, *MAXIMUM CONCENTRATION 90%, WITH AT LEAST 10% WATER*	5.2	UN2119	ORGANIC PEROXIDE
	CYCLOHEXANONE PEROXIDES, *MAXIMUM CONCENTRATION 72% AND CONTAINING NOT MORE THAN 9% AVAILABLE OXYGEN, IN SOLUTION*	5.2	UN2118	ORGANIC PEROXIDE
	CYCLOHEXANONE PEROXIDES, *MAXIMUM CONCENTRATION 72% AND CONTAINING NOT MORE THAN 9% AVAILABLE OXYGEN, AS A PASTE*	5.2	UN2896	ORGANIC PEROXIDE
	CYCLOHEXENE	3.1	UN2256	FLAMMABLE LIQUID
	CYCLOHEXENE	3.2	UN2256	FLAMMABLE LIQUID
	CYCLOHEXENYL TRICHLOROSILANE	8	UN1762	CORROSIVE
	CYCLOHEXYL ACETATE	3.3	UN2243	FLAMMABLE LIQUID
	CYCLOHEXYLAMINE	3.2 3.3	UN2357 UN2357	FLAMMABLE LIQUID, CORROSIVE FLAMMABLE LIQUID, CORROSIVE
	CYCLOHEXYL ISOCYANATE	6.1	UN2488	POISON, FLAMMABLE LIQUID
	CYCLOHEXYL TRICHLOROSILANE	8	UN1763	CORROSIVE
	CYCLOOCTADIENES	3.3	UN2520	FLAMMABLE LIQUID
	CYCLOOCTATETRAENE	3.2	UN2358	FLAMMABLE LIQUID
	CYCLOPENTANE	3.1	UN1146	FLAMMABLE LIQUID
	CYCLOPENTANOL	3.3	UN2244	FLAMMABLE LIQUID
	CYCLOPENTANONE	3.3	UN2245	FLAMMABLE LIQUID
	CYCLOPENTENE	3.1	UN2246	FLAMMABLE LIQUID
	CYCLOPROPANE, LIQUEFIED	2.1	UN1027	FLAMMABLE GAS
N	CYCLOTETRAMETHYLENETETRANITRAMINE, WETTED *WITH NOT LESS THAN 25% WATER, BY WEIGHT, OR* DESENSITIZED *WITH NOT LESS THAN 10% PHLEGMATIZER, BY WEIGHT*	1.1D	UN0226	EXPLOSIVE (1.1D)
N	CYCLOTRIMETHYLENETRINITRAMINE AND CYCLOTETRAMETHYLENE-TETRANITRAMINE, MIXTURES, WETTED *WITH NOT LESS THAN 25% WATER, BY WEIGHT, OR* DESENSITIZED *WITH NOT LESS THAN 10% PHLEGMATIZER, BY WEIGHT*	1.1D	UN0391	EXPLOSIVE (1.1D)
N	CYCLOTRIMETHYLENETRINITRAMINE, WETTED *WITH NOT LESS THAN 25%, WATER, BY WEIGHT, OR* DESENSITIZED *WITH NOT LESS THAN 10% PHLEGMATIZER, BY WEIGHT*	1.1D	UN0072	EXPLOSIVE (1.1D), CORROSIVE
	CYMENES	3.3	UN2046	FLAMMABLE LIQUID
	DECABORANE	4.1	UN1868	FLAMMABLE SOLID, POISON
	DECAHYDRONAPHTHALENE	3.3	UN1147	FLAMMABLE LIQUID
	DECALIN. SEE DECAHYDRONAPHTHALENE			
	N-DECANE	3.3	UN2247	FLAMMABLE LIQUID
N	DEFLAGRATING METAL SALTS OF AROMATIC NITRO-DERIVATIVES, N.O.S.	1.3C	UN0132	EXPLOSIVE (1.3C)
	DETONATOR ASSEMBLIES, NON-ELECTRIC, *FOR BLASTING*	1.4B	UN0361	EXPLOSIVE (1.4B)
N	DETONATOR ASSEMBLIES, NON-ELECTRIC, *FOR BLASTING*	1.1B	UN0360	EXPLOSIVE (1.1B)
	DETONATORS, ELECTRIC, *FOR BLASTING*	1.4 B	UN0255	EXPLOSIVE (1.4B)
N	DETONATORS, ELECTRIC, *FOR BLASTING*	1.1B	UN0030	EXPLOSIVE (1.1B)
	DETONATORS FOR AMMUNITION	1.4B	UN0365	EXPLOSIVE (1.4B)
	DETONATORS FOR AMMUNITION	1.4S	UN0366	NONE. PACKAGE TO BE MARKED '1.4S'
N	DETONATORS FOR AMMUNITION	1.1B	UN0073	EXPLOSIVE (1.1B)
N	DETONATORS FOR AMMUNITION	1.2B	UN0364	EXPLOSIVE (,1.2B)
	DETONATORS, NON-ELECTRIC, *FOR BLASTING*	1.4B	UN0267	EXPLOSIVE (1.4B)
N	DETONATORS, NON-ELECTRIC, *FOR BLASTING*	1.1B	UN0029	EXPLOSIVE (1.1B)
	DEUTERIUM	2.1	UN1957	FLAMMABLE GAS
	DIACETONE ALCHOHOL	3.2 3.3	UN1148 UN1148	FLAMMABLE LIQUID FLAMMABLE LIQUID
	DIACETONE ALCOHOL PEROXIDES, *MAXIMUM 57% IN SOLUTION WITH MAXIMUM 9% HYDROGEN PEROXIDE, MINIMUM 26% DIACETONE ALCOHOL AND MINIMUM 9% WATER; TOTAL ACTIVE OXYGEN CONTENT MAXIMUM 10%*	5.2	UN2163	ORGANIC PEROXIDE
	DIALLYLAMINE	3.2	UN2359	FLAMMABLE LIQUID
	DIALLYLETHER	3.2	UN2360	FLAMMABLE LIQUID, POISON
	4,4'-DIAMINODIPHENYL METHANE	6.1	UN2651	ST. ANDREWS CROSS
	DI-N-AMYLAMINE	6.1	UN2841	ST. ANDREWS CROSS, FLAMMABLE LIQUID

§172.102 OPTIONAL HAZARDOUS MATERIALS TABLE

(1) Notes and Symbols	(2) Hazardous Materials and Description and Proper Shipping Names	(3) IMO Class	(4) Identification Number	(5) Label(s) required
N	DIAZODINITROPHENOL, *CONTAINING, BY WEIGHT, AT LEAST 40% WATER OR MIXTURE OF ALCOHOL AND WATER*	1.1A	UN0074	EXPLOSIVE (1.1A)
	DIBENZYLDICHLOROSILANE	8	UN2434	CORROSIVE
	DIBENZYL PEROXYDICARBONATE, *MAXIMUM CONCENTRATION 87% WITH WATER*	5.2	UN2149	ORGANIC PEROXIDE
	DIBORANE	2.1	UN1911	FLAMMABLE GAS, POISON GAS
	DIBROMOBENZENE	3.3	UN2711	FLAMMABLE LIQUID
	1,2-DIBROMOBUTAN-3-ONE	6.1	UN2648	POISON
	1,2-DIBROMO-3-CHLOROPROPANE	6.1	UN2872	ST. ANDREWS CROSS
	DIBROMODIFLUOROMETHANE	9	UN1941	NONE
	DIBROMOMETHANE	6.1	UN2664	ST. ANDREWS CROSS
	DI-(N-BUTYL)AMINE	8	UN2248	CORROSIVE, FLAMMABLE LIQUID
	N,N-DI-N-BUTYLAMINOETHANOL	6.1	UN2873	ST. ANDREWS CROSS
	DI-(4-TERT-BUTYLCYCLOHEXYL) PEROXYDICARBONATE, *MAXIMUM CONCENTRATION 42%, STABLE DISPERSION IN WATER*	5.2	UN2894	ORGANIC PEROXIDE
	DI-(4-TERT-BUTYLCYCLOHEXYL) PEROXYDICARBONATE, *TECHNICAL PURE*	5.2	UN2154	ORGANIC PEROXIDE
	DIBUTYL ETHERS	3.3	UN1149	FLAMMABLE LIQUID
	DI-TERT-BUTYL PEROXIDE, *TECHNICAL PURE*	5.2	UN2102	ORGANIC PEROXIDE, FLAMMABLE LIQUID
	2,2-DI-(TERT-BUTYLPEROXY) BUTANE, *MAXIMUM CONCENTRATION 55% IN SOLUTION*	5.2	UN2111	ORGANIC PEROXIDE
	1,1-DI-(TERT-BUTYLPEROXY) CYCLOHEXANE, *MAXIMUM CONCENTRATION 77% IN SOLUTION*	5.2	UN2180	ORGANIC PEROXIDE
	1,1-DI-(TERT-BUTYLPEROXY) CYCLOHEXANE, *TECHNICAL PURE*	5.2	UN2179	ORGANIC PEROXIDE
	1,1-DI-(TERT-BUTYLPEROXY) CYCLOHEXANE *WITH AT LEAST 13% PHLEGMATIZER AND 47% INERT INORGANIC SOLID*	5.2	UN2885	ORGANIC PEROXIDE
	1,1-DI-(TERT-BUTYLPEROXY) CYCLOHEXANE *WITH AT LEAST 50% PHLEGMATIZER*	5.2	UN2897	ORGANIC PEROXIDE
	DI-SEC-BUTYL PEROXYDICARBONATE, *MAXIMUM CONCENTRATION 52% IN SOLUTION*	5.2	UN2151	ORGANIC PEROXIDE
	DI-SEC-BUTYL PEROXYDICARBONATE, *TECHNICAL PURE*	5.2	UN2150	ORGANIC PEROXIDE
	1,4-DI-(2-TERT-BUTYLPEROXY ISOPROPYL) BENZENE, *OR* 1,3-DI-(2-TERT-BUTYLPEROXY ISOPROPYL) BENZENE, *OR MIXTURES THEREOF, TECHNICAL PURE OR IN A CONCENTRATION OF MORE THAN 40% WITH INERT SOLID*	5.2	UN2112	ORGANIC PEROXIDE
	DI-(TERT-BUTYLPEROXY)PHTHALATE, *MAXIMUM CONCENTRATION 55% AS A PASTE*	5.2	UN2108	ORGANIC PEROXIDE
	DI-(TERT-BUTYLPEROXY)PHTHALATE, *MAXIMUM CONCENTRATION 55% IN SOLUTION*	5.2	UN2107	ORGANIC PEROXIDE
	DI-(TERT-BUTYLPEROXY)PHTHALATE, *TECHNICAL PURE*	5.2	UN2106	ORGANIC PEROXIDE
	2,2-DI-(TERT-BUTYLPEROXY) PROPANE, *WITH AT LEAST 50% PHLEGMATIZER*	5.2	UN2883	ORGANIC PEROXIDE
	2,2-DI-(TERT-BUTYLPEROXY) PROPANE, *WITH AT LEAST 13% PHLEGMATIZER AND 47% INERT INORGANIC SOLID*	5.2	UN2884	ORGANIC PEROXIDE
	1,1-DI-(TERT-BUTYLPEROXY)-3,3,5-TRIMETHYL CYCLOHEXANE, *TECHNICAL PURE*	5.2	UN2145	ORGANIC PEROXIDE
	1,1-DI-(TERT-BUTYLPEROXY)-3,3,5-TRIMETHYL CYCLOHEXANE, *MAXIMUM CONCENTRATION 57% IN SOLVENT*	5.2	UN2146	ORGANIC PEROXIDE
	1,1-DI-(TERT-BUTYLPEROXY)-3,3,5-TRIMETHYL CYCLOHEXANE, *MAXIMUM CONCENTRATION 58% WITH INERT SOLID*	5.2	UN2147	ORGANIC PEROXIDE
	DICETYL PEROXYDICARBONATE *MAXIMUM CONCENTRATION 42%, STABLE DISPERSION IN WATER*	5.2	UN2895	ORGANIC PEROXIDE
	DICETYL PEROXYDICARBONATE, *TECHNICAL PURE*	5.2	UN2164	ORGANIC PEROXIDE
	DICHLOROACETIC ACID	8	UN1764	CORROSIVE
	1,3-DICHLOROACETONE	6.1	UN2649	POISON
	DICHLOROACETYL CHLORIDE	8	UN1765	CORROSIVE
	DICHLOROANILINES	6.1	UN1590	POISON
	O-DICHLOROBENZENE	6.1	UN1591	ST. ANDREWS CROSS
	P-DICHLOROBENZENE	6.1	UN1592	ST. ANDREWS CROSS
	DI-(4-CHLOROBENZOYL) PEROXIDE, *MAXIMUM CONCENTRATION 52% AS A PASTE*	5.2	UN2114	ORGANIC PEROXIDE
	DI-(4-CHLOROBENZOYL) PEROXIDE, *MAXIMUM CONCENTRATION 52% IN SOLUTION*	5.2	UN2115	ORGANIC PEROXIDE

§172.102 OPTIONAL HAZARDOUS MATERIALS TABLE

(1) Notes and Symbols	(2) Hazardous Materials and Description and Proper Shipping Names	(3) IMO Class	(4) Identification Number	(5) Label(s) required
	DI-(4-CHLOROBENZOYL) PEROXIDE, *MAXIMUM CONCENTRATION 75% WITH WATER*	5.2	UN2113	ORGANIC PEROXIDE
	DICHLORODIFLUOROMETHANE	2.2	UN1028	NONFLAMMABLE GAS
	DICHLORODIFLUOROMETHANE AND DIFLUOROETHANE, AZEOTROPIC MIXTURE, *WITH APPROXIMATELY 74% DICHLORODIFLUOROMETHANE*	2.2	UN2602	NONFLAMMABLE GAS
	DICHLORODIMETHYL ETHER, SYMMETRICAL	6.1	UN2249	POISON, FLAMMABLE LIQUID
	1,1-DICHLOROETHANE	3.2	UN2362	FLAMMABLE LIQUID
	DICHLOROETHYLENE	3.2	UN1150	FLAMMABLE LIQUID
	DICHLOROETHYL ETHER	6.1	UN1916	POISON
	DICHLOROISOCYANURIC ACID, DRY *OR* DICHLOROISOCYANURIC ACID SALTS	5.1	UN2465	OXIDIZER
	DICHLOROISOPROPYL ETHER	6.1	UN2490	POISON
	DICHLOROMETHANE	6.1	UN1593	ST. ANDREWS CROSS
	DICHLOROMONOFLUOROMETHANE	2.2	UN1029	NONFLAMMABLE GAS
	1,1-DICHLORO-1-NITROETHANE	6.1	UN2650	POISON
	DICHLOROPENTANES	3.3	UN1152	FLAMMABLE LIQUID
	DICHLOROPHENYL ISOCYANATES	6.1	UN2250	POISON
	DICHLOROPHENYL TRICHLOROSILANE	8	UN1766	CORROSIVE
	1,3-DICHLOROPROPANOL-2	6.1	UN2750	POISON
	DICHLOROPROPENE	3.3	UN2047	FLAMMABLE LIQUID
	DICHLOROSILANE	2.3	UN2189	POISON GAS, FLAMMABLE GAS
	DICHLOROTETRAFLUOROETHANE	2.2	UN1958	NONFLAMMABLE GAS
	DICUMYL PEROXIDE, *TECHNICAL PURE OR WITH INERT SOLID*	5.2	UN2121	ORGANIC PEROXIDE
	DICYCLOHEXYLAMINE	8	UN2565	CORROSIVE
	DICYCLOHEXYLAMMONIUM NITRITE	6.1	UN2687	ST. ANDREWS CROSS
	DICYCLOHEXYL PEROXYDICARBONATE, *MAXIMUM CONCENTRATION 91% WITH WATER*	5.2	UN2153	ORGANIC PEROXIDE
	DICYCLOHEXYL PEROXYDICARBONATE, *TECHNICAL PURE*	5.2	UN2152	ORGANIC PEROXIDE
	DICYCLOPENTADIENE	3.3	UN2048	FLAMMABLE LIQUID
	DIDECANOYL PEROXIDE, *TECHNICAL PURE*	5.2	UN2120	ORGANIC PEROXIDE
	2,2-DI-(4,4-DI-TERT-BUTYLPEROXY CYCLOHEXYL) PROPANE, *MAXIMUM CONCENTRATION 42% WITH INERT SOLID*	5.2	UN2168	ORGANIC PEROXIDE
	DI-2,4-DICHLOROBENZOYL PEROXIDE, *MAXIMUM CONCENTRATION 75% WITH WATER*	5.2	UN2137	ORGANIC PEROXIDE
	DI-2,4-DICHLOROBENZOYL PEROXIDE, *MAXIMUM CONCENTRATION 52% AS A PASTE*	5.2	UN2138	ORGANIC PEROXIDE
	DI-2,4-DICHLOROBENZOYL PEROXIDE, *MAXIMUM CONCENTRATION 52% IN SOLUTION*	5.2	UN2139	ORGANIC PEROXIDE
	1,2-DI-(DIMETHYLAMINO)-ETHANE	3.2	UN2372	FLAMMABLE LIQUID
	DIDYMIUM NITRATE	5.1	UN1465	OXIDIZER
	DIETHOXYMETHANE	3.1	UN2373	FLAMMABLE LIQUID
	3,3-DIETHOXYPROPENE	3.2	UN2374	FLAMMABLE LIQUID
	DIETHYLALUMINIUM CHLORIDE	4.2	UN1101	SPONTANEOUSLY COMBUSTIBLE
	DIETHYLAMINE	3.1	UN1154	FLAMMABLE LIQUID
	DIETHYLAMINOETHANOL	3.3	UN2686	FLAMMABLE LIQUID
	3-(DIETHYLAMINO) PROPYLAMINE	8	UN2684	CORROSIVE, FLAMMABLE LIQUID
	N,N-DIETHYLANILINE	6.1	UN2432	ST. ANDREWS CROSS
	DIETHYLBENZENE	3.3	UN2049	FLAMMABLE LIQUID
	DIETHYLCARBINOL	3.3	UN2706	FLAMMABLE LIQUID
	DIETHYL CARBONATE	3.3	UN2366	FLAMMABLE LIQUID
	DIETHYL DICHLOROSILANE	8	UN1767	CORROSIVE, FLAMMABLE LIQUID
	N,N-DIETHYLETHYLENE DIAMINE	8	UN2685	CORROSIVE, FLAMMABLE LIQUID
N	DIETHYLENEGLYCOL DINITRATE, DESENSITIZED *WITH NOT LESS THAN 25% NON-VOLATILE WATER-INSOLUBLE PHLEGMATIZER, BY WEIGHT*	1.1D	UN0075	EXPLOSIVE (1.1D)
	DIETHYLENETRIAMINE	8	UN2079	CORROSIVE
	DIETHYL ETHER	3.1	UN1155	FLAMMABLE LIQUID
	DI-(2-ETHYLHEXYL) PEROXYDICARBONATE, *MAXIMUM CONCENTRATION 77% IN SOLUTION*	5.2	UN2123	ORGANIC PEROXIDE
	DI-(2-ETHYLHEXYL) PEROXYDICARBONATE, *MAXIMUM CONCENTRATION 32%, STABLE DISPERSION IN WATER*	5.2	UN2960	ORGANIC PEROXIDE
	DI-(2-ETHYLHEXYL) PEROXYDICARBONATE, *TECHNICAL PURE*	5.2	UN2122	ORGANIC PEROXIDE
	DIETHYL KETONE	3.2	UN1156	FLAMMABLE LIQUID

§172.102 OPTIONAL HAZARDOUS MATERIALS TABLE

(1) Notes and Symbols	(2) Hazardous Materials and Description and Proper Shipping Names	(3) IMO Class	(4) Identi-fication Number	(5) Label(s) required
	DIETHYLMAGNESIUM	4.2	UN1367	SPONTANEOUSLY COMBUSTIBLE
	DIETHYL-P-NITROSOANILINE	4.2	--	SPONTANEOUSLY COMBUSTIBLE
	DIETHYL PEROXYDICARBONATE, *MAXIMUM CONCENTRATION 27% IN SOLUTION*	5.2	UN2175	ORGANIC PEROXIDE
	DIETHYL SULFIDE	3.2	UN2375	FLAMMABLE LIQUID, POISON
	DIETHYL SULPHATE	6.1	UN1594	POISON
	DIETHYLTHIOPHOSPHORYL CHLORIDE	8	UN2751	CORROSIVE, FLAMMABLE LIQUID
	DIETHYLZINC	4.2	UN1366	SPONTANEOUSLY COMBUSTIBLE
	1,1-DIFLUOROETHANE	2.1	UN1030	FLAMMABLE GAS
	1,1-DIFLUOROETHYLENE	2.1	UN1959	FLAMMABLE GAS
	DIFLUOROMONOCHLOROETHANE. *SEE* CHLORODIFLUOROETHANES			
	DIFLUOROPHOSPHORIC ACID, *ANHYDROUS*	8	UN1768	CORROSIVE
	2,2-DIHYDROPEROXY PROPANE, *MAXIMUM CONCENTRATION 25% WITH INERT INORGANIC SOLID*	5.2	UN2178	ORGANIC PEROXIDE
	2,3-DIHYDROPYRAN	3.2	UN2376	FLAMMABLE LIQUID
	DI-(1-HYDROXY CYCLOHEXYL) PEROXIDE, *TECHNICAL PURE*	5.2	UN2148	ORGANIC PEROXIDE
	DIISOBUTYLAMINE	3.3	UN2361	FLAMMABLE LIQUID
	DIISOBUTYLENE, *(ISOMERIC COMPOUNDS)*	3.2	UN2050	FLAMMABLE LIQUID
	DIISOBUTYL KETONE	3.3	UN1157	FLAMMABLE LIQUID
	DIISOOCTYL ACID PHOSPHATE	8	UN1902	CORROSIVE
	DIISOPROPYLAMINE	3.2	UN1158	FLAMMABLE LIQUID
	DIISOPROPYLBENZENE HYDROPEROXIDE, *MAXIMUM CONCENTRATION 72% IN SOLUTION*	5.2	UN2171	ORGANIC PEROXIDE
	N,N-DIISOPROPYL ETHANOLAMINE	8	UN2825	CORROSIVE
	DIISOPROPYL ETHER	3.1	UN1159	FLAMMABLE LIQUID
	DIISOPROPYL PEROXYDICARBONATE, *TECHNICAL PURE*	5.2	UN2133	ORGANIC PEROXIDE
	DIISOPROPYL PEROXYDICARBONATE, *MAXIMUM CONCENTRATION 52% IN SOLUTION*	5.2	UN2134	ORGANIC PEROXIDE
	DIISOTRIDECYL PEROXIDICARBONATE, *TECHNICAL PURE*	5.2	UN2889	ORGANIC PEROXIDE
	DIKETENE, INHIBITED	3.3	UN2521	FLAMMABLE LIQUID
	DILAUROYL PEROXIDE, *MAXIMUM CONCENTRATION 42%, STABLE DISPERSION IN WATER*	5.2	UN2893	ORGANIC PEROXIDE
	DILAUROYL PEROXIDE, *TECHNICAL PURE*	5.2	UN2124	ORGANIC PEROXIDE
	1,1-DIMETHOXYETHANE	3.1 3.2	UN2377 UN2377	FLAMMABLE LIQUID FLAMMABLE LIQUID
	1,2-DIMETHOXYETHANE	3.2	UN2252	FLAMMABLE LIQUID
	DIMETHYLAMINE, *ANHYDROUS*	2.1	UN1032	FLAMMABLE GAS
	DIMETHYLAMINE, SOLUTION	3.2	UN1160	FLAMMABLE LIQUID
	2-DIMETHYLAMINOACETONITRILE	3.3	UN2378	FLAMMABLE LIQUID, POISON
	DIMETHYLAMINOETHYL METHACRYLATE	6.1	UN2522	POISON
	N,N-DIMETHYLANILINE	6.1	UN2253	POISON
	DI-(2-METHYLBENZOYL)PEROXIDE, *WITH AT LEAST 15% WATER*	5.2	UN2593	ORGANIC PEROXIDE
	2,3-DIMETHYLBUTANE	3.1	UN2457	FLAMMABLE LIQUID
	1,3-DIMETHYLBUTYLAMINE	3.2	UN2379	FLAMMABLE LIQUID
	N,N-DIMETHYLCARBAMOYL CHLORIDE	8	UN2262	CORROSIVE
	DIMETHYL CARBONATE	3.2	UN1161	FLAMMABLE LIQUID
	DIMETHYLCYCLOHEXANES	3.2	UN2263	FLAMMABLE LIQUID
	N,N-DIMETHYLCYCLOHEXYLAMINE	8	UN2264	CORROSIVE, FLAMMABLE LIQUID
	2,5-DIMETHYL-2,5-DI-(BENZOYLPEROXY) HEXANE, *MAXIMUM CONCENTRATION 82% WITH WATER*	5.2	UN2959	ORGANIC PEROXIDE
	2,5-DIMETHYL-2,5-DI-(BENZOYLPEROXY) HEXANE, *MAXIMUM CONCENTRATION 82% WITH INERT SOLID*	5.2	UN2173	ORGANIC PEROXIDE
	2,5-DIMETHYL-2,5-DI-(BENZOYLPEROXY) HEXANE, *TECHNICAL PURE*	5.2	UN2172	ORGANIC PEROXIDE
	2,5-DIMETHYL-2,5-DI-(TERT-BUTYLPEROXY) HEXANE, *MAXIMUM CONCENTRATION 52% WITH INERT SOLID*	5.2	UN2156	ORGANIC PEROXIDE
	2,5-DIMETHYL-2,5-DI-(TERT-BUTYLPEROXY) HEXANE, *TECHNICAL PURE*	5.2	UN2155	ORGANIC PEROXIDE
	2,5-DIMETHYL-2,5-DI-(TERT-BUTYLPEROXY) HEXYNE-3, *MAXIMUM CONCENTRATION 52% WITH INERT SOLID*	5.2	UN2159	ORGANIC PEROXIDE
	2,5-DIMETHYL-2,5-DI-(TERT-BUTYLPEROXY) HEXYNE-3, *TECHNICAL PURE*	5.2	UN2158	ORGANIC PEROXIDE
	DIMETHYLDICHLOROSILANE	3.2	UN1162	FLAMMABLE LIQUID, CORROSIVE
	DIMETHYLDIETHOXYSILANE	3.2	UN2380	FLAMMABLE LIQUID

§172.102 OPTIONAL HAZARDOUS MATERIALS TABLE

(1) Notes and Symbols	(2) Hazardous Materials and Description and Proper Shipping Names	(3) IMO Class	(4) Identification Number	(5) Label(s) required
	2,5-DIMETHYL-2,5-DI -(2-ETHYLHEXANOYLPEROXY) HEXANE, *TECHNICAL PURE*	5.2	UN2157	ORGANIC PEROXIDE
	2,5-DIMETHYL-2,5-DIHYDROPEROXY HEXANE, *MAXIMUM CONCENTRATION 82% WITH WATER*	5.2	UN2174	ORGANIC PEROXIDE
	DIMETHYLDIOXANES	3.2	UN2707	FLAMMABLE LIQUID
	DIMETHYLDIOXANES	3.3	UN2707	FLAMMABLE LIQUID
	DIMETHYL DISULFIDE	3.2	UN2381	FLAMMABLE LIQUID
	DIMETHYLETHANOLAMINE	3.3	UN2051	FLAMMABLE LIQUID
	DIMETHYL ETHER	2.1	UN1033	FLAMMABLE GAS
	N,N-DIMETHYLFORMAMIDE	3.3	UN2265	FLAMMABLE LIQUID
	DIMETHYLHYDRAZINE, *SYMMETRICAL*	3.2	UN2382	FLAMMABLE LIQUID, POISON
	DIMETHYLHYDRAZINE, UNSYMMETRICAL	3.2	UN1163	FLAMMABLE LIQUID, CORROSIVE
	DIMETHYLMAGNESIUM	4.2	UN1368	SPONTANEOUSLY COMBUSTIBLE
	DIMETHYL-P-NITROSOANILINE. *SEE* P-NITROSODIMETHYLANILINE			
	2,2-DIMETHYLPROPANE *OTHER THAN PENTANE AND ISOPENTANE*	2.1	UN2044	FLAMMABLE GAS
	DIMETHYL-N-PROPYLAMINE	3.2	UN2266	FLAMMABLE LIQUID, CORROSIVE
	DIMETHYL SULPHATE	6.1	UN1595	POISON
	DIMETHYL SULPHIDE	3.1	UN1164	FLAMMABLE LIQUID
	DIMETHYL THIOPHOSPHORYL CHLORIDE	8	UN2267	CORROSIVE
	DIMETHYLZINC	4.2	UN1370	SPONTANEOUSLY COMBUSTIBLE
	DIMYRISTYLPEROXYDICARBONATE, *MAXIMUM CONCENTRATION 22%, STABLE DISPERSION IN WATER*	5.2	UN2892	ORGANIC PEROXIDE
	DIMYRISTYLPEROXYDICARBONATE, *TECHNICAL PURE*	5.2	UN2595	ORGANIC PEROXIDE
	DINITROANILINES	6.1	UN1596	POISON
	DINITROBENZENES *(O-, M-, P-)*	6.1	UN1597	POISON
	DINITRO-O-CRESOL, *LIQUID*	6.1 6.1 6.1	UN1598 UN1598 UN1598	POISON POISON ST. ANDREWS CROSS
	DINITRO-O-CRESOL, *SOLID*	6.1 6.1	UN1598 UN1598	POISON ST. ANDREWS CROSS
	DINITROGEN TETROXIDE. *SEE* NITROGEN DIOXIDE			
N	DINITROPHENATES *ALKALI METALS, DRY OR WETTED WITH LESS THAN 15% WATER, BY WEIGHT*	1.3C	UN0077	EXPLOSIVE (1.3C), POISON
	DINITROPHENOLATES, WETTED *WITH, BY WEIGHT, AT LEAST 15% WATER*	4.1	UN1321	FLAMMABLE SOLID, POISON
N	DINITROPHENOL, *DRY OR WETTED WITH LESS THAN 15% WATER, BY WEIGHT*	1.1D	UN0076	EXPLOSIVE (1.1D), POISON
	DINITROPHENOL, SOLUTION *IN WATER OR FLAMMABLE LIQUID*	6.1	UN1599	POISON, FLAMMABLE LIQUID (ONLY IF FLASHPOINT BETWEEN 23 DEG C AND 61 DEG C)
	DINITROPHENOL, WETTED *WITH, BY WEIGHT, AT LEAST 15% WATER*	4.1	UN1320	FLAMMABLE SOLID, POISON
N	DINITRORESORCINOL, *DRY OR WETTED WITH LESS THAN 15% WATER*	1.1D	UN0078	EXPLOSIVE (1.1D)
	DINITRORESORCINOL, WETTED *WITH, BY WEIGHT, AT LEAST 15% WATER*	4.1	UN1322	FLAMMABLE SOLID
N	DINITROSOBENZENE	1.3C	UN0406	EXPLOSIVE (1.3C)
	DINITROTOLUENES, MOLTEN	6.1	UN1600	POISON
	DINITROTOLUENES, SOLID	6.1	UN2038	POISON
	DI-N-NONANOYL PEROXIDE, *TECHNICAL PURE*	5.2	UN2130	ORGANIC PEROXIDE
	DI-N-OCTANOYL PEROXIDE, *TECHNICAL PURE*	5.2	UN2129	ORGANIC PEROXIDE
	DIOXANE	3.2	UN1165	FLAMMABLE LIQUID
	DIOXOLANE	3.2	UN1166	FLAMMABLE LIQUID
	DIPENTENE	3.3	UN2052	FLAMMABLE LIQUID
	DIPEROXYAZELAIC ACID, *MAXIMUM CONCENTRATION 27%, WITH NOT LESS THAN 13% AZELAIC ACID AND NOT LESS THAN 53% SODIUM SULPHATE*	5.2	UN2958	ORGANIC PEROXIDE
	DIPHENYLAMINECHLOROARSINE	6.1	UN1698	POISON
	DIPHENYLCHLOROARSINE	6.1	UN1699	POISON
	DIPHENYL DICHLOROSILANE	8	UN1769	CORROSIVE
	DIPHENYLMETHANE-4,4'DIISOCYANATE	6.1	UN2489	ST. ANDREWS CROSS
	DIPHENYLMETHYL BROMIDE	8	UN1770	CORROSIVE
N	DIPICRYL SULFIDE, *DRY OR WETTED WITH LESS THAN 10% WATER, BY WEIGHT*	1.1D	UN0401	EXPLOSIVE (1.1D)
	DIPICRYL SULPHIDE, WETTED *WITH, BY WEIGHT, AT LEAST 10% WATER*	4.1	UN2852	FLAMMABLE SOLID
	DIPROPIONYL PEROXIDE, *MAXIMUM CONCENTRATION 28% IN SOLUTION*	5.2	UN2132	ORGANIC PEROXIDE
	DIPROPYLAMINE	3.2	UN2383	FLAMMABLE LIQUID

§172.102 OPTIONAL HAZARDOUS MATERIALS TABLE

(1) Notes and Symbols	(2) Hazardous Materials and Description and Proper Shipping Names	(3) IMO Class	(4) Identi- fication Number	(5) Label(s) required
	DIPROPYL ETHER	3.1	UN2384	FLAMMABLE LIQUID
	DIPROPYLKETONE	3.3	UN2710	FLAMMABLE LIQUID
	DI-N-PROPYL PEROXYDICARBONATE, *TECHNICAL PURE*	5.2	UN2176	ORGANIC PEROXIDE
	DISINFECTANTS, CORROSIVE, LIQUID, N.O.S.	8	UN1903	CORROSIVE
	DISINFECTANTS, N.O.S., *POISONOUS, LIQUID OR SOLID*	6.1	UN1601	POISON
		6.1	UN1601	POISON
		6.1	UN1601	ST. ANDREWS CROSS
	DISTEARYLPEROXYDICARBONATE, *WITH 15% STEARYL ALCOHOL*	5.2	UN2592	ORGANIC PEROXIDE
	DISUCCINIC ACID PEROXIDE, *MAXIMUM CONCENTRATION 72%, WETTED WITH WATER*	5.2	UN2962	ORGANIC PEROXIDE
	DISUCCINIC ACID PEROXIDE, *TECHNICAL PURE*	5.2	UN2135	ORGANIC PEROXIDE
	DITHIOCARBAMATE PESTICIDES, LIQUID, FLAMMABLE, TOXIC, N.O.S., *FLASHPOINT LESS THAN 23 DEG C*	3.2	UN2772	FLAMMABLE LIQUID AND POISON OR ST. ANDREWS CROSS (ACCORDING TO TOXICITY)
	DITHIOCARBAMATE PESTICIDES, LIQUID, TOXIC, FLAMMABLE, N.O.S., *FLASHPOINT BETWEEN 23 DEG C AND 61 DEG C*	6.1	UN3005	POISON, FLAMMABLE LIQUID
		6.1	UN3005	POISON, FLAMMABLE LIQUID
		6.1	UN3005	ST. ANDREWS CROSS, FLAMMABLE LIQUID
	DITHIOCARBAMATE PESTICIDES, LIQUID, TOXIC, N.O.S.	6.1	UN3006	POISON
		6.1	UN3006	POISON
		6.1	UN3006	ST. ANDREWS CROSS
	DITHIOCARBAMATE PESTICIDES, SOLID, TOXIC, N.O.S.	6.1	UN2771	POISON
		6.1	UN2771	ST. ANDREWS CROSS
	DI-(3,5,5-TRIMETHYL-1,2-DIOXOLANYL-3)PEROXIDE, *AS A PASTE WITH AT LEAST 50% PHLEGMATIZER*	5.2	UN2597	ORGANIC PEROXIDE
	DI-(3,5,5-TRIMETHYLHEXANOYL)PEROXIDE, *TECHNICAL PURE OR IN SOLUTION*	5.2	UN2128	ORGANIC PEROXIDE
	DIVINYL ETHER, INHIBITED	3.1	UN1167	FLAMMABLE LIQUID
	DODECYL TRICHLOROSILANE	8	UN1771	CORROSIVE
	DRESSING, LEATHER. *SEE* FLAMMABLE LIQUID PREPARATIONS, N.O.S.			
	DRIERS, *PAINT OR VARNISH*, LIQUID, N.O.S.	3.2	UN1168	FLAMMABLE LIQUID
		3.3	UN1168	FLAMMABLE LIQUID
	DRIERS, *PAINT OR VARNISH*, SOLID, N.O.S.	4.1	UN1371	FLAMMABLE SOLID
	DYES, N.O.S. *OR* DYE INTERMEDIATES, N.O.S., *LIQUID OR SOLID, CORROSIVE*	8	UN2801	CORROSIVE, FLAMMABLE LIQUID (ONLY IF FLASHPOINT BETWEEN 23 AND 61 DEG C)
	DYES, N.O.S. *OR* DYE INTERMEDIATES, N.O.S., *LIQUID OR SOLID, POISONOUS*	6.1	UN1602	POISON
		6.1	UN1602	POISON
		6.1	UN1602	ST. ANDREWS CROSS
	ENAMELS. *SEE* PAINTS, *ETC.*			
	ENGINE STARTING FLUID, *WITH FLAMMABLE GAS*	2.1	UN1960	FLAMMABLE GAS
	EPIBROMOHYDRIN	6.1	UN2558	POISON, FLAMMABLE LIQUID
	EPICHLOROHYDRIN	6.1	UN2023	POISON, FLAMMABLE LIQUID
	1,2-EPOXY-3-ETHYLOXY PROPANE	3.3	UN2752	FLAMMABLE LIQUID
	ERADICATORS, *PAINT OR GREASE, LIQUID. TO BE CLASSIFIED AND LABELED ACCORDING TO THE PRINCIPLE HAZARDOUS CONSTITUENT*	--	UN1850	
	ESSENCES. *SEE* EXTRACTS, AROMATIC, LIQUID			
	ETHANE, *COMPRESSED*	2.1	UN1035	FLAMMABLE GAS
	ETHANE, *REFRIGERATED LIQUID*	2.1	UN1961	FLAMMABLE GAS
	ETHANOLAMINE *OR* ETHANOLAMINE SOLUTIONS	8	UN2491	CORROSIVE
	ETHANOL *OR* ETHANOL SOLUTIONS *INCLUDING* ALCOHOLIC BEVERAGES	3.2	UN1170	FLAMMABLE LIQUID
		3.3	UN1170	FLAMMABLE LIQUID
	ETHER. *SEE* DIETHYL ETHER			
	ETHYL ACETATE	3.2	UN1173	FLAMMABLE LIQUID
	ETHYL ACETYLENE, INHIBITED	2.1	UN2452	FLAMMABLE GAS
	ETHYL ACRYLATE, *INHIBITED*	3.2	UN1917	FLAMMABLE LIQUID
	ETHYL ALCOHOL. *SEE* ETHANOL			
	ETHYL ALDEHYDE. *SEE* ACETALDEHYDE			
	ETHYL ALUMINIUM DICHLORIDE	4.2	UN1924	SPONTANEOUSLY COMBUSTIBLE
	ETHYL ALUMINIUM SESQUICHLORIDE	4.2	UN1925	SPONTANEOUSLY COMBUSTIBLE
	ETHYLAMINE	2.1	UN1036	FLAMMABLE GAS
	ETHYLAMINE SOLUTION *IN WATER, CONCENTRATIONS UP TO 70%*	3.1	UN2270	FLAMMABLE LIQUID
		3.2	UN2270	FLAMMABLE LIQUID
		3.3	UN2270	FLAMMABLE LIQUID

§172.102 OPTIONAL HAZARDOUS MATERIALS TABLE

(1) Notes and Symbols	(2) Hazardous Materials and Description and Proper Shipping Names	(3) IMO Class	(4) Identification Number	(5) Label(s) required
	ETHYL AMYL KETONE	3.3	UN2271	FLAMMABLE LIQUID
	N-ETHYLANILINE	6.1	UN2272	ST. ANDREWS CROSS
	2-ETHYLANILINE	6.1	UN2273	ST. ANDREWS CROSS
	ETHYLBENZENE	3.2	UN1175	FLAMMABLE LIQUID
	N-ETHYL-N-BENZYLANILINE	6.1	UN2274	ST. ANDREWS CROSS
	N-ETHYLBENZYLTOLUIDINES	6.1	UN2753	ST. ANDREWS CROSS
	ETHYL BORATE	3.2	UN1176	FLAMMABLE LIQUID
	ETHYL BROMIDE	6.1	UN1891	POISON
	ETHYL BROMOACETATE	6.1	UN1603	POISON, FLAMMABLE LIQUID
	2-ETHYLBUTANOL	3.3	UN2275	FLAMMABLE LIQUID
	ETHYLBUTYL ACETATE	3.3	UN1177	FLAMMABLE LIQUID
	ETHYL BUTYL ETHER	3.2	UN1179	FLAMMABLE LIQUID
	2-ETHYLBUTYRALDEHYDE	3.2	UN1178	FLAMMABLE LIQUID
	ETHYL BUTYRATE	3.3	UN1180	FLAMMABLE LIQUID
	ETHYL CHLORIDE	2.1	UN1037	FLAMMABLE GAS
	ETHYL CHLOROACETATE	6.1	UN1181	POISON, FLAMMABLE LIQUID
	ETHYL CHLOROCARBONATE. SEE ETHYL CHLOROFORMATE			
	ETHYL CHLOROFORMATE	3.2	UN1182	FLAMMABLE LIQUID, POISON CORROSIVE
	ETHYL-2-CHLOROPROPIONATE	3.3	UN2935	FLAMMABLE LIQUID
	ETHYL CHLOROTHIOFORMATE	8	UN2826	CORROSIVE, FLAMMABLE LIQUID
	ETHYL CROTONATE	3.2	UN1862	FLAMMABLE LIQUID
	ETHYL CYANOACETATE	6.1	UN2666	ST. ANDREWS CROSS
	ETHYL-3,3-DI-(TERT-BUTYLPEROXY) BUTYRATE, *MAXIMUM CONCENTRATION 77% IN SOLUTION*	5.2	UN2185	ORGANIC PEROXIDE
	ETHYL-3,3-DI-(TERT-BUTYLPEROXY) BUTYRATE, *TECHNICAL PURE*	5.2	UN2184	ORGANIC PEROXIDE, EXPLOSIVE (MAY NOT BE REQUIRED UNDER CERTAIN CONDITIONS. SEE IMDG CODE)
	ETHYL-3,3-DI(TERT-BUTYLPEROXY)BUTYRATE, *WITH AT LEAST 50% INERT, INORGANIC SOLID*	5.2	UN2598	ORGANIC PEROXIDE
	ETHYL DICHLOROARSINE	6.1	UN1892	POISON
	ETHYLDICHLOROSILANE	3.2	UN1183	FLAMMABLE LIQUID, CORROSIVE
	ETHYLENE CHLOROHYDRIN	6.1	UN1135	POISON, FLAMMABLE LIQUID
	ETHYLENE, *COMPRESSED*	2.1	UN1962	FLAMMABLE GAS
	ETHYLENEDIAMINE	8	UN1604	CORROSIVE, FLAMMABLE LIQUID
	ETHYLENE DIBROMIDE	6.1	UN1605	POISON
	ETHYLENE DICHLORIDE	3.2	UN1184	FLAMMABLE LIQUID, POISON
	ETHYLENE GLYCOL DIETHYL ETHER	3.3	UN1153	FLAMMABLE LIQUID
	ETHYLENE GLYCOL MONOBUTYL ETHER	6.1	UN2369	ST. ANDREWS CROSS, FLAMMABLE LIQUID (ONLY IF FLASHPOINT BELOW 61 DEG C)
	ETHYLENE GLYCOL MONOETHYL ETHER	3.3	UN1171	FLAMMABLE LIQUID
	ETHYLENE GLYCOL MONOETHYL ETHER ACETATE	3.3	UN1172	FLAMMABLE LIQUID
	ETHYLENE GLYCOL MONOMETHYL ETHER	3.3	UN1188	FLAMMABLE LIQUID
	ETHYLENE GLYCOL MONOMETHYL ETHER ACETATE	3.3	UN1189	FLAMMABLE LIQUID
	ETHYLENEIMINE, *INHIBITED*	3.2	UN1185	FLAMMABLE LIQUID, POISON
	ETHYLENE OXIDE AND CARBON DIOXIDE, MIXTURES. SEE CARBON DIOXIDE AND ETHYLENE OXIDE MIXTURES			
	ETHYLENE OXIDE AND PROPYLENE OXIDE MIXTURES, *NOT MORE THAN 30% ETHYLENE OXIDE*	3.1	UN2983	FLAMMABLE LIQUID, POISON
	ETHYLENE OXIDE, *CONTAINING NOT MORE THAN 0.2% OF NITROGEN*	2.1	UN1040	FLAMMABLE GAS, POISON GAS
	ETHYLENE, *REFRIGERATED LIQUID*	2.1	UN1038	FLAMMABLE GAS
	ETHYL ETHER. *SEE* DIETHYL ETHER			
	ETHYL FLUID. *SEE* MOTOR FUEL ANTIKNOCK MIXTURES.			
	ETHYL FLUORIDE	2.1	UN2453	FLAMMABLE GAS
	ETHYL FORMATE	3.1	UN1190	FLAMMABLE LIQUID
	ETHYL HEXALDEHYDE	3.3	UN1191	FLAMMABLE LIQUID
	2-ETHYLHEXYLAMINE	8	UN2276	CORROSIVE, FLAMMABLE LIQUID
	2-ETHYLHEXYLCHLOROFORMATE	6.1	UN2748	POISON CORROSIVE
	ETHYLISOBUTYRATE	3.2	UN2385	FLAMMABLE LIQUID
	ETHYL ISOCYANATE	3.2	UN2481	FLAMMABLE LIQUID, POISON

§172.102 OPTIONAL HAZARDOUS MATERIALS TABLE

(1) Notes and Symbols	(2) Hazardous Materials and Description and Proper Shipping Names	(3) IMO Class	(4) Identification Number	(5) Label(s) required
	ETHYL LACTATE	3.3	UN1192	FLAMMABLE LIQUID
	ETHYL MERCAPTAN	3.1	UN2363	FLAMMABLE LIQUID, POISON
	ETHYL METHACRYLATE, INHIBITED	3.2	UN2277	FLAMMABLE LIQUID
	ETHYL METHYL ETHER	2.1	UN1039	FLAMMABLE GAS
	ETHYL METHYL KETONE	3.2	UN1193	FLAMMABLE LIQUID
	ETHYL METHYL KETONE PEROXIDE(S), *MAXIMUM CONCENTRATION 60%*	5.2	UN2127	ORGANIC PEROXIDE
	ETHYL METHYL KETONE PEROXIDE(S), *MAXIMUM CONCENTRATION 50%, CONTAINING NOT MORE THAN 10% AVAILABLE OXYGEN*	5.2	UN2550	ORGANIC PEROXIDE
N	ETHYL METHYL KETONE PEROXIDE(S), *MAXIMUM CONCENTRATION 50%, CONTAINING MORE THAN 10% AVAILABLE OXYGEN*			
	ETHYL NITRITE, SOLUTIONS	3.1	UN1194	FLAMMABLE LIQUID
	ETHYL ORTHOFORMATE	3.3	UN2524	FLAMMABLE LIQUID
	ETHYL OXALATE	6.1	UN2525	ST. ANDREWS CROSS
	ETHYLPHENYLDICHLOROSILANE	8	UN2435	CORROSIVE
	1-ETHYL PIPERIDINE	3.2	UN2386	FLAMMABLE LIQUID
	ETHYL PROPIONATE	3.2	UN1195	FLAMMABLE LIQUID
	ETHYL PROPYL ETHER	3.1	UN2615	FLAMMABLE LIQUID
	ETHYL SILICATE. *SEE* TETRAETHYL SILICATE			
	ETHYLSULPHURIC ACID	8	UN2571	CORROSIVE
	N-ETHYLTOLUIDINES *(O,M,P)*	6.1	UN2754	POISON, FLAMMABLE LIQUID (ONLY IF FLASHPOINT BELOW 61 DEG C)
	ETHYLTRICHLOROSILANE	3.2	UN1196	FLAMMABLE LIQUID, CORROSIVE
N	EXPLOSIVES, BLASTING, TYPE A	1.1D	UN0081	EXPLOSIVE (1.1D)
	EXPLOSIVES, BLASTING, TYPE B	1.5 D	UN0331	EXPLOSIVE (1.5D)
N	EXPLOSIVES, BLASTING, TYPE B	1.1D	UN0082	EXPLOSIVE (1.1D)
N	EXPLOSIVES, BLASTING, TYPE C	1.1D	UN0083	EXPLOSIVE (1.1D)
N	EXPLOSIVES, BLASTING, TYPE D	1.1D	UN0084	EXPLOSIVE (1.1D)
	EXPLOSIVES, BLASTING, TYPE E	1.5 D	UN0332	EXPLOSIVE (1.5D)
N	EXPLOSIVES, BLASTING, TYPE E	1.1D	UN0241	EXPLOSIVE (1.1D)
	EXTRACTS, AROMATIC, LIQUID	3.2 3.3	UN1169 UN1169	FLAMMABLE LIQUID FLAMMABLE LIQUID
	EXTRACTS, FLAVOURING, LIQUID	3.2	UN1197	FLAMMABLE LIQUID
	FABRIC, ANIMAL OR VEGETABLE. *SEE* FIBRES, ANIMAL OR VEGETABLE	3.3	UN1197	FLAMMABLE LIQUID
	FERRIC ARSENATE	6.1	UN1606	POISON
	FERRIC ARSENITE	6.1	UN1607	POISON
	FERRIC CHLORIDE, ANHYDROUS	8	UN1773	CORROSIVE
	FERRIC CHLORIDE, SOLUTION	8	UN2582	CORROSIVE
	FERRIC NITRATE	5.1	UN1466	OXIDIZER
	FERROCERIUM	4.1	UN1323	FLAMMABLE SOLID
	FERROSILICON, *CONTAINING MORE THAN 30% BUT LESS THAN 90% SILICON.*	4.3	UN1408	DANGEROUS WHEN WET POISON
	FERROUS ARSENATE	6.1	UN1608	POISON
	FERROUS METAL BORINGS, SHAVINGS, TURNINGS, *OR* CUTTINGS *IN A FORM LIABLE TO SELF HEATING*	4.2	UN2793	SPONTANEOUSLY COMBUSTIBLE
	FERTILIZER AMMONIATING SOLUTION, *CONTAINING FREE AMMONIA IN EXCESS OF 35% AMMONIA*	2.2	UN1043	NONFLAMMABLE GAS
	FIBRES, ANIMAL OR VEGETABLE, *BURNT, WET OR DAMP*	4.2	UN1372	SPONTANEOUSLY COMBUSTIBLE
	FIBRES *OR* FABRIC, ANIMAL *OR* VEGETABLE, *CONTAINING MORE THAN 5% OF ANIMAL OR VEGETABLE OIL*	4.2	UN1373	SPONTANEOUSLY COMBUSTIBLE
	FIBRES, *VEGETABLE, DRY*	4.1	--	NONE
	FILLERS, *LIQUID. SEE* PAINTS, *ETC.*			
	FILM, MOTION PICTURE, *NITROCELLULOSE BASE, EXPOSED AND DEVELOPED FILM, OR UNEXPOSED OR UNDEVELOPED FILM*	4.1	UN1324	FLAMMABLE SOLID
	FILM, MOTION PICTURE, *NITROCELLULOSE BASE, OLD FILM*	4.1	UN1324	FLAMMABLE SOLID
	FIRE EXTINGUISHER CHARGES, *CORROSIVE LIQUID*	8	UN1774	CORROSIVE
	FIRE EXTINGUISHERS, *CONTAINING COMPRESSED OR LIQUEFIED GAS*	2.2	UN1044	NONFLAMMABLE GAS
	FIRELIGHTERS, *SOLID, WITH A FLAMMABLE LIQUID*	4.1	UN2623	FLAMMABLE SOLID
N	FIREWORKS, TYPE A	1.1G	UN0333	EXPLOSIVE (1.1G)
N	FIREWORKS, TYPE B	1.2G	UN0334	EXPLOSIVE (1.2G)
N	FIREWORKS, TYPE C	1.3G	UN0335	EXPLOSIVE (1.3G)
	FIREWORKS, TYPE D	1.4 G	UN0336	EXPLOSIVE (1.4G)

§172.102 OPTIONAL HAZARDOUS MATERIALS TABLE

(1) Notes and Symbols	(2) Hazardous Materials and Description and Proper Shipping Names	(3) IMO Class	(4) Identification Number	(5) Label(s) required
	FIREWORKS, TYPE D	1.4 S	UN0337	NONE. PACKAGE TO BE MARKED '1.4 S'
	FISHMEAL *OR* FISHSCRAP, *ANTIOXIDANT TREATED, MOISTURE CONTENT GREATER THAN 5% BUT NOT EXCEEDING 12%, BY WEIGHT, FAT CONTENT NOT MORE THAN 15% BY WEIGHT*	9	UN2216	NONE
	FISHMEAL *OR* FISHSCRAP, *HIGH HAZARD, UNRESTRICTED MOISTURE CONTENT, UNRESTRICTED FAT CONTENT IN EXCESS OF 12% BY WEIGHT; UNRESTRICTED FAT CONTENT IN EXCESS OF 15%, BY WEIGHT, IN THE CASE OF ANTI-OXIDANT TREATED FISHMEAL OR FISHSCRAP*	4.2	UN1374	SPONTANEOUSLY COMBUSTIBLE
	FISHMEAL *OR* FISHSCRAP, *NOT ANTIOXIDANT TREATED, MOISTURE CONTENT GREATER THAN 5% BUT NOT EXCEEDING 12% BY WEIGHT, FAT CONTENT NOT EXCEEDING 12% BY WEIGHT*	4.2	UN1374	NONE. PACKAGE TO BE MARKED 'CLASS 4.2'
	FLAMMABLE GAS IN LIGHTERS. SEE LIGHTERS FOR CIGARS AND CIGARETTES, *ETC. WITH FLAMMABLE GAS*			
	FLAMMABLE LIQUID PREPARATIONS, N.O.S.	3.2 3.3	UN1142 UN1142	FLAMMABLE LIQUID FLAMMABLE LIQUID
	FLAMMABLE LIQUIDS, CORROSIVE, N.O.S.	3.1 3.2	UN2924 UN2924	FLAMMABLE LIQUID, CORROSIVE FLAMMABLE LIQUID, CORROSIVE
	FLAMMABLE LIQUIDS, N.O.S.	3.1 3.2 3.3	UN1993 UN1993 UN1993	FLAMMABLE LIQUID FLAMMABLE LIQUID FLAMMABLE LIQUID
	FLAMMABLE LIQUIDS, POISONOUS, N.O.S.	3.1 3.2 3.3	UN1992 UN1992 UN1992	FLAMMABLE LIQUID, POISON FLAMMABLE LIQUID, POISON FLAMMABLE LIQUID, POISON
	FLAMMABLE SOLIDS, CORROSIVE, N.O.S.	4.1	UN2925	FLAMMABLE SOLID, CORROSIVE
	FLAMMABLE SOLIDS, N.O.S.	4.1 4.1	UN1325 UN1325	FLAMMABLE SOLID FLAMMABLE SOLID
	FLAMMABLE SOLIDS, POISONOUS, N.O.S.	4.1 4.1	UN2926 UN2926	FLAMMABLE SOLID FLAMMABLE SOLID
N	FLARES, AERIAL	1.1G	UN0420	EXPLOSIVE (1.1G)
N	FLARES, AERIAL	1.2G	UN0421	EXPLOSIVE (1.2G)
N	FLARES, AERIAL	1.3G	UN0093	EXPLOSIVE (1.3G)
	FLARES, AERIAL	1.4G	UN0403	EXPLOSIVE (1.4G)
	FLARES, AERIAL	1.4S	UN0404	NONE. PACKAGE TO BE MARKED '1.4S'
N	FLARES, SURFACE *(OTHER THAN WATER-ACTIVATED FLARES)*	1.1G	UN0418	EXPLOSIVE (1.1G)
N	FLARES, SURFACE *(OTHER THAN WATER-ACTIVATED FLARES)*	1.2G	UN0419	EXPLOSIVE (1.2G)
N	FLARES, SURFACE *(OTHER THAN WATER-ACTIVATED FLARES)*	1.3G	UN0092	EXPLOSIVE (1.3G)
	FLAX, DRY. SEE FIBRES, *VEGETABLE, DRY*			
	FLUOBORIC ACID	8	UN1775	CORROSIVE
	FLUORIC ACID. *SEE* HYDROFLUORIC ACID, SOLUTION			
	FLUORINE, *COMPRESSED*	2.3	UN1045	POISON GAS, OXIDIZER
	FLUOROACETIC ACID	6.1	UN2642	POISON
	4-FLUOROANILINE	6.1	UN2944	ST. ANDREWS CROSS
	2-FLUOROANILINE	6.1	UN2941	ST. ANDREWS CROSS, FLAMMABLE LIQUID
	FLUOROBENZENE	3.2	UN2387	FLAMMABLE LIQUID
	FLUOROPHOSPHORIC ACID, *ANHYDROUS*	8	UN1776	CORROSIVE
	FLUOROSULPHONIC ACID	8	UN1777	CORROSIVE
	FLUOROTOLUENES	3.2 3.3	UN2388 UN2388	FLAMMABLE LIQUID FLAMMABLE LIQUID
	FLUOSILICIC ACID	8	UN1778	CORROSIVE
	FORMALDEHYDE SOLUTIONS	3.3	UN1198	FLAMMABLE LIQUID
	FORMALDEHYDE, SOLUTIONS *WITH A FLASHPOINT ABOVE 61 DEGREES C*	9	UN2209	NONE
	FORMALIN. *SEE* FORMALDEHYDE			
	FORMIC ACID	8	UN1779	CORROSIVE
N	FRACTURING DEVICES, EXPLOSIVE, *FOR OIL WELLS*	1.1D	UN0099	EXPLOSIVE (1.1D)
	FUEL, AVIATION, *TURBINE ENGINE*	3.2	UN1863	FLAMMABLE LIQUID
	FUEL, PYROPHORIC, N.O.S.	4.2	UN1375	SPONTANEOUSLY COMBUSTIBLE
	FUMARYL CHLORIDE	8	UN1780	CORROSIVE
	FURAN	3.1	UN2389	FLAMMABLE LIQUID
	FURFURAL	3.3	UN1199	FLAMMABLE LIQUID
	FURFURYL ALCOHOL	6.1	UN2874	ST. ANDREWS CROSS
	FURFURYLAMINE	3.3	UN2526	FLAMMABLE LIQUID

§172.102 OPTIONAL HAZARDOUS MATERIALS TABLE

(1) Notes and Symbols	(2) Hazardous Materials and Description and Proper Shipping Names	(3) IMO Class	(4) Identification Number	(5) Label(s) required
	FUSE, IGNITER, *TUBULAR, METAL CLAD*	1.4 G	UN0103	EXPLOSIVE (1.4G)
N	FUSE, INSTANTANEOUS, NON-DETONATING	1.3G	UN0101	EXPLOSIVE (1.3G)
	FUSEL OIL	3.2	UN1201	FLAMMABLE LIQUID
	FUSE, SAFETY	1.4 S	UN0105	NONE. PACKAGE TO BE MARKED '1.4 S'
	FUZES, DETONATING	1.4S	UN0367	NONE. PACKAGE TO BE MARKED '1.4S'
	FUZES, DETONATING	1.4 B	UN0257	EXPLOSIVE (1.4B)
N	FUZES, DETONATING	1.1B	UN0106	EXPLOSIVE (1.1B)
N	FUZES, DETONATING	1.2B	UN0107	EXPLOSIVE (1.2B)
N	FUZES, DETONATING, *WITH PROTECTIVE FEATURES*	1.1D	UN0408	EXPLOSIVE (1.1D)
N	FUZES, DETONATING, *WITH PROTECTIVE FEATURES*	1.2D	UN0409	EXPLOSIVE (1.2D)
N	FUZES, DETONATING, *WITH PROTECTIVE FEATURES*	1.4D	UN0410	EXPLOSIVE (1.4D)
	FUZES, IGNITING	1.4S	UN0368	NONE. PACKAGE TO BE MARKED '1.4S'
N	FUZES, IGNITING	1.3G	UN0316	EXPLOSIVE (1.3G)
N	FUZES, IGNITING	1.4G	UN0317	EXPLOSIVE (1.4G)
	GALLIUM, *METAL*	8	UN2803	CORROSIVE
	GAS DRIPS, *HYDROCARBON*	3.2	UN1864	FLAMMABLE LIQUID
	GAS OIL	3.3	UN1202	FLAMMABLE LIQUID
	GASOLINE	3.1	UN1203	FLAMMABLE LIQUID
	GERMANE	2.3	UN2192	POISON GAS, FLAMMABLE GAS
	GLYCEROL-ALPHA-MONOCHLOROHYDRIN	6.1	UN2689	ST. ANDREWS CROSS
	GLYCIDALDEHYDE	3.3	UN2622	FLAMMABLE LIQUID, POISON
N	GRENADES, *HAND OR RIFLE, WITH BURSTING CHARGE*	1.1D	UN0284	EXPLOSIVE (1.1D)
N	GRENADES, *HAND OR RIFLE, WITH BURSTING CHARGE*	1.2D	UN0285	EXPLOSIVE (1.2D)
N	GRENADES, *HAND OR RIFLE, WITH BURSTING CHARGE*	1.1F	UN0292	EXPLOSIVE (1.1F)
N	GRENADES, *HAND OR RIFLE, WITH BURSTING CHARGE*	1.2F	UN0293	EXPLOSIVE (1.2F)
	GRENADES, PRACTICE, *HAND OR RIFLE*	1.4 S	UN0110	NONE. PACKAGE TO BE MARKED '1.4 S'
N	GRENADES, PRACTICE, *HAND OR RIFLE*	1.2G	UN0372	EXPLOSIVE (1.2G)
N	GRENADES, PRACTICE, *HAND OR RIFLE*	1.3G	UN0318	EXPLOSIVE (1.3G)
	GUANIDINE NITRATE	5.1	UN1467	OXIDIZER
N	GUANYL NITROSAMINO GUANYLIDENE HYDRAZINE, *CONTAINING, BY WEIGHT, AT LEAST 30% WATER*	1.1A	UN0113	EXPLOSIVE (1.1A)
N	GUANYL NITROSAMINO GUANYL TETRAZENE, *CONTAINING, BY WEIGHT, AT LEAST 30% WATER OR MIXTURE OF ALCOHOL AND WATER*	1.1A	UN0114	EXPLOSIVE (1.1A)
	GUTTA PERCHA SOLUTION	3.2 3.3	UN1205 UN1205	FLAMMABLE LIQUID FLAMMABLE LIQUID
	HAFNIUM *METAL* POWDER, DRY	4.2	UN2545	SPONTANEOUSLY COMBUSTIBLE
	HAFNIUM *METAL* POWDER, WETTED, *WITH NOT LESS THAN 25% WATER (A VISIBLE EXCESS OF WATER MUST BE PRESENT) MECHANICALLY PRODUCED, PARTICLE SIZE LESS THAN 53 MICRONS, OR CHEMICALLY PRODUCED, PARTICLE SIZE LESS THAN 840 MICRONS*	4.1	UN1326	FLAMMABLE SOLID
	HALOGENATED IRRITATING LIQUIDS, N.O.S.	6.1 6.1	UN1610 UN1610	POISON, FLAMMABLE LIQUID (ONLY IF FLASHPOINT BETWEEN 23 AND 61 DEG C) ST. ANDREWS CROSS, FLAMMABLE LIQUID (ONLY IF FLASHPOINT BETWEEN 23 AND 61 DEG C)
	HAY	4.1	UN1327	NONE
	HELIUM, *COMPRESSED*	2.2	UN1046	NONFLAMMABLE GAS
	HELIUM, REFRIGERATED LIQUID	2.2	UN1963	NONFLAMMABLE GAS
	HEMP, DRY. SEE FIBRES, *VEGETABLE, DRY*			
	HEPTANE, *AND ITS ISOMERS*	3.2	UN1206	FLAMMABLE LIQUID
	N-HEPTENE	3.2	UN2278	FLAMMABLE LIQUID
	HEXACHLOROACETONE	6.1	UN2661	POISON
	HEXACHLOROBENZENE	6.1	UN2729	ST. ANDREWS CROSS
	HEXACHLOROBUTADIENE	6.1	UN2279	ST. ANDREWS CROSS
	HEXACHLOROCYCLOPENTADIENE	6.1	UN2646	POISON
	HEXACHLOROPHENE	6.1	UN2875	ST. ANDREWS CROSS
	HEXADECYL TRICHLOROSILANE	8	UN1781	CORROSIVE
	HEXADIENE	3.1	UN2458	FLAMMABLE LIQUID

§172.102 OPTIONAL HAZARDOUS MATERIALS TABLE

(1) Notes and Symbols	(2) Hazardous Materials and Description and Proper Shipping Names	(3) IMO Class	(4) Identification Number	(5) Label(s) required
	HEXAETHYL TETRAPHOSPHATE	6.1 6.1	UN1611 UN1611	POISON ST. ANDREWS CROSS
	HEXAETHYL TETRAPHOSPHATE AND COMPRESSED GAS *MIXTURE*	2.3	UN1612	POISON GAS
	HEXAFLUOROACETONE	2.3	UN2420	POISON GAS
	HEXAFLUOROACETONE HYDRATE	6.1	UN2552	POISON
	HEXAFLUOROETHANE	2.2	UN2193	NONFLAMMABLE GAS
	HEXAFLUOROPHOSPHORIC ACID	8	UN1782	CORROSIVE
	HEXAFLUOROPROPYLENE	2.2	UN1858	NONFLAMMABLE GAS
	HEXALDEHYDE	3.3	UN1207	FLAMMABLE LIQUID
	HEXAMETHYLENEDIAMINE, SOLID	8	UN2280	CORROSIVE
	HEXAMETHYLENEDIAMINE, SOLUTION	8	UN1783	CORROSIVE POISON
	HEXAMETHYLENEDIISOCYANATE	6.1	UN2281	POISON
	HEXAMETHYLENEIMINE	3.2	UN2493	FLAMMABLE LIQUID, CORROSIVE
	3,3,6,6,9,9-HEXAMETHYL-1,2,4,5-TETRAOXOCYCLONONANE, *MAXIMUM CONCENTRATION 52% WITH INERT SOLID*	5.2	UN2166	ORGANIC PEROXIDE
	3,3,6,6,9,9-HEXAMETHYL-1,2,4,5-TETRAOXOCYCLONONANE, *MAXIMUM CONCENTRATION 52% IN SOLUTION*	5.2	UN2167	ORGANIC PEROXIDE
	3,3,6,6,9,9-HEXAMETHYL-1,2,4,5-TETRAOXOCYCLONONANE, *TECHNICAL PURE.*	5.2	UN2165	ORGANIC PEROXIDE
	HEXAMINE	4.1	UN1328	FLAMMABLE SOLID
	HEXANE, *AND ITS ISOMERS*	3.1	UN1208	FLAMMABLE LIQUID
N	HEXANITRODIPHENYLAMINE	1.1D	UN0079	EXPLOSIVE (1.1D), POISON
N	HEXANITROSTILBENE	1.1D	UN0392	EXPLOSIVE (1.1D)
	HEXANOLS	3.3	UN2282	FLAMMABLE LIQUID
N	HEXATONAL, CAST	1.1D	UN0393	EXPLOSIVE (1.1D)
	HEX-1-ENE	3.1	UN2370	FLAMMABLE LIQUID
	1-HEXENE. *SEE* HEX-1-ENE			
N	HEXOLITE, *DRY OR WETTED WITH LESS THAN 15% WATER, BY WEIGHT*	1.1D	UN0118	EXPLOSIVE (1.1D)
	HEXYL TRICHLOROSILANE	8	UN1784	CORROSIVE
	HYDRAZINE, ANHYDROUS *OR* HYDRAZINE AQUEOUS SOLUTIONS *CONTAINING MORE THAN 64%, BY WEIGHT, HYDRAZINE*	3.3	UN2029	FLAMMABLE LIQUID, POISON CORROSIVE
	HYDRAZINE HYDRATE *OR* HYDRAZINE AQUEOUS SOLUTIONS *WITH NOT MORE THAN 64%, BY WEIGHT, HYDRAZINE*	8	UN2030	CORROSIVE, POISON
	HYDRIDES, *METAL*, N.O.S.	4.3	UN1409	DANGEROUS WHEN WET
	HYDRIODIC ACID, SOLUTION	8	UN1787	CORROSIVE
	HYDROBROMIC ACID, SOLUTION	8	UN1788	CORROSIVE
	HYDROCARBON GASES, COMPRESSED, N.O.S. *OR* HYDROCARBON GASES, MIXTURES, COMPRESSED, N.O.S.	2.1	UN1964	FLAMMABLE GAS
	HYDROCARBON GASES, LIQUEFIED, N.O.S. *OR* HYDROCARBON GASES, MIXTURES, LIQUEFIED, N.O.S.	2.1	UN1965	FLAMMABLE GAS
	HYDROCHLORIC ACID, SOLUTION	8	UN1789	CORROSIVE
	HYDROCYANIC ACID, ANHYDROUS, STABILIZED. SEE HYDROGEN CYANIDE, ANHYDROUS, STABILIZED			
	HYDROCYANIC ACID, AQUEOUS SOLUTIONS, *WITH NOT MORE THAN 20% HYDROCYANIC ACID*	6.1	UN1613	POISON
	HYDROFLUORIC ACID AND SULPHURIC ACID MIXTURES	8	UN1786	CORROSIVE POISON
	HYDROFLUORIC ACID, SOLUTION	8	UN1790	CORROSIVE POISON
	HYDROFLUOSILICIC ACID. *SEE* FLUOSILICIC ACID			
	HYDROGEN AND METHANE MIXTURES, COMPRESSED	2.1	UN2034	FLAMMABLE GAS
	HYDROGEN BROMIDE, ANHYDROUS	2.3	UN1048	POISON GAS, CORROSIVE
	HYDROGEN CHLORIDE, ANHYDROUS	2.2	UN1050	NONFLAMMABLE GAS, CORROSIVE
	HYDROGEN, *COMPRESSED*	2.1	UN1049	FLAMMABLE GAS
	HYDROGEN CYANIDE, ANHYDROUS, STABILIZED	6.1	UN1051	POISON, FLAMMABLE LIQUID
	HYDROGEN CYANIDE, ANHYDROUS, STABILIZED, *ABSORBED IN A POROUS INERT MATERIAL*	6.1	UN1614	POISON
	HYDROGEN FLUORIDE, ANHYDROUS	2.3	UN1052	POISON GAS, CORROSIVE
	HYDROGEN IODIDE, ANHYDROUS	2.2	UN2197	NONFLAMMABLE GAS, CORROSIVE
	HYDROGEN IODIDE, SOLUTION. *SEE* HYDRIODIC ACID SOLUTION			
	HYDROGEN PEROXIDE, AQUEOUS SOLUTIONS *WITH NOT LESS THAN 8% BUT LESS THAN 20% HYDROGEN PEROXIDE (STABILIZED AS NECESSARY)*	5.1	UN2984	OXIDIZER
	HYDROGEN PEROXIDE, AQUEOUS SOLUTIONS *WITH NOT LESS THAN 20% BUT NOT MORE THAN 60% HYDROGEN PEROXIDE (STABILIZED AS NECESSARY)*	5.1	UN2014	OXIDIZER, CORROSIVE

§172.102 OPTIONAL HAZARDOUS MATERIALS TABLE

(1) Notes and Symbols	(2) Hazardous Materials and Description and Proper Shipping Names	(3) IMO Class	(4) Identification Number	(5) Label(s) required
	HYDROGEN PEROXIDE, STABILIZED, *OR* HYDROGEN PEROXIDE, AQUEOUS SOLUTIONS, STABILIZED, *WITH MORE THAN 60% HYDROGEN PEROXIDE*	5.1	UN2015	OXIDIZER, CORROSIVE
	HYDROGEN, *REFRIGERATED LIQUID*	2.1	UN1966	FLAMMABLE GAS
	HYDROGEN SELENIDE, *ANHYDROUS*	2.3	UN2202	POISON GAS, FLAMMABLE GAS
	HYDROGEN SULPHIDE	2.1	UN1053	FLAMMABLE GAS, POISON GAS
	HYDROQUINONE	6.1	UN2662	ST. ANDREWS CROSS
	HYDROSILICOFLUORIC ACID. *SEE* FLUOSILICIC ACID			
	HYDROXYLAMINE SULPHATE	8	UN2865	CORROSIVE
	HYPOCHLORITE, SOLUTIONS *WITH MORE THAN 5% AVAILABLE CHLORINE*	8	UN1791	CORROSIVE
	IGNITER FOR AIRCRAFT THRUST DEVICE *FOR ASSISTED TAKE-OFF*	4.1	UN2792	FLAMMABLE SOLID
	IGNITERS	1.4 G	UN0325	EXPLOSIVE (1.4G)
N	IGNITERS	1.1G	UN0121	EXPLOSIVE (1.1G)
N	IGNITERS	1.2G	UN0314	EXPLOSIVE (1.2G)
N	IGNITERS	1.3G	UN0315	EXPLOSIVE (1.3G)
	3,3'-IMINOBISPROPYLAMINE	8	UN2269	CORROSIVE
	INFLAMMABLE GAS IN LIGHTERS. *SEE* LIGHTERS, FOR CIGARS AND CIGARETTES, *ETC. WITH FLAMMABLE GAS*			
	INFLAMMABLE LIQUID PREPARATIONS, N.O.S. *SEE* FLAMMABLE LIQUID PREPARATIONS, N.O.S.			
	INFLAMMABLE LIQUIDS, CORROSIVE, N.O.S. *SEE* FLAMMABLE LIQUIDS, CORROSIVE, N.O.S.			
	INFLAMMABLE LIQUIDS, N.O.S. *SEE* FLAMMABLE LIQUIDS, N.O.S.			
	INFLAMMABLE LIQUIDS, POISONOUS, N.O.S. *SEE* FLAMMABLE LIQUIDS, POISONOUS, N.O.S.			
	INFLAMMABLE SOLIDS, CORROSIVE, N.O.S. *SEE* FLAMMABLE SOLIDS, , CORROSIVE, N.O.S.			
	INFLAMMABLE SOLIDS, N.O.S. *SEE* FLAMMABLE SOLIDS, N.O.S.			
	INFLAMMABLE SOLIDS, POISONOUS, N.O.S. *SEE* FLAMMABLE SOLIDS, POISONOUS, N.O.S.			
	INK, PRINTERS	3.2 3.3	UN1210 UN1210	FLAMMABLE LIQUID FLAMMABLE LIQUID
	INSECTICIDE GASES, *NON-TOXIC*, N.O.S.	2.1 2.2	UN1968 UN1968	FLAMMABLE GAS NONFLAMMABLE GAS
	INSECTICIDE GASES, *TOXIC*, N.O.S.	2.3	UN1967	POISON GAS
	IODINE MONOCHLORIDE	8	UN1792	CORROSIVE
	IODINE PENTAFLUORIDE	5.1	UN2495	OXIDIZER, POISON
	2-IODOBUTANE	3.2	UN2390	FLAMMABLE LIQUID
	IODOMETHYLPROPANES	3.2	UN2391	FLAMMABLE LIQUID
	IODOPROPANES	3.2 3.3	UN2392 UN2392	FLAMMABLE LIQUID FLAMMABLE LIQUID
	IRON CARBONYL. *SEE* IRON PENTACARBONYL			
	IRON CHLORIDE. SEE FERRIC CHLORIDE, ANHYDROUS			
	IRON OXIDE, *SPENT*	4.2	UN1376	SPONTANEOUSLY COMBUSTIBLE
	IRON PENTACARBONYL	6.1	UN1994	POISON, FLAMMABLE LIQUID
	IRON SESQUICHLORIDE. *SEE* FERRIC CHLORIDE			
	IRON SPONGE, *SPENT. SEE* IRON OXIDE, *SPENT*			
	ISOBUTANE *OR* ISOBUTANE MIXTURES	2.1	UN1969	FLAMMABLE GAS
	ISOBUTANOL	3.3	UN1212	FLAMMABLE LIQUID
	ISOBUTYL ACETATE	3.2	UN1213	FLAMMABLE LIQUID
	ISOBUTYL ACRYLATE, *INHIBITED*	3.3	UN2527	FLAMMABLE LIQUID
	ISOBUTYL ALCOHOL. *SEE* ISOBUTANOL			
	ISOBUTYL ALDEHYDE. *SEE* ISOBUTYRALDEHYDE			
	ISOBUTYLAMINE	3.2	UN1214	FLAMMABLE LIQUID
	ISOBUTYLENE	2.1	UN1055	FLAMMABLE GAS
	ISOBUTYL FORMATE	3.2	UN2393	FLAMMABLE LIQUID
	ISOBUTYL ISOBUTYRATE	3.3	UN2528	FLAMMABLE LIQUID
	ISOBUTYL ISOCYANATE	3.2	UN2486	FLAMMABLE LIQUID, POISON
	ISOBUTYL METHACRYLATE, *INHIBITED*	3.3	UN2283	FLAMMABLE LIQUID
	ISOBUTYL METHYL KETONE PEROXIDE, *MAXIMUM CONCENTRATION 62% IN SOLUTION*	5.2	UN2126	ORGANIC PEROXIDE
	ISOBUTYL PROPIONATE	3.2	UN2394	FLAMMABLE LIQUID
	ISOBUTYRALDEHYDE	3.1	UN2045	FLAMMABLE LIQUID

§172.102 OPTIONAL HAZARDOUS MATERIALS TABLE

(1) Notes and Symbols	(2) Hazardous Materials and Description and Proper Shipping Names	(3) IMO Class	(4) Identification Number	(5) Label(s) required
	ISOBUTYRIC ACID	3.3	UN2529	FLAMMABLE LIQUID
	ISOBUTYRIC ANHYDRIDE	3.3	UN2530	FLAMMABLE LIQUID
	ISOBUTYRONITRILE	3.2	UN2284	FLAMMABLE LIQUID, POISON
	ISOBUTYRYL CHLORIDE	3.2	UN2395	FLAMMABLE LIQUID, POISON
	ISOBUTYRYL PEROXIDE, *MAXIMUM CONCENTRATION 52% IN SOLUTION*	5.2	UN2182	ORGANIC PEROXIDE
	ISOCYANATES, N.O.S. *OR* ISOCYANATE SOLUTIONS, N.O.S., *BOILING POINT NOT LESS THAN 300 DEG C*	6.1	UN2207	ST. ANDREWS CROSS
	ISOCYANATES, N.O.S. *OR* ISOCYANATE SOLUTIONS, N.O.S., *FLASHPOINT NOT LESS THAN 23 DEG C, BOILING POINT LESS THAN 300 DEG C*	6.1	UN2206	POISON, FLAMMABLE LIQUID (ONLY IF FLASHPOINT IS BELOW 61 DEG C.)
	ISOCYANATES, N.O.S. *OR* ISOCYANATE SOLUTIONS, N.O.S. *FLASHPOINT NOT LESS THAN -18 DEG C BUT LESS THAN 23 DEG C*	3.2	UN2478	FLAMMABLE LIQUID, POISON
	ISOCYANATES, N.O.S. *OR* ISOCYANATE SOLUTIONS, N.O.S. *FLASHPOINT LESS THAN -18 DEG C*	3.1	UN2478	FLAMMABLE LIQUID, POISON
	ISOCYANATOBENZOTRIFLUORIDES	6.1	UN2285	POISON
	ISOHEPTENE	3.1	UN2287	FLAMMABLE LIQUID
	ISOHEXENE	3.1	UN2288	FLAMMABLE LIQUID
	ISONONANOYL PEROXIDE. *SEE* DI-(3,3,5-TRIMETHYLHEXANOYL) PEROXIDE			
	ISOOCTENE	3.2	UN1216	FLAMMABLE LIQUID
	ISOPENTANE. *SEE* PENTANES			
	ISOPENTENES	3.1	UN2371	FLAMMABLE LIQUID
	ISOPHORONEDIAMINE	8	UN2289	CORROSIVE
	ISOPHORONE DIISOCYANATE	6.1	UN2290	POISON
	ISOPRENE, INHIBITED	3.1	UN1218	FLAMMABLE LIQUID
	ISOPROPANOL	3.2	UN1219	FLAMMABLE LIQUID
	ISOPROPENYL ACETATE	3.2	UN2403	FLAMMABLE LIQUID
	ISOPROPENYLBENZENE	3.3	UN2303	FLAMMABLE LIQUID
	ISOPROPYL ACETATE	3.2	UN1220	FLAMMABLE LIQUID
	ISOPROPYL ACID PHOSPHATIC	8	UN1793	CORROSIVE
	ISOPROPYL ALCOHOL. *SEE* ISOPROPANOL			
	ISOPROPYLAMINE	3.1	UN1221	FLAMMABLE LIQUID
	ISOPROPYLBENZENE	3.3	UN1918	FLAMMABLE LIQUID
	ISOPROPYL BUTYRATE	3.3	UN2405	FLAMMABLE LIQUID
	ISOPROPYL CHLOROACETATE	3.3	UN2947	FLAMMABLE LIQUID
	ISOPROPYL CHLOROFORMATE	3.2	UN2407	FLAMMABLE LIQUID, CORROSIVE
	ISOPROPYL-2-CHLOROPROPIONATE	3.3	UN2934	FLAMMABLE LIQUID
	ISOPROPYL ISOBUTYRATE	3.2	UN2406	FLAMMABLE LIQUID
	ISOPROPYL ISOCYANATE	3.2	UN2483	FLAMMABLE LIQUID, POISON
	ISOPROPYL NITRATE	3.2	UN1222	FLAMMABLE LIQUID
	ISOPROPYL PROPIONATE	3.2	UN2409	FLAMMABLE LIQUID
	ISOSORBIDE DINITRATE MIXTURE *WITH NOT LESS THAN 60% LACTOSE, MANNOSE, STARCH, OR CALCIUM HYDROGEN PHOSPHATE*	4.1	UN2907	FLAMMABLE SOLID
N	JET PERFORATING GUNS, CHARGED, *OIL WELL, WITHOUT DETONATOR*	1.1D	UN0124	EXPLOSIVE (1.1D)
	JUTE, DRY. SEE FIBRES, *VEGETABLE, DRY*			
	KAPOK, DRY. SEE FIBRES, *VEGETABLE, DRY*			
	KEROSENE	3.3	UN1223	FLAMMABLE LIQUID
	KETONES, LIQUID, N.O.S.	3.1 3.2 3.3	UN1224 UN1224 UN1224	FLAMMABLE LIQUID FLAMMABLE LIQUID FLAMMABLE LIQUID
	KRYPTON, *COMPRESSED*	2.2	UN1056	NONFLAMMABLE GAS
	KRYPTON, REFRIGERATED LIQUID	2.2	UN1970	NONFLAMMABLE GAS
	LACQUER BASE. *SEE* PAINTS, *ETC.*			
	LACQUER CHIPS. *SEE* NITROCELLULOSE, *WETTED WITH, BY WEIGHT, MORE THAN 40% FLAMMABLE LIQUIDS.*			
	LACQUERS. *SEE* PAINTS, *ETC.*			
	LAUROYL PEROXIDE. *SEE* DILAUROYL PEROXIDE			
	LEAD ACETATE	6.1	UN1616	ST. ANDREWS CROSS
	LEAD ARSENATES	6.1	UN1617	POISON
	LEAD ARSENITES	6.1	UN1618	POISON
N	LEAD AZIDE, *CONTAINING, BY WEIGHT, AT LEAST 20% WATER OR MIXTURE OF ALCOHOL AND WATER*	1.1A	UN0129	EXPLOSIVE (1.1A)

§172.102 OPTIONAL HAZARDOUS MATERIALS TABLE

(1) Notes and Symbols	(2) Hazardous Materials and Description and Proper Shipping Names	(3) IMO Class	(4) Identification Number	(5) Label(s) required
	LEAD COMPOUNDS, SOLUBLE, N.O.S.	6.1	UN2291	ST. ANDREWS CROSS
	LEAD CYANIDE	6.1	UN1620	POISON
	LEAD DIOXIDE	5.1	UN1872	OXIDIZER
	LEAD DROSS. *SEE* LEAD SULPHATE, *WITH MORE THAN 3% FREE ACID*			
	LEAD NITRATE	5.1	UN1469	OXIDIZER POISON
	LEAD PERCHLORATE	5.1	UN1470	OXIDIZER POISON
	LEAD PEROXIDE. *SEE* LEAD DIOXIDE			
	LEAD PHOSPHITE DIBASIC	4.1	UN2989	FLAMMABLE SOLID
N	LEAD STYPHNATE, *CONTAINING, BY WEIGHT, AT LEAST 20% WATER OR MIXTURE OF ALCOHOL AND WATER*	1.1A	UN0130	EXPLOSIVE (1.1A)
	LEAD SULPHATE, *WITH MORE THAN 3% FREE ACID*	8	UN1794	CORROSIVE
	LIFE-RAFTS, *INFLATABLE*	9	--	NONE
	LIGHTER FUELS, *CIGAR AND CIGARETTE*	3.2	UN1226	FLAMMABLE LIQUID
	LIGHTERS FOR CIGARS AND CIGARETTES, *ETC., WITH FLAMMABLE GAS*	2.1	UN1057	FLAMMABLE GAS
	LIGHTERS *FOR CIGARS AND CIGARETTES, ETC., WITH* FUEL	3.2	UN1226	FLAMMABLE LIQUID
	LIGHTERS, FUSE	1.4 S	UN0131	NONE. PACKAGE TO BE MARKED '1.4 S'
	LIQUEFIED GASES, N.O.S. SEE COMPRESSED OR LIQUEFIED GASES, N.O.S.			
	LIQUEFIED NON-FLAMMABLE GASES *CHARGED WITH* NITROGEN, CARBON DIOXIDE *OR* AIR	2.2	UN1058	NONFLAMMABLE GAS
	LITHIUM ALKYLS	4.2	UN2445	SPONTANEOUSLY COMBUSTIBLE
	LITHIUM ALUMINIUM HYDRIDE	4.3	UN1410	DANGEROUS WHEN WET
	LITHIUM ALUMINIUM HYDRIDE, *ETHEREAL*	4.3	UN1411	DANGEROUS WHEN WET, FLAMMABLE LIQUID
	LITHIUM AMIDE	4.3	UN1412	DANGEROUS WHEN WET
	LITHIUM BOROHYDRIDE	4.3	UN1413	DANGEROUS WHEN WET
	LITHIUM FERROSILICON	4.3	UN2830	DANGEROUS WHEN WET
	LITHIUM HYDRIDE	4.3	UN1414	DANGEROUS WHEN WET
	LITHIUM HYDRIDE, FUSED SOLID	4.3	UN2805	DANGEROUS WHEN WET
	LITHIUM HYDROXIDE MONOHYDRATE	8	UN2680	CORROSIVE
	LITHIUM HYDROXIDE, SOLUTION	8	UN2679	CORROSIVE
	LITHIUM HYPOCHLORITE, DRY *OR* LITHIUM HYPOCHLORITE MIXTURES *CONTAINING MORE THAN 39% AVAILABLE CHLORINE (8.8% AVAILABLE OXYGEN)*	5.1	UN1471	OXIDIZER
	LITHIUM, *METAL*	4.3	UN1415	DANGEROUS WHEN WET
	LITHIUM NITRATE	5.1	UN2722	OXIDIZER
	LITHIUM NITRIDE	4.3	UN2806	DANGEROUS WHEN WET
	LITHIUM PEROXIDE	5.1	UN1472	OXIDIZER
	LITHIUM SILICON	4.3	UN1417	DANGEROUS WHEN WET
	LONDON PURPLE	6.1	UN1621	POISON
	LYE. *SEE* SODIUM HYDROXIDE			
	MAGNESIUM ALUMINIUM PHOSPHIDE	4.3	UN1419	DANGEROUS WHEN WET
	MAGNESIUM ARSENATE	6.1	UN1622	POISON
	MAGNESIUM BROMATE	5.1	UN1473	OXIDIZER
	MAGNESIUM CHLORATE	5.1	UN2723	OXIDIZER
	MAGNESIUM DIAMIDE	4.2	UN2004	SPONTANEOUSLY COMBUSTIBLE
	MAGNESIUM DIPHENYL	4.2	UN2005	SPONTANEOUSLY COMBUSTIBLE
	MAGNESIUM GRANULES, COATED *CONTAINING MORE THAN 50% MAGNESIUM, PARTICLE SIZE NOT LESS THAN 149 MICRONS*	4.3	UN2950	DANGEROUS WHEN WET
	MAGNESIUM HYDRIDE	4.3	UN2010	DANGEROUS WHEN WET
	MAGNESIUM NITRATE	5.1	UN1474	OXIDIZER
	MAGNESIUM *OR* MAGNESIUM ALLOYS *WITH MORE THAN 50% MAGNESIUM, IN PELLETS, TURNINGS OR RIBBON*	4.1	UN1869	FLAMMABLE SOLID
	MAGNESIUM PERCHLORATE	5.1	UN1475	OXIDIZER
	MAGNESIUM PEROXIDE	5.1	UN1476	OXIDIZER
	MAGNESIUM PHOSPHIDE	4.3	UN2011	DANGEROUS WHEN WET, POISON
	MAGNESIUM POWDER *OR* MAGNESIUM ALLOYS, POWDER *CONTAINING MORE , THAN 50% MAGNESIUM, NON-PYROPHORIC*	4.3	UN1418	DANGEROUS WHEN WET
	MAGNESIUM SILICIDE	4.3	UN2624	DANGEROUS WHEN WET
	MAGNESIUM SILICOFLUORIDE	6.1	UN2853	ST. ANDREWS CROSS
	MALEIC ANHYDRIDE, *SOLID OR MOLTEN*	8	UN2215	NONE. PACKAGE TO BE MARKED 'CLASS 8'

§172.102 OPTIONAL HAZARDOUS MATERIALS TABLE

(1) Notes and Symbols	(2) Hazardous Materials and Description and Proper Shipping Names	(3) IMO Class	(4) Identification Number	(5) Label(s) required
	MALONONITRILE	6.1	UN2647	POISON
	MANEB, *OR* MANEB PREPARATION(S), *STABILIZED AGAINST SELF-HEATING*	4.3	UN2968	DANGEROUS WHEN WET
	MANEB, *OR* MANEB PREPARATION(S) *WITH NOT LESS THAN 60% MANEB*	4.2	UN2210	SPONTANEOUSLY COMBUSTIBLE DANGEROUS WHEN WET
	MANGANESE ETHYLENE-BIS-DITHIOCARBAMATE. SEE MANEB			
	MANGANESE NITRATE	5.1	UN2724	OXIDIZER
	MANGANESE RESINATE	4.1	UN1330	FLAMMABLE SOLID
N	MANNITOL HEXANITRATE, *CONTAINING, BY WEIGHT, AT LEAST 40% WATER OR MIXTURE OF ALCOHOL AND WATER*	1.1D	UN0133	EXPLOSIVE (1.1D)
	MATCHES, *FUSEE*	4.1	UN2254	FLAMMABLE SOLID
	MATCHES, SAFETY	4.1	UN1944	FLAMMABLE SOLID
	MATCHES, *STRIKE ANYWHERE*	4.1	UN1331	FLAMMABLE SOLID
	MATCHES, *WAX 'VESTA'*	4.1	UN1945	FLAMMABLE SOLID
N	MEDICINES, N.O.S. *TO BE CLASSIFIED AND LABELED ACCORDING TO THE PRINCIPLE HAZARDOUS CONSTITUENT*	--	UN1851	
	MEKP. SEE ETHYL METHYL KETONE PEROXIDE			
	P-MENTHANE HYDROPEROXIDE, *TECHNICAL PURE*	5.2	UN2125	ORGANIC PEROXIDE
	MERCAPTANS, LIQUID, N.O.S. *OR* MERCAPTAN MIXTURES, LIQUID, N.O.S.	3.1 3.2 3.3	UN1228 UN1228 UN1228	FLAMMABLE LIQUID FLAMMABLE LIQUID FLAMMABLE LIQUID
	MERCURIC ACETATE. *SEE* MERCURY ACETATE			
	MERCURIC ARSENATE	6.1	UN1623	POISON
	MERCURIC BROMIDE. *SEE* MERCURY BROMIDES			
	MERCURIC CHLORIDE	6.1	UN1624	POISON
	MERCURIC CYANIDE. *SEE* MERCURY CYANIDE			
	MERCURIC NITRATE	6.1	UN1625	POISON
	MERCURIC OXYCYANIDE. *SEE* MERCURY OXYCYANIDE			
	MERCURIC POTASSIUM CYANIDE	6.1	UN1626	POISON
	MERCURIC SULPHATE	6.1	UN1645	POISON
	MERCUROL. *SEE* MERCURY NUCLEATE			
	MERCUROUS ACETATE. *SEE* MERCURY ACETATE			
	MERCUROUS BROMIDE. *SEE* MERCURY BROMIDES			
	MERCUROUS NITRATE	6.1	UN1627	POISON
	MERCUROUS SULPHATE	6.1	UN1628	POISON
	MERCURY ACETATE	6.1	UN1629	POISON
	MERCURY AMMONIUM CHLORIDE	6.1	UN1630	POISON
	MERCURY BASED PESTICIDES, LIQUID, FLAMMABLE, TOXIC, N.O.S., *FLASHPOINT BELOW 23 DEG C*	3.2	UN2778	FLAMMABLE LIQUID AND POISON OR ST. ANDREWS CROSS (ACCORDING TO TOXICITY)
	MERCURY BASED PESTICIDES, LIQUID, TOXIC, FLAMMABLE, N.O.S., *FLASHPOINT BETWEEN 23 DEG C AND 61 DEG C*	6.1 6.1 6.1	UN3011 UN3011 UN3011	POISON FLAMMABLE LIQUID POISON FLAMMABLE LIQUID ST. ANDREWS CROSS FLAMMABLE LIQUID
	MERCURY BASED PESTICIDES, LIQUID, TOXIC, N.O.S.	6.1 6.1 6.1	UN3012 UN3012 UN3012	POISON POISON ST. ANDREWS CROSS
	MERCURY BASED PESTICIDES, SOLID, TOXIC, N.O.S.	6.1 6.1	UN2777 UN2777	POISON ST. ANDREWS CROSS
	MERCURY BENZOATE	6.1	UN1631	POISON
	MERCURY BISULPHATE	6.1	UN1633	POISON
	MERCURY BROMIDES	6.1	UN1634	POISON
	MERCURY COMPOUNDS, LIQUID, N.O.S.	6.1 6.1	UN2024 UN2024	POISON ST. ANDREWS CROSS
	MERCURY COMPOUNDS, SOLID, N.O.S.	6.1 6.1	UN2025 UN2025	POISON ST. ANDREWS CROSS
	MERCURY CYANIDE	6.1	UN1636	POISON
N	MERCURY FULMINATE, *CONTAINING, BY WEIGHT, AT LEAST 20% WATER OR MIXTURE OF ALCOHOL AND WATER*	1.1A	UN0135	EXPLOSIVE (1.1A)
	MERCURY GLUCONATE	6.1	UN1637	POISON
	MERCURY IODIDE	6.1	UN1638	POISON
	MERCURY, *METAL*	8	UN2809	CORROSIVE
	MERCURY NUCLEATE	6.1	UN1639	POISON

§172.102 OPTIONAL HAZARDOUS MATERIALS TABLE

(1) Notes and Symbols	(2) Hazardous Materials and Description and Proper Shipping Names	(3) IMO Class	(4) Identi-fication Number	(5) Label(s) required
	MERCURY OLEATE	6.1	UN1640	POISON
	MERCURY OXIDE	6.1	UN1641	POISON
	MERCURY OXYCYANIDE	6.1	UN1642	POISON
	MERCURY POTASSIUM IODIDE	6.1	UN1643	POISON
	MERCURY SALICYLATE	6.1	UN1644	POISON
	MERCURY THIOCYANATE	6.1	UN1646	POISON
	MESITYL OXIDE	3.3	UN1229	FLAMMABLE LIQUID
	METAL ALKYLS, N.O.S.	4.2	UN2003	SPONTANEOUSLY COMBUSTIBLE
	METALDEHYDE	4.1	UN1332	FLAMMABLE SOLID
	METHACRALDEHYDE	3.2	UN2396	FLAMMABLE LIQUID, POISON
	METHACRYLIC ACID, INHIBITED	8	UN2531	CORROSIVE
	METHALLYL ALCOHOL	3.3	UN2614	FLAMMABLE LIQUID
	METHANE *OR* NATURAL GASES *WITH A HIGH* METHANE *CONTENT, COMPRESSED*	2.1	UN1971	FLAMMABLE GAS
	METHANE *OR* NATURAL GASES *WITH A HIGH* METHANE *CONTENT, REFRIGERATED LIQUID*	2.1	UN1972	FLAMMABLE GAS
	METHANOL	3.2	UN1230	FLAMMABLE LIQUID, POISON
	METHOXYMETHYL ISOCYANATE	3.2	UN2605	FLAMMABLE LIQUID, POISON
	4-METHOXY-4-METHYLPENTAN-2-ONE	3.3	UN2293	FLAMMABLE LIQUID
	METHYL ACETATE	3.2	UN1231	FLAMMABLE LIQUID
	METHYL ACETONE	3.2	UN1232	FLAMMABLE LIQUID
	METHYL ACETYLENE AND PROPADIENE *MIXTURE, STABILIZED*	2.1	UN1060	FLAMMABLE GAS
	BETA-METHYL ACROLEIN. *SEE* CROTONALDEHYDE, *INHIBITED*			
	METHYL ACRYLATE, *INHIBITED*	3.2	UN1919	FLAMMABLE LIQUID
	METHYLAL	3.1	UN1234	FLAMMABLE LIQUID
	METHYL ALCOHOL. *SEE* METHANOL			
	METHYL ALLYL CHLORIDE	3.1	UN2554	FLAMMABLE LIQUID
	METHYL ALUMINIUM SESQUIBROMIDE	4.2	UN1926	SPONTANEOUSLY COMBUSTIBLE
	METHYL ALUMINIUM SESQUICHLORIDE	4.2	UN1927	SPONTANEOUSLY COMBUSTIBLE
	METHYLAMINE, *ANHYDROUS*	2.1	UN1061	FLAMMABLE GAS
	METHYLAMINE, AQUEOUS SOLUTION	3.1	UN1235	FLAMMABLE LIQUID
	METHYLAMYL ACETATE	3.3	UN1233	FLAMMABLE LIQUID
	METHYL AMYL KETONE. *SEE* AMYL METHYL KETONE			
	N-METHYLANILINE	6.1	UN2294	ST. ANDREWS CROSS
	METHYLBENZOATE	6.1	UN2938	ST. ANDREWS CROSS
	ALPHA-METHYLBENZYL ALCOHOL	6.1	UN2937	ST. ANDREWS CROSS
	METHYL BROMIDE	2.3	UN1062	POISON GAS
	METHYL BROMIDE AND CHLOROPICRIN, MIXTURES. *SEE* CHLOROPICRIN AND METHYL BROMIDE, MIXTURES			
	METHYL BROMIDE AND ETHYLENE DIBROMIDE MIXTURES, LIQUID	6.1	UN1647	POISON
	METHYL BROMOACETATE	6.1	UN2643	POISON
	2-METHYL-1,3-BUTADIENE. *SEE* ISOPRENE			
	3-METHYL BUTAN-2-ONE	3.2	UN2397	FLAMMABLE LIQUID
	3-METHYL-1-BUTENE	3.2	UN2561	FLAMMABLE LIQUID
	2-METHYL-1-BUTENE	3.1	UN2459	FLAMMABLE LIQUID
	2-METHYL-2-BUTENE	3.1	UN2460	FLAMMABLE LIQUID
	N-METHYLBUTYLAMINE	3.2	UN2945	FLAMMABLE LIQUID
	METHYL-TERT-BUTYL ETHER	3.2	UN2398	FLAMMABLE LIQUID
	METHYL BUTYRATE	3.2	UN1237	FLAMMABLE LIQUID
	METHYL CHLORIDE	2.3	UN1063	POISON GAS, FLAMMABLE GAS
	METHYL CHLORIDE AND CHLOROPICRIN, MIXTURES. SEE CHLOROPICRIN , AND METHYL CHLORIDE, MIXTURES			
	METHYL CHLORIDE AND METHYLENE CHLORIDE, MIXTURES	2.1	UN1912	FLAMMABLE GAS
	METHYL CHLOROACETATE	3.3	UN2295	FLAMMABLE LIQUID
	METHYL CHLOROCARBONATE. *SEE* METHYL CHLOROFORMATE			
	METHYL CHLOROFORMATE	3.2	UN1238	FLAMMABLE LIQUID, POISON, CORROSIVE
	METHYLCHLOROMETHYL ETHER	3.1	UN1239	FLAMMABLE LIQUID
	METHYL-2-CHLOROPROPIONATE	3.3	UN2933	FLAMMABLE LIQUID
	METHYL CHLOROSILANE	3.2	UN2534	FLAMMABLE LIQUID, CORROSIVE

§172.102 OPTIONAL HAZARDOUS MATERIALS TABLE

(1) Notes and Symbols	(2) Hazardous Materials and Description and Proper Shipping Names	(3) IMO Class	(4) Identification Number	(5) Label(s) required
	METHYL CYANIDE	3.2	UN1648	FLAMMABLE LIQUID, POISON
	METHYL CYCLOHEXANE	3.2	UN2296	FLAMMABLE LIQUID
	METHYL CYCLOHEXANOL	3.3	UN2617	FLAMMABLE LIQUID
	METHYL CYCLOHEXANONE	3.3	UN2297	FLAMMABLE LIQUID
	METHYL CYCLOPENTANE	3.1	UN2298	FLAMMABLE LIQUID
	METHYL DICHLOROACETATE	6.1	UN2299	ST. ANDREWS CROSS
	METHYLDICHLOROSILANE	3.2	UN1242	FLAMMABLE LIQUID, CORROSIVE
	METHYLENE CHLORIDE. *SEE* DICHLOROMETHANE			
	METHYLENE DI (PHENYLENE ISOCYANATE). SEE DIPHENYLMETHANE DIISOCYANATE			
	METHYL ETHYL ETHER. *SEE* ETHYL METHYL ETHER			
	METHYL ETHYL KETONE PEROXIDE(S). *SEE* ETHYL METHYL KETONE PEROXIDE			
	METHYL ETHYL KETONE. *SEE* ETHYL METHYL KETONE			
	2-METHYL-5-ETHYLPYRIDINE	6.1	UN2300	ST. ANDREWS CROSS
	METHYL FLUORIDE	2.1	UN2454	FLAMMABLE GAS
	METHYL FORMATE	3.1	UN1243	FLAMMABLE LIQUID
	2-METHYLFURAN	3.1	UN2301	FLAMMABLE LIQUID
	5-METHYLHEXAN-2-ONE	3.3	UN2302	FLAMMABLE LIQUID
	METHYLHYDRAZINE	3.2	UN1244	FLAMMABLE LIQUID, CORROSIVE
	METHYL IODIDE	6.1	UN2644	POISON
	METHYL ISOBUTYL CARBINOL	3.3	UN2053	FLAMMABLE LIQUID
	METHYL ISOBUTYL KETONE	3.2	UN1245	FLAMMABLE LIQUID
	METHYL ISOCYANATE *OR* METHYL ISOCYANATE SOLUTIONS	3.2	UN2480	FLAMMABLE LIQUID, POISON
	METHYL ISOPROPENYL KETONE, *INHIBITED*	3.2	UN1246	FLAMMABLE LIQUID
	METHYL ISOTHIOCYANATE	3.2	UN2477	FLAMMABLE LIQUID, POISON
	METHYLISOVALERATE	3.2	UN2400	FLAMMABLE LIQUID
	METHYL MAGNESIUM BROMIDE, IN ETHYL ETHER	4.2	UN1928	SPONTANEOUSLY COMBUSTIBLE
	METHYLMERCAPTAN	2.1	UN1064	FLAMMABLE GAS
	METHYL METHACRYLATE, *MONOMER, INHIBITED*	3.2	UN1247	FLAMMABLE LIQUID
	METHYLMORPHOLINE	3.2 3.3	UN2535 UN2535	FLAMMABLE LIQUID, CORROSIVE FLAMMABLE LIQUID, CORROSIVE
	METHYL NITRITE - CARRIAGE FORBIDDEN			
	METHYL ORTHOSILICATE	3.2	UN2606	FLAMMABLE LIQUID, POISON
	METHYLPENTADIENE	3.1	UN2461	FLAMMABLE LIQUID
	METHYLPENTANES	3.1	UN2462	FLAMMABLE LIQUID
	2-METHYLPENTAN-2-OL	3.3	UN2560	FLAMMABLE LIQUID
	METHYLPHENYLDICHLOROSILANE	3.3	UN2437	FLAMMABLE LIQUID, CORROSIVE
	1-METHYLPIPERIDINE	3.2	UN2399	FLAMMABLE LIQUID
	METHYL PROPIONATE	3.2	UN1248	FLAMMABLE LIQUID
	METHYL PROPYL ETHER	3.1	UN2612	FLAMMABLE LIQUID
	METHYL PROPYL KETONE	3.2	UN1249	FLAMMABLE LIQUID
	METHYL SULPHIDE. *SEE* DIMETHYL SULFIDE			
	METHYLTETRAHYDROFURAN	3.2	UN2536	FLAMMABLE LIQUID
	METHYL TRICHLOROACETATE	6.1	UN2533	ST. ANDREWS CROSS
	METHYLTRICHLOROSILANE	3.2	UN1250	FLAMMABLE LIQUID, CORROSIVE
	ALPHA-METHYL VALERALDEHYDE	3.3	UN2367	FLAMMABLE LIQUID
	METHYL VINYL KETONE	3.2	UN1251	FLAMMABLE LIQUID
N	MINES, *WITH BURSTING CHARGE*	1.1D	UN0137	EXPLOSIVE (1.1D)
N	MINES, *WITH BURSTING CHARGE*	1.2D	UN0138	EXPLOSIVE (1.2D)
N	MINES, *WITH BURSTING CHARGE*	1.1F	UN0136	EXPLOSIVE (1.1F)
N	MINES, *WITH BURSTING CHARGE*	1.2F	UN0294	EXPLOSIVE (1.2F)
	MIXED ACID, SPENT. SEE ACID MIXTURES, SPENT, NITRATING			
	MOLYBDENUM PENTACHLORIDE	8	UN2508	CORROSIVE
	MONOCHLORODIFLUOROMETHANE. *SEE* CHLORODIFLUOROMETHANE			
	MONOCHLOROPENTAFLUOROETHANE. *SEE* CHLOROPENTAFLUOROETHANE.			
	MONOCHLOROTETRAFLUOROETHANE. *SEE* CHLOROTETRAFLUOROETHANE.			
	MONOCHLOROTRIFLUOROMETHANE. *SEE* CHLOROTRIFLUOROMETHANE			
	MONOETHANOLAMINE. *SEE* ETHANOLAMINE *AND SOLUTIONS THEREOF.*			
	MONOETHYLAMINE. *SEE* ETHYLAMINE			

§172.102 OPTIONAL HAZARDOUS MATERIALS TABLE

(1) Notes and Symbols	(2) Hazardous Materials and Description and Proper Shipping Names	(3) IMO Class	(4) Identification Number	(5) Label(s) required
	MONOMETHYLAMINE, *ANHYDROUS. SEE* METHYLAMINE, *ANHYDROUS*			
	MONOMETHYLAMINE, *AQUEOUS SOLUTION. SEE* METHYLAMINE, AQUEOUS SOLUTION			
	MONONITROTOLUIDINES	6.1	UN2660	ST. ANDREWS CROSS
	MONOPROPYLAMINE	3.1	UN1277	FLAMMABLE LIQUID
	MORPHOLINE	3.3	UN2054	FLAMMABLE LIQUID
	MOTOR FUEL ANTI-KNOCK MIXTURES	6.1	UN1649	POISON, FLAMMABLE LIQUID (ONLY IF FLASHPOINT BELOW 61 DEG C)
	MOTOR FUEL, N.O.S.	3.1	(UN1203)	FLAMMABLE LIQUID
	MOTOR SPIRIT. *SEE* GASOLINE			
	MURIATIC ACID. *SEE* HYDROCHLORIC ACID			
	NAPHTHALENE, CRUDE *OR* REFINED	4.1	UN1334	FLAMMABLE SOLID
	NAPHTHA, PETROLEUM	3.2	UN1255	FLAMMABLE LIQUID
	NAPHTHA, SOLVENT	3.2	UN1256	FLAMMABLE LIQUID
	NAPHTHYLAMINE *(ALPHA)*	6.1	UN2077	ST. ANDREWS CROSS
	NAPHTHYLAMINE *(BETA)*	6.1	UN1650	POISON
	ALPHA-NAPHTHYLTHIOUREA	6.1	UN1651	POISON
	NAPHTHYLUREA	6.1	UN1652	POISON
	NAPTHALENE, MOLTEN	4.1	UN2304	FLAMMABLE SOLID
	NATURAL GASES *WITH A HIGH METHANE CONTENT. SEE* METHANE *OR* NATURAL GASES, *ETC.*			
	NATURAL GASOLINE. *SEE* CASINGHEAD GASOLINE			
	NEOHEXANE. *SEE* DIMETHYL BUTANE			
	NEON, *COMPRESSED*	2.2	UN1065	NONFLAMMABLE GAS
	NEON, REFRIGERATED LIQUID	2.2	UN1913	NONFLAMMABLE GAS
	NICKEL CARBONYL	6.1	UN1259	POISON, FLAMMABLE LIQUID
	NICKEL CATALYST, DRY *PRECIPITATED ON A CARRIER WITH A SPECIAL ACTIVATOR*	4.2	UN2881	SPONTANEOUSLY COMBUSTIBLE
	NICKEL CATALYST, WETTED *WITH NOT LESS THAN 40% WATER OR OTHER SUITABLE LIQUID, BY WEIGHT, FINELY DIVIDED, ACTIVATED, OR SPENT*	4.2	UN1378	SPONTANEOUSLY COMBUSTIBLE
	NICKEL CYANIDE	6.1	UN1653	POISON
	NICKEL NITRATE	5.1	UN2725	OXIDIZER
	NICKEL NITRITE	5.1	UN2726	OXIDIZER
	NICOTINE *(AND SALTS OF)*, LIQUID	6.1 6.1 6.1	UN1654 UN1654 UN1654	POISON POISON ST. ANDREWS CROSS
	NICOTINE *(AND SALTS OF)*, SOLID	6.1 6.1	UN1654 UN1654	POISON ST. ANDREWS CROSS
	NICOTINE COMPOUNDS, N.O.S. *OR* NICOTINE PREPARATIONS, N.O.S.	6.1 6.1	UN1655 UN1655	POISON ST. ANDREWS CROSS
	NICOTINE HYDROCHLORIDE, *OR* NICOTINE HYDROCHLORIDE SOLUTIONS	6.1	UN1656	POISON
	NICOTINE SALICYLATE	6.1	UN1657	POISON
	NICOTINE SULPHATE, SOLID *OR* SOLUTION	6.1	UN1658	POISON
	NICOTINE TARTRATE	6.1	UN1659	POISON
	NITRATE OF SODA AND POTASH, MIXTURE. *SEE* SODIUM NITRATE AND POTASH, *MIXTURE*			
	NITRATES, INORGANIC, N.O.S.	5.1	UN1477	OXIDIZER
	NITRATING ACID, MIXTURES	8	UN1796	CORROSIVE, OXIDIZER (NOT REQUIRED IF NITRIC ACID CONTENT NOT MORE THAN 50%)
	NITRIC ACID, *OTHER THAN RED FUMING, ALL CONCENTRATIONS*	8	UN2031	CORROSIVE
	NITRIC ACID, RED FUMING	8	UN2032	CORROSIVE, OXIDIZER, POISON
	NITRIC OXIDE	2.3	UN1660	POISON GAS
	NITRIC OXIDE AND NITROGEN TETROXIDE MIXTURES	2.3	UN1975	POISON GAS, OXIDIZER
	NITRITES, INORGANIC, N.O.S.	5.1	UN2627	OXIDIZER
	NITROANILINES *(O-, M-, P-)*	6.1	UN1661	POISON
	NITROANISOLES	6.1	UN2730	ST. ANDREWS CROSS
	NITROBENZENE	6.1	UN1662	POISON
	NITROBENZENESULPHONIC ACID	8	UN2305	CORROSIVE
	NITROBENZOL. *SEE* NITROBENZENE			

§172.102 OPTIONAL HAZARDOUS MATERIALS TABLE

(1) Notes and Symbols	(2) Hazardous Materials and Description and Proper Shipping Names	(3) IMO Class	(4) Identification Number	(5) Label(s) required
N	5-NITROBENZOTRIAZOL	1.1D	UN0385	EXPLOSIVE (1.1D)
	NITROBENZOTRIFLUORIDES	6.1	UN2306	POISON
	NITROBROMOBENZENES	6.1	UN2732	ST. ANDREWS CROSS
	NITROCELLULOSE, *CONTAINING AT LEAST 25% ALCOHOL, BY WEIGHT, AND NOT EXCEEDING 12.6% NITROGEN BY DRY WEIGHT*	4.1	UN2556	FLAMMABLE SOLID
	NITROCELLULOSE, *CONTAINING AT LEAST 25%, BY WEIGHT, WATER*	4.1	UN2555	FLAMMABLE SOLID
	NITROCELLULOSE, *CONTAINING AT LEAST 18% PLASTICIZING SUBSTANCE, BY WEIGHT, AND NOT EXCEEDING 12.6% NITROGEN BY DRY WEIGHT*	4.1	UN2557	FLAMMABLE SOLID
N	NITROCELLULOSE *DRY OR WETTED WITH LESS THAN 25% ALCOHOL, BY WEIGHT*	1.1D	UN0340	EXPLOSIVE (1.1D)
N	NITROCELLULOSE, *DRY OR WETTED WITH LESS THAN 25% WATER, BY WEIGHT*	1.1D	UN0340	EXPLOSIVE (1.1D)
	NITROCELLULOSE, *IN SOLUTION IN FLAMMABLE LIQUIDS*	3.2 3.3	UN2059 UN2060	FLAMMABLE LIQUID FLAMMABLE LIQUID
N	NITROCELLULOSE, PLASTICIZED *WITH NOT LESS THAN 18% PLASTICIZING SUBSTANCE, BY WEIGHT*	1.3C	UN0343	EXPLOSIVE (1.3C)
N	NITROCELLULOSE *UNMODIFIED OR UNPLASTICIZED WITH LESS THAN 18% PLASTICIZING SUBSTANCE, BY WEIGHT*	1.1D	UN0341	EXPLOSIVE (1.1D)
	NITROCELLULOSE, *WETTED WITH, BY WEIGHT, MORE THAN 40% FLAMMABLE LIQUIDS*	3.2 3.3	(UN2556) (UN2556)	FLAMMABLE LIQUID FLAMMABLE LIQUID
N	NITROCELLULOSE, WETTED *WITH NOT LESS THAN 25% ALCOHOL, BY WEIGHT*	1.3C	UN0342	EXPLOSIVE (1.3C)
	3-NITRO-4-CHLOROBENZOTRIFLUORIDE	6.1	UN2307	POISON
	NITROCRESOLS	6.1	UN2446	ST. ANDREWS CROSS
	NITROETHANE	3.3	UN2842	FLAMMABLE LIQUID
	NITROGEN, *COMPRESSED*	2.2	UN1066	NONFLAMMABLE GAS
	NITROGEN DIOXIDE, *LIQUEFIED*	2.3	UN1067	POISON GAS, OXIDIZER
	NITROGEN, REFRIGERATED LIQUID	2.2	UN1977	NONFLAMMABLE GAS
	NITROGEN TETROXIDE. *SEE* NITROGEN DIOXIDE			
	NITROGEN TRIFLUORIDE	2.3	UN2451	POISON GAS
	NITROGEN TRIOXIDE	2.3	UN2421	POISON GAS
N	NITROGLYCERINE, DESENSITIZED, *WITH NOT LESS THAN 40% NON-VOLATILE WATER-INSOLUBLE PHLEGMATIZER, BY WEIGHT*	1.1D	UN0143	EXPLOSIVE (1.1D), POISON
N	NITROGLYCERINE, SPIRIT OF, *WITH MORE THAN 1% BUT NOT MORE THAN 10% NITROGLYCERINE IN SOLUTION IN ALCOHOL*	1.1D	UN0144	EXPLOSIVE (1.1D)
	NITROGLYCERIN SOLUTION, IN ALCOHOL *WITH NOT MORE THAN 1% NITROGLYCERIN*	3.2	UN1204	FLAMMABLE LIQUID
N	NITROGUANIDINE, *DRY OR CONTAINING, BY WEIGHT, LESS THAN 20% WATER*	1.1D	UN0282	EXPLOSIVE (1.1D)
	NITROGUANIDINE, WETTED, *WITH, BY WEIGHT, AT LEAST 20% WATER*	4.1	UN1336	FLAMMABLE SOLID
	NITROHYDROCHLORIC ACID	8	UN1798	CORROSIVE
	NITROMETHANE	3.3	UN1261	FLAMMABLE LIQUID
	NITROMURIATIC ACID. *SEE* NITROHYDROCHLORIC ACID			
	NITRONAPTHALENE	4.1	UN2538	FLAMMABLE SOLID
	NITROPHENOLS *(O-, M-, P-)*	6.1	UN1663	ST. ANDREWS CROSS
	NITROPROPANES	3.3	UN2608	FLAMMABLE LIQUID
	P-NITROSODIMETHYLANILINE	4.2	UN1369	SPONTANEOUSLY COMBUSTIBLE
N	NITROSTARCH, *DRY OR CONTAINING, BY WEIGHT, LESS THAN 20% WATER*	1.1D	UN0146	EXPLOSIVE (1.1D)
	NITROSTARCH, WETTED *WITH, BY WEIGHT, AT LEAST 20% WATER*	4.1	UN1337	FLAMMABLE SOLID
	NITROSYL CHLORIDE	2.3	UN1069	POISON GAS, CORROSIVE
	NITROSYLSULPHURIC ACID	8	UN2308	CORROSIVE
	NITROTOLUENES *(O-, M-, P-)*	6.1	UN1664	POISON
N	NITRO UREA	1.1D	UN0147	EXPLOSIVE (1.1D)
	NITROUS OXIDE, *COMPRESSED*	2.2	UN1070	NONFLAMMABLE GAS, OXIDIZER
	NITROUS OXIDE, REFRIGERATED LIQUID	2.2	UN2201	NONFLAMMABLE GAS
	NITROXYLENES *(O-, M-, P-)*	6.1	UN1665	POISON
	NONANE, *AND ITS ISOMERS*	3.3	UN1920	FLAMMABLE LIQUID
	NONYL TRICHLOROSILANE	8	UN1799	CORROSIVE
	OCTADECYL TRICHLOROSILANE	8	UN1800	CORROSIVE
	OCTADIENE	3.2	UN2309	FLAMMABLE LIQUID
	OCTAFLUOROBUT-2-ENE	2.2	UN2422	NONFLAMMABLE GAS
	OCTAFLUOROCYCLOBUTANE	2.2	UN1976	NONFLAMMABLE GAS

(1) Notes and Symbols	(2) Hazardous Materials and Description and Proper Shipping Names	(3) IMO Class	(4) Identification Number	(5) Label(s) required
	OCTAFLUOROPROPANE	2.2	UN2424	NONFLAMMABLE GAS
	OCTANE *AND ITS ISOMERS*	3.2	UN1262	FLAMMABLE LIQUID
	N-OCTANOYL PEROXIDE. *SEE* DI-N-OCTANOYL PEROXIDE			
N	OCTOLITE, *DRY OR WETTED WITH LESS THAN 15% WATER, BY WEIGHT*	1.1D	UN0266	EXPLOSIVE (1.1D)
	OCTYL TRICHLOROSILANE	8	UN1801	CORROSIVE
	OIL GAS	2.1	UN1071	FLAMMABLE GAS, POISON GAS
	OLEUM. *SEE* SULPHURIC ACID, FUMING			
	ORGANIC PEROXIDES, MIXTURE (*THIS DESCRIPTION MUST BE SUPPLEMENTED WITH THE NAME OF THE PRIMARY CONSTITUENT OF THE MIXTURE*)	5.2	UN2756	ORGANIC PEROXIDE
	ORGANIC PEROXIDES, N.O.S., SAMPLES	5.2	UN2255	ORGANIC PEROXIDE
	ORGANIC PEROXIDES, N.O.S., TRIAL QUANTITIES	5.2	UN2899	ORGANIC PEROXIDE
	ORGANOCHLORINE PESTICIDES, LIQUID, FLAMMABLE, TOXIC, N.O.S., *FLASHPOINT LESS THAN 23 DEG C*	3.2	UN2762	FLAMMABLE LIQUID AND POISON OR ST. ANDREWS CROSS (ACCORDING TO TOXICITY)
	ORGANOCHLORINE PESTICIDES, LIQUID, TOXIC, FLAMMABLE, N.O.S., *FLASHPOINT BETWEEN 23 DEG C AND 61 DEG C*	6.1 6.1 6.1	UN2995 UN2995 UN2995	POISON, FLAMMABLE LIQUID POISON, FLAMMABLE LIQUID ST. ANDREWS CROSS, FLAMMABLE LIQUID
	ORGANOCHLORINE PESTICIDES, LIQUID, TOXIC, N.O.S.	6.1 6.1 6.1	UN2996 UN2996 UN2996	POISON POISON ST. ANDREWS CROSS
	ORGANOCHLORINE PESTICIDES, SOLID, TOXIC, N.O.S.	6.1 6.1	UN2761 UN2761	POISON ST. ANDREWS CROSS
	ORGANOPHOSPHORUS PESTICIDES, LIQUID, FLAMMABLE, TOXIC, N.O.S., *FLASHPOINT LESS THAN 23 DEG C*	3.2	UN2784	FLAMMABLE LIQUID AND POISON OR ST. ANDREWS CROSS (ACCORDING TO TOXICITY)
	ORGANOPHOSPHORUS PESTICIDES, LIQUID, TOXIC, FLAMMABLE, N.O.S., *FLASHPOINT BETWEEN 23 DEG C AND 61 DEG C*	6.1 6.1 6.1	UN3017 UN3017 UN3017	POISON, FLAMMABLE LIQUID POISON, FLAMMABLE LIQUID ST. ANDREWS CROSS, FLAMMABLE LIQUID
	ORGANOPHOSPHORUS PESTICIDES, LIQUID, TOXIC, N.O.S.	6.1 6.1 6.1	UN3018 UN3018 UN3018	POISON POISON ST. ANDREWS CROSS
	ORGANOPHOSPHORUS PESTICIDES SOLID, TOXIC, N.O.S.	6.1 6.1	UN2783 UN2783	POISON ST. ANDREWS CROSS
	ORGANOTIN COMPOUNDS, N.O.S.	6.1 6.1	UN2788 UN2788	POISON ST. ANDREWS CROSS
	ORGANOTIN PESTICIDES, LIQUID, FLAMMABLE, TOXIC, N.O.S., *FLASHPOINT BELOW 23 DEG C*	3.2	UN2787	FLAMMABLE LIQUID AND POISON OR ST. ANDREWS CROSS (ACCORDING TO TOXICITY)
	ORGANOTIN PESTICIDES, LIQUID, TOXIC, FLAMMABLE, N.O.S., *FLASHPOINT BETWEEN 23 DEG C AND 61 DEG C*	6.1 6.1 6.1	UN3019 UN3019 UN3019	POISON, FLAMMABLE LIQUID POISON, FLAMMABLE LIQUID ST. ANDREWS CROSS, FLAMMABLE LIQUID
	ORGANOTIN PESTICIDES, LIQUID, TOXIC, N.O.S.	6.1 6.1 6.1	UN3020 UN3020 UN3020	POISON POISON ST. ANDREWS CROSS
	ORGANOTIN PESTICIDES, SOLID, TOXIC, N.O.S.	6.1 6.1	UN2786 UN2786	POISON ST. ANDREWS CROSS
	OSMIUM TETROXIDE	6.1	UN2471	POISON
	OXALATES, *WATER SOLUBLE*	6.1	UN2449	ST. ANDREWS CROSS
	OXIDIZING SUBSTANCES, N.O.S.	5.1	UN1479	OXIDIZER
	OXYGEN AND CARBON DIOXIDE, MIXTURES. SEE CARBON DIOXIDE AND OXYGEN *MIXTURES*			
	OXYGEN, *COMPRESSED*	2.2	UN1072	NONFLAMMABLE GAS, OXIDIZER
	OXYGEN DIFLUORIDE	2.3	UN2190	POISON GAS
	OXYGEN, *REFRIGERATED LIQUID*	2.2	UN1073	NONFLAMMABLE GAS, OXIDIZER
	PAINTS, ENAMELS, LACQUERS, STAINS, SHELLAC *SOLUTIONS,* VARNISH, POLISHES, FILLERS *LIQUID,* LACQUER BASE *OR* THINNERS	3.2 3.3	UN1263 UN1263	FLAMMABLE LIQUID FLAMMABLE LIQUID
	PAPER, UNSATURATED OIL TREATED, *INCOMPLETELY DRIED*	4.2	UN1379	SPONTANEOUSLY COMBUSTIBLE
	PARAFORMALDEHYDE	4.1	UN2213	NONE. PACKAGE TO BE MARKED 'CLASS 4.1'
	PARALDEHYDE	3.3	UN1264	FLAMMABLE LIQUID
	PENTABORANE	4.2	UN1380	SPONTANEOUSLY COMBUSTIBLE, POISON

§172.102 OPTIONAL HAZARDOUS MATERIALS TABLE

(1) Notes and Symbols	(2) Hazardous Materials and Description and Proper Shipping Names	(3) IMO Class	(4) Identification Number	(5) Label(s) required
	PENTACHLOROETHANE	6.1	UN1669	POISON
N	PENTAERYTHRITE TETRANITRATE, WETTED *WITH NOT LESS THAN 25% WATER, BY WEIGHT,* OR DESENSITIZED *WITH NOT LESS THAN 15% PHLEGMATIZER, BY WEIGHT*	1.1D	UN0150	EXPLOSIVE (1.1D)
N	PENTAERYTHRITE TETRANITRATE, *WITH NOT LESS THAN 7% WAX, BY WEIGHT*	1.1D	UN0411	EXPLOSIVE (1.1D)
	PENTAMETHYLHEPTANE	3.3	UN2286	FLAMMABLE LIQUID
	2,4-PENTANEDIONE	3.3	UN2310	FLAMMABLE LIQUID
	PENTANES	3.1	UN1265	FLAMMABLE LIQUID
	1-PENTOL	8	UN2705	CORROSIVE
N	PENTOLITE, *DRY OR WETTED WITH LESS THAN 15% WATER, BY WEIGHT*	1.1D	UN0151	EXPLOSIVE (1.1D)
	PERCHLORATES, INORGANIC, N.O.S.	5.1	UN1481	OXIDIZER
	PERCHLORIC ACID, *NOT EXCEEDING 50%, BY WEIGHT, OF ACID*	8	UN1802	CORROSIVE, OXIDIZER
	PERCHLORIC ACID, *OVER 50% AND NOT EXCEEDING 72% OF ACID*	5.1	UN1873	OXIDIZER, CORROSIVE
	PERCHLOROETHYLENE. *SEE* TETRACHLOROETHYLENE			
	PERCHLOROMETHYL MERCAPTAN	6.1	UN1670	POISON
	PERFUMERY PRODUCTS, *WITH FLAMMABLE SOLVENTS*	3.2 3.3	UN1266 UN1266	FLAMMABLE LIQUID FLAMMABLE LIQUID
	PERMANGANATES, INORGANIC, N.O.S.	5.1	UN1482	OXIDIZER
	PEROXIDES, INORGANIC, N.O.S.	5.1	UN1483	OXIDIZER
	PEROXYACETIC ACID, *MAXIMUM CONCENTRATION 43% IN ACETIC ACID OR IN A MIXTURE OF ACETIC ACID AND WATER, WITH IN EITHER CASE NOT MORE THAN 6% HYDROGEN PEROXIDE AND NOT MORE THAN 1% SULPHURIC ACID*	5.2	UN2131	ORGANIC PEROXIDE, CORROSIVE
	PESTICIDES, LIQUID, FLAMMABLE, TOXIC, N.O.S. *FLASHPOINT BELOW 23 DEG C*	3.2	UN3021	FLAMMABLE LIQUID AND POISON OR ST. ANDREWS CROSS (ACCORDING TO TOXICITY)
	PESTICIDES, LIQUID, TOXIC, FLAMMABLE, N.O.S. *FLASHPOINT BETWEEN 23 DEG C AND 61 DEG C*	6.1 6.1 6.1	UN2903 UN2903 UN2903	POISON, FLAMMABLE LIQUID POISON, FLAMMABLE LIQUID ST. ANDREWS CROSS, FLAMMABLE LIQUID
	PESTICIDES, LIQUID, TOXIC, N.O.S.	6.1 6.1 6.1	UN2902 UN2902 UN2902	POISON POISON ST. ANDREWS CROSS
	PESTICIDES, SOLID, TOXIC, N.O.S.	6.1 6.1	UN2588 UN2588	POISON ST. ANDREWS CROSS
	PETROLEUM CRUDE OIL	3.1 3.2 3.3	UN1267 UN1267 UN1267	FLAMMABLE LIQUID FLAMMABLE LIQUID FLAMMABLE LIQUID
	PETROLEUM DISTILLATES, N.O.S.	3.1 3.2 3.3	UN1268 UN1268 UN1268	FLAMMABLE LIQUID FLAMMABLE LIQUID FLAMMABLE LIQUID
	PETROLEUM ETHER. *SEE* PETROLEUM SPIRIT			
	PETROLEUM GASES, LIQUEFIED	2.1	UN1075	FLAMMABLE GAS
	PETROLEUM OIL	3.1 3.2 3.3	UN1270 UN1270 UN1270	FLAMMABLE LIQUID FLAMMABLE LIQUID FLAMMABLE LIQUID
	PETROLEUM SPIRIT	3.1 3.2 3.3	UN1271 UN1271 UN1271	FLAMMABLE LIQUID FLAMMABLE LIQUID FLAMMABLE LIQUID
	PETROL. *SEE* GASOLINE			
	PHENACYL BROMIDE	6.1	UN2645	POISON
	PHENETIDINES (*O-, P-*)	6.1	UN2311	ST. ANDREWS CROSS
	PHENOL	6.1	UN1671	POISON
	PHENOL, MOLTEN	6.1	UN2312	POISON
	PHENOL SOLUTIONS	6.1	UN2821	POISON
	PHENOLSULPHONIC ACID, *LIQUID*	8	UN1803	CORROSIVE
	PHENOXY PESTICIDES, SOLID, TOXIC, N.O.S.	6.1 6.1	UN2765 UN2765	POISON ST. ANDREWS CROSS
	PHENOXY PESTICIDES, LIQUID, FLAMMABLE, TOXIC, N.O.S., *FLASHPOINT LESS THAN 23 DEG C*	3.2	UN2766	FLAMMABLE LIQUID AND POISON OR ST. ANDREWS CROSS (ACCORDING TO TOXICITY)
	PHENOXY PESTICIDES, LIQUID, TOXIC, FLAMMABLE, N.O.S., *FLASHPOINT BETWEEN 23 DEG C AND 61 DEG C*	6.1 6.1 6.1	UN2999 UN2999 UN2999	POISON, FLAMMABLE LIQUID POISON, FLAMMABLE LIQUID ST. ANDREWS CROSS, FLAMMABLE LIQUID

§172.102 OPTIONAL HAZARDOUS MATERIALS TABLE

(1) Notes and Symbols	(2) Hazardous Materials and Description and Proper Shipping Names	(3) IMO Class	(4) Identification Number	(5) Label(s) required
	PHENOXY PESTICIDES, LIQUID, TOXIC, N.O.S.	6.1	UN3000	POISON
		6.1	UN3000	POISON
		6.1	UN3000	ST. ANDREWS CROSS
	PHENYLACETONITRILE, LIQUID	6.1	UN2470	ST. ANDREWS CROSS
	PHENYLACETYL CHLORIDE	8	UN2577	CORROSIVE
	PHENYLCARBYLAMINE CHLORIDE	6.1	UN1672	POISON
	PHENYLCHLOROFORMATE	6.1	UN2746	POISON, CORROSIVE
	PHENYLENEDIAMINES *(O-, M-, P-)*	6.1	UN1673	ST. ANDREWS CROSS
	PHENYLHYDRAZINE	6.1	UN2572	POISON
	PHENYL ISOCYANATE	6.1	UN2487	POISON, FLAMMABLE LIQUID
	PHENYL MERCAPTAN	6.1	UN2337	POISON, FLAMMABLE LIQUID
	PHENYLMERCURIC ACETATE	6.1	UN1674	POISON
	PHENYLMERCURIC COMPOUNDS, N.O.S.	6.1	UN2026	POISON
		6.1	UN2026	ST. ANDREWS CROSS
	PHENYLMERCURIC HYDROXIDE	6.1	UN1894	POISON
	PHENYLMERCURIC NITRATE	6.1	UN1895	POISON
	PHENYL PHOSPHORUS DICHLORIDE	8	UN2798	CORROSIVE
	PHENYL PHOSPHORUS THIODICHLORIDE	8	UN2799	CORROSIVE
	PHENYL TRICHLOROSILANE	8	UN1804	CORROSIVE
	PHENYL UREA PESTICIDES, LIQUID, FLAMMABLE, TOXIC, N.O.S., *FLASHPOINT LESS THAN 23 DEG C*	3.2	UN2768	FLAMMABLE LIQUID AND POISON OR ST. ANDREWS CROSS (ACCORDING TO TOXICITY)
	PHENYL UREA PESTICIDES, LIQUID, TOXIC, FLAMMABLE, N.O.S., *FLASHPOINT BETWEEN 23 DEG C AND 61 DEG C*	6.1	UN3001	POISON, FLAMMABLE LIQUID
		6.1	UN3001	POISON, FLAMMABLE LIQUID
		6.1	UN3001	ST. ANDREWS CROSS, FLAMMABLE LIQUID
	PHENYL UREA PESTICIDES, LIQUID, TOXIC, N.O.S.	6.1	UN3002	POISON
		6.1	UN3002	POISON
		6.1	UN3002	ST. ANDREWS CROSS
	PHENYL UREA PESTICIDES, SOLID, TOXIC, N.O.S.	6.1	UN2767	POISON
		6.1	UN2767	ST. ANDREWS CROSS
	PHOSGENE	2.3	UN1076	POISON GAS, CORROSIVE
	9-PHOSPHABICYONONANES	4.2	UN2940	SPONTANEOUSLY COMBUSTIBLE
	PHOSPHINE	2.3	UN2199	POISON GAS, FLAMMABLE GAS
	PHOSPHORIC ACID, *LIQUID*	8	UN1805	CORROSIVE
	PHOSPHORIC ACID, *SOLID*	8	UN1805	CORROSIVE
	PHOSPHORIC ANHYDRIDE. *SEE* PHOSPHORUS PENTOXIDE			
	PHOSPHOROUS ACID, ORTHO	8	UN2834	CORROSIVE
	PHOSPHOROUS BROMIDE. *SEE* PHOSPHORUS TRIBROMIDE			
	PHOSPHOROUS CHLORIDE. *SEE* PHOSPHORUS TRICHLORIDE			
	PHOSPHORUS, AMORPHOUS	4.1	UN1338	FLAMMABLE SOLID
	PHOSPHORUS HEPTASULPHIDE, *FREE FROM YELLOW OR WHITE PHOSPHORUS*	4.1	UN1339	FLAMMABLE SOLID
	PHOSPHORUS OXYBROMIDE, MOLTEN	8	UN2576	CORROSIVE
	PHOSPHORUS OXYBROMIDE, SOLID	8	UN1939	CORROSIVE
	PHOSPHORUS OXYCHLORIDE	8	UN1810	CORROSIVE
	PHOSPHORUS PENTABROMIDE	8	UN2691	CORROSIVE
	PHOSPHORUS PENTACHLORIDE	8	UN1806	CORROSIVE
	PHOSPHORUS PENTAFLUORIDE	2.3	UN2198	POISON GAS
	PHOSPHORUS PENTASULPHIDE, *FREE FROM YELLOW OR WHITE PHOSPHORUS*	4.1	UN1340	FLAMMABLE SOLID
	PHOSPHORUS PENTOXIDE	8	UN1807	CORROSIVE
	PHOSPHORUS SESQUISULPHIDE, *FREE FROM YELLOW OR WHITE PHOSPHORUS*	4.1	UN1341	FLAMMABLE SOLID
	PHOSPHORUS TRIBROMIDE	8	UN1808	CORROSIVE
	PHOSPHORUS TRICHLORIDE	8	UN1809	CORROSIVE
	PHOSPHORUS TRIOXIDE	8	UN2578	CORROSIVE
	PHOSPHORUS TRISULPHIDE, *FREE FROM YELLOW OR WHITE PHOSPHORUS*	4.1	UN1343	FLAMMABLE SOLID
	PHOSPHORUS, WHITE, MOLTEN	4.2	UN2447	SPONTANEOUSLY COMBUSTIBLE, POISON
	PHOSPHORUS, *WHITE OR YELLOW, DRY*	4.2	UN1381	SPONTANEOUSLY COMBUSTIBLE
	PHOSPHORUS, *WHITE OR YELLOW, IN WATER*	4.2	UN1381	SPONTANEOUSLY COMBUSTIBLE
N	PHOTO-FLASH POWDER *IN UNITS*	1.1G	UN0094	EXPLOSIVE (1.1G)

§172.102 OPTIONAL HAZARDOUS MATERIALS TABLE

(1) Notes and Symbols	(2) Hazardous Materials and Description and Proper Shipping Names	(3) IMO Class	(4) Identification Number	(5) Label(s) required
N	PHOTO-FLASH POWDER *IN UNITS*	1.2G	UN0096	EXPLOSIVE (1.2G)
N	PHOTO-FLASH POWDER *IN UNITS*	1.3G	UN0305	EXPLOSIVE (1.3G)
	PHTHALIC ANHYDRIDE, *SOLID OR MOLTEN*	8	UN2214	NONE. PACKAGE TO BE MARKED 'CLASS 8'
	PHTHALIMIDE DERIVATIVE PESTICIDES, LIQUID, FLAMMABLE, TOXIC, N.O.S., *FLASHPOINT BELOW 23 DEG C*	3.2	UN2774	FLAMMABLE LIQUID AND POISON OR ST. ANDREWS CROSS (ACCORDING TO TOXICITY)
	PHTHALIMIDE DERIVATIVE PESTICIDES, LIQUID, TOXIC, FLAMMABLE, N.O.S., *FLASHPOINT BETWEEN 23 DEG C AND 61 DEG C*	6.1 6.1 6.1	UN3007 UN3007 UN3007	POISON FLAMMABLE LIQUID POISON FLAMMABLE LIQUID ST. ANDREWS CROSS FLAMMABLE LIQUID
	PHTHALIMIDE DERIVATIVE PESTICIDES, LIQUID, TOXIC, N.O.S.	6.1 6.1 6.1	UN3008 UN3008 UN3008	POISON POISON ST. ANDREWS CROSS
	PHTHALIMIDE DERIVATIVE PESTICIDES, SOLID, TOXIC, N.O.S.	6.1 6.1	UN2773 UN2773	POISON ST. ANDREWS CROSS
	PICOLINES	3.3	UN2313	FLAMMABLE LIQUID
	PICRIC ACID, WETTED. *SEE* TRINITROPHENOL, WETTED			
	PINANYL HYDROPEROXIDE *TECHNICAL PURE*	5.2	UN2162	ORGANIC PEROXIDE
	PINDONE *(AND SALTS OF), LIQUID*	6.1 6.1 6.1	UN2472 UN2472 UN2472	POISON POISON ST. ANDREWS CROSS
	PINDONE *(AND SALTS OF), SOLID*	6.1 6.1	UN2472 UN2472	POISON ST. ANDREWS CROSS
	ALPHA-PINENE	3.3	UN2368	FLAMMABLE LIQUID
	PINE OIL	3.3	UN1272	FLAMMABLE LIQUID
	PIPERAZINE	8	UN2579	CORROSIVE
	PIPERIDINE	3.2	UN2401	FLAMMABLE LIQUID
	PLASTICS MOULDING MATERIALS EVOLVING FLAMMABLE VAPOURS	9	UN2211	NONE
	PLASTICS, NITROCELLULOSE-BASED, SPONTANEOUSLY COMBUSTIBLE, N.O.S.	4.2	UN2006	SPONTANEOUSLY COMBUSTIBLE
	POISONOUS LIQUIDS, CORROSIVE, N.O.S.	6.1	UN2927	POISON, CORROSIVE
	POISONOUS LIQUIDS, FLAMMABLE, N.O.S.	6.1	UN2929	POISON, FLAMMABLE LIQUID
	POISONOUS LIQUIDS, N.O.S.	6.1 6.1	UN2810 UN2810	POISON ST. ANDREWS CROSS
	POISONOUS SOLIDS, CORROSIVE, N.O.S.	6.1	UN2928	POISON, CORROSIVE
	POISONOUS SOLIDS, FLAMMABLE, N.O.S.	6.1	UN2930	POISON, FLAMMABLE SOLID
	POISONOUS SOLIDS, N.O.S.	6.1 6.1	UN2811 UN2811	POISON ST. ANDREWS CROSS
	POLISHES. *SEE* PAINTS, *ETC.*			
	POLISHING FLUID. *SEE* FLAMMABLE LIQUID PREPARATIONS, N.O.S.			
	POLYCHLORINATED BIPHENYLS,	9	UN2315	NONE[3]
	POLYSTYRENE BEADS, EXPANDABLE, IMPREGNATED WITH FLAMMABLE LIQUID. SEE PLASTICS MOULDING MATERIALS EVOLVING FLAMMABLE VAPOURS			
	POTASSIUM ARSENATE	6.1	UN1677	POISON
	POTASSIUM ARSENITE	6.1	UN1678	POISON
	POTASSIUM BIFLUORIDE, *SOLID*	8	UN1811	CORROSIVE, POISON
	POTASSIUM BIFLUORIDE, *SOLUTION*	8	UN1811	CORROSIVE, POISON
	POTASSIUM BOROHYDRIDE	4.3	UN1870	DANGEROUS WHEN WET
	POTASSIUM BROMATE	5.1	UN1484	OXIDIZER
	POTASSIUM CHLORATE	5.1	UN1485	OXIDIZER
	POTASSIUM CHLORATE, *AQUEOUS SOLUTION*	5.1	UN2427	OXIDIZER
	POTASSIUM CUPROCYANIDE	6.1	UN1679	POISON
	POTASSIUM CYANIDE	6.1	UN1680	POISON
	POTASSIUM FLUORIDE	6.1	UN1812	ST. ANDREWS CROSS
	POTASSIUM FLUOROACETATE	6.1	UN2628	POISON
	POTASSIUM HYDROGEN FLUORIDE. *SEE* POTASSIUM BIFLUORIDE			
	POTASSIUM HYDROGEN SULPHATE	8	UN2509	CORROSIVE
	POTASSIUM HYDROSULPHITE	4.2	UN1929	SPONTANEOUSLY COMBUSTIBLE
	POTASSIUM HYDROXIDE, SOLID	8	UN1813	CORROSIVE
	POTASSIUM HYDROXIDE, SOLUTION	8	UN1814	CORROSIVE
	POTASSIUM HYPOCHLORITE, *SOLUTION. SEE* HYPOCHLORITE, SOLUTIONS, *ETC.*			

[3] EPA REGULATIONS REQUIRE THE DISPLAY OF THE LARGE PCB LABEL ON THE FRONT, REAR, AND EACH SIDE OF THE VEHICLE

(1) Notes and Symbols	(2) Hazardous Materials and Description and Proper Shipping Names	(3) IMO Class	(4) Identification Number	(5) Label(s) required
	POTASSIUM METAL	4.2	UN2257	SPONTANEOUSLY COMBUSTIBLE
	POTASSIUM, METAL ALLOYS	4.3	UN1420	DANGEROUS WHEN WET
	POTASSIUM METAVANADATE	6.1	UN2864	POISON
	POTASSIUM NITRATE	5.1	UN1486	OXIDIZER
	POTASSIUM NITRATE AND SODIUM NITRITE, MIXTURE	5.1	UN1487	OXIDIZER
	POTASSIUM NITRATE BAGS, *EMPTY. SEE* BAGS, *EMPTY AND UNWASHED, ETC.*			
	POTASSIUM NITRITE	5.1	UN1488	OXIDIZER
	POTASSIUM OXIDE	8	UN2033	CORROSIVE
	POTASSIUM PERCHLORATE	5.1	UN1489	OXIDIZER
	POTASSIUM PERMANGANATE	5.1	UN1490	OXIDIZER
	POTASSIUM PEROXIDE	5.1	UN1491	OXIDIZER
	POTASSIUM PERSULPHATE	5.1	UN1492	OXIDIZER
	POTASSIUM PHOSPHIDE	4.3	UN2012	DANGEROUS WHEN WET, POISON
N	POTASSIUM SALTS OF NITRO-AROMATIC DERIVATIVES, *EXPLOSIVE*	1.3C	UN0158	EXPLOSIVE (1.3C)
	POTASSIUM SILICOFLUORIDE	6.1	UN2655	ST. ANDREWS CROSS
	POTASSIUM-SODIUM, *ALLOY*	4.3	UN1422	DANGEROUS WHEN WET
	POTASSIUM SULPHIDE, ANHYDROUS *OR* POTASSIUM SULPHIDE *WITH LESS THAN 30% WATER OF CRYSTALLIZATION*	4.2	UN1382	SPONTANEOUSLY COMBUSTIBLE
	POTASSIUM SULPHIDE, HYDRATED, *WITH NOT LESS THAN 30% WATER OF CRYSTALLIZATION*	8	UN1847	CORROSIVE
	POTASSIUM SUPEROXIDE	5.1	UN2466	OXIDIZER
N	POWDER CAKE, WETTED *WITH NOT LESS THAN 35% WATER, BY WEIGHT*	1.3C	UN0159	EXPLOSIVE (1.3C)
N	POWDER, SMOKELESS	1.1C	UN0160	EXPLOSIVE (1.1C)
N	POWDER, SMOKELESS	1.3C	UN0161	EXPLOSIVE (1.3C)
	PRIMERS, CAP TYPE	1.4B	UN0378	EXPLOSIVE (1.4B)
	PRIMERS, CAP TYPE	1.4S	UN0044	NONE. PACKAGE TO BE MARKED '1.4S'
N	PRIMERS, CAP TYPE	1.1B	UN0377	EXPLOSIVE (1.1B)
	PRIMERS, TUBULAR	1.4G	UN0320	EXPLOSIVE (1.4G)
	PRIMERS, TUBULAR	1.4S	UN0376	NONE. PACKAGE TO BE MARKED '1.4S'
N	PRIMERS, TUBULAR	1.3G	UN0319	EXPLOSIVE (1.3G)
	PROJECTILES, *INERT, WITH TRACER*	1.4 S	UN0345	NONE. PACKAGE TO BE MARKED '1.4 S'
N	PROJECTILES, *INERT, WITH TRACER*	1.3G	UN0424	EXPLOSIVE (1.3G)
	PROJECTILES, *INERT, WITH TRACER*	1.4G	UN0425	EXPLOSIVE (1.4G)
N	PROJECTILES, *WITH BURSTER OR EXPELLING CHARGE*	1.2D	UN0346	EXPLOSIVE (1.2D)
N	PROJECTILES, *WITH BURSTER OR EXPELLING CHARGE*	1.2F	UN0426	EXPLOSIVE (1.2F)
N	PROJECTILES, *WITH BURSTER OR EXPELLING CHARGE*	1.4F	UN0427	EXPLOSIVE (1.4F)
N	PROJECTILES, *WITH BURSTER OR EXPELLING CHARGE*	1.4D	UN0347	EXPLOSIVE (1.4D)
N	PROJECTILES, *WITH BURSTING CHARGE*	1.1D	UN0168	EXPLOSIVE (1.1D)
N	PROJECTILES, *WITH BURSTING CHARGE*	1.2D	UN0169	EXPLOSIVE (1.2D)
N	PROJECTILES, *WITH BURSTING CHARGE*	1.4D	UN0344	EXPLOSIVE (1.4D)
N	PROJECTILES, *WITH BURSTING CHARGE*	1.1F	UN0167	EXPLOSIVE (1.1F)
N	PROJECTILES, *WITH BURSTING CHARGE*	1.2F	UN0324	EXPLOSIVE (1.2F)
	PROPADIENE, INHIBITED	2.1	UN2200	FLAMMABLE GAS
	PROPANE	2.1	UN1978	FLAMMABLE GAS
	PROPANETHIOLS	3.1	UN2402	FLAMMABLE LIQUID
	PROPANOL	3.2	UN1274	FLAMMABLE LIQUID
	PROPIONALDEHYDE	3.1 3.2	UN1275 UN1275	FLAMMABLE LIQUID FLAMMABLE LIQUID
	PROPIONIC ACID, *SOLUTION CONTAINING NOT LESS THAN 80% ACID*	8	UN1848	CORROSIVE, FLAMMABLE LIQUID (ONLY IF FLASHPOINT 61 DEG C OR BELOW)
	PROPIONIC ANHYDRIDE	8	UN2496	CORROSIVE
	PROPIONITRILE	3.2	UN2404	FLAMMABLE LIQUID, POISON
	PROPIONYL CHLORIDE	3.2	UN1815	FLAMMABLE LIQUID, CORROSIVE
	N-PROPYL ACETATE	3.2	UN1276	FLAMMABLE LIQUID
	SEC-PROPYL ALCOHOL. *SEE* ISOPROPANOL			
	N-PROPYL ALCOHOL. *SEE* PROPANOL			
	PROPYLAMINE. *SEE* MONOPROPYLAMINE			

§172.102 OPTIONAL HAZARDOUS MATERIALS TABLE

(1) Notes and Symbols	(2) Hazardous Materials and Description and Proper Shipping Names	(3) IMO Class	(4) Identification Number	(5) Label(s) required
	PROPYL BENZENE	3.3	UN2364	FLAMMABLE LIQUID
	PROPYL CHLORIDE	3.1	UN1278	FLAMMABLE LIQUID
	N-PROPYL CHLOROFORMATE	3.3	UN2740	FLAMMABLE LIQUID, POISON, CORROSIVE
	PROPYLENE	2.1	UN1077	FLAMMABLE GAS
	PROPYLENE CHLOROHYDRIN	6.1	UN2611	POISON, FLAMMABLE LIQUID
	PROPYLENEDIAMINE	3.2 3.3	UN2258 UN2258	FLAMMABLE LIQUID, CORROSIVE FLAMMABLE LIQUID, CORROSIVE
	PROPYLENE DICHLORIDE	3.2	UN1279	FLAMMABLE LIQUID
	PROPYLENEIMINE, *INHIBITED*	3.2	UN1921	FLAMMABLE LIQUID
	PROPYLENE OXIDE	3.1	UN1280	FLAMMABLE LIQUID
	PROPYLENE TETRAMER	3.3	UN2850	FLAMMABLE LIQUID
	PROPYL FORMATES	3.2	UN1281	FLAMMABLE LIQUID
	N-PROPYL ISOCYANATE	3.2	UN2482	FLAMMABLE LIQUID, POISON
	N-PROPYL NITRATE	3.2	UN1865	FLAMMABLE LIQUID
	PROPYL TRICHLOROSILANE	8	UN1816	CORROSIVE, FLAMMABLE LIQUID
	PYRIDINE	3.2	UN1282	FLAMMABLE LIQUID, POISON
	PYROPHORIC FUEL, N.O.S. *SEE* FUEL, PYROPHORIC, N.O.S.			
	PYROPHORIC LIQUIDS, N.O.S.	4.2	UN2845	SPONTANEOUSLY COMBUSTIBLE
	PYROPHORIC METALS, N.O.S. *OR* PYROPHORIC ALLOYS, N.O.S.	4.2	UN1383	SPONTANEOUSLY COMBUSTIBLE
	PYROPHORIC SOLIDS, N.O.S.	4.2	UN2846	SPONTANEOUSLY COMBUSTIBLE
	PYROSULPHURYL CHLORIDE	8	UN1817	CORROSIVE
	PYROXYLIN. *SEE* NITROCELLULOSE			
	PYRROLIDINE	3.2	UN1922	FLAMMABLE LIQUID
	QUINOLINE	6.1	UN2656	ST. ANDREWS CROSS, FLAMMABLE LIQUID (ONLY IF FLASHPOINT BELOW 61 DEG C)
	RAGS, *OILY*	4.2	UN1856	SPONTANEOUSLY COMBUSTIBLE
	RARE GASES, *MIXTURES*	2.2	UN1979	NONFLAMMABLE GAS
	RARE GASES, *MIXTURES WITH* NITROGEN	2.2	UN1981	NONFLAMMABLE GAS
	RARE GASES, *MIXTURES WITH* OXYGEN	2.2	UN1980	NONFLAMMABLE GAS
	RECEPTACLES, *SMALL, CONTAINING* FLAMMABLE COMPRESSED GAS, *NOT FITTED WITH A DISPERSION DEVICE, NOT REFILLABLE*	2.1	UN2037	FLAMMABLE GAS
	REDUCING LIQUID. *SEE* FLAMMABLE LIQUID PREPARATION, N.O.S.			
N	REFRIGERANT GASES, N.O.S.	2.1	UN1078	FLAMMABLE GAS
	REFRIGERANT GASES, N.O.S.	2.2	UN1078	NONFLAMMABLE GAS
	REFRIGERATING MACHINES *CONTAINING NON-FLAMMABLE, NON-TOXIC LIQUEFIED GASES*	2.2	UN2857	NONFLAMMABLE GAS
	RELEASE DEVICES, EXPLOSIVE	1.4S	UN0173	NONE. PACKAGE TO BE MARKED '1.4 S'
	REMOVING LIQUID. *SEE* FLAMMABLE LIQUID PREPARATIONS, N.O.S.			
	RESIN SOLUTION *IN FLAMMABLE LIQUID*	3.2 3.3	UN1866 UN1866	FLAMMABLE LIQUID FLAMMABLE LIQUID
	RESIN SOLUTION, POISONOUS	6.1	UN1896 UN1896	POISON FLAMMABLE LIQUID (ONLY IF FLASHPOINT BETWEEN 23 AND 61 DEG C) ST. ANDREWS CROSS FLAMMABLE LIQUID (ONLY IF FLASHPOINT BETWEEN 23 AND AND 61 DEG C)
	RESORCINOL	6.1	UN2876	ST. ANDREWS CROSS
	RIVETS, EXPLOSIVE	1.4 S	UN0174	NONE. PACKAGE TO BE MARKED '1.4 S'
	ROAD ASPHALT, *LIQUID, TARS OR OIL. SEE* CUT-BACKS, *ASPHALT OR BITUMEN*			
N	ROCKET MOTORS	1.1C	UN0280	EXPLOSIVE (1.1C)
N	ROCKET MOTORS	1.2C	UN0281	EXPLOSIVE (1.2C)
N	ROCKET MOTORS	1.3C	UN0186	EXPLOSIVE (1.3C)
N	ROCKET MOTORS, *CONTAINING HYPERGOLIC LIQUIDS, WITH OR WITHOUT , EXPELLING CHARGE*	1.2L	UN0322	EXPLOSIVE (1.2L)
N	ROCKET MOTORS, *CONTAINING HYPERGOLIC LIQUIDS, WITH OR WITHOUT EXPELLING CHARGE*	1.3L	UN0250	EXPLOSIVE (1.3L)
N	ROCKET MOTORS, LIQUID FUELED	1.2J	UN0395	EXPLOSIVE (1.2J)
N	ROCKET MOTORS, LIQUID FUELED	1.3J	UN0396	EXPLOSIVE (1.3J)

§172.102 OPTIONAL HAZARDOUS MATERIALS TABLE

(1) Notes and Symbols	(2) Hazardous Materials and Description and Proper Shipping Names	(3) IMO Class	(4) Identification Number	(5) Label(s) required
N	ROCKETS, LINE THROWING	1.2G	UN0238	EXPLOSIVE (1.2G)
N	ROCKETS, LINE THROWING	1.3G	UN0240	EXPLOSIVE (1.3G)
N	ROCKETS, LIQUID FUELED, *WITH BURSTING CHARGE*	1.1J	UN0397	EXPLOSIVE (1.1J)
N	ROCKETS, LIQUID FUELED, *WITH BURSTING CHARGE*	1.2J	UN0398	EXPLOSIVE (1.2J)
N	ROCKETS, *WITH BURSTING CHARGE*	1.1E	UN0181	EXPLOSIVE (1.1E)
N	ROCKETS, *WITH BURSTING CHARGE*	1.2E	UN0182	EXPLOSIVE (1.2E)
N	ROCKETS, *WITH BURSTING CHARGE*	1.1F	UN0180	EXPLOSIVE (1.1F)
N	ROCKETS, *WITH BURSTING CHARGE*	1.2F	UN0295	EXPLOSIVE (1.2F)
N	ROCKETS, *WITH INERT HEAD*	1.3C	UN0183	EXPLOSIVE (1.3C)
	RODENTICIDES, N.O.S., *LIQUID*	6.1 6.1 6.1	UN1681 UN1681 UN1681	POISON POISON ST. ANDREWS CROSS
	RODENTICIDES, N.O.S., *SOLID*	6.1 6.1	UN1681 UN1681	POISON ST. ANDREWS CROSS
	ROSIN OIL	3.2 3.3	UN1286 UN1286	FLAMMABLE LIQUID FLAMMABLE LIQUID
	R 22. SEE CHLORODIFLUOROMETHANE			
	R 12. SEE DICHLORODIFLUOROMETHANE			
	RUBBER SCRAP, *POWDERED OR GRANULATED*	4.1	UN1345	FLAMMABLE SOLID
	RUBBER SHODDY. *SEE* RUBBER SCRAP			
	RUBBER SOLUTION	3.2 3.3	UN1287 UN1287	FLAMMABLE LIQUID FLAMMABLE LIQUID
	RUBIDIUM HYDROXIDE, SOLID	8	UN2678	CORROSIVE
	RUBIDIUM HYDROXIDE, SOLUTION	8	UN2677	CORROSIVE
	RUBIDIUM, *METAL*	4.3	UN1423	DANGEROUS WHEN WET
N	SAMPLES, EXPLOSIVE, *OTHER THAN INITIATING EXPLOSIVE*	1	UN0190	AS DIRECTED BY THE COMPETENT AUTHORITY
	SAND ACID. *SEE* FLUOSILICIC ACID			
	SEED CAKE, *CONTAINING VEGETABLE OIL, MECHANICALLY EXPELLED SEEDS, CONTAINING MORE THAN 10% OF OIL OR MORE THAN 20% OF OIL AND MOISTURE COMBINED*	4.2	UN1386	NONE. PACKAGE TO BE MARKED 'CLASS 4.2'
	SEED CAKE, *CONTAINING VEGETABLE OIL, SOLVENT EXTRACTIONS AND EXPELLED SEEDS, CONTAINING NOT MORE THAN 10% OF OIL AND, WHEN THE AMOUNT OF MOISTURE IS HIGHER THAN 10%, NOT MORE THAN 20% OF OIL AND MOISTURE COMBINED*	4.2	UN1386	NONE. PACKAGE TO BE MARKED 'CLASS 4.2'
	SEED CAKE, *CONTAINING VEGETABLE OIL, SOLVENT EXTRACTIONS CONTAINING NOT MORE THAN 1.5% OF OIL AND 11% OF MOISTURE*	4.2	UN2217	NONE. PACKAGE TO BE MARKED 'CLASS 4.2'
	SELENATES, N.O.S. *OR* SELENITES, N.O.S.	6.1	UN2630	POISON
	SELENIC ACID	8	UN1905	CORROSIVE
	SELENIUM DISULPHIDE	6.1	UN2657	POISON
	SELENIUM HEXAFLUORIDE	2.3	UN2194	POISON GAS
	SELENIUM *METAL* POWDER, *NON-PYROPHORIC*	6.1	UN2658	ST. ANDREWS CROSS
	SELENIUM OXYCHLORIDE	8	UN2879	CORROSIVE POISON
	SHALE OIL	3.2 3.3	UN1288 UN1288	FLAMMABLE LIQUID FLAMMABLE LIQUID
	SHELLAC. *SEE* PAINTS, *ETC.*			
	SIGNAL DEVICES, HAND	1.4S	UN0373	NONE. PACKAGE TO BE MARKED '1.4S'
	SIGNAL DEVICES, HAND	1.4 G	UN0191	EXPLOSIVE (1.4G)
N	SIGNALS, DISTRESS, *SHIP (OTHER THAN WATER-ACTIVATED)*	1.1G	UN0194	EXPLOSIVE (1.1G)
N	SIGNALS, DISTRESS, *SHIP (OTHER THAN WATER-ACTIVATED)*	1.3G	UN0195	EXPLOSIVE (1.3G)
N	SIGNALS, RAILWAY TRACK, EXPLOSIVE	1.1G	UN0192	EXPLOSIVE (1.1G)
N	SIGNALS, RAILWAY TRACK, EXPLOSIVE	1.4S	UN0193	NONE. PACKAGE TO BE MARKED '1.4S'
N	SIGNALS, SMOKE, *WITH EXPLOSIVE SOUND UNIT*	1.1G	UN0196	EXPLOSIVE (1.1G)
N	SIGNALS, SMOKE, *WITH EXPLOSIVE SOUND UNIT*	1.2G	UN0313	EXPLOSIVE (1.2G)
	SIGNALS, SMOKE, *WITHOUT EXPLOSIVE SOUND UNIT*	1.4 G	UN0197	EXPLOSIVE (1.4G)
	SILANE	2.1	UN2203	FLAMMABLE GAS
	SILICOFLUORIC ACID. *SEE* FLUOSILICIC ACID			
	SILICOFLUORIDES, N.O.S.	6.1	UN2856	ST. ANDREWS CROSS
	SILICON CHLORIDE. *SEE* SILICON TETRACHLORIDE			
	SILICON POWDER, *AMORPHOUS*	4.1	UN1346	FLAMMABLE SOLID
	SILICON TETRACHLORIDE	8	UN1818	CORROSIVE
	SILICON TETRAFLUORIDE	2.3	UN1859	POISON GAS, CORROSIVE

§172.102 OPTIONAL HAZARDOUS MATERIALS TABLE

(1) Notes and Symbols	(2) Hazardous Materials and Description and Proper Shipping Names	(3) IMO Class	(4) Identification Number	(5) Label(s) required
	SILVER ARSENITE	6.1	UN1683	POISON
	SILVER CYANIDE	6.1	UN1684	POISON
	SILVER NITRATE	5.1	UN1493	OXIDIZER
	SILVER PICRATE, WETTED *WITH, BY WEIGHT, AT LEAST 30% WATER*	4.1	UN1347	FLAMMABLE SOLID
	SISAL, DRY. SEE FIBRE, *VEGETABLE, DRY*			
	SLUDGE ACID	8	UN1906	CORROSIVE
	SODA LIME	8	UN1907	CORROSIVE
	SODIUM ALUMINATE, SOLUTION	8	UN1819	CORROSIVE
	SODIUM ALUMINIUM HYDRIDE	4.3	UN2835	DANGEROUS WHEN WET
	SODIUM AMALGAM	4.3	UN1424	DANGEROUS WHEN WET
	SODIUM AMIDE	4.3	UN1425	DANGEROUS WHEN WET
	SODIUM AMMONIUM VANADATE	6.1	UN2863	POISON
	SODIUM ARSANILATE	6.1	UN2473	ST. ANDREWS CROSS
	SODIUM ARSENATE	6.1	UN1685	POISON
	SODIUM ARSENITE, AQUEOUS SOLUTIONS	6.1 6.1	UN1686 UN1686	POISON ST. ANDREWS CROSS
	SODIUM ARSENITE, SOLID	6.1	UN2027	POISON
	SODIUM AZIDE	6.1	UN1687	POISON
	SODIUM BISULPHATE, SOLID. *SEE* SODIUM HYDROGEN SULPHATE, SOLID			
	SODIUM BISULPHITE, SOLUTION. *SEE* SODIUM HYDROGEN SULPHITE, SOLUTION			
	SODIUM BOROHYDRIDE	4.3	UN1426	DANGEROUS WHEN WET
	SODIUM BROMATE	5.1	UN1494	OXIDIZER
	SODIUM CACODYLATE	6.1	UN1688	POISON
	SODIUM CHLORATE	5.1	UN1495	OXIDIZER
	SODIUM CHLORATE, *AQUEOUS* SOLUTION	5.1	UN2428	OXIDIZER
	SODIUM CHLORITE	5.1	UN1496	OXIDIZER
	SODIUM CHLORITE, SOLUTION *WITH MORE THAN 5% AVAILABLE CHLORINE*	8	UN1908	CORROSIVE
	SODIUM CHLOROACETATE	6.1	UN2659	ST. ANDREWS CROSS
	SODIUM CUPROCYANIDE, *SOLID*	6.1	UN2316	POISON
	SODIUM CUPROCYANIDE SOLUTION	6.1	UN2317	POISON
	SODIUM CYANIDE	6.1	UN1689	POISON
N	SODIUM DINITRO-O-CRESOLATE, *DRY OR WETTED WITH LESS THAN 15% , WATER, BY WEIGHT*	1.3C	UN0234	EXPLOSIVE (1.3C)
	SODIUM DINITRO-O-CRESOLATE, WETTED *WITH, BY WEIGHT, AT LEAST 15% WATER*	4.1	UN1348	FLAMMABLE SOLID, POISON
	SODIUM DINITRO-O-CRESOLATE, WETTED *WITH, BY WEIGHT, AT LEAST 10% WATER*	4.1	UN1348	FLAMMABLE SOLID, POISON
	SODIUM DITHIONITE	4.2	UN1384	SPONTANEOUSLY COMBUSTIBLE
	SODIUM FLUORIDE, SOLID	6.1	UN1690	ST. ANDREWS CROSS
	SODIUM FLUORIDE, SOLUTION	6.1	UN1690	ST. ANDREWS CROSS
	SODIUM FLUOROACETATE	6.1	UN2629	POISON
	SODIUM HYDRATE. *SEE* SODIUM HYDROXIDE, *SOLUTION*			
	SODIUM HYDRIDE	4.3	UN1427	DANGEROUS WHEN WET
	SODIUM HYDROGEN FLUORIDE	8	UN2439	CORROSIVE
	SODIUM HYDROGEN SULPHATE, SOLID	8	UN1821	CORROSIVE
	SODIUM HYDROGEN SULPHATE, SOLUTION	8	UN2837	CORROSIVE
	SODIUM HYDROGEN SULPHITE, SOLUTION. *SEE* BISULPHITES, INORGANIC, AQUEOUS SOLUTIONS, N.O.S.			
	SODIUM HYDROSULPHIDE, *WITH LESS THAN 25% WATER OF CRYSTALLIZATION*	4.2	UN2318	SPONTANEOUSLY COMBUSTIBLE
	SODIUM HYDROSULPHIDE, *WITH NOT LESS THAN 25% WATER OF CRYSTALLIZATION*	8	UN2949	CORROSIVE
	SODIUM HYDROSULPHIDE, *WITH NOT LESS THAN 25% WATER OF CRYSTALLIZATION*	8	UN2949	CORROSIVE
	SODIUM HYDROSULPHITE. *SEE* SODIUM DITHIONITE			
	SODIUM HYDROXIDE, SOLID	8	UN1823	CORROSIVE
	SODIUM HYDROXIDE, SOLUTION	8	UN1824	CORROSIVE
	SODIUM, *METAL*	4.3	UN1428	DANGEROUS WHEN WET
	SODIUM *METAL, DISPERSION IN ORGANIC LIQUIDS*	4.3	UN1429	DANGEROUS WHEN WET
	SODIUM METHYLATE	4.3	UN1431	DANGEROUS WHEN WET
	SODIUM METHYLATE, SOLUTIONS *IN ALCOHOL*	3.2 3.3	UN1289 UN1289	FLAMMABLE LIQUID FLAMMABLE LIQUID

§172.102 OPTIONAL HAZARDOUS MATERIALS TABLE

(1) Notes and Symbols	(2) Hazardous Materials and Description and Proper Shipping Names	(3) IMO Class	(4) Identification Number	(5) Label(s) required
	SODIUM MONOXIDE	8	UN1825	CORROSIVE
	SODIUM NITRATE	5.1	UN1498	OXIDIZER
	SODIUM NITRATE AND POTASH, MIXTURES	5.1	UN1478	OXIDIZER
	SODIUM NITRATE AND POTASSIUM NITRATE, *MIXTURES*	5.1	UN1499	OXIDIZER
	SODIUM NITRATE BAGS, *EMPTY. SEE* BAGS, *EMPTY AND UNWASHED, ETC.*			
	SODIUM NITRITE	5.1	UN1500	OXIDIZER
	SODIUM PENTACHLOROPHENATE	6.1	UN2567	POISON
	SODIUM PERCARBONATES	5.1	UN2467	OXIDIZER
	SODIUM PERCHLORATE	5.1	UN1502	OXIDIZER
	SODIUM PERMANGANATE	5.1	UN1503	OXIDIZER
	SODIUM PEROXIDE	5.1	UN1504	OXIDIZER
	SODIUM PERSULPHATE	5.1	UN1505	OXIDIZER
	SODIUM PHENOLATE, SOLID	8	UN2497	CORROSIVE
	SODIUM PHOSPHIDE	4.3	UN1432	DANGEROUS WHEN WET, POISON
N	SODIUM PICRAMATE, *DRY OR WETTED WITH LESS THAN 20% WATER, BY WEIGHT*	1.3C	UN0235	EXPLOSIVE (1.3C)
	SODIUM PICRAMATE, WETTED *WITH, BY WEIGHT, AT LEAST 20% WATER*	4.1	UN1349	FLAMMABLE SOLID
	SODIUM-POTASSIUM, *ALLOY. SEE* POTASSIUM-SODIUM, *ALLOY*			
N	SODIUM SALTS OF NITRO-AROMATIC DERIVATIVES, N.O.S., *EXPLOSIVE*	1.3C	UN0203	EXPLOSIVE (1.3C)
	SODIUM SILICOFLUORIDE	6.1	UN2674	ST. ANDREWS CROSS
	SODIUM SULPHIDE, ANHYDROUS *OR* SODIUM SULPHIDE *WITH LESS THAN 30% WATER OF CRYSTALLIZATION*	4.2	UN1385	SPONTANEOUSLY COMBUSTIBLE
	SODIUM SULPHIDE, HYDRATED, *WITH AT LEAST 30% WATER*	8	UN1849	CORROSIVE
	SODIUM SUPEROXIDE	5.1	UN2547	OXIDIZER
N	SOUNDING DEVICES, EXPLOSIVE	1.1E	UN0374	EXPLOSIVE (1.1E)
N	SOUNDING DEVICES, EXPLOSIVE	1.2E	UN0375	EXPLOSIVE (1.2E)
N	SOUNDING DEVICES, EXPLOSIVE	1.1F	UN0296	EXPLOSIVE (1.1F)
N	SOUNDING DEVICES, EXPLOSIVE	1.2F	UN0204	EXPLOSIVE (1.2F)
	SPENT MIXED ACID. SEE ACID MIXTURES, *SPENT*			
	SPIRITS OF SALTS. *SEE* HYDROCHLORIC ACID			
	SQUIBS	1.4S	UN0206	NONE. PACKAGE TO BE MARKED '1.4S'
	SQUIBS	1.4B	UN0422	EXPLOSIVE (1.4B)
	SQUIBS	1.4G	UN0423	EXPLOSIVE (1.4G)
	STAINS. *SEE* PAINTS, *ETC.*			
	STANNIC CHLORIDE, ANHYDROUS	8	UN1827	CORROSIVE
	STANNIC CHLORIDE PENTAHYDRATE	8	UN2440	CORROSIVE
	STANNIC PHOSPHIDES	4.3	UN1433	DANGEROUS WHEN WET
	STIBINE	2.3	UN2676	POISON GAS, FLAMMABLE GAS
	STRAW	4.1	UN1327	NONE
	STRIKE ANYWHERE MATCHES. *SEE* MATCHES, *STRIKE ANYWHERE*			
	STRONTIUM, *ALLOYS, NON-PYROPHORIC*	4.3	UN1434	DANGEROUS WHEN WET
	STRONTIUM ARSENITE	6.1	UN1691	POISON
	STRONTIUM CHLORATE	5.1	UN1506	OXIDIZER
	STRONTIUM NITRATE	5.1	UN1507	OXIDIZER
	STRONTIUM PERCHLORATE	5.1	UN1508	OXIDIZER
	STRONTIUM PEROXIDE	5.1	UN1509	OXIDIZER
	STRONTIUM PHOSPHIDE	4.3	UN2013	DANGEROUS WHEN WET, POISON
	STRYCHNINE *(AND SALTS OF), LIQUID*	6.1 6.1 6.1	UN1692 UN1692 UN1692	POISON POISON ST. ANDREWS CROSS
	STRYCHNINE *(AND SALTS OF), SOLID*	6.1 6.1	UN1692 UN1692	POISON ST. ANDREWS CROSS
	STYRENE MONOMER, INHIBITED	3.3	UN2055	FLAMMABLE LIQUID
N	SUBSTANCES, EXPLOSIVE, N.O.S.	1.1L	UN0357	EXPLOSIVE (1.1L)
N	SUBSTANCES, EXPLOSIVE, N.O.S.	1.2L	UN0358	EXPLOSIVE (1.2L)
N	SUBSTANCES, EXPLOSIVE, N.O.S.	1.3L	UN0359	EXPLOSIVE (1.3L)
	SUBSTANCES WHICH, IN CONTACT WITH WATER, EMIT FLAMMABLE GASES, N.O.S.	4.3	UN2813	DANGEROUS WHEN WET
	SUBSTITUTED NITROPHENOL PESTICIDES, LIQUID, FLAMMABLE, TOXIC, N.O.S., *FLASHPOINT BELOW 23 DEG C*	3.2	UN2780	FLAMMABLE LIQUID AND POISON OR ST. ANDREWS CROSS (ACCORDING TO TOXICITY)

§172.102 OPTIONAL HAZARDOUS MATERIALS TABLE

(1) Notes and Symbols	(2) Hazardous Materials and Description and Proper Shipping Names	(3) IMO Class	(4) Identification Number	(5) Label(s) required
	SUBSTITUTED NITROPHENOL PESTICIDES, LIQUID, TOXIC, FLAMMABLE, N.O.S., *FLASHPOINT BETWEEN 23 DEG C AND 61 DEG C*	6.1 6.1 6.1	UN3013 UN3013 UN3013	POISON, FLAMMABLE LIQUID POISON, FLAMMABLE LIQUID ST. ANDREWS CROSS, FLAMMABLE LIQUID
	SUBSTITUTED NITROPHENOL PESTICIDES, LIQUID, TOXIC, N.O.S.	6.1 6.1 6.1	UN3014 UN3014 UN3014	POISON POISON ST. ANDREWS CROSS
	SUBSTITUTED NITROPHENOL PESTICIDES, SOLID, TOXIC, N.O.S.	6.1 6.1	UN2779 UN2779	POISON ST. ANDREWS CROSS
	SUCCINIC ACID PEROXIDE. *SEE* DISUCCINIC ACID PEROXIDE			
	SULPHAMIC ACID	8	UN2967	CORROSIVE
	SULPHUR CHLORIDES	8	UN1828	CORROSIVE
	SULPHUR DICHLORIDE. SEE SULPHUR CHLORIDES			
	SULPHUR DIOXIDE, *LIQUEFIED*	2.3	UN1079	POISON GAS
	SULPHUR HEXAFLUORIDE	2.2	UN1080	NONFLAMMABLE GAS
	SULPHURIC ACID, FUMING	8	UN1831	CORROSIVE , POISON
	SULPHURIC ACID, SPENT	8	UN1832	CORROSIVE
	SULPHURIC ACID, *WITH MORE THAN 51% ACID*	8	UN1830	CORROSIVE
	SULPHURIC ACID, *WITH NOT MORE THAN 51% ACID*	8	UN2796	CORROSIVE
	SULPHURIC AND HYDROFLUORIC ACID, MIXTURES. SEE HYDROFLUORIC ACID AND SULPHURIC ACID MIXTURES			
	SULPHURIC ANHYDRIDE. *SEE* SULPHUR TRIOXIDE, *STABILIZED*			
	SULPHUR, *LUMP AND COARSE GRAINED POWDER, OR FINE GRAINED POWDER*	4.1	UN1350	FLAMMABLE SOLID
	SULPHUR, MOLTEN	4.1	UN2448	FLAMMABLE SOLID
	SULPHUROUS ACID	8	UN1833	CORROSIVE
	SULPHUR TETRAFLUORIDE	2.3	UN2418	POISON GAS
	SULPHUR TRIOXIDE, *STABILIZED*	8	UN1829	CORROSIVE
	SULPHURYL CHLORIDE	8	UN1834	CORROSIVE
	SULPHURYL FLUORIDE	2.3	UN2191	POISON GAS
	TARS, *LIQUID. SEE* CUT-BACKS, *ASPHALT OR BITUMEN*			
	TEAR GAS CANDLES, *NON-EXPLOSIVE*	6.1	UN1700	POISON, FLAMMABLE SOLID
	TEAR GAS GRENADES, *NON-EXPLOSIVE. SEE* TEAR GAS CANDLES			
	TEAR GAS, *(IRRITATING SUBSTANCES, LIQUID OR SOLID)*, N.O.S.	6.1 6.1	UN1693 UN1693	POISON ST. ANDREWS CROSS
	TELLURIUM HEXAFLUORIDE	2.3	UN2195	POISON GAS
	T.E.L. SEE MOTOR FUEL ANTI-KNOCK MIXTURES			
	TERPENE HYDROCARBONS N.O.S.	3.3	UN2319	FLAMMABLE LIQUID
	TERPINOLENE	3.3	UN2541	FLAMMABLE LIQUID
	TETRABROMOETHANE	6.1	UN2504	ST. ANDREWS CROSS
	1,1,2,2-TETRACHLOROETHANE	6.1	UN1702	POISON
	TETRACHLOROETHYLENE	6.1	UN1897	ST. ANDREWS CROSS
	TETRAETHYL DITHIOPYROPHOSPHATE, LIQUID *OR* MIXTURES	6.1 6.1	UN1704 UN1704	POISON ST. ANDREWS CROSS
	TETRAETHYL DITHIOPYROPHOSPHATE WITH GASES, *INCLUDING SOLUTIONS AND MIXTURES THEREOF*	2.3	UN1703	POISON GAS
	TETRAETHYLENEPENTAMINE	8	UN2320	CORROSIVE
	TETRAETHYL LEAD. *SEE* MOTOR FUEL ANTI-KNOCK MIXTURES			
	TETRAETHYL PYROPHOSPHATE AND COMPRESSED GAS, *MIXTURE*	2.3	UN1705	POISON GAS
	TETRAETHYL SILICATE	3.3	UN1292	FLAMMABLE LIQUID
	TETRAFLUOROETHYLENE, INHIBITED	2.1	UN1081	NON FLAMMABLE GAS
	TETRAFLUOROHYDRAZINE	2.3	(UN1955)	POISON GAS
	TETRAFLUOROMETHANE	2.2	UN1982	NONFLAMMABLE GAS
	1,2,3,6-TETRAHYDROBENZALDEHYDE	3.3	UN2498	FLAMMABLE LIQUID
	TETRAHYDROFURAN	3.1	UN2056	FLAMMABLE LIQUID
	TETRAHYDROFURFURYLAMINE	3.3	UN2943	FLAMMABLE LIQUID
	TETRAHYDROPHTHALIC ANHYDRIDES	8	UN2698	NONE. PACKAGE TO BE AS MARKED 'CLASS 8'
	1,2,3,6-TETRAHYDROPYRIDINE	3.2	UN2410	FLAMMABLE LIQUID
	TETRAHYDROTHIOPHENE	3.2	UN2412	FLAMMABLE LIQUID
	TETRALIN HYDROPEROXIDE, *TECHNICAL PURE*	5.2	UN2136	ORGANIC PEROXIDE
	TETRAMETHYLAMMONIUM HYDROXIDE	8	UN1835	CORROSIVE
	1,1,3,3-TETRAMETHYL BUTYL HYDROPEROXIDE, *TECHNICAL PURE*	5.2	UN2160	ORGANIC PEROXIDE

§172.102 OPTIONAL HAZARDOUS MATERIALS TABLE

(1) Notes and Symbols	(2) Hazardous Materials and Description and Proper Shipping Names	(3) IMO Class	(4) Identification Number	(5) Label(s) required
	1,1,3,3-TETRAMETHYL BUTYL PEROXY-2-ETHYL HEXANOATE, *TECHNICAL PURE*	5.2	UN2161	ORGANIC PEROXIDE
	TETRAMETHYLSILANE	3.1	UN2749	FLAMMABLE LIQUID
N	TETRANITRO-ANILINE	1.1D	UN0207	EXPLOSIVE (1.1D)
	TETRANITROMETHANE	5.1	UN1510	OXIDIZER
	TETRAPROPYLORTHOTITANATE	3.3	UN2413	FLAMMABLE LIQUID
N	TETRAZOL-1-ACETIC ACID	1.4C	UN0407	EXPLOSIVE (1.4C)
	TEXTILE WASTE, WET, N.O.S	4.2	UN1857	SPONTANEOUSLY COMBUSTIBLE
	THALLIUM CHLORATE	5.1	UN2573	OXIDIZER, POISON
	THALLIUM COMPOUNDS, N.O.S. *LIQUID*	6.1	UN1707	POISON
		6.1	UN1707	POISON
		6.1	UN1707	ST. ANDREWS CROSS
	THALLIUM COMPOUNDS, N.O.S. *SOLID*	6.1	UN1707	POISON
		6.1	UN1707	ST. ANDREWS CROSS
	THALLIUM NITRATE	6.1	UN2727	POISON
	4-THIAPENTANAL	6.1	UN2785	ST. ANDREWS CROSS, FLAMMABLE LIQUID (ONLY IF FLASHPOINT BELOW 61 DEG C)
	THINNERS. *SEE* PAINTS, *ETC.*			
	THINNING LIQUID. *SEE* FLAMMABLE LIQUID PREPARATIONS, N.O.S.			
	THIOACETIC ACID	3.2	UN2436	FLAMMABLE LIQUID
	THIOCARBONYL CHLORIDE. SEE THIOPHOSGENE			
	THIOGLYCOL	6.1	UN2966	POISON
	THIOGLYCOLIC ACID	8	UN1940	CORROSIVE
	THIOLACTIC ACID	6.1	UN2936	POISON
	THIONYL CHLORIDE	8	UN1836	CORROSIVE
	THIOPHENE	3.2	UN2414	FLAMMABLE LIQUID
	THIOPHOSGENE	6.1	UN2474	POISON
	THIOPHOSPHORYL CHLORIDE	8	UN1837	CORROSIVE
	THIOUREA	6.1	UN2877	ST. ANDREWS CROSS
	TIN CHLORIDE, *FUMING. SEE* STANNIC CHLORIDE, ANHYDROUS			
	TINCTURES, MEDICINAL	3.2	UN1293	FLAMMABLE LIQUID
		3.3	UN1293	FLAMMABLE LIQUID
	TIN TETRACHLORIDE. *SEE* STANNIC CHLORIDE, ANHYDROUS			
	TITANIUM HYDRIDE	4.1	UN1871	FLAMMABLE SOLID
	TITANIUM *METAL* POWDER, DRY *MECHANICALLY PRODUCED, PARTICLE SIZE BETWEEN 3 AND 53 MICRONS; OR CHEMICALY PRODUCED, PARTICLE SIZE LESS THAN 840 MICRONS*	4.2	UN2546	SPONTANEOUSLY COMBUSTIBLE
	TITANIUM *METAL* POWDER, WETTED *WITH NOT LESS THAN 25% WATER(A VISIBLE EXCESS OF WATER MUST BE PRESENT) MECHANICALLY PRODUCED, PARTICLE SIZE BETWEEN 3 AND 53 MICRONS; OR CHEMICALLY PRODUCED, PARTICLE SIZE LESS THAN 840 MICRONS*	4.1	UN1352	FLAMMABLE SOLID
	TITANIUM SPONGE GRANULES *OR* TITANIUM SPONGE POWDERS	4.1	UN2878	FLAMMABLE SOLID
	TITANIUM TETRACHLORIDE	8	UN1838	CORROSIVE
	TITANIUM TRICHLORIDE MIXTURES, *NON-PYROPHORIC*	8	UN2869	CORROSIVE
	TITANIUM TRICHLORIDE, PYROPHORIC *OR* TITANIUM TRICHLORIDE MIXTURES, PYROPHORIC	4.2	UN2441	SPONTANEOUSLY COMBUSTIBLE, CORROSIVE
	TOE PUFFS, *NITROCELLULOSE BASE*	4.1	UN1353	FLAMMABLE SOLID
	TOLUENE	3.2	UN1294	FLAMMABLE LIQUID
	TOLUENE DIISOCYANATE *(T.D.I.)*	6.1	UN2078	POISON
	TOLUIDINES *(O-, M-, P-)*	6.1	UN1708	POISON
	2,4-TOLUYLENEDIAMINE	6.1	UN1709	ST. ANDREWS CROSS
N	TORPEDOES, *WITH BURSTING CHARGE*	1.1E	UN0329	EXPLOSIVE (1.1E)
N	TORPEDOES, *WITH BURSTING CHARGE*	1.1F	UN0330	EXPLOSIVE (1.1F)
	TRACERS FOR AMMUNITION	1.4 G	UN0306	EXPLOSIVE (1.4G)
N	TRACERS FOR AMMUNITION	1.3G	UN0212	EXPLOSIVE (1.3G)
	TRIALLYLAMINE	3.3	UN2610	FLAMMABLE LIQUID
	TRIALLYL BORATE	6.1	UN2609	ST. ANDREWS CROSS, FLAMMABLE LIQUID (ONLY IF FLASHPOINT BELOW 61 DEG C)
	TRIAZINE PESTICIDES, LIQUID, FLAMMABLE, TOXIC, N.O.S., *FLASHPOINT BELOW 23 DEG C*	3.2	UN2764	FLAMMABLE LIQUID AND POISON OR ST. ANDREWS CROSS (ACCORDING TO TOXICITY)

§172.102 OPTIONAL HAZARDOUS MATERIALS TABLE

(1) Notes and Symbols	(2) Hazardous Materials and Description and Proper Shipping Names	(3) IMO Class	(4) Identi-fication Number	(5) Label(s) required
	TRIAZINE PESTICIDES, LIQUID, TOXIC, FLAMMABLE, N.O.S., *FLASHPOINT BETWEEN 23 DEG C AND 61 DEG C*	6.1 6.1 6.1	UN2997 UN2997 UN2997	POISON, FLAMMABLE LIQUID POISON, FLAMMABLE LIQUID ST. ANDREWS CROSS, FLAMMABLE LIQUID
	TRIAZINE PESTICIDES, LIQUID, TOXIC, N.O.S.	6.1 6.1 6.1	UN2998 UN2998 UN2998	POISON POISON ST. ANDREWS CROSS
	TRIAZINE PESTICIDES, SOLID, TOXIC, N.O.S.	6.1 6.1	UN2763 UN2763	POISON ST. ANDREWS CROSS
	TRIBUTYLAMINE	8	UN2542	CORROSIVE
	TRICHLOROACETIC ACID, SOLID	8	UN1839	CORROSIVE
	TRICHLOROACETIC ACID, *SOLUTIONS*	8	UN2564	CORROSIVE
	TRICHLOROACETYL CHLORIDE	8	UN2442	CORROSIVE
	TRICHLOROBENZENES, *LIQUID*	6.1	UN2321	ST. ANDREWS CROSS
	TRICHLOROBUTENE	6.1	UN2322	POISON
	1,1,1,-TRICHLOROETHANE	6.1	UN2831	ST. ANDREWS CROSS
	TRICHLOROETHYLENE	6.1	UN1710	ST. ANDREWS CROSS
	TRICHLOROISOCYANURIC ACID, DRY	5.1	UN2468	OXIDIZER
	TRICHLOROSILANE	4.3	UN1295	DANGEROUS WHEN WET, FLAMMABLE LIQUID, CORROSIVE
	TRICRESYLPHOSPHATE, *WITH MORE THAN 3% ORTHO ISOMER*	6.1	UN2574	POISON
	TRIETHYLALUMINIUM	4.2	UN1102	SPONTANEOUSLY COMBUSTIBLE
	TRIETHYLAMINE	3.2	UN1296	FLAMMABLE LIQUID
	TRIETHYLENETETRAMINE	8	UN2259	CORROSIVE
	TRIETHYL PHOSPHITE	3.3	UN2323	FLAMMABLE LIQUID
	TRIFLUOROACETIC ACID	8	UN2699	CORROSIVE
	TRIFLUOROBROMOMETHANE. SEE BROMOTRIFLUOROMETHANE			
	TRIFLUOROCHLOROETHANE. SEE CHLOROTRIFLUOROETHANE			
	TRIFLUOROCHLOROETHYLENE, INHIBITED	2.1	UN1082	FLAMMABLE GAS
	TRIFLUOROCHLOROMETHANE. SEE CHLOROTRIFLUOROMETHANE			
	TRIFLUOROETHANE	2.1	UN2035	FLAMMABLE GAS
	TRIFLUOROMETHANE	2.2	UN1984	NONFLAMMABLE GAS
	3-TRIFLUOROMETHYL ANILINE	6.1	UN2948	POISON
	2-TRIFLUOROMETHYL ANILINE	6.1	UN2942	ST. ANDREWS CROSS
	TRIISOBUTYL ALUMINIUM	4.2	UN1930	SPONTANEOUSLY COMBUSTIBLE
	TRIISOBUTYLENE	3.3	UN2324	FLAMMABLE LIQUID
	TRIISOCYANATOISOCYANURATE OF ISOPHORONEDIISOCYANATE, SOLUTION *70%, BY WEIGHT*	3.3	UN2906	FLAMMABLE LIQUID
	TRIISOPROPYL BORATE	3.3	UN2616	FLAMMABLE LIQUID
	TRIMETHYLACETYL CHLORIDE	8	UN2438	CORROSIVE, FLAMMABLE LIQUID
	TRIMETHYLALUMINIUM	4.2	UN1103	SPONTANEOUSLY COMBUSTIBLE
	TRIMETHYLAMINE, ANHYDROUS	2.1	UN1083	FLAMMABLE GAS
	TRIMETHYLAMINE, *AQUEOUS SOLUTIONS CONTAINING NOT MORE THAN 30% OF TRIMETHYLAMINE*	3.2	UN1297	FLAMMABLE LIQUID
	1,3,5-TRIMETHYLBENZENE	3.3	UN2325	FLAMMABLE LIQUID
	TRIMETHYL BORATE	3.2 3.3	UN2416 UN2416	FLAMMABLE LIQUID FLAMMABLE LIQUID
	TRIMETHYLCHLOROSILANE	3.2	UN1298	FLAMMABLE LIQUID, CORROSIVE
	TRIMETHYLCYCLOHEXYLAMINE	8	UN2326	CORROSIVE
	TRIMETHYLHEXAMETHYLENE DIAMINES	8	UN2327	CORROSIVE
	TRIMETHYLHEXAMETHYLENE DIISOCYANATE	6.1	UN2328	POISON
	2,4,4-TRIMETHYLPENTYL-2-PEROXY PHENOXY ACETATE, *MAXIMUM CONCENTRATION 37%, IN SOLUTION*	5.2	UN2961	ORGANIC PEROXIDE
	TRIMETHYL PHOSPHITE	3.3	UN2329	FLAMMABLE LIQUID
N	TRINITRO-ANILINE	1.1D	UN0153	EXPLOSIVE (1.1D)
N	TRINITROANISOLE	1.1D	UN0213	EXPLOSIVE (1.1D)
N	TRINITROBENZENE, *DRY OR WETTED WITH LESS THAN 30% WATER, BY WEIGHT*	1.1D	UN0214	EXPLOSIVE (1.1D)
N	TRINITROBENZENESULFONIC ACID	1.1L	UN0386	EXPLOSIVE (1.1L)
	TRINITROBENZENE, WETTED *WITH, BY WEIGHT, AT LEAST 10% WATER*	4.1	UN1354	FLAMMABLE SOLID
	TRINITROBENZENE, WETTED *WITH NOT LESS THAN 30% WATER, BY WEIGHT*	4.1	UN1354	FLAMMABLE SOLID

§172.102 OPTIONAL HAZARDOUS MATERIALS TABLE

(1) Notes and Symbols	(2) Hazardous Materials and Description and Proper Shipping Names	(3) IMO Class	(4) Identification Number	(5) Label(s) required
N	TRINITROBENZOIC ACID, *DRY OR CONTAINING, BY WEIGHT, LESS THAN 30% WATER*	1.1D	UN0215	EXPLOSIVE (1.1D)
	TRINITROBENZOIC ACID, WETTED *WITH, BY WEIGHT, AT LEAST 10% WATER.*	4.1	UN1355	FLAMMABLE SOLID
	TRINITROBENZOIC ACID, WETTED *WITH NOT LESS THAN 30% WATER, BY WEIGHT*	4.1	UN1355	FLAMMABLE SOLID
N	TRINITROCHLOROBENZENE	1.1D	UN0155	EXPLOSIVE (1.1D)
N	TRINITROFLUORENONE	1.1D	UN0387	EXPLOSIVE (1.1D)
N	TRINITRO-M-CRESOL	1.1D	UN0216	EXPLOSIVE (1.1D)
N	TRINITRONAPHTHALENE	1.1D	UN0217	EXPLOSIVE (1.1D)
N	TRINITROPHENETOLE	1.1D	UN0218	EXPLOSIVE (1.1D)
N	TRINITROPHENOL, *DRY OR WETTED WITH LESS THAN 30% WATER, BY WEIGHT*	1.1D	UN0154	EXPLOSIVE (1.1D)
	TRINITROPHENOL, WETTED, *WITH NOT LESS THAN 30% WATER, BY WEIGHT*	4.1	UN1344	FLAMMABLE SOLID
	TRINITROPHENOL, WETTED *WITH NOT LESS THAN 10% WATER, BY WEIGHT.*	4.1	UN1344	FLAMMABLE SOLID
N	TRINITROPHENYLMETHYLNITRAMINE	1.1D	UN0208	EXPLOSIVE (1.1D)
N	TRINITRORESORCINOL, *CONTAINING, BY WEIGHT, NOT LESS THAN 20% WATER OR MIXTURE OF ALCOHOL AND WATER*	1.1D	UN0394	EXPLOSIVE (1.1D)
N	TRINITRORESORCINOL, *DRY OR WETTED WITH LESS THAN 20% WATER, BY WEIGHT (OR MIXTURE OF ALCOHOL AND WATER)*	1.1D	UN0219	EXPLOSIVE (1.1D)
N	TRINITROTOLUENE AND HEXANITROSTILBENE MIXTURES *OR* TRINITROTOLUENE AND TRINITROBENZENE MIXTURES	1.1D	UN0388	EXPLOSIVE (1.1D)
N	TRINITROTOLUENE, *DRY OR WETTED WITH LESS THAN 30% WATER, BY WEIGHT*	1.1D	UN0209	EXPLOSIVE (1.1D)
N	TRINITROTOLUENE MIXTURES CONTAINING TRINITROBENZENE AND HEXANITROSTILBENE	1.1D	UN0389	EXPLOSIVE (1.1D)
	TRINITROTOLUENE, WETTED *WITH, BY WEIGHT, AT LEAST 10% WATER*	4.1	UN1356	FLAMMABLE SOLID
	TRINITROTOLUENE, WETTED *WITH NOT LESS THAN 30% WATER, BY WEIGHT*	4.1	UN1356	FLAMMABLE SOLID
	TRIPROPYLALUMINUM	4.2	UN2718	SPONTANEOUSLY COMBUSTIBLE
	TRIPROPYLAMINE	3.3	UN2260	FLAMMABLE LIQUID, CORROSIVE
	TRIPROPYLENE	3.2 3.3	UN2057 UN2057	FLAMMABLE LIQUID FLAMMABLE LIQUID
	TRIS-(1-AZIRIDINYL)PHOSPHINE OXIDE, SOLUTION	6.1	UN2501	POISON
N	TRITONAL	1.1D	UN0390	EXPLOSIVE (1.1D)
	TUNGSTEN HEXAFLUORIDE	2.3	UN2196	POISON GAS
	TURPENTINE	3.3	UN1299	FLAMMABLE LIQUID
	TURPENTINE SUBSTITUTE	3.2 3.3	UN1300 UN1300	FLAMMABLE LIQUID FLAMMABLE LIQUID
	U D M H. SEE DIMETHYLHYDRAZINE, UNSYMMETRICAL			
	UNDECANE	3.3	UN2330	FLAMMABLE LIQUID
	UREA HYDROGEN PEROXIDE	5.1	UN1511	OXIDIZER
N	UREA NITRATE, *DRY OR CONTAINING, BY WEIGHT, LESS THAN 20% WATER*	1.1D	UN0220	EXPLOSIVE (1.1D)
	UREA NITRATE, WETTED *WITH NOT LESS THAN 10% WATER, BY WEIGHT*	4.1	UN1357	FLAMMABLE SOLID
	UREA NITRATE, WETTED *WITH NOT LESS THAN 20% WATER, BY WEIGHT*	4.1	UN1357	FLAMMABLE SOLID
	VALERALDEHYDE	3.2	UN2058	FLAMMABLE LIQUID
	VALERYL CHLORIDES	8	UN2502	CORROSIVE
	VANADIUM OXITRICHLORIDE	8	UN2443	CORROSIVE
	VANADIUM PENTOXIDE, *NON-FUSED FORM*	6.1	UN2862	POISON
	VANADIUM TETRACHLORIDE	8	UN2444	CORROSIVE
	VANADIUM TRICHLORIDE	8	UN2475	CORROSIVE
	VANADIUM TRIOXIDE, *NON-FUSED FORM*	6.1	UN2860	POISON
	VANADYL SULPHATE	6.1	UN2931	POISON
	VARNISH. *SEE* PAINTS, *ETC.*			
	VINYL ACETATE, *INHIBITED*	3.2	UN1301	FLAMMABLE LIQUID
	VINYL BROMIDE, *INHIBITED*	2.1	UN1085	FLAMMABLE GAS
	VINYL BUTYRATE, INHIBITED	3.2	UN2838	FLAMMABLE LIQUID
	VINYL CHLORIDE, *INHIBITED*	2.1	UN1086	FLAMMABLE GAS
	VINYL CHLOROACETATE	6.1	UN2589	POISON, FLAMMABLE LIQUID
	VINYL ETHYL ETHER, *INHIBITED*	3.1	UN1302	FLAMMABLE LIQUID
	VINYL FLUORIDE, *INHIBITED*	2.1	UN1860	FLAMMABLE GAS
	VINYLIDENE CHLORIDE, *INHIBITED*	3.1	UN1303	FLAMMABLE LIQUID
	VINYL ISOBUTYL ETHER, *INHIBITED*	3.2	UN1304	FLAMMABLE LIQUID

§172.102 OPTIONAL HAZARDOUS MATERIALS TABLE

(1) Notes and Symbols	(2) Hazardous Materials and Description and Proper Shipping Names	(3) IMO Class	(4) Identification Number	(5) Label(s) required
	VINYL METHYL ETHER, *INHIBITED*	2.1	UN1087	FLAMMABLE GAS
	VINYL TOLUENES *(MIXED ISOMERS), INHIBITED*	3.3	UN2618	FLAMMABLE LIQUID
	VINYL TRICHLOROSILANE, INHIBITED	3.2	UN1305	FLAMMABLE LIQUID, CORROSIVE
N	WARHEADS, ROCKET, *WITH BURSTER OR EXPELLING CHARGE*	1.4D	UN0370	EXPLOSIVE (1.4D)
N	WARHEADS, ROCKET, *WITH BURSTER OR EXPELLING CHARGE*	1.4F	UN0371	EXPLOSIVE (1.4F)
N	WARHEADS, ROCKET, *WITH BURSTING CHARGE*	1.1D	UN0286	EXPLOSIVE (1.1D)
N	WARHEADS, ROCKET, *WITH BURSTING CHARGE*	1.2D	UN0287	EXPLOSIVE (1.2D)
N	WARHEADS, ROCKET, *WITH BURSTING CHARGE*	1.1F	UN0369	EXPLOSIVE (1.1F)
N	WARHEADS, TORPEDO, *WITH BURSTING CHARGE*	1.1D	UN0221	EXPLOSIVE (1.1D)
	WHITE ASBESTOS. *SEE* ASBESTOS, WHITE			
	WHITE PHOSPHORUS, *DRY. SEE* PHOSPHORUS, *WHITE OR YELLOW, DRY*			
	WHITE PHOSPHORUS, *WET. SEE* PHOSPHORUS, *WHITE OR YELLOW, INWATER*			
	WOOD ALCOHOL. SEE METHANOL			
	WOOL WASTE, *WET*	4.2	UN1387	SPONTANEOUSLY COMBUSTIBLE
	XENON	2.2	UN2036	NONFLAMMABLE GAS
	XENON, REFRIGERATED LIQUID	2.2	UN2591	NONFLAMMABLE GAS
	XYLENES	3.2 3.3	UN1307 UN1307	FLAMMABLE LIQUID FLAMMABLE LIQUID
	XYLENOLS	6.1	UN2261	POISON
	XYLIDINES	6.1	UN1711	POISON
	XYLOLS. SEE XYLENES			
	XYLYL BROMIDE	6.1	UN1701	POISON
	YELLOW PHOSPHORUS, *DRY. SEE* PHOSPHORUS, *WHITE OR YELLOW, DRY*			
	YELLOW PHOSPHORUS, *WET. SEE* PHOSPHORUS, *WHITE OR YELLOW, IN WATER*			
	ZINC ARSENATE *OR* ZINC ARSENITE, *OR* MIXTURES *THEREOF, SOLID*	6.1	UN1712	POISON
	ZINC ASHES	4.3	UN1435	DANGEROUS WHEN WET
	ZINC BROMATE	5.1	UN2469	OXIDIZER
	ZINC CHLORATE	5.1	UN1513	OXIDIZER
	ZINC CHLORIDE, ANHYDROUS	8	UN2331	CORROSIVE
	ZINC CHLORIDE, SOLUTION	8	UN1840	CORROSIVE
	ZINC CYANIDE	6.1	UN1713	POISON
	ZINC HYDROSULPHITE	9	UN1931	NONE
	ZINC NITRATE	5.1	UN1514	OXIDIZER
	ZINC PERMANGANATE	5.1	UN1515	OXIDIZER
	ZINC PEROXIDE	5.1	UN1516	OXIDIZER
	ZINC PHOSPHIDE	4.3	UN1714	DANGEROUS WHEN WET
	ZINC, *POWDER OR DUST, PYROPHORIC. SEE* PYROPHORIC METALS			
	ZINC POWDER *OR* ZINC DUST, *NON-PYROPHORIC*	4.3	UN1436	DANGEROUS WHEN WET
	ZINC RESINATE	4.1	UN2714	FLAMMABLE SOLID
	ZINC SILICOFLUORIDE	6.1	UN2855	ST. ANDREWS CROSS
	ZIRCONIUM HYDRIDE	4.1	UN1437	FLAMMABLE SOLID
	ZIRCONIUM, *METAL*, DRY, *FINISHED METAL SHEETS, STRIP OR COILED WIRE (THINNER THAN 18 MICRONS)*	4.2	UN2009	SPONTANEOUSLY COMBUSTIBLE
	ZIRCONIUM, *METAL*, DRY, *FINISHED SHEETS, STRIP OR COILED WIRE (THINNER THAN 254 MICRONS BUT NOT THINNER THAN 18 MICRONS)*	4.1	UN2858	FLAMMABLE SOLID
	ZIRCONIUM *METAL* POWDER, DRY *MECHANICALLY PRODUCED, PARTICLE SIZE BETWEEN 3 AND 53 MICRONS; OR CHEMICALLY PRODUCED, PARTICLE SIZE BETWEEN 10 AND 840 MICRONS*	4.2	UN2008	SPONTANEOUSLY COMBUSTIBLE
	ZIRCONIUM *METAL* POWDER, WETTED *WITH NOT LESS THAN 25% WATER (A VISIBLE EXCESS OF WATER MUST BE PRESENT), MECHANICALLY PRODUCED, HAVING A PARTICLE SIZE LESS THAN 53 MICRONS; OR CHEMICALLY PRODUCED, HAVING A PARTICLE SIZE LESS THAN 840 MICRONS*	4.1	UN1358	FLAMMABLE SOLID
	ZIRCONIUM NITRATE	5.1	UN2728	OXIDIZER
N	ZIRCONIUM PICRAMATE, *DRY OR WETTED WITH LESS THAN 20% WATER, BY WEIGHT*	1.3C	UN0236	EXPLOSIVE (1.3C)
	ZIRCONIUM PICRAMATE, WETTED *WITH, BY WEIGHT, AT LEAST 20% WATER.*	5.1	UN1517	OXIDIZER
	ZIRCONIUM, *SCRAP*	4.2	UN1932	SPONTANEOUSLY COMBUSTIBLE
	ZIRCONIUM, *SUSPENDED IN FLAMMABLE LIQUID*	3.1	UN1308	FLAMMABLE LIQUID
	ZIRCONIUM TETRACHLORIDE	8	UN2503	CORROSIVE

APPENDIX A TO SUBPART B—IDENTIFICATION NUMBER CROSS REFERENCE TO PROPER SHIPPING NAMES IN §§172.101 AND 172.102

This listing is provided for information purposes only.

(1)—Identification Number	(2)—Source 172.***	(3)—Description
UN 0001	102	Alarm devices, explosive
UN 0004	102	Ammonium picrate
UN 0005	102	Cartridges for weapons
UN 0006	102	Cartridges for weapons
UN 0007	102	Cartridges for weapons
UN 0009	102	Ammunition, incendiary
UN 0010	102	Ammunition, incendiary
UN 0012	102	Cartridges for weapons
UN 0014	102	Cartridges for weapons, blank *or* Cartridges, safety, blank
UN 0015	102	Ammunition, smoke
UN 0016	102	Ammunition, smoke
UN 0018	102	Ammunition, tear producing
UN 0019	102	Ammunition, tear producing
UN 0020	102	Ammunition, toxic
UN 0021	102	Ammunition, toxic
UN 0027	102	Black powder
UN 0028	102	Black powder, compressed
UN 0029	102	Detonators, non-electric
UN 0030	102	Detonators, electric
UN 0033	102	Bombs
UN 0034	102	Bombs
UN 0035	102	Bombs
UN 0037	102	Bombs, photo-flash
UN 0038	102	Bombs, photo-flash
UN 0039	102	Bombs, photo-flash
UN 0042	102	Boosters
UN 0043	102	Bursters
UN 0044	102	Primers, cap type
UN 0048	102	Charges, demolition
UN 0049	102	Cartridges, flash
UN 0050	102	Cartridges, flash
UN 0054	102	Cartridges, signal
UN 0055	102	Cases, cartridges, empty, with primer
UN 0056	102	Charges, depth
UN 0059	102	Charges, shaped, commercial
UN 0060	102	Charges, supplementary, explosive
UN 0065	102	Cord, detonating
UN 0066	102	Cord, igniter
UN 0070	102	Cutters, cable, explosive
UN 0072	102	Cyclotrimethylenetrinitramine, wetted
UN 0073	102	Detonators for ammunition
UN 0074	102	Diazodinitrophenol
UN 0075	102	Diethyleneglycol dinitrate, desensitized
UN 0076	102	Dinitrophenol
UN 0077	102	Dinitrophenates
UN 0078	102	Dinitroresorcinol
UN 0079	102	Hexanitrodiphenylamine
UN 0081	102	Explosives, blasting, Type A
UN 0082	102	Explosives, blasting, Type B
UN 0083	102	Explosives, blasting, Type C
UN 0084	102	Explosives, blasting, Type D
UN 0092	102	Flares, surface
UN 0093	102	Flares, aerial
UN 0094	102	Photo-flash powder
UN 0096	102	Photo-flash powder
UN 0099	102	Fracturing devices, explosive
UN 0101	102	Fuse, instantaneous, non-detonating
UN 0102	102	Cord, detonating
UN 0103	102	Fuse, igniter
UN 0104	102	Cord, detonating, mild effect
UN 0105	102	Fuse, safety
UN 0106	102	Fuzes, detonating
UN 0107	102	Fuzes, detonating
UN 0110	102	Grenades, practice
UN 0113	102	Guanyl nitrosamino guanylidene hydrazine
UN 0114	102	Guanyl nitrosamino guanyl tetrazene
UN 0118	102	Hexolite
UN 0121	102	Igniters
UN 0124	102	Jet perforating guns, charged
UN 0129	102	Lead azide
UN 0130	102	Lead styphnate
UN 0131	102	Lighters, fuse

(1)—Identification Number	(2)—Source 172.***	(3)—Description
UN 0132	102	Deflagrating metal salts of aromatic nitro-derivatives, n.o.s.
UN 0133	102	Mannitol hexanitrate
UN 0135	102	Mercury fulminate
UN 0136	102	Mines
UN 0137	102	Mines
UN 0138	102	Mines
UN 0143	102	Nitroglycerine, desensitized
UN 0144	102	Nitroglycerine, spirit of
UN 0146	102	Nitrostarch
UN 0147	102	Nitro urea
UN 0150	102	Pentaerythrite tetranitrate, wetted
UN 0151	102	Pentolite
UN 0153	102	Trinitro-aniline
UN 0154	102	Trinitrophenol
UN 0155	102	Trinitrochlorobenzene
UN 0158	102	Potassium salts of nitro-aromatic derivatives
UN 0159	102	Powder cake, wetted
UN 0160	102	Powder, smokeless
UN 0161	102	Powder, smokeless
UN 0167	102	Projectiles
UN 0168	102	Projectiles
UN 0169	102	Projectiles
UN 0171	102	Ammunition, illuminating
UN 0173	102	Release devices, explosive
UN 0174	102	Rivets, explosive
UN 0180	102	Rockets
UN 0181	102	Rockets
UN 0182	102	Rockets
UN 0183	102	Rockets
UN 0186	102	Rocket motors
UN 0190	102	Samples, explosive
UN 0191	102	Signal devices, hand
UN 0192	102	Signals, railway track, explosive
UN 0193	102	Signals, railway track, explosive
UN 0194	102	Signals, distress
UN 0195	102	Signals, distress
UN 0196	102	Signals, smoke
UN 0197	102	Signals, smoke
UN 0203	102	Sodium salts of nitro-aromatic derivatives, n.o.s.
UN 0204	102	Sounding devices, explosive
UN 0206	102	Squibs
UN 0207	102	Tetranitro-aniline
UN 0208	102	Trinitrophenylmethylnitramine
UN 0209	102	Trinitrotoluene
UN 0212	102	Tracers for ammunition
UN 0213	102	Trinitroanisole
UN 0214	102	Trinitrobenzene
UN 0215	102	Trinitrobenzoic acid
UN 0216	102	Trinitro-m-cresol
UN 0217	102	Trinitronaphthalene
UN 0218	102	Trinitrophenetole
UN 0219	102	Trinitroresorcinol
UN 0220	102	Urea nitrate
UN 0221	102	Warheads, torpedo
UN 0222	102	Ammonium nitrate
UN 0223	102	Ammonium nitrate fertilizers
UN 0224	102	Barium azide
UN 0225	102	Boosters, with detonator
UN 0226	102	Cyclotetramethylenetetranitramine, wetted
UN 0234	102	Sodium dinitro-o-cresolate
UN 0235	102	Sodium picramate
UN 0236	102	Zirconium picramate
UN 0237	102	Charges, shaped, flexible, linear
UN 0238	102	Rockets, line throwing
UN 0240	102	Rockets, line throwing
UN 0241	102	Explosives, blasting, Type E
UN 0242	102	Charges, propelling, for cannon
UN 0243	102	Ammunition, incendiary, white phosphorus
UN 0244	102	Ammunition, incendiary, white phosphorus
UN 0245	102	Ammunition, smoke, white phosphorus
UN 0246	102	Ammunition, smoke, white phosphorus
UN 0247	102	Ammunition, incendiary
UN 0248	102	Contrivances, water-activated
UN 0249	102	Contrivances, water-activated
UN 0250	102	Rocket motors
UN 0254	102	Ammunition, illuminating
UN 0255	102	Detonators, electric
UN 0257	102	Fuzes, detonating
UN 0266	102	Octolite
UN 0267	102	Detonators, non-electric

(1)—Identification Number	(2)—Source 172.***	(3)—Description
UN 0268	102	Boosters, with detonator
UN 0271	102	Charges, propelling, for rocket motors
UN 0272	102	Charges, propelling, for rocket motors
UN 0273	102	Charges, propelling, for rocket motors
UN 0274	102	Charges, propelling, for rocket motors
UN 0275	102	Cartridges, power device
UN 0276	102	Cartridges, power device
UN 0277	102	Cartridges, oil well
UN 0278	102	Cartridges, oil well
UN 0279	102	Charges, propelling, for cannon
UN 0280	102	Rocket motors
UN 0281	102	Rocket motors
UN 0282	102	Nitroguanidine
UN 0283	102	Boosters
UN 0284	102	Grenades
UN 0285	102	Grenades
UN 0286	102	Warheads, rocket
UN 0287	102	Warheads, rocket
UN 0288	102	Charges, shaped, flexible, linear
UN 0289	102	Cord, detonating
UN 0290	102	Cord, detonating
UN 0291	102	Bombs
UN 0292	102	Grenades
UN 0293	102	Grenades
UN 0294	102	Mines
UN 0295	102	Rockets
UN 0296	102	Sounding devices, explosive
UN 0297	102	Ammunition, illuminating
UN 0299	102	Bombs, photo-flash
UN 0300	102	Ammunition, incendiary
UN 0301	102	Ammunition, tear-producing
UN 0303	102	Ammunition, smoke
UN 0305	102	Photo-flash powder
UN 0306	102	Tracers for ammunition
UN 0312	102	Cartridges, signal
UN 0313	102	Signals, smoke
UN 0314	102	Igniters
UN 0315	102	Igniters
UN 0316	102	Fuzes, igniting
UN 0317	102	Fuzes, igniting
UN 0318	102	Grenades, practice
UN 0319	102	Primers, tubular
UN 0320	102	Primers, tubular
UN 0321	102	Cartridges for weapons
UN 0322	102	Rocket motors
UN 0323	102	Cartridges, power device
UN 0324	102	Projectiles
UN 0325	102	Igniters
UN 0326	102	Cartridges for weapons, blank
UN 0327	102	Cartridges for weapons, blank
UN 0328	102	Cartridges for weapons, with inert projectile
UN 0329	102	Torpedoes
UN 0330	102	Torpedoes
UN 0331	102	Explosives, blasting, Type B
UN 0332	102	Explosives, blasting, Type E
UN 0333	102	Fireworks, Type A
UN 0334	102	Fireworks, Type B
UN 0335	102	Fireworks, Type C
UN 0336	102	Fireworks, Type D
UN 0337	102	Fireworks, Type D
UN 0338	102	Cartridges for weapons, blank
UN 0339	102	Cartridges for weapons, with inert projectile
UN 0340	102	Nitrocellulose
UN 0341	102	Nitrocellulose
UN 0342	102	Nitrocellulose, wetted
UN 0343	102	Nitrocellulose, plasticized
UN 0344	102	Projectiles
UN 0345	102	Projectiles
UN 0346	102	Projectiles
UN 0347	102	Projectiles
UN 0348	102	Cartridges for weapons
UN 0349	102	Articles, explosive, n.o.s.
UN 0350	102	Articles, explosive, n.o.s.
UN 0351	102	Articles, explosive, n.o.s.
UN 0352	102	Articles, explosive, n.o.s.
UN 0353	102	Articles, explosive, n.o.s.
UN 0354	102	Articles, explosive, n.o.s.
UN 0355	102	Articles, explosive, n.o.s.
UN 0356	102	Articles, explosive, n.o.s.
UN 0357	102	Substances, explosive, n.o.s.
UN 0358	102	Substances, explosive, n.o.s.
UN 0359	102	Substances, explosive, n.o.s.

(1)—Identification Number	(2)—Source 172.***	(3)—Description
UN 0360	102	Detonator assemblies, non-electric
UN 0361	102	Detonator assemblies, non-electric
UN 0362	102	Ammunition, practice
UN 0363	102	Ammunition, proof
UN 0364	102	Detonators for ammunition
UN 0365	102	Detonators for ammunition
UN 0366	102	Detonators for ammunition
UN 0367	102	Fuzes, detonating
UN 0368	102	Fuzes, igniting
UN 0369	102	Warheads, rocket
UN 0370	102	Warheads, rocket
UN 0371	102	Warheads, rocket
UN 0372	102	Grenades, practice
UN 0373	102	Signal devices, hand
UN 0374	102	Sounding devices, explosive
UN 0375	102	Sounding devices, explosive
UN 0376	102	Primers, tubular
UN 0377	102	Primers, cap type
UN 0378	102	Primers, cap type
UN 0379	102	Cases, cartridge, empty, with primer
UN 0380	102	Articles, pyrophoric
UN 0381	102	Cartridges, power device
UN 0382	102	Components, explosive train, n.o.s.
UN 0383	102	Components, explosive train, n.o.s.
UN 0384	102	Components, explosive train, n.o.s.
UN 0385	102	5-Nitrobenzotriazol
UN 0386	102	Trinitrobenzenesulfonic acid
UN 0387	102	Trinitrofluorenone
UN 0388	102	Trinitrotoluene and hexanitrostilbene mixtures *or* Trinitrotoluene and trinitrobenzene mixtures
UN 0389	102	Trinitrotoluene mixtures containing trinitrobenzene and hexanitrostilbene
UN 0390	102	Tritonal
UN 0391	102	Cyclotrimethylenetrinitramine and cyclotetramethylenetetranitramine, mixtures, wetted
UN 0392	102	Hexanitrostilbene
UN 0393	102	Hexatonal, cast
UN 0394	102	Trinitroresorcinol
UN 0395	102	Rocket motors, liquid fueled
UN 0396	102	Rocket motors, liquid fueled
UN 0397	102	Rockets, liquid fueled
UN 0398	102	Rockets, liquid fueled
UN 0399	102	Bombs, containing flammable liquid
UN 0400	102	Bombs, containing flammable liquid
UN 0401	102	Dipicryl sulfide
UN 0402	102	Ammonium perchlorate
UN 0403	102	Flares, aerial
UN 0404	102	Flares, aerial
UN 0405	102	Cartridges, signal
UN 0406	102	Dinitrosobenzene
UN 0407	102	Tetrazol-1-acetic acid
UN 0408	102	Fuzes, detonating
UN 0409	102	Fuzes, detonating
UN 0410	102	Fuzes, detonating
UN 0411	102	Pentaerythrite tetranitrate
UN 0412	102	Cartridges for weapons
UN 0413	102	Cartridges for weapons, blank
UN 0414	102	Charges, propelling, for cannon
UN 0415	102	Charges, propelling, for rocket motors
UN 0416	102	Charges, propelling, for rocket motors
UN 0417	102	Cartridges for weapons, with inert projectile
UN 0418	102	Flares, surface
UN 0419	102	Flares, surface
UN 0420	102	Flares, aerial
UN 0421	102	Flares, aerial
UN 0422	102	Squibs
UN 0423	102	Squibs
UN 0424	102	Projectiles
UN 0425	102	Projectiles
UN 0426	102	Projectiles
UN 0427	102	Projectiles
UN 1001	101	Acetylene
UN 1002	102	Air
UN 1002	101	Air, compressed
UN 1003	102	Air
UN 1005	102	Ammonia
UN 1005	101	Ammonia, anhydrous
UN 1006	102	Argon
UN 1006	101	Argon *or* Argon, compressed
UN 1008	101	Boron trifluoride
UN 1009	101	Bromotrifluoromethane
UN 1010	102	Butadiene
UN 1010	101	Butadiene, inhibited
UN 1011	102	Butane *or* butane mixtures
UN 1012	102	Butylene
UN 1013	101	Carbon dioxide
UN 1014	102	Carbon dioxide and oxygen
UN 1014	101	Carbon dioxide-oxygen mixture
UN 1015	102	Carbon dioxide and nitrous oxide
UN 1015	101	Carbon dioxide-nitrous oxide mixture
UN 1016	101	Carbon monoxide
UN 1017	101	Chlorine
UN 1018	101	Chlorodifluoromethane
UN 1020	101	Chloropentafluoroethane
UN 1021	101	Chlorotetrafluoroethane
UN 1022	101	Chlorotrifluoromethane
UN 1023	102	Coal gas
UN 1026	102	Cyanogen
UN 1026	101	Cyanogen gas
UN 1027	101	Cyclopropane
UN 1027	102	Cyclopropane, liquefied
UN 1028	101	Dichlorodifluoromethane
UN 1029	102	Dichloromonofluoromethane
UN 1030	102	1,1-Difluoroethane
UN 1030	101	Difluoroethane
UN 1032	102	Dimethylamine
UN 1032	101	Dimethylamine, anhydrous
UN 1033	101	Dimethyl ether
UN 1035	102	Ethane
UN 1035	101	Ethane *or* Ethane, compressed
UN 1036	102	Ethylamine
UN 1036	101	Monoethylamine
UN 1037	101	Ethyl chloride
UN 1038	102	Ethylene
UN 1038	101	Ethylene, refrigerated liquid
UN 1039	101	Ethyl methyl ether
UN 1040	101	Ethylene oxide
UN 1041	102	Carbon dioxide and ethylene oxide mixtures
NA 1043	101	Crude nitrogen fertilizer solution
UN 1043	101	Fertilizer ammoniating solution
NA 1043	101	Nitrogen fertilizer solution
UN 1044	101	Fire extinguisher
UN 1044	102	Fire extinguishers
UN 1045	101	Fluorine
UN 1046	102	Helium
UN 1046	101	Helium *or* Helium, compressed
UN 1048	101	Hydrogen bromide
UN 1048	102	Hydrogen bromide, anhydrous
UN 1049	102	Hydrogen
UN 1049	101	Hydrogen *or* Hydrogen, compressed
UN 1050	101	Hydrogen chloride *or* Hydrogen chloride, anhydrous
UN 1050	102	Hydrogen chloride, anhydrous
NA 1051	101	Hydrocyanic acid, liquefied
UN 1051	102	Hydrogen cyanide, anhydrous, stabilized
UN 1052	101	Hydrogen fluoride
UN 1052	102	Hydrogen fluoride, anhydrous
UN 1053	101	Hydrogen sulfide
UN 1053	102	Hydrogen sulphide
UN 1055	102	Isobutylene
UN 1056	102	Krypton
UN 1057	101	Cigarette lighter
UN 1057	102	Lighters for cigars and cigarettes
NA 1058	101	Liquefied nonflammable gas
UN 1058	102	Liquefied non-flammable gases
UN 1060	102	Methyl acetylene and propadiene
UN 1060	101	Methylacetylene-propadiene, stabilized
UN 1061	102	Methylamine
UN 1061	101	Methylamine, anhydrous
UN 1062	102	Methyl bromide
UN 1062	101	Methyl bromide, liquid
UN 1063	101	Methyl chloride
UN 1064	101	Methyl mercaptan
UN 1064	102	Methylmercaptan
UN 1065	102	Neon
UN 1065	101	Neon *or* Neon, compressed
UN 1066	102	Nitrogen
UN 1066	101	Nitrogen *or* Nitrogen, compressed
UN 1067	102	Nitrogen dioxide
UN 1067	101	Nitrogen dioxide, liquid
NA 1067	101	Nitrogen peroxide, liquid
NA 1067	101	Nitrogen tetroxide, liquid
UN 1069	101	Nitrosyl chloride
UN 1070	102	Nitrous oxide
UN 1070	101	Nitrous oxide *or* Nitrous oxide, compressed
UN 1071	102	Oil gas
UN 1072	102	Oxygen
UN 1072	101	Oxygen *or* Oxygen, compressed
UN 1073	102	Oxygen
UN 1073	101	Oxygen, refrigerated liquid
UN 1075	101	Liquefied petroleum gas
UN 1075	102	Petroleum gases, liquefied
UN 1076	101	Phosgene
UN 1077	102	Propylene
UN 1078	102	Refrigerant gases, n.o.s.
UN 1078	101	Refrigerant gas, n.o.s. *or* Dispersant gas, n.o.s.
UN 1079	101	Sulfur dioxide
UN 1079	102	Sulphur dioxide
UN 1080	101	Sulfur hexafluoride
UN 1080	102	Sulphur hexafluoride
UN 1081	101	Tetrafluoroethylene, inhibited
UN 1082	101	Trifluorochloroethylene
UN 1082	102	Trifluorochloroethylene, inhibited
UN 1083	101	Trimethylamine, anhydrous
UN 1085	102	Vinyl bromide
UN 1086	101	Vinyl chloride
UN 1087	101	Vinyl methyl ether
UN 1088	101	Acetal
UN 1089	101	Acetaldehyde
UN 1090	101	Acetone
UN 1091	101	Acetone oil
UN 1091	102	Acetone oils
UN 1092	101	Acrolein, inhibited
UN 1093	101	Acrylonitrile
UN 1093	102	Acrylonitrile, inhibited
UN 1098	101	Allyl alcohol
UN 1099	101	Allyl bromide
UN 1100	101	Allyl chloride
UN 1101	102	Diethylaluminium chloride
UN 1102	102	Triethylaluminium
UN 1103	102	Trimethylaluminium
UN 1104	101	Amyl acetate
UN 1104	102	Amyl acetates
UN 1105	102	Amyl alcohols
UN 1106	101	Amylamine
UN 1107	101	Amyl chloride
UN 1108	101	Amylene
UN 1108	102	n-Amylene
UN 1109	101	Amyl formate
UN 1109	102	Amyl formates
UN 1110	102	Amyl methyl ketone
UN 1110	101	Methyl amyl ketone
UN 1111	101	Amyl mercaptan
UN 1112	102	Amyl nitrate
UN 1113	101	Amyl nitrite
UN 1114	101	Benzene
UN 1115	101	Benzine
UN 1118	102	Brake fluid
UN 1120	102	Butanol
NA 1120	101	Butyl alcohol
UN 1120	102	sec-Butanol
UN 1120	102	tert-Butanol
UN 1123	101	Butyl acetate
UN 1123	102	Butyl acetates
UN 1125	101	Butylamine
UN 1125	102	n-Butylamine
UN 1126	101	Butyl bromide
UN 1126	102	n-Butyl bromide
UN 1127	101	Butyl chloride
UN 1127	102	Chlorobutanes
UN 1128	101	Butyl formate
UN 1128	102	n-Butyl formate
UN 1129	101	Butyraldehyde
UN 1130	101	Camphor oil
UN 1131	101	Carbon bisulfide, *or* Carbon disulfide
UN 1131	102	Carbon disulphide
UN 1132	102	Carbon remover
UN 1132	101	Carbon remover, liquid
UN 1133	101	Adhesive
UN 1133	102	Adhesives
NA 1133	101	Cement
NA 1133	101	Cement, container, linoleum, tile, *or* wallboard, liquid
NA 1133	101	Cement, leather
NA 1133	101	Cement, pyroxylin
NA 1133	101	Cement, roofing, liquid
NA 1133	101	Cement, rubber
UN 1134	101	Chlorobenzene
UN 1135	101	Ethylene chlorohydrin
UN 1136	101	Coal tar distillate
NA 1136	101	Coal tar light oil

(1)—Identification Number	(2)—Source 172.***	(3)—Description
NA 1136	101	Coal tar oil
UN 1137	101	Coal tar distillate
NA 1137	101	Coal tar light oil
NA 1137	101	Coal tar oil
UN 1139	101	Coating solution
NA 1142	101	Antifreeze compound, liquid
NA 1142	101	Antifreeze preparation, liquid
NA 1142	101	Compound, polishing, liquid
NA 1142	101	Compound, vulcanizing, liquid
NA 1142	101	Dressing, leather
UN 1142	102	Flammable liquid preparations, n.o.s.
NA 1142	101	Leather bleach *or* dressing
NA 1142	101	Polish, metal, stove, furniture *or* wood, liquid
NA 1142	101	Rust preventive coating
UN 1143	101	Crotonaldehyde
UN 1143	102	Crotonaldehyde, inhibited
UN 1144	101	Crotonylene
UN 1145	101	Cyclohexane
UN 1146	101	Cyclopentane
UN 1147	101	Decahydronaphthalene
UN 1148	102	Diacetone alcohol
UN 1148	101	Diacetone alcohol
UN 1149	101	Butyl ether
UN 1149	102	Dibutyl ethers
UN 1150	101	Dichloroethylene
UN 1152	101	Dichloropentane
UN 1152	102	Dichloropentanes
UN 1153	101	Ethylene glycol diethyl ether
UN 1154	101	Diethylamine
UN 1155	102	Diethyl ether
UN 1155	101	Ethyl ether
UN 1156	101	Diethyl ketone
UN 1157	101	Diisobutyl ketone
UN 1158	101	Diisopropylamine
UN 1159	101	Diisopropyl ether
UN 1160	101	Dimethylamine, aqueous solution
UN 1160	102	Dimethylamine, solution
UN 1161	101	Dimethyl carbonate
UN 1162	101	Dimethyldichlorosilane
UN 1163	101	Dimethylhydrazine, unsymmetrical
UN 1164	101	Dimethyl sulfide
UN 1164	102	Dimethyl sulphide
UN 1165	101	Dioxane
UN 1166	101	Dioxolane
UN 1167	101	Divinyl ether
UN 1167	102	Divinyl ether, inhibited
UN 1168	102	Driers
UN 1168	101	Driers, paint *or* varnish, liquid, n.o.s.
UN 1169	102	Extracts, aromatic, liquid
UN 1170	101	Alcoholic beverage
NA 1170	101	Cologne spirits
UN 1170	102	Ethanol *or* Ethanol solutions
UN 1170	101	Ethyl alcohol
UN 1171	101	Ethylene glycol monoethyl ether
UN 1172	101	Ethylene glycol monoethyl ether acetate
UN 1173	101	Ethyl acetate
UN 1175	101	Ethyl benzene
UN 1175	102	Ethylbenzene
UN 1176	101	Ethyl borate
UN 1177	102	Ethylbutyl acetate
UN 1177	101	Ethyl butyl acetate
UN 1178	102	2-Ethylbutyraldehyde
UN 1178	101	Ethyl butyraldehyde
UN 1179	101	Ethyl butyl ether
UN 1180	101	Ethyl butyrate
UN 1181	101	Ethyl chloroacetate
UN 1182	101	Ethyl chloroformate
UN 1183	101	Ethyl dichlorosilane
UN 1183	102	Ethyldichlorosilane
UN 1184	101	Ethylene dichloride
UN 1185	102	Ethyleneimine
UN 1185	101	Ethylene imine, inhibited
UN 1188	101	Ethylene glycol monomethyl ether
UN 1189	101	Ethylene glycol monomethyl ether acetate
UN 1190	101	Ethyl formate
UN 1191	102	Ethyl hexaldehyde
UN 1191	101	Ethylhexaldehyde
UN 1192	101	Ethyl lactate
UN 1193	101	Ethyl methyl ketone
UN 1193	101	Methyl ethyl ketone
UN 1194	101	Ethyl nitrite (nitrous ether)
UN 1194	102	Ethyl nitrite, solutions
UN 1195	101	Ethyl propionate
UN 1196	101	Ethyl trichlorosilane
UN 1196	102	Ethyltrichlorosilane
UN 1197	101	Extract, liquid, flavoring
UN 1197	102	Extracts, flavouring, liquid
UN 1198	101	Formaldehyde solution
UN 1198	102	Formaldehyde solutions
UN 1199	101	Furfural
UN 1201	101	Fusel oil
UN 1202	102	Gas oil
UN 1203	101	Gasoline
(UN 1203)	102	Motor fuel, n.o.s.
NA 1203	101	Motor fuel, n.o.s.
UN 1204	102	Nitroglycerin solution, in alcohol
NA 1204	101	Spirits of nitroglycerin
NA 1204	101	Spirits of nitroglycerin, not exceeding 1% nitroglycerin by weight
UN 1205	102	Gutta percha solution
UN 1206	101	Heptane
UN 1207	101	Hexaldehyde
UN 1208	101	Hexane
UN 1208	101	Neohexane
UN 1210	101	Ink
UN 1210	102	Ink, printers
UN 1212	102	Isobutanol
UN 1213	101	Isobutyl acetate
UN 1214	101	Isobutylamine
UN 1216	101	Isooctene
UN 1218	101	Isoprene
UN 1218	102	Isoprene, inhibited
UN 1219	101	Isopropanol
UN 1220	101	Isopropyl acetate
UN 1221	101	Isopropylamine
UN 1222	101	Isopropyl nitrate
UN 1223	101	Kerosene
UN 1224	102	Ketones, liquid, n.o.s.
UN 1226	101	Cigarette lighter
UN 1226	101	Lighter fluid
UN 1226	102	Lighter fuels
UN 1226	102	Lighters
NA 1228	101	Mercaptan mixture, aliphatic
UN 1228	102	Mercaptans, liquid, n.o.s. *or* Mercaptan mixtures, liquid, n.o.s.
UN 1229	101	Mesityl oxide
NA 1230	101	Columbian spirits
UN 1230	102	Methanol
UN 1230	101	Methyl alcohol
UN 1231	101	Methyl acetate
UN 1232	101	Methyl acetone
UN 1233	101	Methylamyl acetate
UN 1234	101	Methylal
UN 1235	101	Methylamine, aqueous solution
UN 1237	101	Methyl butyrate
UN 1238	101	Methyl chloroformate
UN 1239	102	Methylchloromethyl ether
UN 1239	101	Methylchloromethyl ether, anhydrous
UN 1242	101	Methyl dichlorosilane
UN 1242	102	Methyldichlorosilane
UN 1243	101	Methyl formate
UN 1244	101	Methylhydrazine
UN 1245	102	Methyl isobutyl ketone
UN 1246	102	Methyl isopropenyl ketone
UN 1246	101	Methyl isopropenyl ketone, inhibited
UN 1247	102	Methyl methacrylate
UN 1247	101	Methyl methacrylate monomer, inhibited
NA 1247	101	Methyl methacrylate monomer, uninhibited
UN 1248	101	Methyl propionate
UN 1249	101	Methyl propyl ketone
UN 1250	101	Methyltrichlorosilane
UN 1251	102	Methyl vinyl ketone
UN 1251	101	Methyl vinyl ketone, inhibited
UN 1255	102	Naphtha, petroleum
UN 1255	101	Petroleum naphtha
UN 1256	101	Naphtha, solvent
UN 1257	102	Casinghead gasoline
UN 1259	101	Nickel carbonyl
UN 1261	101	Nitromethane
UN 1262	101	Isooctane
UN 1262	101	Octane
UN 1263	101	Lacquer base *or* Lacquer chips, plastic
UN 1263	101	Paint
NA 1263	101	Paint related material
UN 1263	102	Paints, enamels, lacquers, stains, shellac
UN 1264	101	Paraldehyde
UN 1265	101	Isopentane
UN 1265	101	Pentane
UN 1265	102	Pentanes
UN 1266	102	Perfumery products
UN 1267	101	Crude oil, petroleum
UN 1267	102	Petroleum crude oil
NA 1268	101	Naphtha distillate
UN 1268	101	Petroleum distillate
UN 1268	102	Petroleum distillates, n.o.s.
NA 1268	101	Road oil
NA 1270	101	Oil
UN 1270	102	Petroleum oil
UN 1271	101	Petroleum ether
UN 1271	102	Petroleum spirit
UN 1272	101	Pine oil
UN 1274	102	Propanol
UN 1274	101	Propyl alcohol
UN 1275	101	Propionaldehyde
UN 1276	102	n-Propyl acetate
UN 1276	101	Propyl acetate
UN 1277	102	Monopropylamine
UN 1277	101	Propylamine
UN 1278	101	Propyl chloride
UN 1279	101	Propylene dichloride
UN 1280	101	Propylene oxide
UN 1281	101	Propyl formate
UN 1281	102	Propyl formates
UN 1282	101	Pyridine
UN 1286	102	Rosin oil
UN 1287	102	Rubber solution
UN 1288	102	Shale oil
NA 1289	101	Sodium methylate, alcohol mixture
UN 1289	102	Sodium methylate, solutions
UN 1292	101	Ethyl silicate
UN 1292	102	Tetraethyl silicate
UN 1293	102	Tinctures, medicinal
UN 1294	101	Toluene
UN 1295	101	Trichlorosilane
UN 1296	101	Triethylamine
UN 1297	102	Trimethylamine
UN 1297	101	Trimethylamine, aqueous solution
UN 1298	101	Trimethylchlorosilane
UN 1299	101	Turpentine
UN 1300	101	Turpentine substitute
UN 1301	101	Vinyl acetate
UN 1302	102	Vinyl ethyl ether
UN 1302	101	Vinyl ethyl ether, inhibited
UN 1303	102	Vinylidene chloride
UN 1303	101	Vinylidene chloride, inhibited
UN 1304	101	Vinyl isobutyl ether
UN 1305	101	Vinyl trichlorosilane
UN 1305	102	Vinyl trichlorosilane, inhibited
UN 1307	101	Xylene
UN 1307	102	Xylenes
UN 1308	102	Zirconium
UN 1308	101	Zirconium, metal, liquid, suspensions
UN 1309	102	Aluminium powder, coated
UN 1310	101	Ammonium picrate, wet
UN 1310	102	Ammonium picrate, wetted
UN 1312	102	Borneol
UN 1313	101	Calcium resinate
UN 1314	101	Calcium resinate, fused
UN 1318	102	Cobalt resinate
UN 1318	101	Cobalt resinate, precipitated
UN 1320	102	Dinitrophenol, wetted
UN 1321	102	Dinitrophenolates, wetted
UN 1322	102	Dinitroresorcinol, wetted
UN 1323	102	Ferrocerium
NA 1324	101	Film
UN 1324	102	Film, motion picture
NA 1325	101	Antimony sulfide, solid
NA 1325	101	Burnt cotton, not repicked
NA 1325	101	Cosmetics, n.o.s.
NA 1325	101	Drugs, n.o.s.
UN 1325	101	Flammable solid, n.o.s.
UN 1325	102	Flammable solids, n.o.s.
NA 1325	101	Fusee
NA 1325	101	Garbage tankage
NA 1325	101	N-Methyl-N'-nitro-N-nitrosoguanidine
NA 1325	101	Paper stock, wet
NA 1325	101	Pyroxylin plastic, rods, sheets, rolls, *or* tubes
NA 1325	101	Pyroxylin plastic, scrap
NA 1325	101	Rags, wet
NA 1325	101	Rough ammoniate tankage
NA 1325	101	Smokeless powder for small arms

(1)—Identification Number	(2)—Source 172.***	(3)—Description
NA 1325	101	Tankage fertilizer
NA 1325	101	Tankage, rough ammoniate
NA 1325	101	Waste paper, wet
UN 1326	102	Hafnium
UN 1326	101	Hafnium metal, wet
UN 1327	102	Bhusa
UN 1327	101	Hay
UN 1327	101	Hay or straw
UN 1327	102	Straw
UN 1328	102	Hexamine
UN 1330	102	Manganese resinate
UN 1331	102	Matches
UN 1331	101	Matches, strike anywhere
UN 1332	102	Metaldehyde
UN 1333	102	Cerium, crude
UN 1334	102	Naphthalene, crude *or* refined
UN 1334	101	Naphthalene *or* Naphthalin
UN 1336	102	Nitroguanidine, wetted
UN 1336	101	Nitroguanidine, wet with not less than 20% water
UN 1337	102	Nitrostarch, wetted
UN 1337	101	Nitrostarch, wet with not less than 20% water
NA 1337	101	Nitrostarch, wet with not less than 30% alcohol *or* solvent
UN 1338	102	Phosphorus, amorphous
UN 1338	101	Phosphorus, amorphous, red
UN 1339	101	Phosphorus heptasulfide
UN 1339	102	Phosphorus heptasulphide
UN 1340	101	Phosphorus pentasulfide
UN 1340	102	Phosphorus pentasulphide
UN 1341	101	Phosphorus sesquisulfide
UN 1341	102	Phosphorus sesquisulphide
UN 1343	101	Phosphorus trisulfide
UN 1343	102	Phosphorus trisulphide
NA 1344	101	Picric acid, wet, with not less than 10% water
UN 1344	102	Trinitrophenol, wetted
UN 1345	102	Rubber scrap
UN 1345	101	Rubber scrap *or* Rubber buffings
UN 1345	101	Rubber shoddy *or* Rubber, regenerated *or* Rubber, reclaimed
UN 1346	102	Silicon powder
UN 1347	102	Silver picrate, wetted
UN 1348	102	Sodium dinitro-o-cresolate, wetted
UN 1349	101	Sodium picramate, wet
UN 1349	102	Sodium picramate, wetted
UN 1350	101	Sulfur, solid
UN 1350	102	Sulphur
UN 1352	102	Titanium
UN 1352	101	Titanium metal powder, wet with 20% or more water
UN 1353	102	Toe puffs
UN 1354	101	Trinitrobenzene, wet
UN 1354	102	Trinitrobenzene, wetted
UN 1355	101	Trinitrobenzoic acid, wet
UN 1355	102	Trinitrobenzoic acid, wetted
UN 1356	101	Trinitrotoluene, wet
UN 1356	102	Trinitrotoluene, wetted
UN 1357	101	Urea nitrate, wet
UN 1357	102	Urea nitrate, wetted
UN 1358	102	Zirconium
UN 1358	101	Zirconium metal, wet
UN 1359	102	Bags
UN 1359	101	Bags, sodium nitrate, empty and unwashed
UN 1360	101	Calcium phosphide
UN 1361	102	Carbon
NA 1361	101	Charcoal briquettes *or* briquets
NA 1361	101	Charcoal screenings, made from 'pinon' wood
NA 1361	101	Charcoal, shell
NA 1361	101	Charcoal, wood, ground, crushed, granulated, *or* pulverized
NA 1361	101	Charcoal, wood, lump
NA 1361	101	Charcoal wood screenings, other than 'pinon' wood screenings
NA 1361	101	Coal, ground bituminous, sea coal, coal facings
UN 1362	102	Carbon, activated
UN 1362	101	Charcoal, activated
UN 1363	101	Copra
UN 1364	101	Cotton waste, oily
UN 1365	102	Cotton
UN 1366	102	Diethylzinc
UN 1367	102	Diethylmagnesium
UN 1368	102	Dimethylmagnesium
UN 1369	102	p-Nitrosodimethylaniline

(1)—Identification Number	(2)—Source 172.***	(3)—Description
UN 1370	102	Dimethylzinc
UN 1371	102	Driers
NA 1372	101	Burnt fiber
NA 1372	101	Fibers
NA 1372	101	Fibers, burnt
UN 1372	102	Fibres, animal or vegetable
NA 1372	101	Hair, wet
NA 1373	101	Fibers *or* fabric, containing more than 5% animal *or* vegetable oil
UN 1373	102	Fibres *or* Fabric, animal *or* vegetable
UN 1374	102	Fishmeal *or* fishscrap
NA 1374	101	Fish meal *or* fish scrap containing less than 6% or more than 12% water
UN 1375	102	Fuel, pyrophoric, n.o.s.
UN 1376	101	Iron mass *or* sponge, spent
UN 1376	102	Iron oxide
UN 1378	101	Nickel catalyst, wet
UN 1378	102	Nickel catalyst, wetted
UN 1379	102	Paper, unsaturated oil treated
UN 1380	101	Pentaborane
UN 1381	102	Phosphorus
UN 1381	101	Phosphorus, white *or* yellow, dry
UN 1381	101	Phosphorus, white *or* yellow, in water
UN 1382	101	Potassium sulfide
UN 1382	102	Potassium sulphide, anhydrous *or* Potassium sulphide
NA 1383	101	Iron mass *or* sponge
UN 1383	102	Pyrophoric metals, n.o.s. *or* Pyrophoric alloys, n.o.s.
UN 1384	102	Sodium dithionite
UN 1384	101	Sodium hydrosulfite
UN 1385	101	Sodium sulfide, anhydrous
UN 1385	102	Sodium sulphide, anhydrous *or* Sodium sulphide
UN 1386	102	Seed cake
UN 1387	101	Waste wool, wet
UN 1387	102	Wool waste
UN 1389	102	Alkali metal amalgams, n.o.s.
UN 1390	102	Alkali metal amides, n.o.s.
UN 1391	102	Alkali metal dispersions, n.o.s. *or* Alkali earth metal dispersions, n.o.s.
UN 1392	102	Alkaline earth metal amalgams, n.o.s.
UN 1393	102	Alkaline earth metal alloys, n.o.s.
UN 1394	102	Aluminium carbide
UN 1395	102	Aluminium ferrosilicon
UN 1396	102	Aluminium, powder, uncoated
UN 1396	101	Aluminum, metallic, powder
UN 1397	102	Aluminium phosphide
UN 1397	101	Aluminum phosphide
UN 1398	102	Aluminium silicon
UN 1399	102	Barium
UN 1400	102	Barium
UN 1401	102	Calcium
UN 1401	101	Calcium, metal
NA 1401	101	Calcium, metal, crystalline
UN 1402	101	Calcium carbide
UN 1403	102	Calcium cyanamide
UN 1403	101	Calcium cyanamide, not hydrated
UN 1404	102	Calcium hydride
UN 1405	102	Calcium silicide
UN 1406	101	Calcium silicon
UN 1407	102	Caesium
UN 1407	101	Cesium metal
UN 1408	101	Ferrosilicon
UN 1409	102	Hydrides
UN 1410	102	Lithium aluminium hydride
UN 1410	101	Lithium aluminum hydride
UN 1411	102	Lithium aluminium hydride
UN 1411	101	Lithium aluminum hydride, ethereal
UN 1412	102	Lithium amide
UN 1412	101	Lithium amide, powdered
UN 1413	101	Lithium borohydride
UN 1414	101	Lithium hydride
UN 1415	102	Lithium
UN 1415	101	Lithium metal
UN 1415	101	Lithium metal, in cartridges
UN 1417	101	Lithium silicon
UN 1418	102	Magnesium powder *or* Magnesium alloys, powder
UN 1419	102	Magnesium aluminium phosphide
UN 1419	101	Magnesium aluminum phosphide
UN 1420	102	Potassium, metal alloys
UN 1420	101	Potassium, metal liquid alloy

(1)—Identification Number	(2)—Source 172.***	(3)—Description
UN 1421	102	Alkali metal alloys, liquid
NA 1421	101	Sodium, metal liquid alloy
UN 1422	102	Potassium-sodium
UN 1422	101	Sodium potassium alloy, liquid
UN 1422	101	Sodium potassium alloy, solid
UN 1423	102	Rubidium
UN 1423	101	Rubidium metal
UN 1423	101	Rubidium metal, in cartridges
UN 1424	102	Sodium amalgam
UN 1425	101	Sodium amide
UN 1426	102	Sodium borohydride
UN 1427	101	Sodium hydride
UN 1428	102	Sodium
UN 1428	101	Sodium, metal *or* metallic
UN 1429	102	Sodium
UN 1429	101	Sodium, metal dispersion in organic solvent
UN 1431	102	Sodium methylate
UN 1431	101	Sodium methylate, dry
UN 1432	101	Sodium phosphide
UN 1433	101	Stannic phosphide
UN 1433	102	Stannic phosphides
UN 1434	102	Strontium
UN 1435	102	Zinc ashes
UN 1436	102	Zinc powder *or* zinc dust
UN 1437	101	Zirconium hydride
UN 1438	102	Aluminium nitrate
UN 1438	101	Aluminum nitrate
UN 1439	101	Ammonium dichromate
UN 1442	101	Ammonium perchlorate
UN 1444	102	Ammonium persulphate
UN 1445	101	Barium chlorate
NA 1445	101	Barium chlorate, wet
UN 1446	101	Barium nitrate
UN 1447	101	Barium perchlorate
UN 1448	101	Barium permanganate
UN 1449	102	Barium peroxide
UN 1449	101	Barium peroxide
UN 1450	102	Bromates, inorganic, n.o.s.
UN 1451	102	Caesium nitrate
UN 1452	101	Calcium chlorate
UN 1453	101	Calcium chlorite
UN 1454	101	Calcium nitrate
UN 1455	102	Calcium perchlorate
UN 1456	101	Calcium permanganate
UN 1457	101	Calcium peroxide
UN 1458	101	Chlorate and borate mixture
UN 1458	102	Chlorate and borate, mixtures
UN 1459	101	Chlorate and magnesium chloride mixture
UN 1459	102	Chlorate and magnesium chloride, mixture
UN 1461	101	Chlorate, n.o.s.
NA 1461	101	Chlorate, n.o.s., wet
UN 1461	102	Chlorates, inorganic, n.o.s.
UN 1462	102	Chlorites, inorganic, n.o.s.
NA 1463	101	Chromic acid mixture, dry
NA 1463	101	Chromic acid, solid
UN 1463	102	Chromium trioxide, anhydrous
UN 1465	102	Didymium nitrate
UN 1466	101	Ferric nitrate
UN 1467	101	Guanidine nitrate
UN 1469	101	Lead nitrate
UN 1470	102	Lead perchlorate
UN 1471	101	Lithium hypochlorite compound, dry
UN 1471	102	Lithium hypochlorite, dry *or* Lithium hypochlorite mixtures
UN 1472	101	Lithium peroxide
UN 1473	102	Magnesium bromate
UN 1474	101	Magnesium nitrate
UN 1475	101	Magnesium perchlorate
UN 1476	102	Magnesium peroxide
UN 1476	101	Magnesium peroxide, solid
NA 1477	101	Ammonium sulfate nitrate
NA 1477	101	Nitrate, n.o.s.
UN 1477	102	Nitrates, inorganic, n.o.s.
UN 1478	102	Sodium nitrate and potash, mixtures
NA 1479	101	Compound, tree *or* weed killing, solid
NA 1479	101	Cosmetics, n.o.s.
NA 1479	101	Cupric nitrate
NA 1479	101	Drugs, n.o.s.
UN 1479	101	Oxidizer, n.o.s. *or* Oxidizing material, n.o.s.
UN 1479	102	Oxidizing substances, n.o.s.
NA 1479	101	Potassium dichromate

(1)—Identification Number	(2)—Source 172.***	(3)—Description
NA 1479	101	Sodium dichromate
NA 1481	101	Perchlorate, n.o.s.
UN 1481	102	Perchlorates, inorganic, n.o.s.
NA 1482	101	Permanganate, n.o.s.
UN 1482	102	Permanganates, inorganic, n.o.s.
UN 1483	102	Peroxides, inorganic, n.o.s.
UN 1484	101	Potassium bromate
UN 1485	101	Potassium chlorate
UN 1486	101	Potassium nitrate
UN 1487	102	Potassium nitrate and sodium nitrite, mixture
UN 1487	101	Sodium nitrite mixed with potassium nitrate
NA 1487	101	Sodium nitrite mixture
UN 1488	101	Potassium nitrite
UN 1489	101	Potassium perchlorate
UN 1490	101	Potassium permanganate
UN 1491	101	Potassium peroxide
UN 1492	102	Potassium persulphate
UN 1493	101	Silver nitrate
UN 1494	101	Sodium bromate
UN 1495	101	Sodium chlorate
UN 1496	101	Sodium chlorite
UN 1498	101	Sodium nitrate
UN 1499	102	Sodium nitrate and potassium nitrate
UN 1500	101	Sodium nitrite
UN 1502	101	Sodium perchlorate
UN 1503	101	Sodium permanganate
UN 1504	101	Sodium peroxide
UN 1505	102	Sodium persulphate
UN 1506	101	Strontium chlorate
UN 1506	101	Strontium chlorate, wet
UN 1507	101	Strontium nitrate
UN 1508	102	Strontium perchlorate
UN 1509	101	Strontium peroxide
UN 1510	101	Tetranitromethane
UN 1511	102	Urea hydrogen peroxide
NA 1511	101	Urea peroxide
UN 1512	101	Zinc ammonium nitrite
UN 1513	101	Zinc chlorate
UN 1514	101	Zinc nitrate
UN 1515	101	Zinc permanganate
UN 1516	101	Zinc peroxide
UN 1517	101	Zirconium picramate, wet
UN 1517	102	Zirconium picramate, wetted
UN 1541	101	Acetone cyanohydrin
UN 1544	102	Alkaloids, n.o.s. *or* Alkaloid salts, n.o.s.
UN 1545	102	Allyl isothiocyanate
UN 1546	102	Ammonium arsenate
UN 1546	101	Ammonium arsenate, solid
UN 1547	102	Aniline
UN 1547	101	Aniline oil, liquid
UN 1548	102	Aniline hydrochloride
UN 1549	102	Antimony compounds, inorganic, n.o.s.
NA 1549	101	Antimony tribromide, solid
NA 1549	101	Antimony tribromide solution
NA 1549	101	Antimony trifluoride, solid
NA 1549	101	Antimony trifluoride solution
UN 1550	102	Antimony lactate
UN 1550	101	Antimony lactate, solid
UN 1551	102	Antimony potassium tartrate
UN 1551	101	Antimony potassium tartrate, solid
UN 1553	102	Arsenic acid, liquid
UN 1553	101	Arsenic acid solution
UN 1554	101	Arsenic acid, solid
UN 1555	102	Arsenic bromide
UN 1555	101	Arsenic bromide, solid
UN 1556	101	Arsenical compound, liquid, n.o.s., *or* Arsenical mixture, liquid, n.o.s.
UN 1556	102	Arsenic compounds, liquid, n.o.s.
NA 1556	101	Methyldichloroarsine
NA 1556	101	Phenyldichloroarsine
UN 1557	101	Arsenical compound, solid, n.o.s., *or* Arsenical mixture, solid, n.o.s.
NA 1557	101	Arsenical dip, liquid
UN 1557	102	Arsenic compounds, solid, n.o.s.
NA 1557	101	Arsenic iodide, solid
NA 1557	101	Arsenic sulfide, solid
NA 1557	101	Arsenic trisulfide
UN 1558	102	Arsenic
UN 1558	101	Arsenic, solid
UN 1559	102	Arsenic pentoxide
UN 1559	101	Arsenic pentoxide, solid
UN 1560	102	Arsenic trichloride
UN 1560	101	Arsenic trichloride, liquid
UN 1561	102	Arsenic trioxide
UN 1561	101	Arsenic trioxide, solid
UN 1562	101	Arsenical dust
UN 1564	102	Barium compounds, n.o.s.
UN 1565	102	Barium cyanide
UN 1565	101	Barium cyanide, solid
NA 1566	101	Beryllium chloride
UN 1566	101	Beryllium compound, n.o.s.
UN 1566	102	Beryllium compounds
NA 1566	101	Beryllium fluoride
UN 1567	102	Beryllium
UN 1569	102	Bromoacetone
UN 1569	101	Bromoacetone, liquid
UN 1570	102	Brucine
UN 1570	101	Brucine, solid
UN 1571	101	Barium azide, wet
UN 1571	102	Barium azide, wetted
UN 1572	102	Cacodylic acid
UN 1573	102	Calcium arsenate
UN 1573	101	Calcium arsenate, solid
UN 1574	102	Calcium arsenate and arsenite
NA 1574	101	Calcium arsenite, solid
UN 1575	102	Calcium cyanide
UN 1575	101	Calcium cyanide, solid *or* Calcium cyanide mixture, solid
UN 1577	102	Chlorodinitrobenzene
UN 1577	101	Dinitrochlorobenzene
UN 1578	102	Chloronitrobenzenes
UN 1578	101	Nitrochlorobenzene, meta *or* para, solid
UN 1578	101	Nitrochlorobenzene, ortho, liquid
UN 1579	101	4-Chloro-o-toluidine hydrochloride
UN 1580	102	Chloropicrin
UN 1580	101	Chloropicrin, liquid
UN 1581	102	Chloropicrin and methyl bromide
NA 1581	101	Methyl bromide and more than 2% chloropicrin mixture, liquid
UN 1582	102	Chloropicrin and methyl chloride
UN 1582	101	Chloropicrin and methyl chloride mixture
NA 1583	101	Chloropicrin, absorbed
UN 1583	101	Chloropicrin mixture
UN 1583	102	Chloropicrin mixtures, n.o.s.
UN 1584	102	Cocculus
UN 1584	101	Cocculus, solid
UN 1585	102	Copper acetoarsenite
UN 1585	101	Copper acetoarsenite, solid
UN 1586	102	Copper arsenite
UN 1586	101	Copper arsenite, solid
UN 1587	101	Copper cyanide
UN 1588	101	Cyanide *or* cyanide mixture, dry
UN 1588	102	Cyanides, inorganic, n.o.s.
UN 1589	101	Cyanogen chloride
UN 1590	102	Dichloroanilines
UN 1591	101	Dichlorobenzene, ortho, liquid
UN 1591	102	o-Dichlorobenzene
UN 1592	101	Dichlorobenzene, para, solid
UN 1592	102	p-Dichlorobenzene
UN 1593	102	Dichloromethane
UN 1593	101	Dichloromethane *or* Methylene chloride
UN 1594	102	Diethyl sulphate
UN 1595	101	Dimethyl sulfate
UN 1595	102	Dimethyl sulphate
UN 1596	102	Dinitroanilines
UN 1597	102	Dinitrobenzenes
UN 1597	101	Dinitrobenzene, solid, *or* Dinitrobenzol, solid
UN 1597	101	Dinitrobenzene solution
UN 1598	102	Dinitro-o-cresol
UN 1599	101	Dinitrophenol solution
UN 1599	102	Dinitrophenol, solution
UN 1600	101	Dinitrotoluene, liquid
UN 1600	102	Dinitrotoluenes, molten
UN 1601	101	Disinfectant, liquid
UN 1601	102	Disinfectants, n.o.s.
UN 1601	101	Disinfectant, solid
UN 1602	102	Dyes, n.o.s. *or* Dye intermediates, n.o.s.
UN 1603	102	Ethyl bromoacetate
UN 1604	101	Ethylenediamine
UN 1605	101	Ethylene dibromide
UN 1606	102	Ferric arsenate
UN 1606	101	Ferric arsenate, solid
UN 1607	102	Ferric arsenite
UN 1607	101	Ferric arsenite, solid
UN 1608	102	Ferrous arsenate
UN 1608	101	Ferrous arsenate, solid
UN 1610	102	Halogenated irritating liquids, n.o.s.
UN 1611	102	Hexaethyl tetraphosphate
UN 1611	101	Hexaethyl tetraphosphate, liquid
UN 1612	102	Hexaethyl tetraphosphate and compressed gas
UN 1612	101	Hexaethyl tetraphosphate and compressed gas mixture
UN 1613	102	Hydrocyanic acid, aqueous solutions
UN 1613	101	Hydrocyanic acid solution
UN 1613	101	Hydrocyanic acid solution, less than 5% hydrocyanic acid
UN 1614	102	Hydrogen cyanide, anhydrous, stabilized
UN 1616	101	Lead acetate
UN 1617	102	Lead arsenates
UN 1617	101	Lead arsenate, solid
UN 1618	102	Lead arsenites
UN 1618	101	Lead arsenite, solid
UN 1620	101	Lead cyanide
UN 1621	102	London purple
UN 1621	101	London purple, solid
UN 1622	102	Magnesium arsenate
UN 1622	101	Magnesium arsenate, solid
UN 1623	102	Mercuric arsenate
UN 1624	102	Mercuric chloride
UN 1624	101	Mercuric chloride, solid
UN 1625	101	Mercuric nitrate
UN 1626	102	Mercuric potassium cyanide
UN 1626	101	Mercuric potassium cyanide, solid
UN 1627	102	Mercurous nitrate
UN 1627	101	Mercurous nitrate, solid
UN 1628	101	Mercurous sulfate, solid
UN 1628	102	Mercurous sulphate
UN 1629	101	Mercuric acetate
UN 1629	101	Mercurous acetate, solid
UN 1629	102	Mercury acetate
UN 1630	101	Mercuric ammonium chloride, solid
UN 1630	102	Mercury ammonium chloride
UN 1631	101	Mercuric benzoate, solid
UN 1631	102	Mercury benzoate
UN 1633	102	Mercury bisulphate
UN 1634	101	Mercuric bromide, solid
UN 1634	101	Mercurous bromide, solid
UN 1634	102	Mercury bromides
UN 1636	101	Mercuric cyanide, solid
UN 1636	102	Mercury cyanide
UN 1637	101	Mercurous gluconate, solid
UN 1637	102	Mercury gluconate
UN 1638	101	Mercuric iodide, solid
UN 1638	101	Mercuric iodide, solution
UN 1638	101	Mercurous iodide, solid
UN 1638	102	Mercury iodide
UN 1639	101	Mercurol *or* Mercury nucleate, solid
UN 1639	102	Mercury nucleate
UN 1640	101	Mercuric oleate, solid
UN 1640	102	Mercury oleate
UN 1641	101	Mercuric oxide, solid
UN 1641	101	Mercurous oxide, black, solid
UN 1641	102	Mercury oxide
UN 1642	101	Mercuric oxycyanide, solid
UN 1642	102	Mercury oxycyanide
UN 1643	101	Mercuric potassium iodide, solid
UN 1643	102	Mercury potassium iodide
UN 1644	101	Mercuric salicylate solid
UN 1644	102	Mercury salicylate
UN 1645	101	Mercuric sulfate, solid
UN 1645	102	Mercuric sulphate
UN 1646	101	Mercuric sulfocyanate, solid *or* Mercuric thiocyanate, solid
UN 1646	102	Mercury thiocyanate
UN 1647	102	Methyl bromide and ethylene dibromide mixtures, liquid
UN 1647	101	Methyl bromide - ethylene dibromide mixture, liquid
NA 1648	101	Acetonitrile
UN 1648	102	Methyl cyanide
UN 1649	101	Motor fuel antiknock compound *or* Antiknock compound
UN 1649	102	Motor fuel anti-knock mixtures
NA 1649	101	Tetraethyl lead, liquid
UN 1650	102	Naphthylamine
UN 1651	102	alpha-Naphthylthiourea
UN 1652	102	Naphthylurea
UN 1653	102	Nickel cyanide
UN 1653	101	Nickel cyanide, solid
UN 1654	102	Nicotine
UN 1654	101	Nicotine, liquid

(1)— Identification Number	(2)— Source 172.***	(3)—Description
UN 1655	102	Nicotine compounds, n.o.s. *or* Nicotine preparations, n.o.s.
UN 1656	101	Nicotine hydrochloride
UN 1656	102	Nicotine hydrochloride, *or* Nicotine hydrochloride solutions
UN 1657	101	Nicotine salicylate
UN 1658	101	Nicotine sulfate, liquid
UN 1658	101	Nicotine sulfate, solid
UN 1658	102	Nicotine sulphate, solid *or* solution
UN 1659	101	Nicotine tartrate
UN 1660	101	Nitric oxide
UN 1661	101	Nitroaniline
UN 1661	102	Nitroanilines
UN 1662	102	Nitrobenzene
UN 1662	101	Nitrobenzene, liquid *or* Nitrobenzol, liquid
UN 1663	101	Nitrophenol
UN 1663	102	Nitrophenols
UN 1664	101	Nitrotoluene
UN 1664	102	Nitrotoluenes
UN 1665	102	Nitroxylenes
NA 1665	101	Nitroxylol
UN 1669	102	Pentachloroethane
UN 1670	101	Perchloromethyl mercaptan
UN 1671	101	Phenol
UN 1672	102	Phenylcarbylamine chloride
UN 1673	101	Phenylenediamine, meta *or* para, solid
UN 1673	102	Phenylenediamines
UN 1674	102	Phenylmercuric acetate
UN 1677	102	Potassium arsenate
UN 1677	101	Potassium arsenate, solid
UN 1678	102	Potassium arsenite
UN 1678	101	Potassium arsenite, solid
UN 1679	102	Potassium cuprocyanide
UN 1680	102	Potassium cyanide
UN 1680	101	Potassium cyanide, solid
UN 1680	101	Potassium cyanide solution
UN 1681	102	Rodenticides, n.o.s.
UN 1683	102	Silver arsenite
UN 1684	101	Silver cyanide
UN 1685	101	Sodium arsenate
UN 1686	102	Sodium arsenite, aqueous solutions
NA 1686	101	Sodium arsenite, liquid
UN 1687	101	Sodium azide
UN 1688	102	Sodium cacodylate
UN 1689	102	Sodium cyanide
UN 1689	101	Sodium cyanide, solid
UN 1689	101	Sodium cyanide solution
UN 1690	101	Sodium fluoride, solid
UN 1690	101	Sodium fluoride, solution
UN 1691	102	Strontium arsenite
UN 1691	101	Strontium arsenite, solid
UN 1692	102	Strychnine
UN 1692	101	Strychnine salt, solid
UN 1692	101	Strychnine, solid
NA 1693	101	Irritating agent, n.o.s.
NA 1693	101	ORM-A, n.o.s.
UN 1693	102	Tear gas
NA 1693	101	Tear gas device
UN 1694	102	Bromobenzyl cyanides
UN 1695	102	Chloroacetone
UN 1695	101	Monochloroacetone, stabilized *or* inhibited
UN 1697	102	Chloroacetophenone
UN 1697	101	Chloroacetophenone, gas, liquid, *or* solid
UN 1698	101	Diphenylaminechloroarsine
UN 1699	102	Diphenylchloroarsine
UN 1700	101	Tear gas candle
UN 1700	102	Tear gas candles
UN 1701	101	Xylyl bromide
UN 1702	102	1,1,2,2-Tetrachloroethane
UN 1702	101	Tetrachloroethane
UN 1703	101	Tetraethyl dithiopyrophosphate and compressed gas mixture
UN 1703	102	Tetraethyl dithiopyrophosphate with gases
UN 1704	101	Tetraethyl dithiopyrophosphate, liquid
UN 1704	102	Tetraethyl dithiopyrophosphate, liquid *or* mixtures
UN 1704	101	Tetraethyl dithiopyrophosphate mixture, dry
UN 1704	101	Tetraethyl dithiopyrophosphate mixture, liquid
UN 1705	102	Tetraethyl pyrophosphate and compressed gas
UN 1705	101	Tetraethyl pyrophosphate and compressed gas mixture
UN 1707	102	Thallium compounds, n.o.s.
NA 1707	101	Thallium salt, solid, n.o.s.
NA 1707	101	Thallium sulfate, solid
UN 1708	102	Toluidines
UN 1709	102	2,4-Toluylenediamine
NA 1709	101	Toluenediamine
UN 1710	101	Trichloroethylene
UN 1711	102	Xylidines
UN 1712	101	Zinc arsenate
UN 1712	102	Zinc arsenate *or* Zinc arsenite, *or* mixtures
UN 1712	101	Zinc arsenite, solid
UN 1713	101	Zinc cyanide
UN 1714	101	Zinc phosphide
UN 1715	101	Acetic anhydride
UN 1716	101	Acetyl bromide
UN 1717	101	Acetyl chloride
UN 1718	101	Acid butyl phosphate
UN 1718	102	Butyl acid phosphate
NA 1719	101	Alkaline liquid, n.o.s.
UN 1719	102	Caustic alkali liquids, n.o.s.
UN 1722	101	Allyl chlorocarbonate
UN 1722	102	Allyl chloroformate
UN 1723	102	Allyl iodide
UN 1724	101	Allyl trichlorosilane
UN 1725	102	Aluminium bromide, anhydrous
UN 1725	101	Aluminum bromide, anhydrous
UN 1726	102	Aluminium chloride, anhydrous
UN 1727	101	Ammonium hydrogen fluoride, solid
UN 1728	102	Amyl trichlorosilane
UN 1728	101	Amyl trichorosilane
UN 1729	101	Anisoyl chloride
UN 1730	101	Antimony pentachloride
UN 1731	102	Antimony pentachloride
UN 1731	101	Antimony pentachloride solution
UN 1732	101	Antimony pentafluoride
UN 1733	102	Antimony trichloride
UN 1733	101	Antimony trichloride, solid
UN 1733	101	Antimony trichloride solution
UN 1736	101	Benzoyl chloride
UN 1737	101	Benzyl bromide
UN 1738	101	Benzyl chloride
UN 1739	101	Benzyl chloroformate
UN 1740	102	Bifluorides, n.o.s.
UN 1741	101	Boron trichloride
UN 1742	102	Boron trifluoride acetic acid complex
UN 1742	101	Boron trifluoride-acetic acid complex
UN 1743	102	Boron trifluoride propionic acid complex
UN 1744	101	Bromine
UN 1745	101	Bromine pentafluoride
UN 1746	101	Bromine trifluoride
UN 1747	101	Butyl trichlorosilane
UN 1748	102	Calcium hypochlorite, dry *or* Calcium hypochlorite mixtures
UN 1748	101	Calcium hypochlorite mixture
UN 1749	101	Chlorine trifluoride
UN 1750	102	Chloroacetic acid
UN 1750	101	Chloroacetic acid, liquid *or* solution
UN 1751	101	Chloroacetic acid, solid
UN 1752	101	Chloroacetyl chloride
UN 1753	102	Chlorophenyl trichlorosilane
UN 1753	101	Chlorophenyltrichlorosilane
UN 1754	101	Chlorosulfonic acid
UN 1754	101	Chlorosulfonic acid-sulfur trioxide mixture
UN 1754	102	Chlorosulphonic acid
UN 1755	102	Chromic acid
UN 1755	101	Chromic acid solution
UN 1756	101	Chromic fluoride, solid
UN 1757	101	Chromic fluoride solution
UN 1757	102	Chromic fluoride, solution
UN 1758	102	Chromium oxychloride
UN 1758	101	Chromium oxychloride *or* Chromyl chloride
UN 1759	101	Corrosive solid, n.o.s.
UN 1759	102	Corrosive solids, n.o.s.
NA 1759	101	Cosmetics, solid, n.o.s.
NA 1759	101	Drugs, solid, n.o.s.
NA 1759	101	Ferrous chloride, solid
NA 1759	101	Stannous chloride, solid
NA 1760	101	2-(2-Aminoethoxy) ethanol
NA 1760	101	2,2-Dichloropropionic acid
NA 1760	101	Acid, liquid, n.o.s.
NA 1760	101	Aluminum phosphate solution
NA 1760	101	Aluminum sulfate solution
NA 1760	101	Aminopropyldiethanolamine
NA 1760	101	bis (Aminopropyl) piperazine
NA 1760	101	Boiler compound, liquid
NA 1760	101	Chemical kit
NA 1760	101	Compound, cleaning, liquid
NA 1760	101	Compound, rust preventing *or* Compound, rust removing
NA 1760	101	Compound, tree *or* weed killing, liquid
NA 1760	101	Compound, vulcanizing, liquid
UN 1760	101	Corrosive liquid, n.o.s.
UN 1760	102	Corrosive liquids, n.o.s.
NA 1760	101	Cosmetics, liquid, n.o.s.
NA 1760	101	Drugs, liquid, n.o.s.
NA 1760	101	Ethyl phosphonothioic dichloride, anhydrous
NA 1760	101	Ethyl phosphorodichloridate
NA 1760	101	Ferrous chloride, solution
NA 1760	101	Flame retardant compound liquid
NA 1760	101	Hexanoic acid
NA 1760	101	Isopentanoic acid
NA 1760	101	Memtetrahydro phthalic anhydride
NA 1760	101	Methyl phosphonothioic dichloride, anhydrous
NA 1760	101	Morpholine, aqueous, mixture
NA 1760	101	N-Aminopropylmorpholine
NA 1760	101	Nitric acid, 40% or less
NA 1760	101	ORM-B, n.o.s.
NA 1760	101	Paint *or* paint related material
NA 1760	101	Textile treating compound *or* mixture, liquid
NA 1760	101	Titanium sulfate solution
NA 1760	101	Valeric acid
NA 1760	101	Water treatment compounds, liquid
NA 1760	101	White acid
UN 1761	101	Cupriethylene-diamine solution
UN 1761	102	Cupriethylenediamine, solution
UN 1762	101	Cyclohexenyl trichlorosilane
UN 1763	101	Cyclohexyl trichlorosilane
UN 1764	101	Dichloroacetic acid
UN 1765	101	Dichloroacetyl chloride
UN 1766	102	Dichlorophenyl trichlorosilane
UN 1766	101	Dichlorophenyltrichlorosilane
UN 1767	101	Diethyl dichlorosilane
UN 1768	102	Difluorophosphoric acid
UN 1768	101	Difluorophosphoric acid, anhydrous
UN 1769	101	Diphenyl dichlorosilane
UN 1770	102	Diphenylmethyl bromide
UN 1770	101	Diphenyl methyl bromide, solid
UN 1770	101	Diphenyl methyl bromide solution
UN 1771	101	Dodecyl trichlorosilane
UN 1773	102	Ferric chloride, anhydrous
UN 1773	101	Ferric chloride, solid
UN 1774	101	Fire extinguisher charge containing sulfuric acid
UN 1774	102	Fire extinguisher charges
UN 1775	101	Fluoboric acid
UN 1776	102	Fluorophosphoric acid
UN 1776	101	Monofluorophosphoric acid, anhydrous
UN 1777	101	Fluorosulfonic acid *or* Fluosulfonic acid
UN 1777	102	Fluorosulphonic acid
UN 1778	102	Fluosilicic acid
NA 1778	101	Hydrofluorosilicic acid
UN 1779	101	Formic acid
UN 1779	101	Formic acid solution
UN 1780	101	Fumaryl chloride
UN 1781	102	Hexadecyl trichlorosilane
UN 1781	101	Hexadecyltrichlorosilane
UN 1782	101	Hexafluorophosphoric acid
UN 1783	101	Hexamethylenediamine, solution
UN 1784	102	Hexyl trichlorosilane
UN 1784	101	Hexyltrichlorosilane
UN 1786	102	Hydrofluoric acid and sulphuric acid mixtures
UN 1786	101	Hydrofluoric and sulfuric acid mixture
UN 1787	101	Hydriodic acid
UN 1787	102	Hydriodic acid, solution
UN 1788	101	Hydrobromic acid

(1)—Identification Number	(2)—Source 172.***	(3)—Description
UN 1788	101	Hydrobromic acid not more than 49% strength
UN 1788	102	Hydrobromic acid, solution
NA 1789	101	Compound, cleaning, liquid (containing hydrochloric (muriatic) acid)
UN 1789	101	Hydrochloric acid
NA 1789	101	Hydrochloric acid mixture
UN 1789	102	Hydrochloric acid, solution
UN 1789	101	Hydrochloric acid solution, inhibited
NA 1790	101	Compound, cleaning, liquid (containing hydrofluoric acid)
NA 1790	101	Etching acid, liquid, n.o.s.
UN 1790	101	Hydrofluoric acid solution
UN 1790	102	Hydrofluoric acid, solution
UN 1791	101	Hypochlorite solution
NA 1791	101	Hypochlorite solution containing not more than 7% available chlorine
UN 1791	102	Hypochlorite, solutions
UN 1792	101	Iodine monochloride
UN 1793	101	Isopropyl acid phosphate, solid
UN 1793	102	Isopropyl acid phosphatic
NA 1794	101	Lead dross
UN 1794	101	Lead sulfate, solid
UN 1794	102	Lead sulphate
UN 1796	102	Nitrating acid, mixtures
UN 1796	101	Nitrating acid, mixture (with more than 50% nitric acid)
UN 1796	101	Nitrating acid, mixture (with not more than 50% nitric acid)
UN 1798	101	Nitrohydrochloric acid
UN 1798	101	Nitrohydrochloric acid, diluted
UN 1799	102	Nonyl trichlorosilane
UN 1799	101	Nonyltrichlorosilane
UN 1800	102	Octadecyl trichlorosilane
UN 1800	101	Octadecyltrichlorosilane
UN 1801	102	Octyl trichlorosilane
UN 1801	101	Octyltrichlorosilane
UN 1802	102	Perchloric acid
UN 1802	101	Perchloric acid, not over 50% acid
UN 1803	102	Phenolsulphonic acid
UN 1804	102	Phenyl trichlorosilane
UN 1804	101	Phenyltrichlorosilane
UN 1805	101	Phosphoric acid
UN 1806	102	Phosphorus pentachloride
UN 1806	101	Phosphorus pentachloride, solid
NA 1807	101	Phosphoric anhydride
UN 1807	102	Phosphorus pentoxide
UN 1808	101	Phosphorus tribromide
UN 1809	101	Phosphorus trichloride
UN 1810	101	Phosphorus oxychloride
UN 1811	102	Potassium bifluoride
NA 1811	101	Potassium hydrogen fluoride solution
UN 1812	101	Potassium fluoride
UN 1812	101	Potassium fluoride solution
NA 1813	101	Battery
UN 1813	101	Potassium hydroxide, dry solid, flake, bead, *or* granular
UN 1813	102	Potassium hydroxide, solid
UN 1814	101	Potassium hydroxide, liquid *or* solution
UN 1814	102	Potassium hydroxide, solution
UN 1815	102	Propionyl chloride
UN 1816	101	Propyl trichlorosilane
UN 1817	101	Pyrosulfuryl chloride
UN 1817	102	Pyrosulphuryl chloride
UN 1818	101	Silicon chloride *or* Silicon tetrachloride
UN 1818	102	Silicon tetrachloride
UN 1819	101	Sodium aluminate solution
UN 1819	102	Sodium aluminate, solution
UN 1821	101	Sodium hydrogen sulfate, solid
UN 1821	102	Sodium hydrogen sulphate, solid
UN 1823	101	Sodium hydroxide, dry solid, flake, bead, *or* granular
UN 1823	102	Sodium hydroxide, solid
UN 1824	101	Sodium hydroxide, liquid *or* solution
UN 1824	102	Sodium hydroxide, solution
UN 1825	102	Sodium monoxide
UN 1825	101	Sodium monoxide, solid
UN 1826	102	Acid mixtures, spent, nitrating
NA 1826	101	Nitrating acid, spent
UN 1827	102	Stannic chloride, anhydrous
UN 1827	101	Tin tetrachloride, anhydrous
UN 1828	101	Sulfur chloride
UN 1828	102	Sulphur chlorides
UN 1829	101	Sulfur trioxide
UN 1829	102	Sulphur trioxide
UN 1830	101	Sulfuric acid
UN 1830	102	Sulphuric acid
NA 1831	101	Oleum
UN 1831	102	Sulphuric acid, fuming
UN 1832	101	Sulfuric acid, spent
UN 1832	102	Sulphuric acid, spent
UN 1833	101	Sulfurous acid
UN 1833	102	Sulphurous acid
UN 1834	101	Sulfuryl chloride
UN 1834	102	Sulphuryl chloride
UN 1835	102	Tetramethylammonium hydroxide
UN 1835	101	Tetramethylammonium hydroxide, liquid
UN 1836	101	Thionyl chloride
UN 1837	101	Thiophosphoryl chloride
UN 1838	101	Titanium tetrachloride
UN 1839	101	Trichloroacetic acid, solid
UN 1840	101	Zinc chloride solution
UN 1840	102	Zinc chloride, solution
UN 1841	101	Acetaldehyde ammonia
UN 1843	102	Ammonium dinitro-o-cresolate
UN 1845	102	Carbon dioxide, solid
UN 1845	101	Carbon dioxide, solid, *or* Dry ice, *or* Carbonice
UN 1846	101	Carbon tetrachloride
UN 1847	102	Potassium sulphide, hydrated
UN 1848	101	Propionic acid
UN 1848	101	Propionic acid, solution
UN 1849	102	Sodium sulphide, hydrated
UN 1850	101	Eradicator, paint *or* grease, liquid
UN 1850	102	Eradicators
UN 1851	101	Medicines, liquid, n.o.s.
UN 1851	101	Medicines, n.o.s.
UN 1851	101	Medicines, solid, n.o.s.
UN 1854	102	Barium alloys, pyrophoric
UN 1855	102	Calcium, pyrophoric *or* Calcium alloys, pyrophoric
UN 1856	102	Rags
UN 1856	101	Rags, oily
UN 1857	101	Textile waste, wet
UN 1857	102	Textile waste, wet, n.o.s.
UN 1857	101	Waste textile, wet
UN 1858	101	Hexafluoropropylene
UN 1859	101	Silicon tetrafluoride
UN 1860	102	Vinyl fluoride
UN 1860	101	Vinyl fluoride, inhibited
UN 1862	101	Ethyl crotonate
UN 1863	102	Fuel, aviation
UN 1863	101	Fuel, aviation, turbine engine
UN 1864	102	Gas drips
UN 1864	101	Gas drips, hydrocarbon
UN 1865	102	n-Propyl nitrate
UN 1866	101	Resin solution
UN 1867	102	Cigarettes
UN 1867	101	Self-lighting cigarette
UN 1868	101	Decaborane
UN 1869	101	Magnesium, metal
UN 1869	102	Magnesium *or* Magnesium alloys
NA 1869	101	Magnesium scrap
UN 1870	102	Potassium borohydride
UN 1871	102	Titanium hydride
UN 1872	102	Lead dioxide
UN 1872	101	Lead peroxide
UN 1873	101	Perchloric acid
UN 1884	101	Barium oxide
UN 1885	101	Benzidine
UN 1886	102	Benzylidene chloride
UN 1887	101	Bromochloromethane
UN 1888	101	Chloroform
UN 1889	101	Cyanogen bromide
UN 1891	102	Ethyl bromide
UN 1892	102	Ethyl dichloroarsine
UN 1894	102	Phenylmercuric hydroxide
UN 1895	102	Phenylmercuric nitrate
UN 1896	102	Resin solution, poisonous
UN 1897	102	Tetrachloroethylene
UN 1897	101	Tetrachloroethylene *or* Perchloroethylene
UN 1898	101	Acetyl iodide
NA 1902	101	Di-(2-ethylhexyl) phosphoric acid
UN 1902	101	Diisooctyl acid phosphate
UN 1903	101	Disinfectant, liquid
UN 1903	102	Disinfectants, corrosive, liquid, n.o.s.
UN 1905	102	Selenic acid
UN 1905	101	Selenic acid, liquid
UN 1906	101	Acid, sludge
UN 1906	102	Sludge acid
UN 1907	102	Soda lime
UN 1907	101	Soda lime, solid
UN 1908	101	Sodium chlorite solution
UN 1908	102	Sodium chlorite, solution
UN 1910	101	Calcium oxide
UN 1911	102	Diborane
UN 1911	101	Diborane *or* diborane mixtures
UN 1912	102	Methyl chloride and methylene chloride, mixtures
UN 1912	101	Methyl chloride-methylene chloride mixture
UN 1913	101	Neon, refrigerated liquid
UN 1913	102	Neon, refrigerated liquid
UN 1914	102	Butyl propionate
UN 1915	102	Cyclohexanone
UN 1916	102	Dichloroethyl ether
UN 1917	102	Ethyl acrylate
UN 1917	101	Ethyl acrylate, inhibited
UN 1918	102	Isopropylbenzene
UN 1919	102	Methyl acrylate
UN 1919	101	Methyl acrylate, inhibited
UN 1920	102	Nonane
UN 1921	102	Propyleneimine
UN 1921	101	Propyleneimine, inhibited
UN 1922	101	Pyrrolidine
UN 1923	102	Calcium hydrosulphite
UN 1924	102	Ethyl aluminium dichloride
UN 1925	102	Ethyl aluminium sesquichloride
UN 1926	102	Methyl aluminium sesquibromide
UN 1927	102	Methyl aluminium sesquichloride
UN 1928	101	Methyl magnesium bromide in ethyl ether
UN 1928	102	Methyl magnesium bromide, in ethyl ether
UN 1929	102	Potassium hydrosulphite
UN 1930	102	Triisobutyl aluminium
UN 1931	101	Zinc hydrosulfite
UN 1931	102	Zinc hydrosulphite
UN 1932	102	Zirconium
UN 1932	101	Zirconium scrap
UN 1935	101	Cyanide solution, n.o.s.
UN 1935	102	Cyanide, solutions
UN 1938	102	Bromoacetic acid
UN 1938	101	Bromoacetic acid, solid
UN 1938	101	Bromoacetic acid solution
UN 1939	101	Phosphorus oxybromide
UN 1939	102	Phosphorus oxybromide, solid
UN 1940	101	Thioglycolic acid
UN 1941	101	Dibromodifluoromethane
UN 1942	102	Ammonium nitrate
UN 1942	101	Ammonium nitrate (no organic coating)
NA 1942	101	Ammonium nitrate (organic coating)
UN 1944	101	Matches, safety
UN 1945	102	Matches
UN 1950	102	Aerosol dispensers
UN 1951	101	Argon, refrigerated liquid
UN 1951	102	Argon, refrigerated liquid
UN 1952	102	Carbon dioxide and ethylene oxide mixtures
UN 1953	102	Compressed *or* liquefied gases, flammable, toxic, n.o.s.
NA 1953	101	Poisonous liquid *or* gas, flammable, n.o.s.
UN 1954	101	Compressed gas, n.o.s.
UN 1954	102	Compressed *or* liquefied gases, flammable
NA 1954	101	Refrigerant gas, n.o.s. *or* Dispersant gas, n.o.s.
NA 1954	101	Refrigerating machine
NA 1955	101	Chloropicrin and nonflammable, nonliquefied compressed gas mixture
UN 1955	102	Compressed *or* liquefied gases
NA 1955	101	Methyl bromide and nonflammable, nonliquefied compressed gas mixture, liquid
NA 1955	101	Organic phosphate, Organic phosphate compound, *or* Organic phosphorus compound; mixed with compressed gas
NA 1955	101	Poisonous liquid *or* gas, n.o.s.
(UN 1955)	102	Tetrafluorohydrazine
NA 1956	101	Accumulator, pressurized
UN 1956	101	Compressed gas, n.o.s.
UN 1956	102	Compressed *or* liquefied gases
NA 1956	101	Hexafluoropropylene oxide
NA 1956	101	Mine rescue equipment containing carbon dioxide

(1)—Identification Number	(2)—Source 172.***	(3)—Description
NA 1956	101	Water pump system
UN 1957	102	Deuterium
UN 1958	102	Dichlorotetrafluoroethane
UN 1959	102	1,1-Difluoroethylene
UN 1960	101	Engine starting fluid
UN 1961	102	Ethane
NA 1961	101	Ethane-Propane mixture, refrigerated liquid
UN 1961	101	Ethane, refrigerated liquid
UN 1962	102	Ethylene
UN 1962	101	Ethylene *or* Ethylene, compressed
UN 1963	101	Helium, refrigerated liquid
UN 1963	102	Helium, refrigerated liquid
UN 1964	102	Hydrocarbon gases, compressed, n.o.s. *or* Hydrocarbon gases, mixtures, compressed, n.o.s.
UN 1964	101	Hydrocarbon gas, nonliquefied
UN 1965	102	Hydrocarbon gases, liquefied, n.o.s. *or* Hydrocarbon gases, mixtures, liquefied, n.o.s.
UN 1965	101	Hydrocarbon gas, liquefied
UN 1966	102	Hydrogen
UN 1966	101	Hydrogen, refrigerated liquid
UN 1967	102	Insecticide gases
NA 1967	101	Insecticide, liquefied gas, containing Poison A material *or* Poison B material
NA 1967	101	Parathion and compressed gas mixture
UN 1968	102	Insecticide gases
NA 1968	101	Insecticide, liquefied gas
UN 1969	102	Isobutane *or* Isobutane mixtures
UN 1970	102	Krypton, refrigerated liquid
UN 1971	101	Methane *or* Methane, compressed
UN 1971	102	Methane *or* Natural gases
UN 1972	102	Methane *or* Natural gases
UN 1972	101	Methane, refrigerated liquid
UN 1972	101	Natural gas, refrigerated liquid
UN 1973	102	Chlorodifluoromethane and chloropentafluoroethane
UN 1974	102	Chlorodifluorobromomethane
UN 1975	102	Nitric oxide and nitrogen tetroxide mixtures
UN 1976	102	Octafluorocyclobutane
UN 1977	101	Nitrogen, refrigerated liquid
UN 1977	102	Nitrogen, refrigerated liquid
UN 1978	102	Propane
UN 1979	102	Rare gases
NA 1980	101	Helium-oxygen mixture
UN 1980	102	Rare gases
UN 1981	102	Rare gases
UN 1982	102	Tetrafluoromethane
UN 1983	102	Chlorotrifluoroethane
UN 1984	102	Trifluoromethane
UN 1986	102	Alcohols, toxic, n.o.s.
NA 1986	101	Denatured alcohol
NA 1986	101	Propargyl alcohol
NA 1986	101	Rum, denatured
UN 1987	101	Alcohol, n.o.s.
UN 1987	102	Alcohols, n.o.s.
UN 1988	102	Aldehydes, toxic, n.o.s.
UN 1989	102	Aldehydes, n.o.s.
NA 1989	101	Benzaldehyde
UN 1991	101	Chloroprene, inhibited
UN 1992	101	Flammable liquid, poisonous, n.o.s.
UN 1992	102	Flammable liquids, poisonous, n.o.s.
NA 1993	101	Combustible liquid, n.o.s.
NA 1993	101	Compound, cleaning, liquid
NA 1993	101	Compound, tree *or* weed killing, liquid
NA 1993	101	Cosmetics, n.o.s.
NA 1993	101	Creosote, coal tar
NA 1993	101	Disinfectant, liquid, n.o.s.
NA 1993	101	Drugs, n.o.s.
NA 1993	101	Ethyl nitrate
UN 1993	101	Flammable liquid, n.o.s.
UN 1993	102	Flammable liquids, n.o.s.
NA 1993	101	Fuel oil
NA 1993	101	Fuel oil, No. 1, 2, 4, 5 *or* 6
NA 1993	101	Heater for refrigerator car, liquid fuel type
NA 1993	101	Insecticide, liquid, n.o.s.
NA 1993	101	Organic peroxide, liquid *or* solution, n.o.s.
NA 1993	101	Plastic solvent, n.o.s.
NA 1993	101	Refrigerating machine
NA 1993	101	Solvent, n.o.s.
NA 1993	101	Wax, liquid
UN 1994	102	Iron pentacarbonyl
NA 1999	101	Asphalt
NA 1999	101	Asphalt, cut back
UN 1999	102	Cut-backs
UN 1999	101	Tar, liquid
UN 2000	102	Celluloid
UN 2001	102	Cobalt naphthenates
UN 2002	102	Celluloid, scrap
UN 2003	102	Metal alkyls, n.o.s.
UN 2004	102	Magnesium diamide
UN 2005	102	Magnesium diphenyl
UN 2006	102	Plastics, nitrocellulose-based, spontaneously combustible, n.o.s.
UN 2008	102	Zirconium
UN 2008	101	Zirconium metal, dry
UN 2009	102	Zirconium
UN 2010	102	Magnesium hydride
UN 2011	102	Magnesium phosphide
UN 2012	102	Potassium phosphide
UN 2013	102	Strontium phosphide
UN 2014	102	Hydrogen peroxide, aqueous solutions
UN 2014	101	Hydrogen peroxide solution (8% to 40% peroxide)
UN 2014	101	Hydrogen peroxide solution (40% to 52% peroxide)
UN 2015	101	Hydrogen peroxide solution (over 52% peroxide)
UN 2015	102	Hydrogen peroxide, stabilized, *or* Hydrogen peroxide, aqueous solutions, stabilized
UN 2016	102	Ammunition
UN 2016	101	Chemical ammunition, nonexplosive
NA 2016	101	Grenade
UN 2017	102	Ammunition, tear producing, nonexplosive
UN 2017	101	Chemical ammunition, nonexplosive
NA 2017	101	Grenade, tear gas
UN 2018	102	Chloroanilines
UN 2019	102	Chloroanilines
UN 2020	102	Chlorophenols, solid
NA 2020	101	Pentachlorophenol
NA 2020	101	Trichlorophenol
UN 2021	102	Chlorophenols, liquid
UN 2022	102	Cresylic acid
NA 2022	101	Mining reagent, liquid
UN 2023	101	Epichlorohydrin
UN 2024	102	Mercury compounds, liquid, n.o.s.
NA 2025	101	Mercuric subsulfate, solid
UN 2025	101	Mercury compound, solid, n.o.s.
UN 2025	102	Mercury compounds, solid, n.o.s.
UN 2026	102	Phenylmercuric compounds, n.o.s.
UN 2027	102	Sodium arsenite, solid
UN 2028	102	Bombs, smoke
UN 2029	101	Hydrazine, anhydrous
UN 2029	102	Hydrazine, anhydrous *or* Hydrazine aqueous solutions
UN 2030	101	Hydrazine, aqueous solution
UN 2030	102	Hydrazine hydrate *or* Hydrazine aqueous solutions
UN 2031	101	Nitric acid
UN 2032	101	Nitric acid, fuming
UN 2032	102	Nitric acid, red fuming
UN 2033	102	Potassium oxide
UN 2034	102	Hydrogen and methane mixtures, compressed
UN 2035	102	Trifluoroethane
UN 2036	101	Xenon
UN 2037	102	Receptacles
UN 2038	101	Dinitrotoluene, solid
UN 2038	102	Dinitrotoluenes, solid
UN 2044	102	2,2-Dimethylpropane
UN 2045	102	Isobutyraldehyde
UN 2046	102	Cymenes
UN 2047	101	Dichloropropene
NA 2047	101	Dichloropropene and propylene dichloride mixture
UN 2048	102	Dicyclopentadiene
UN 2049	102	Diethylbenzene
UN 2050	102	Diisobutylene
UN 2051	102	Dimethylethanolamine
UN 2052	102	Dipentene
UN 2053	102	Methyl isobutyl carbinol
UN 2054	101	Morpholine
NA 2054	101	Morpholine, aqueous, mixture
UN 2055	101	Styrene monomer, inhibited
UN 2056	101	Tetrahydrofuran
UN 2057	102	Tripropylene
UN 2058	102	Valeraldehyde
UN 2059	101	Box toe gum
NA 2059	101	Collodion
UN 2059	102	Nitrocellulose
NA 2059	101	Nitrocellulose, colloided, granular *or* flake, wet with not less than 20% alcohol *or* solvent, *or* block, wet with not less than 25% alcohol
NA 2059	101	Pyroxylin solution
NA 2059	101	Pyroxylin solvent, n.o.s.
UN 2060	101	Box toe gum
UN 2067	101	Ammonium nitrate fertilizer
UN 2067	102	Ammonium nitrate fertilizers
UN 2068	101	Ammonium nitrate-carbonate mixture
UN 2069	101	Ammonium nitrate mixed fertilizer
UN 2070	101	Ammonium nitrate-phosphate
UN 2071	102	Ammonium nitrate fertilizers
UN 2072	102	Ammonium nitrate fertilizers, n.o.s.
UN 2073	101	Ammonia solution
UN 2073	102	Ammonia solutions
UN 2074	102	Acrylamide
UN 2075	102	Chloral
UN 2076	101	Cresol
UN 2076	102	Cresols
UN 2077	102	Naphthylamine
UN 2078	101	Toluene diisocyanate
UN 2079	102	Diethylenetriamine
UN 2080	101	Acetyl acetone peroxide
UN 2081	101	Acetyl benzoyl peroxide
UN 2081	101	Acetyl benzoyl peroxide solution
UN 2082	101	Acetyl cyclohexanesulphonyl peroxide
UN 2082	102	Acetyl cyclohexane sulphonyl peroxide
UN 2083	101	Acetyl cyclohexanesulphonyl peroxide
UN 2083	102	Acetyl cyclohexane sulphonyl peroxide
UN 2084	101	Acetyl peroxide
UN 2084	101	Acetyl peroxide solution
NA 2085	101	Benzoyl peroxide
UN 2085	101	Benzoyl peroxide
UN 2086	102	Benzoyl peroxide
UN 2087	101	Benzoyl peroxide
UN 2088	101	Benzoyl peroxide
UN 2089	101	Benzoyl peroxide
UN 2090	101	Benzoyl peroxide
UN 2091	101	tert-Butyl cumyl peroxide
NA 2091	101	tert-Butyl isopropyl benzene hydroperoxide
UN 2092	101	tert-Butyl hydroperoxide
UN 2093	101	tert-Butyl hydroperoxide
UN 2094	101	tert-Butyl hydroperoxide
UN 2095	101	tert-Butyl peroxyacetate
UN 2096	101	tert-Butyl peroxyacetate
UN 2097	101	tert-Butyl peroxybenzoate
UN 2098	101	tert-Butyl peroxybenzoate
UN 2099	102	tert-Butyl monoperoxymaleate
UN 2099	101	tert-Butyl peroxymaleate
UN 2100	102	tert-Butyl monoperoxymaleate
UN 2100	101	tert-Butyl peroxymaleate
UN 2101	102	tert-Butyl monoperoxymaleate
UN 2101	101	tert-Butyl peroxymaleate
UN 2102	101	Di-tert-butyl peroxide
UN 2103	101	tert-Butyl peroxyisopropyl carbonate
UN 2103	102	tert-Butyl peroxy isopropyl carbonate
UN 2104	102	tert-Butyl peroxy-3,5,5-trimethyl hexanoate
UN 2104	101	tert-Butyl peroxy-3,5,5-trimethylhexanoate *or* tert-Butyl peroxyisononanoate
UN 2105	102	tert-Butyl monoperoxyphthalate
UN 2105	101	tert-Butyl peroxyphthalate
UN 2106	101	Di-(tert-butylperoxy)phthalate
UN 2107	101	Di-(tert-butylperoxy)phthalate
UN 2108	101	Di-(tert-butylperoxy)phthalate
UN 2110	101	tert-Butyl peroxypivalate
UN 2111	101	2,2-Di-(tert-butylperoxy)butane
UN 2111	102	2,2-Di-(tert-butylperoxy) butane
UN 2112	101	1,3-Di-(2-tert-butylperoxyisopropyl) benzene
UN 2112	101	1,3-Di-(2-tert-butylperoxyisopropyl) benzene and 1,4-Di-(2-tert-butylperoxyisopropyl) benzene mixture

(1)—Identification Number	(2)—Source 172.***	(3)—Description
UN 2112	101	1,4-Di-(2-tert-butylperoxyisopropyl) benzene
UN 2112	102	1,4-Di-(2-tert-butylperoxy isopropyl) benzene, *or* ,1,3-di-(2-tert-butylperoxy isopropyl) benzene, *or* mixtures
UN 2113	102	Di-(4-chlorobenzoyl) peroxide
UN 2113	101	p-Chlorobenzoyl peroxide
UN 2114	102	Di-(4-chlorobenzoyl) peroxide
UN 2114	101	p-Chlorobenzoyl peroxide
UN 2115	102	Di-(4-chlorobenzoyl) peroxide
UN 2115	101	p-Chlorobenzoyl peroxide
UN 2116	101	Cumene hydroperoxide
UN 2116	102	Cumyl hydroperoxide
UN 2117	102	Cyclohexanone peroxides
UN 2118	101	Cyclohexanone peroxide
UN 2118	102	Cyclohexanone peroxides
UN 2119	101	Cyclohexanone peroxide
UN 2119	102	Cyclohexanone peroxides
UN 2120	101	Decanoyl peroxide
UN 2120	102	Didecanoyl peroxide
NA 2121	101	Dicumyl peroxide
UN 2121	101	Dicumyl peroxide
UN 2121	101	Dicumyl peroxide, dry
UN 2122	101	Di-(2-ethylhexyl) peroxydicarbonate
UN 2123	101	Di-(2-ethylhexyl) peroxydicarbonate
UN 2124	102	Dilauroyl peroxide
UN 2124	101	Lauroyl peroxide
UN 2125	101	Paramenthane hydroperoxide
UN 2125	101	p-Menthane hydroperoxide
UN 2126	102	Isobutyl methyl ketone peroxide
UN 2126	101	Methyl isobutyl ketone peroxide
UN 2127	102	Ethyl methyl ketone peroxide(s)
UN 2128	102	Di-(3,5,5-trimethylhexanoyl)peroxide
UN 2128	101	Isononanyl peroxide
NA 2129	101	Caprylyl peroxide solution
UN 2129	102	Di-n-octanoyl peroxide
UN 2129	101	n-Octanoyl peroxide
UN 2130	102	Di-n-nonanoyl peroxide
UN 2130	101	Pelargonyl peroxide
NA 2131	101	Peracetic acid solution
UN 2131	101	Peroxyacetic acid
UN 2132	102	Dipropionyl peroxide
UN 2132	101	Propionyl peroxide
UN 2133	102	Diisopropyl peroxydicarbonate
NA 2133	101	Isopropyl percarbonate, unstabilized
UN 2133	101	Isopropyl peroxydicarbonate
UN 2134	102	Diisopropyl peroxydicarcbonate
NA 2134	101	Isopropyl percarbonate, stabilized
UN 2134	101	Isopropyl peroxydicarbonate
UN 2135	102	Disuccinic acid peroxide
UN 2135	101	Succinic acid peroxide
UN 2136	101	Tetralin hydroperoxide
UN 2137	101	2,4-Dichlorobenzoyl peroxide
UN 2137	102	Di-2,4-dichlorobenzoyl peroxide
UN 2138	101	2,4-Dichlorobenzoyl peroxide
UN 2138	102	Di-2,4-dichlorobenzoyl peroxide
UN 2139	101	2,4-Dichlorobenzoyl peroxide
UN 2139	102	Di-2,4-dichlorobenzoyl peroxide
UN 2140	101	n-Butyl-4,4-di-(tert-butylperoxy)valerate
UN 2140	102	n-Butyl-4,4-di-(tert-butyl-peroxy) valerate
UN 2141	101	n-Butyl-4,4-di-(tert-butylperoxy)valerate
UN 2141	102	n-Butyl-4,4-di-(tert-butyl-peroxy) valerate
UN 2142	101	tert-Butyl peroxyisobutyrate
UN 2143	101	tert-Butyl peroxy-2-ethylhexanoate
UN 2143	102	tert-Butyl peroxy-2-ethyl hexanoate
UN 2144	101	tert-Butyl peroxydiethylacetate
UN 2145	101	1,1-Di-(tert-butylperoxy)-3,3,5-trimethyl cyclohexane
UN 2146	101	1,1-Di-(tert-butylperoxy)-3,3,5-trimethyl cyclohexane
UN 2147	101	1,1-Di-(tert-butylperoxy)-3,3,5-trimethyl cyclohexane
UN 2148	102	Di-(1-hydroxy cyclohexyl) peroxide
UN 2148	101	Di-(1-hydroxycyclohexyl) peroxide
UN 2149	101	Dibenzyl peroxydicarbonate
UN 2150	101	Di-sec-butyl peroxydicarbonate
UN 2151	101	Di-sec-butyl peroxydicarbonate
UN 2152	101	Dicyclohexyl peroxydicarbonate
UN 2153	101	Dicyclohexyl peroxydicarbonate
UN 2154	101	Di-(4-tert-butylcyclohexyl)peroxydicarbonate
UN 2154	102	Di-(4-tert-butylcyclohexyl) peroxydicarbonate
UN 2155	101	2,5-Dimethyl-2,5-di-(tert-butylperoxy)hexane
UN 2156	101	2,5-Dimethyl-2,5-di-(tert-butylperoxy)hexane
UN 2157	102	2,5-Dimethyl-2,5-di -(2-ethylhexanoylperoxy) hexane
UN 2157	101	2,5-Dimethyl-2,5-di-(2-ethylhexanoylperoxy)hexane
UN 2158	101	2,5-Dimethyl-2,5-di-(tert-butylperoxy)hexyne-3
UN 2159	101	2,5-Dimethyl-2,5-di-(tert-butylperoxy)hexyne-3
UN 2160	101	1,1,3,3-Tetramethylbutyl hydroperoxide
UN 2160	102	1,1,3,3-Tetramethyl butyl hydroperoxide
UN 2161	102	1,1,3,3-Tetramethyl butyl peroxy-2-ethyl hexanoate
UN 2161	101	1,1,3,3-Tetramethylbutyl peroxy-2-ethylhexanoate
UN 2162	101	Pinane hydroperoxide
UN 2162	101	Pinane hydroperoxide solution
UN 2162	102	Pinanyl hydroperoxide
UN 2163	101	Diacetone alcohol peroxide
UN 2163	102	Diacetone alcohol peroxides
UN 2164	101	Dicetyl peroxydicarbonate
UN 2165	101	3,3,6,6,9,9-Hexamethyl-1,2,4,5-tetraoxocyclononane
UN 2166	101	3,3,6,6,9,9-Hexamethyl-1,2,4,5-tetraoxocyclononane
UN 2167	101	3,3,6,6,9,9-Hexamethyl-1,2,4,5-tetraoxocyclononane
UN 2168	101	2,2-Di-(4,4-di-tert-butylperoxycyclohexyl)propane
UN 2169	101	n-Butyl peroxydicarbonate
UN 2170	101	n-Butyl peroxydicarbonate
UN 2171	101	Diisopropylbenzene hydroperoxide
UN 2171	101	Diisopropylbenzene hydroperoxide solution
UN 2172	101	2,5-Dimethyl-2,5-di-(benzoylperoxy)hexane
UN 2173	101	2,5-Dimethyl-2,5-di-(benzoylperoxy)hexane
UN 2174	101	2,5-Dimethyl-2,5-dihydroperoxy hexane
UN 2174	101	Dimethylhexane dihydroperoxide, (with 18% or more water)
UN 2175	101	Diethyl peroxydicarbonate
UN 2176	101	Di-n-propyl peroxydicarbonate
UN 2177	101	tert-Butyl peroxyneodecanoate
UN 2178	101	2,2-Dihydroperoxy propane
UN 2179	101	1,1-Di-(tert-butylperoxy)cyclohexane
UN 2179	102	1,1-Di-(tert-butylperoxy) cyclohexane
UN 2180	101	1,1-Di-(tert-butylperoxy)cyclohexane
UN 2180	102	1,1-Di-(tert-butylperoxy) cyclohexane
UN 2181	101	1,2-Di-(tert-butylperoxy)cyclohexane
UN 2182	101	Diisobutyryl peroxide
UN 2182	102	Isobutyryl peroxide
UN 2183	101	tert-Butyl peroxycrotonate
UN 2184	101	Ethyl-3,3-di-(tert-butylperoxy)butyrate
UN 2184	102	Ethyl-3,3-di-(tert-butylperoxy) butyrate
UN 2185	101	Ethyl-3,3-di-(tert-butylperoxy)butyrate
UN 2185	102	Ethyl-3,3-di-(tert-butylperoxy) butyrate
UN 2186	101	Hydrogen chloride, refrigerated liquid
UN 2187	101	Carbon dioxide, refrigerated liquid
UN 2187	102	Carbon dioxide, refrigerated liquid
UN 2188	101	Arsine
UN 2189	102	Dichlorosilane
UN 2190	102	Oxygen difluoride
UN 2191	101	Sulfuryl fluoride
UN 2191	102	Sulphuryl fluoride
UN 2192	101	Germane
UN 2193	102	Hexafluoroethane
UN 2194	102	Selenium hexafluoride
UN 2195	102	Tellurium hexafluoride
UN 2196	101	Tungsten hexafluoride
UN 2197	102	Hydrogen iodide, anhydrous
UN 2198	102	Phosphorus pentafluoride
UN 2199	101	Phosphine
UN 2200	102	Propadiene, inhibited
UN 2201	101	Nitrous oxide, refrigerated liquid
UN 2201	102	Nitrous oxide, refrigerated liquid
UN 2202	101	Hydrogen selenide
UN 2203	102	Silane
UN 2204	102	Carbonyl sulfide
UN 2205	102	Adiponitrile
UN 2206	102	Isocyanates, n.o.s. *or* Isocyanate solutions, n.o.s.
UN 2207	102	Isocyanates, n.o.s. *or* Isocyanate solutions, n.o.s.
UN 2208	101	Bleaching powder
UN 2208	102	Calcium hypochlorite mixtures, dry
UN 2209	101	Formaldehyde solution
UN 2209	102	Formaldehyde, solutions
UN 2210	102	Maneb, *or* maneb preparation(s)
NA 2210	101	Pesticide, water reactive
UN 2211	102	Plastics moulding materials evolving flammable vapours
UN 2212	102	Asbestos, blue
UN 2213	101	Paraformaldehyde
UN 2214	102	Phthalic anhydride
NA 2215	101	Maleic acid
UN 2215	101	Maleic anhydride
UN 2216	102	Fishmeal *or* fishscrap
NA 2216	101	Fish meal *or* fish scrap containing 6% to 12% water
UN 2217	102	Seed cake
UN 2218	101	Acrylic acid
UN 2218	102	Acrylic acid, inhibited
UN 2219	102	Allyl glycidyl ether
UN 2220	102	Aluminium alkyl halides
UN 2221	102	Aluminium alkyl halides
UN 2222	102	Anisole
UN 2224	101	Benzonitrile
UN 2225	102	Benzene sulphonyl chloride
UN 2226	102	Benzotrichloride
UN 2227	102	n-Butyl methacrylate
UN 2228	102	Butylphenols, liquid
UN 2229	102	Butylphenols, solid
UN 2232	102	Chloroacetaldehyde
UN 2233	102	p-Chloro-o-anisidine
UN 2234	102	Chlorobenzotrifluorides
UN 2235	102	p-Chlorobenzyl chloride
UN 2236	102	3-Chloro-4-methylphenyl isocyanate
UN 2237	102	Chloronitroanilines
UN 2238	102	Chlorotoluenes
UN 2239	102	Chlorotoluidines
UN 2240	102	Chromosulphuric acid
UN 2241	102	Cycloheptane
UN 2242	102	Cycloheptene
UN 2243	102	Cyclohexyl acetate
UN 2244	102	Cyclopentanol
UN 2245	102	Cyclopentanone
UN 2246	102	Cyclopentene
UN 2247	102	n-Decane
UN 2248	102	Di-(n-butyl)amine
UN 2249	102	Dichlorodimethyl ether, symmetrical
UN 2250	102	Dichlorophenyl isocyanates
UN 2252	102	1,2-Dimethoxyethane
UN 2253	102	N,N-Dimethylaniline
UN 2254	102	Matches
UN 2255	101	Organic peroxide, sample, n.o.s.
UN 2255	102	Organic peroxides, n.o.s., samples
UN 2256	102	Cyclohexene
UN 2257	102	Potassium metal
UN 2257	101	Potassium, metal *or* metallic
UN 2258	101	Propylenediamine
UN 2259	102	Triethylenetetramine
UN 2260	102	Tripropylamine
UN 2261	101	Xylenol
UN 2261	102	Xylenols
UN 2262	102	N,N-Dimethylcarbamoyl chloride
UN 2263	101	1,4-Dimethylcyclohexane
UN 2263	102	Dimethylcyclohexanes
UN 2264	102	N,N-Dimethylcyclohexylamine
UN 2265	102	N,N-Dimethylformamide
UN 2266	102	Dimethyl-N-propylamine
UN 2267	102	Dimethyl thiophosphoryl chloride
UN 2269	102	3,3'-Iminobispropylamine
UN 2269	101	Iminobispropylamine
UN 2270	102	Ethylamine solution
UN 2271	102	Ethyl amyl ketone
UN 2272	102	N-Ethylaniline
UN 2273	102	2-Ethylaniline
UN 2274	102	N-Ethyl-n-benzylaniline
UN 2275	102	2-Ethylbutanol
UN 2276	102	2-Ethylhexylamine
UN 2277	102	Ethyl methacrylate, inhibited
UN 2278	102	n-Heptene

(1)—Identification Number	(2)—Source 172.***	(3)—Description
UN 2279	102	Hexachlorobutadiene
UN 2280	101	Hexamethylenediamine, solid
UN 2281	102	Hexamethylenediisocyanate
UN 2282	102	Hexanols
UN 2283	102	Isobutyl methacrylate
UN 2284	102	Isobutyronitrile
UN 2285	102	Isocyanatobenzotrifluorides
UN 2286	102	Pentamethylheptane
UN 2287	102	Isoheptene
UN 2288	102	Isohexene
UN 2289	102	Isophoronediamine
UN 2290	102	Isophorone diisocyanate
NA 2291	101	Lead chloride
UN 2291	102	Lead compounds, soluble, n.o.s.
NA 2291	101	Lead fluoborate
NA 2291	101	Lead sulfate
NA 2291	101	Lead sulfide
NA 2291	101	Lead thiocyanate
UN 2293	102	4-Methoxy-4-methylpentan-2-one
UN 2294	102	N-Methylaniline
UN 2295	102	Methyl chloroacetate
UN 2296	102	Methyl cyclohexane
UN 2296	101	Methylcyclohexane
UN 2297	102	Methyl cyclohexanone
UN 2298	101	Cyclopentane, methyl
UN 2298	102	Methyl cyclopentane
UN 2298	101	Methylcyclopentane
UN 2299	101	Methyl dichloroacetate
UN 2300	102	2-Methyl-5-ethylpyridine
UN 2300	101	Methyl ethyl pyridine
UN 2301	102	2-Methylfuran
UN 2301	101	Methylfuran
UN 2302	102	5-Methylhexan-2-one
UN 2303	102	Isopropenylbenzene
UN 2304	102	Napthalene, molten
UN 2305	102	Nitrobenzenesulphonic acid
UN 2306	102	Nitrobenzotrifluorides
UN 2307	102	3-Nitro-4-chlorobenzotrifluoride
UN 2308	102	Nitrosylsulphuric acid
UN 2309	102	Octadiene
UN 2310	102	2,4-Pentanedione
UN 2311	102	Phenetidines
UN 2312	102	Phenol, molten
UN 2313	102	Picolines
UN 2315	101	Polychlorinated biphenyls
UN 2316	102	Sodium cuprocyanide
UN 2317	102	Sodium cuprocyanide solution
UN 2318	101	Sodium hydrosulfide, solid
UN 2318	102	Sodium hydrosulphide
UN 2319	102	Terpene hydrocarbons n.o.s.
UN 2320	102	Tetraethylenepentamine
UN 2321	102	Trichlorobenzenes
UN 2322	102	Trichlorobutene
UN 2323	102	Triethyl phosphite
UN 2324	102	Triisobutylene
UN 2325	102	1,3,5-Trimethylbenzene
UN 2326	102	Trimethylcyclohexylamine
UN 2327	102	Trimethylhexamethylene diamines
UN 2328	102	Trimethylhexamethylene diisocyanate
UN 2329	102	Trimethyl phosphite
UN 2330	102	Undecane
UN 2331	102	Zinc chloride, anhydrous
UN 2331	101	Zinc chloride, solid
UN 2332	102	Acetaldehyde oxime
UN 2333	102	Allyl acetate
UN 2334	102	Allylamine
UN 2335	102	Allyl ethyl ether
UN 2336	102	Allyl formate
UN 2337	101	Phenyl mercaptan
UN 2338	102	Benzotrifluoride
UN 2339	102	2-Bromobutane
UN 2340	102	2-Bromoethyl ethyl ether
UN 2341	102	1-Bromo-3-methylbutane
UN 2342	102	Bromomethylpropanes
UN 2343	102	2-Bromopentane
UN 2344	102	Bromopropanes
UN 2345	102	3-Bromopropyne
UN 2346	102	Butanedione
UN 2346	101	Diacetyl
UN 2347	101	Butyl mercaptan
UN 2348	102	Butylacrylate, inhibited
UN 2350	102	Butyl methyl ether
UN 2351	102	Butyl nitrite
UN 2352	102	Butyl vinyl ether
UN 2353	102	Butyryl chloride
UN 2354	102	Chloromethyl ethyl ether
UN 2356	102	2-Chloropropane

(1)—Identification Number	(2)—Source 172.***	(3)—Description
UN 2357	101	Cyclohexylamine
UN 2358	102	Cyclooctatetraene
UN 2359	102	Diallylamine
UN 2360	102	Diallylether
UN 2361	102	Diisobutylamine
UN 2362	102	1,1-Dichloroethane
UN 2363	101	Ethyl mercaptan
UN 2364	102	Propyl benzene
UN 2366	102	Diethyl carbonate
UN 2367	102	alpha-Methyl valeraldehyde
UN 2368	102	alpha-Pinene
UN 2368	101	Pinene
UN 2369	102	Ethylene glycol monobutyl ether
UN 2370	102	Hex-1-ene
UN 2371	102	Isopentenes
UN 2372	102	1,2-Di-(dimethylamino)-ethane
UN 2373	102	Diethoxymethane
UN 2374	102	3,3-Diethoxypropene
UN 2375	102	Diethyl sulfide
UN 2376	102	2,3-Dihydropyran
UN 2376	101	Dihydropyran
UN 2377	102	1,1-Dimethoxyethane
UN 2378	102	2-Dimethylaminoacetonitrile
UN 2379	102	1,3-Dimethylbutylamine
UN 2380	102	Dimethyldiethoxysilane
UN 2381	102	Dimethyl disulfide
UN 2382	102	Dimethylhydrazine
UN 2383	102	Dipropylamine
UN 2384	102	Dipropyl ether
UN 2385	102	Ethylisobutyrate
UN 2386	102	1-Ethyl piperidine
UN 2387	102	Fluorobenzene
UN 2388	102	Fluorotoluenes
UN 2389	101	Furan
UN 2390	102	2-Iodobutane
UN 2391	102	Iodomethylpropanes
UN 2392	102	Iodopropanes
UN 2393	102	Isobutyl formate
UN 2394	102	Isobutyl propionate
UN 2395	102	Isobutyryl chloride
UN 2396	102	Methacraldehyde
UN 2397	102	3-Methyl butan-2-one
UN 2398	102	Methyl-tert-butyl ether
UN 2399	102	1-Methylpiperidine
UN 2400	102	Methylisovalerate
UN 2401	102	Piperidine
UN 2402	102	Propanethiols
UN 2403	102	Isopropenyl acetate
UN 2404	102	Propionitrile
UN 2405	102	Isopropyl butyrate
UN 2406	102	Isopropyl isobutyrate
UN 2407	102	Isopropyl chloroformate
UN 2409	102	Isopropyl propionate
UN 2410	102	1,2,3,6-Tetrahydropyridine
UN 2411	102	Butyronitrile
UN 2412	102	Tetrahydrothiophene
UN 2413	102	Tetrapropylorthotitanate
UN 2414	102	Thiophene
UN 2416	102	Trimethyl borate
UN 2417	102	Carbonyl fluoride
UN 2418	102	Sulphur tetrafluoride
UN 2419	102	Bromotrifluoroethylene
UN 2420	102	Hexafluoroacetone
UN 2421	102	Nitrogen trioxide
UN 2422	102	Octafluorobut-2-ene
NA 2422	101	Perfluoro-2-butene
UN 2424	102	Octafluoropropane
UN 2426	101	Ammonium nitrate, solution
UN 2427	102	Potassium chlorate
UN 2428	102	Sodium chlorate
UN 2429	102	Calcium chlorate
UN 2430	102	Alkyl phenols, n.o.s.
UN 2431	102	o-Anisidine
UN 2432	102	N,N-Diethylaniline
UN 2433	102	Chloro-o-nitrotoluene
UN 2434	102	Dibenzyldichlorosilane
UN 2435	101	Ethyl phenyl dichlorosilane
UN 2435	102	Ethylphenyldichlorosilane
UN 2436	102	Thioacetic acid
UN 2437	102	Methylphenyldichlorosilane
UN 2438	101	Trimethylacetyl chloride
UN 2439	101	Sodium bifluoride, solid
UN 2439	101	Sodium bifluoride, solution
UN 2439	102	Sodium hydrogen fluoride
UN 2440	102	Stannic chloride pentahydrate
UN 2441	102	Titanium trichloride, pyrophoric *or* Titanium trichloride mixtures, pyrophoric

(1)—Identification Number	(2)—Source 172.***	(3)—Description
UN 2442	102	Trichloroacetyl chloride
UN 2443	102	Vanadium oxitrichloride
UN 2443	101	Vanadium oxytrichloride
NA 2443	101	Vanadium oxytrichloride and titanium tetrachloride mixture
UN 2444	101	Vanadium tetrachloride
UN 2445	102	Lithium alkyls
UN 2446	102	Nitrocresols
UN 2447	102	Phosphorus, white, molten
UN 2448	102	Sulphur, molten
NA 2449	101	Ammonium oxalate
NA 2449	101	Cupric oxalate
UN 2449	102	Oxalates
UN 2451	101	Nitrogen trifluoride
UN 2452	102	Ethyl acetylene, inhibited
UN 2453	102	Ethyl fluoride
UN 2454	102	Methyl fluoride
UN 2456	101	2-Chloropropene
UN 2457	101	2,3-Dimethylbutane
UN 2458	101	Hexadiene
UN 2459	102	2-Methyl-1-butene
UN 2460	102	2-Methyl-2-butene
UN 2460	101	Methyl butene
UN 2461	101	Methylpentadiene
UN 2462	101	Methyl pentane
UN 2462	102	Methylpentanes
UN 2463	102	Aluminium hydride
UN 2463	101	Aluminum hydride
UN 2464	101	Beryllium nitrate
UN 2465	102	Dichloroisocyanuric acid, dry *or* Dichloroisocyanuric acid salts
NA 2465	101	Potassium dichloro-s-triazinetrione
UN 2465	101	Sodium dichloro-s-triazinetrione
UN 2466	101	Potassium superoxide
UN 2467	102	Sodium percarbonates
NA 2468	101	(mono-(Trichloro) tetra-(monopotassium dichloro)-penta-s-triazinetrione, dry
UN 2468	102	Trichloroisocyanuric acid, dry
UN 2468	101	Trichloro-s-triazinetrione
UN 2469	102	Zinc bromate
UN 2470	102	Phenylacetonitrile, liquid
UN 2471	102	Osmium tetroxide
UN 2472	102	Pindone
UN 2473	102	Sodium arsanilate
UN 2474	101	Thiophosgene
UN 2475	102	Vanadium trichloride
UN 2477	102	Methyl isothiocyanate
UN 2478	102	Isocyanates, n.o.s. *or* Isocyanate solutions, n.o.s.
UN 2480	101	Methyl isocyanate
UN 2480	102	Methyl isocyanate *or* Methyl isocyanate solutions
UN 2481	102	Ethyl isocyanate
UN 2482	102	n-Propyl isocyanate
UN 2483	102	Isopropyl isocyanate
UN 2484	102	tert-Butyl isocyanate
UN 2485	101	n-Butyl isocyanate
UN 2486	102	Isobutyl isocyanate
UN 2487	102	Phenyl isocyanate
UN 2488	102	Cyclohexyl isocyanate
UN 2489	102	Diphenylmethane-4,4'diisocyanate
UN 2490	101	Dichloroisopropyl ether
UN 2491	102	Ethanolamine *or* Ethanolamine solutions
UN 2491	101	Monoethanolamine
UN 2493	101	Hexamethyleneimine
UN 2495	101	Iodine pentafluoride
UN 2496	101	Propionic anhydride
UN 2497	101	Sodium phenolate, solid
UN 2498	101	1,2,3,6-Tetrahydrobenzaldehyde
UN 2501	101	Tris-(1-aziridinyl) phosphine oxide
UN 2501	102	Tris-(1-aziridinyl)phosphine oxide, solution
UN 2502	101	Valeryl chloride
UN 2502	102	Valeryl chlorides
UN 2503	102	Zirconium tetrachloride
UN 2503	101	Zirconium tetrachloride, solid
UN 2504	101	Acetylene tetrabromide
UN 2504	102	Tetrabromoethane
UN 2505	101	Ammonium fluoride
UN 2506	101	Ammonium hydrogen sulfate
UN 2506	102	Ammonium hydrogen sulphate
UN 2507	101	Chloroplatinic acid, solid
UN 2508	101	Molybdenum pentachloride
UN 2509	101	Potassium hydrogen sulfate, solid
UN 2509	102	Potassium hydrogen sulphate
UN 2511	102	Chloropropionic acid

(1)—Identification Number	(2)—Source 172.***	(3)—Description
UN 2512	102	Aminophenols
UN 2513	102	Bromoacetyl bromide
UN 2514	101	Bromobenzene
UN 2515	102	Bromoform
UN 2516	102	Carbon tetrabromide
UN 2517	101	Chlorodifluoroethane
UN 2517	102	Chlorodifluoroethanes
UN 2518	102	1,5,9-Cyclododecatriene
UN 2520	102	Cyclooctadienes
UN 2521	102	Diketene, inhibited
UN 2522	102	Dimethylaminoethyl methacrylate
UN 2524	102	Ethyl orthoformate
UN 2525	102	Ethyl oxalate
UN 2526	102	Furfurylamine
UN 2527	102	Isobutyl acrylate
UN 2528	102	Isobutyl isobutyrate
UN 2529	101	Isobutyric acid
UN 2530	101	Isobutyric anhydride
UN 2531	102	Methacrylic acid, inhibited
UN 2533	102	Methyl trichloroacetate
UN 2534	102	Methyl chlorosilane
UN 2535	102	Methylmorpholine
UN 2536	102	Methyltetrahydrofuran
UN 2538	102	Nitronapthalene
UN 2541	102	Terpinolene
UN 2542	102	Tributylamine
UN 2545	102	Hafnium
UN 2545	101	Hafnium metal, dry.
UN 2546	102	Titanium
UN 2546	101	Titanium metal powder, dry *or* wet
UN 2547	101	Sodium superoxide
UN 2548	102	Chlorine pentafluoride
UN 2550	102	Ethyl methyl ketone peroxide(s)
UN 2550	101	Methyl ethyl ketone peroxide
UN 2551	102	tert-Butyl peroxydiethylacetate
UN 2551	101	tert-Butyl peroxydiethylacetate, with tert-Butyl peroxybenzoate
UN 2552	102	Hexafluoroacetone hydrate
UN 2553	102	Coal tar naphtha
NA 2553	101	Coal tar naptha
UN 2553	101	Naphtha
UN 2554	102	Methyl allyl chloride
UN 2555	102	Nitrocellulose
NA 2555	101	Nitrocellulose, colloided, granular *or* flake, wet with not less than 20% water
NA 2555	101	Nitrocellulose, wet with not less than 20% water
(UN 2556)	102	Nitrocellulose
UN 2556	102	Nitrocellulose
NA 2556	101	Nitrocellulose, wet with not less than 30% alcohol *or* solvent
NA 2557	101	Lacquer base, *or* Lacquer chips, dry
UN 2557	102	Nitrocellulose
UN 2558	102	Epibromohydrin
UN 2560	102	2-Methylpentan-2-ol
UN 2561	102	3-Methyl-1-butene
UN 2562	101	tert-Butyl peroxyisobutyrate
UN 2564	102	Trichloroacetic acid
UN 2564	101	Trichloroacetic acid solution
UN 2565	102	Dicyclohexylamine
UN 2567	101	Sodium pentachlorophenate
NA 2570	101	Cadmium acetate
NA 2570	101	Cadmium bromide
NA 2570	101	Cadmium chloride
UN 2570	102	Cadmium compounds
UN 2571	102	Ethylsulphuric acid
UN 2572	102	Phenylhydrazine
UN 2573	102	Thallium chlorate
UN 2574	102	Tricresylphosphate
UN 2576	102	Phosphorus oxybromide, molten
UN 2577	102	Phenylacetyl chloride
UN 2578	102	Phosphorus trioxide
UN 2579	102	Piperazine
UN 2580	102	Aluminium bromide solution
UN 2581	102	Aluminium chloride solution
UN 2582	101	Ferric chloride solution
UN 2582	102	Ferric chloride, solution
UN 2583	102	Alkyl, Aryl *or* Toluene sulphonic acid, solid
UN 2584	101	Alkanesulfonic acid
UN 2584	102	Alkyl, Aryl *or* Toluene sulphonic acid, liquid
NA 2584	101	Dodecylbenzenesulfonic acid
UN 2584	101	Toluene sulfonic acid, liquid
UN 2585	102	Alkyl, Aryl *or* Toluene sulphonic acid, solid
UN 2586	102	Alkyl, Aryl *or* Toluene sulphonic acid, liquid
UN 2587	102	Benzoquinone
NA 2588	101	Insecticide, dry, n.o.s.
UN 2588	102	Pesticides, solid, toxic, n.o.s.
UN 2589	102	Vinyl chloroacetate
UN 2590	102	Asbestos, white
UN 2591	102	Xenon, refrigerated liquid
UN 2592	101	Distearyl peroxydicarbonate
UN 2592	102	Distearylperoxydicarbonate
UN 2593	101	Di-(2-methylbenzoyl)peroxide
UN 2594	101	tert-Butyl peroxyneodecanoate
UN 2595	101	Dimyristyl peroxydicarbonate
UN 2595	102	Dimyristylperoxydicarbonate
UN 2596	101	3-tert-Butyl peroxy-3-phenylphthalide
UN 2596	102	3-tert-Butylperoxy-3-phenyl phthalide
UN 2597	101	Di-(3,5,5-trimethyl-1,2-dioxolanyl-3)peroxide
UN 2598	102	Ethyl-3,3-di(tert-butylperoxy)butyrate
UN 2598	101	Ethyl-3,3-di-(tert-butylperoxy)butyrate
UN 2599	102	Chlorotrifluoromethane and trifluormethane azeotropic mixture
UN 2600	102	Carbon monoxide and hydrogen mixture
UN 2601	102	Cyclobutane
UN 2602	102	Dichlorodifluoromethane and difluoroethane, azeotropic mixture
UN 2603	102	Cycloheptatriene
UN 2604	102	Boron trifluoride diethyletherate
UN 2605	102	Methoxymethyl isocyanate
UN 2606	102	Methyl orthosilicate
UN 2607	102	Acrolein dimer
UN 2608	102	Nitropropanes
UN 2609	102	Triallyl borate
UN 2610	102	Triallylamine
UN 2611	102	Propylene chlorohydrin
UN 2612	102	Methyl propyl ether
UN 2614	102	Methallyl alcohol
UN 2615	102	Ethyl propyl ether
UN 2616	102	Triisopropyl borate
UN 2617	102	Methyl cyclohexanol
UN 2618	102	Vinyl toluenes
UN 2619	102	Benzyl dimethylamine
UN 2620	102	Amyl butyrates
UN 2621	102	Acetyl methyl carbinol
UN 2622	102	Glycidaldehyde
UN 2623	102	Firelighters
UN 2624	102	Magnesium silicide
NA 2626	101	Chloric acid
UN 2626	102	Chloric acid solution
UN 2627	102	Nitrites, inorganic, n.o.s.
UN 2628	102	Potassium fluoroacetate
UN 2629	102	Sodium fluoroacetate
UN 2630	102	Selenates, n.o.s. *or* Selenites, n.o.s.
UN 2630	101	Sodium selenite
UN 2642	102	Fluoroacetic acid
UN 2643	102	Methyl bromoacetate
UN 2644	102	Methyl iodide
UN 2645	102	Phenacyl bromide
UN 2646	101	Hexachlorocyclopentadiene
UN 2647	102	Malononitrile
UN 2648	102	1,2-Dibromobutan-3-one
UN 2649	102	1,3-Dichloroacetone
UN 2650	102	1,1-Dichloro-1-nitroethane
UN 2651	102	4,4'-Diaminodiphenyl methane
UN 2653	102	Benzyl iodide
UN 2655	102	Potassium silicofluoride
UN 2656	101	Quinoline
UN 2657	102	Selenium disulphide
UN 2658	102	Selenium
UN 2659	102	Sodium chloroacetate
UN 2660	102	Mononitrotoluidines
UN 2661	102	Hexachloroacetone
UN 2662	102	Hydroquinone
UN 2664	102	Dibromomethane
UN 2666	102	Ethyl cyanoacetate
UN 2667	102	Butyl toluenes
UN 2668	102	Chloroacetonitrile
UN 2669	102	Chlorocresols
UN 2670	102	Cyanuric chloride
UN 2671	102	Aminopyridines
UN 2672	102	Ammonia solutions
NA 2672	101	Ammonium hydroxide
UN 2673	102	2-Amino-4-chlorophenol
UN 2674	102	Sodium silicofluoride
UN 2676	102	Stibine
UN 2677	102	Rubidium hydroxide, solution
UN 2678	102	Rubidium hydroxide, solid
UN 2679	102	Lithium hydroxide, solution
UN 2680	102	Lithium hydroxide monohydrate
UN 2681	102	Caesium hydroxide, solution
UN 2682	102	Caesium hydroxide, solid
NA 2683	101	Ammonium hydrosulfide solution
UN 2683	101	Ammonium sulfide solution
UN 2683	102	Ammonium sulphide, solution
UN 2684	102	3-(Diethylamino) propylamine
UN 2685	102	N,N-Diethylethylene diamine
UN 2686	102	Diethylaminoethanol
UN 2687	102	Dicyclohexylammonium nitrite
UN 2688	102	1-Chloro-3-bromopropane
UN 2689	102	Glycerol-alpha-monochlorohydrin
UN 2690	102	N-n-Butyl imidazole
UN 2691	102	Phosphorus pentabromide
UN 2692	101	Boron tribromide
NA 2693	101	Ammonium bisulfite, solid
NA 2693	101	Ammonium bisulfite solution
UN 2693	102	Bisulphites, inorganic, aqueous solution, n.o.s.
NA 2693	101	Calcium hydrogen sulfite solution
NA 2693	101	Potassium metabisulfite
NA 2693	101	Sodium hydrogen sulfite, solid
NA 2693	101	Sodium hydrogen sulfite, solution
NA 2693	101	Sodium metabisulfite
UN 2698	102	Tetrahydrophthalic anhydrides
UN 2699	102	Trifluoroacetic acid
UN 2703	101	Isopropyl mercaptan
UN 2704	101	Propyl mercaptan
UN 2705	102	1-Pentol
UN 2706	102	Diethylcarbinol
UN 2707	102	Dimethyldioxanes
UN 2708	102	Butoxyl
UN 2709	102	Butyl benzenes
UN 2710	102	Dipropylketone
UN 2711	102	Dibromobenzene
UN 2713	102	Acridine
UN 2714	102	Zinc resinate
UN 2715	102	Aluminium resinate
UN 2716	102	1,4-Butynediol
UN 2717	102	Camphor
UN 2718	102	Tripropylaluminum
UN 2719	102	Barium bromate
UN 2720	102	Chromium nitrate
UN 2721	102	Copper chlorate
UN 2722	102	Lithium nitrate
UN 2723	102	Magnesium chlorate
UN 2724	102	Manganese nitrate
UN 2725	101	Nickel nitrate
UN 2726	102	Nickel nitrite
UN 2727	102	Thallium nitrate
UN 2728	101	Zirconium nitrate
UN 2729	102	Hexachlorobenzene
UN 2730	102	Nitroanisoles
UN 2732	102	Nitrobromobenzenes
UN 2733	102	Alkylamines and polyamines
UN 2734	102	Alkylamines and polyamines
UN 2735	102	Alkylamines and polyamines
UN 2738	102	N-n-Butylaniline
UN 2739	102	Butyric anhydride
UN 2740	102	n-Propyl chloroformate
UN 2741	102	Barium hypochlorite
UN 2742	102	Chloroformates, n.o.s.
UN 2743	102	n-Butylchloroformate
UN 2744	102	Cyclobutylchloroformate
UN 2745	102	Chloromethylchloroformate
UN 2746	102	Phenylchloroformate
UN 2747	102	tert-Butylcyclohexylchloroformate
UN 2748	102	2-Ethylhexylchloroformate
UN 2749	102	Tetramethylsilane
UN 2750	102	1,3-Dichloropropanol-2
UN 2751	102	Diethylthiophosphoryl chloride
UN 2752	102	1,2-Epoxy-3-ethyloxy propane
UN 2753	102	N-ethylbenzyltoluidines
UN 2754	102	N-Ethyltoluidines
UN 2755	101	3-Chloroperoxybenzoic acid
UN 2756	101	Organic peroxide, mixture.
UN 2756	102	Organic peroxides, mixture
UN 2757	101	Carbamate pesticide, liquid, n.o.s.
UN 2757	101	Carbamate pesticide, solid, n.o.s.
UN 2757	102	Carbamate pesticides, solid, toxic, n.o.s.
NA 2757	101	Carbaryl
NA 2757	101	Carbofuran
NA 2757	101	Carbofuran mixture, liquid

(1)—Identification Number	(2)—Source 172.***	(3)—Description
NA 2757	101	Mercaptodimethur
NA 2757	101	Mexacarbate
UN 2758	101	Carbamate pesticide, liquid, n.o.s.
UN 2758	102	Carbamate pesticides, liquid, flammable, toxic, n.o.s.
UN 2759	101	Arsenical pesticide, liquid, n.o.s.
UN 2759	101	Arsenical pesticide, solid, n.o.s.
UN 2759	102	Arsenical pesticides, solid, toxic, n.o.s.
NA 2759	101	Bordeaux arsenite, liquid
NA 2759	101	Bordeaux arsenite, solid
UN 2760	101	Arsenical pesticide, liquid, n.o.s.
UN 2760	102	Arsenical pesticides, liquid, flammable, toxic, n.o.s.
NA 2761	101	Aldrin
NA 2761	101	Aldrin, cast solid
NA 2761	101	Aldrin mixture, dry
NA 2761	101	Aldrin mixture, dry, with 65% or less aldrin
NA 2761	101	DDT *or* Dichlorodiphenyltrichloroethane
NA 2761	101	Dichlone
NA 2761	101	Dieldrin
NA 2761	101	Endosulfan
NA 2761	101	Endosulfan mixture, liquid
NA 2761	101	Endrin
NA 2761	101	Endrin mixture, liquid
NA 2761	101	Heptachlor
NA 2761	101	Kelthane
NA 2761	101	Kepone
NA 2761	101	Lindane
NA 2761	101	Methoxychlor
UN 2761	101	Organochlorine pesticide, liquid, n.o.s.
UN 2761	101	Organochlorine pesticide, solid, n.o.s.
UN 2761	102	Organochlorine pesticides, solid, toxic, n.o.s.
NA 2761	101	TDE
NA 2761	101	Toxaphene
NA 2762	101	Aldrin mixture, liquid
NA 2762	101	Aldrin mixture, liquid, with 60% or less aldrin
NA 2762	101	Chlordane, liquid
UN 2762	101	Organochlorine pesticide, liquid, n.o.s.
UN 2762	102	Organochlorine pesticides, liquid, flammable, toxic, n.o.s.
UN 2763	101	Triazine pesticide, liquid, n.o.s.
UN 2763	101	Triazine pesticide, solid, n.o.s.
UN 2763	102	Triazine pesticides, solid, toxic, n.o.s.
UN 2764	101	Triazine pesticide, liquid, n.o.s.
UN 2764	102	Triazine pesticides, liquid, flammable, toxic, n.o.s.
NA 2765	101	2,4,5-Trichlorophenoxyacetic acid
NA 2765	101	2,4,5-Trichlorophenoxyacetic acid amine, ester, *or* salt
NA 2765	101	2,4,5-Trichlorophenoxypropionic acid
NA 2765	101	2,4,5-Trichlorophenoxypropionic acid ester
NA 2765	101	2,4-Dichlorophenoxyacetic acid
NA 2765	101	2,4-Dichlorophenoxyacetic acid ester
UN 2765	102	Phenoxy pesticides, solid, toxic, n.o.s.
UN 2765	101	Phenoxy pesticide, liquid, n.o.s.
UN 2765	101	Phenoxy pesticide, solid, n.o.s.
NA 2765	101	Propargite
UN 2766	101	Phenoxy pesticide, liquid, n.o.s.
UN 2766	102	Phenoxy pesticides, liquid, flammable, toxic, n.o.s.
NA 2767	101	Diuron
UN 2767	101	Phenylurea pesticide, liquid, n.o.s.
UN 2767	101	Phenylurea pesticide, solid, n.o.s.
UN 2767	102	Phenyl urea pesticides, solid, toxic, n.o.s.
UN 2768	101	Phenylurea pesticide, liquid, n.o.s.
UN 2768	102	Phenyl urea pesticides, liquid, flammable, toxic, n.o.s.
UN 2769	101	Benzoic derivative pesticide, liquid, n.o.s.
UN 2769	101	Benzoic derivative pesticide, solid, n.o.s.
UN 2769	102	Benzoic derivative pesticides, solid, toxic, n.o.s.
NA 2769	101	Dicamba
NA 2769	101	Dichlobenil
UN 2770	101	Benzoic derivative pesticide, liquid, n.o.s.
UN 2770	102	Benzoic derivative pesticides, liquid, flammable, toxic, n.o.s.
UN 2771	101	Dithiocarbamate pesticide, liquid, n.o.s.
UN 2771	101	Dithiocarbamate pesticide, solid, n.o.s.
UN 2771	102	Dithiocarbamate pesticides, solid, toxic, n.o.s.
NA 2771	101	Thiram
UN 2772	101	Dithiocarbamate pesticide, liquid, n.o.s.
UN 2772	102	Dithiocarbamate pesticides, liquid, flammable, toxic, n.o.s.
UN 2773	101	Phthalimide derivative pesticide, liquid, n.o.s.
UN 2773	101	Phthalimide derivative pesticide, solid, n.o.s.
UN 2773	102	Phthalimide derivative pesticides, solid, toxic, n.o.s.
UN 2774	101	Phthalimide derivative pesticide, liquid, n.o.s.
UN 2774	102	Phthalimide derivative pesticides, liquid, flammable, toxic, n.o.s.
UN 2775	101	Copper based pesticide, liquid, n.o.s.
UN 2775	101	Copper based pesticide, solid, n.o.s.
UN 2775	102	Copper based pesticides, solid, toxic, n.o.s.
UN 2776	101	Copper based pesticide, liquid, n.o.s.
UN 2776	102	Copper based pesticides, liquid, flammable, toxic, n.o.s.
UN 2777	101	Mercury based pesticide, liquid, n.o.s.
UN 2777	101	Mercury based pesticide, solid, n.o.s.
UN 2777	102	Mercury based pesticides, solid, toxic, n.o.s.
UN 2778	101	Mercury based pesticide, liquid, n.o.s.
UN 2778	102	Mercury based pesticides, liquid, flammable, toxic, n.o.s.
UN 2779	101	Substituted nitrophenol pesticide, liquid, n.o.s.
UN 2779	101	Substituted nitrophenol pesticide, solid, n.o.s.
UN 2779	102	Substituted nitrophenol pesticides, solid, toxic, n.o.s.
UN 2780	101	Substituted nitrophenol pesticide, liquid, n.o.s.
UN 2780	102	Substituted nitrophenol pesticides, liquid, flammable, toxic, n.o.s.
UN 2781	101	Bipyridilium pesticide, liquid, n.o.s.
UN 2781	101	Bipyridilium pesticide, solid, n.o.s.
UN 2781	102	Bipyridilium pesticides, solid, toxic, n.o.s.
NA 2781	101	Diquat
UN 2782	101	Bipyridilium pesticide, liquid, n.o.s.
UN 2782	102	Bipyridilium pesticides, liquid, flammable, toxic, n.o.s.
NA 2783	101	Azinphos methyl
NA 2783	101	Azinphos methyl mixture, liquid
NA 2783	101	Chlorpyrifos
NA 2783	101	Coumaphos
NA 2783	101	Coumaphos mixture, liquid
NA 2783	101	Diazinon
NA 2783	101	Dichlorvos
NA 2783	101	Dichlorvos mixture, dry
NA 2783	101	Disulfoton
NA 2783	101	Disulfoton mixture, dry
NA 2783	101	Disulfoton mixture, liquid
NA 2783	101	Ethion
NA 2783	101	Ethion mixture, dry
NA 2783	101	Hexaethyl tetraphosphate mixture, dry
NA 2783	101	Hexaethyl tetraphosphate mixture, liquid
UN 2783	101	Hexaethyl tetraphosphate mixture, liquid
NA 2783	101	Malathion
NA 2783	101	Methyl parathion, liquid
NA 2783	101	Methyl parathion mixture, dry
NA 2783	101	Methyl parathion mixture, liquid
NA 2783	101	Methyl parathion mixture, liquid, (containing 25% or less methyl parathion)
NA 2783	101	Mevinphos
NA 2783	101	Mevinphos mixture, dry
NA 2783	101	Mevinphos mixture, liquid
UN 2783	101	Mipafox
NA 2783	101	Naled
NA 2783	101	Organic phosphate mixture, Organic phosphate compound mixture, *or* Organic phosphorus compound mixture; liquid
NA 2783	101	Organic phosphate mixture, Organic phosphate compound mixture, *or* Organic phosphorus compound mixture; solid *or* dry
NA 2783	101	Organic phosphate, Organic phosphate compound, *or* Organic phosphorus compound; liquid
NA 2783	101	Organic phosphate, Organic phosphate compound, *or* Organic phosphorus compound; solid *or* dry
UN 2783	101	Organophosphorus pesticide, liquid, n.o.s.
UN 2783	101	Organophosphorus pesticide, solid, n.o.s.
UN 2783	102	Organophosphorus pesticides solid, toxic, n.o.s.
NA 2783	101	Parathion, liquid
NA 2783	101	Parathion mixture, dry
NA 2783	101	Parathion mixture, liquid
NA 2783	101	Phencapton
NA 2783	101	Tetraethyl pyrophosphate, liquid
NA 2783	101	Tetraethyl pyrophosphate mixture, dry
NA 2783	101	Tetraethyl pyrophosphate mixture, liquid
NA 2783	101	Trichlorfon
UN 2784	101	Organophosphorus pesticide, liquid, n.o.s.
UN 2784	102	Organophosphorus pesticides, liquid, flammable, toxic, n.o.s.
UN 2785	102	4-Thiapentanal
UN 2786	101	Organotin pesticide, liquid, n.o.s.
UN 2786	101	Organotin pesticide, solid, n.o.s.
UN 2786	102	Organotin pesticides, solid, toxic, n.o.s.
UN 2787	101	Organotin pesticide, liquid, n.o.s.
UN 2787	102	Organotin pesticides, liquid, flammable, toxic, n.o.s.
UN 2788	102	Organotin compounds, n.o.s.
UN 2789	101	Acetic acid, glacial
UN 2789	102	Acetic acid, glacial *or* Acetic acid solution
UN 2790	101	Acetic acid
UN 2790	102	Acetic acid solution
NA 2791	101	Aircraft rocket engine
NA 2791	102	Aircraft thrust device
UN 2792	101	Aircraft rocket engine igniter
UN 2792	102	Igniter for aircraft thrust device
UN 2793	102	Ferrous metal borings, shavings, turnings, *or* cuttings
UN 2793	101	Metal borings, shavings, turnings, *or* cuttings
UN 2794	102	Batteries, wet, filled with acid
NA 2794	101	Battery
UN 2794	101	Battery
UN 2795	102	Batteries, wet, filled with alkali
NA 2795	101	Battery
UN 2795	101	Battery
UN 2796	101	Battery fluid, acid
NA 2796	101	Battery fluid, acid, with battery
NA 2796	101	Battery fluid, acid, with electronic equipment *or* actuating device
UN 2796	102	Sulphuric acid
UN 2797	101	Battery fluid, alkali
UN 2797	101	Battery fluid, alkali, with battery
NA 2797	101	Battery fluid, alkali, with electronic equipment *or* actuating device
UN 2798	101	Benzene phosphorus dichloride
UN 2798	102	Phenyl phosphorus dichloride
UN 2799	101	Benzene phosphorus thiodichloride
UN 2799	102	Phenyl phosphorus thiodichloride
UN 2800	102	Batteries, wet, non-spillable
NA 2801	101	Coal tar dye, liquid
UN 2801	101	Dye intermediate, liquid

(1)—Identification Number	(2)—Source 172.***	(3)—Description
UN 2801	102	Dyes, n.o.s. *or* Dye intermediates, n.o.s.
UN 2802	101	Copper chloride
UN 2803	102	Gallium
UN 2803	101	Gallium metal, liquid
UN 2803	101	Gallium metal, solid
UN 2805	102	Lithium hydride, fused solid
UN 2805	101	Lithium hydride in fused solid form
UN 2806	101	Lithium nitride
UN 2807	101	Magnetized material
UN 2809	102	Mercury
NA 2809	101	Mercury, metallic
NA 2810	101	Arsenious and mercuric iodide solution
NA 2810	101	Compound, tree *or* weed killing, liquid
NA 2810	101	Drugs, liquid, n.o.s.
UN 2810	101	Poisonous liquid, n.o.s. *or* Poison B, liquid, n.o.s.
UN 2810	102	Poisonous liquids, n.o.s.
NA 2811	101	Drugs, solid, n.o.s.
NA 2811	101	Flue dust, poisonous
NA 2811	101	Lead fluoride
NA 2811	101	Lead iodide
NA 2811	101	Lead stearate
UN 2811	101	Poisonous solid, n.o.s. *or* Poison B, solid, n.o.s.
UN 2811	102	Poisonous solids, n.o.s.
NA 2811	101	Selenium oxide
UN 2812	101	Sodium aluminate, solid
NA 2813	101	Lithium acetylide-ethylene diamine complex
UN 2813	102	Substances which, in contact with water, emit flammable gases, n.o.s.
UN 2813	101	Water reactive solid, n.o.s.
NA 2814	101	Etiologic agent, n.o.s.
UN 2814	101	Infectious substance, human, n.o.s.
UN 2815	101	N-Aminoethylpiperazine
UN 2817	101	Ammonium hydrogen fluoride solution
UN 2817	102	Ammonium hydrogen fluoride, solution
UN 2818	101	Ammonium polysulfide solution
UN 2818	102	Ammonium polysulphide, solution
UN 2819	101	Amyl acid phosphate
UN 2820	101	Butyric acid
UN 2820	102	n-Butyric acid
NA 2821	101	Phenol, liquid *or* solution
UN 2821	102	Phenol solutions
UN 2822	102	2-Chloropyridine
UN 2823	101	Crotonic acid
UN 2825	102	N,N-Diisopropyl ethanolamine
UN 2826	101	Ethyl chlorothioformate
UN 2830	101	Lithium ferrosilicon
UN 2831	101	1,1,1-Trichloroethane
UN 2831	102	1,1,1,-Trichloroethane
UN 2834	102	Phosphorous acid, ortho
UN 2835	102	Sodium aluminium hydride
UN 2835	101	Sodium aluminum hydride
UN 2837	101	Sodium hydrogen sulfate solution
UN 2837	102	Sodium hydrogen sulphate, solution
UN 2838	102	Vinyl butyrate, inhibited
UN 2839	102	Aldol
UN 2840	102	Butyraldoxime
UN 2841	102	Di-n-amylamine
UN 2842	102	Nitroethane
UN 2844	102	Calcium manganese silicon
UN 2845	101	Pyrophoric liquid, n.o.s. *or* Pyroforic liquid, n.o.s.
UN 2845	102	Pyrophoric liquids, n.o.s.
UN 2846	102	Pyrophoric solids, n.o.s.
UN 2849	102	3-Chloropropanol-1
UN 2850	102	Propylene tetramer
UN 2851	102	Boron trifluoride dihydrate
UN 2852	102	Dipicryl sulphide, wetted
UN 2853	102	Magnesium silicofluoride
UN 2854	101	Ammonium silicofluoride
UN 2855	101	Zinc silicofluoride
UN 2856	102	Silicofluorides, n.o.s.
UN 2857	101	Refrigerating machine
UN 2857	102	Refrigerating machines
UN 2858	102	Zirconium
UN 2859	102	Ammonium metavanadate
UN 2860	102	Vanadium trioxide
UN 2861	102	Ammonium polyvanadate
UN 2862	101	Vanadium pentoxide
UN 2863	102	Sodium ammonium vanadate
UN 2864	102	Potassium metavanadate
UN 2865	102	Hydroxylamine sulphate
UN 2867	101	Ink
UN 2868	101	Resin solution
UN 2869	102	Titanium trichloride mixtures
UN 2870	102	Aluminium borohydride *or* Aluminium borohydride in devices
UN 2871	102	Antimony powder
UN 2872	102	1,2-Dibromo-3-chloropropane
UN 2873	102	N,N-Di-n-butylaminoethanol
UN 2874	102	Furfuryl alcohol
UN 2875	102	Hexachlorophene
UN 2876	101	Resorcinol
UN 2877	102	Thiourea
UN 2878	102	Titanium sponge granules *or* Titanium sponge powders
UN 2879	102	Selenium oxychloride
UN 2880	102	Calcium hypochlorate, hydrated *or* Calcium hypochlorite, hydrated mixtures
UN 2880	101	Calcium hypochlorite, hydrated
UN 2881	102	Nickel catalyst, dry
UN 2883	101	2,2-Di-(tert-butylperoxy)propane
UN 2883	102	2,2-Di-(tert-butylperoxy) propane
UN 2884	101	2,2-Di-(tert-butylperoxy)propane
UN 2884	102	2,2-Di-(tert-butylperoxy) propane
UN 2885	101	1,1-Di-(tert-butylperoxy)cyclohexane
UN 2885	102	1,1-Di-(tert-butylperoxy) cyclohexane
UN 2886	102	tert-Butyl peroxy-2-ethylhexanoate
UN 2886	101	tert-Butyl peroxy-2-ethylhexanoate, with 2,2-Di-(tert-butylperoxy)butane
UN 2887	101	tert-Butyl peroxy-2-ethylhexanoate
UN 2888	101	tert-Butyl peroxy-2-ethylhexanoate
UN 2889	102	Diisotridecyl peroxydicarbonate
UN 2889	101	Diisotridecyl peroxydicarbonate
UN 2890	101	tert-Butyl peroxybenzoate
UN 2891	102	tert-Amyl peroxyneodecanoate
UN 2891	101	tert- Amyl peroxyneodecanoate
UN 2892	101	Dimyristyl peroxydicarbonate
UN 2892	102	Dimyristylperoxydicarbonate
UN 2893	102	Dilauroyl peroxide
UN 2893	101	Lauroyl peroxide
UN 2894	101	Di-(4-tert-butylcyclohexyl)peroxydicarbonate
UN 2894	102	Di-(4-tert-butylcyclohexyl) peroxydicarbonate
UN 2895	101	Dicetyl peroxydicarbonate
UN 2895	102	Dicetyl peroxydicarbonate
UN 2896	101	Cyclohexanone peroxide
UN 2896	102	Cyclohexanone peroxides
UN 2897	101	1,1-Di-(tert-butylperoxy)cyclohexane
UN 2897	102	1,1-Di-(tert-butylperoxy) cyclohexane
UN 2898	101	tert-Amyl peroxy-2-ethylhexanoate
UN 2899	102	Organic peroxides, n.o.s., trial quantities
UN 2899	101	Organic peroxide, trial quantity, n.o.s.
UN 2901	102	Bromine chloride
NA 2902	101	Allethrin
NA 2902	101	Insecticide, liquid, n.o.s.
UN 2902	102	Pesticides, liquid, toxic, n.o.s.
UN 2903	102	Pesticides, liquid, toxic, flammable, n.o.s.
UN 2904	102	Chlorophenates, liquid
UN 2905	102	Chlorophenates, solid
UN 2906	102	Triisocyanatoisocyanurate of isophoronediisocyanate, solution
UN 2907	102	Isosorbide dinitrate mixture
UN 2908	101	Radioactive material, empty packages
UN 2909	101	Radioactive material, articles, manufactured from natural *or* depleted uranium *or* natural thorium
UN 2910	101	Radioactive material, limited quantity, n.o.s.
UN 2911	101	Radioactive material, instruments *and* articles
UN 2912	101	Radioactive material, low specific activity *or* LSA, n.o.s.
UN 2918	101	Radioactive material, fissile, n.o.s.
UN 2920	102	Corrosive liquids, flammable, n.o.s.
UN 2921	102	Corrosive solids, flammable, n.o.s.
UN 2922	101	Corrosive liquid, poisonous, n.o.s.
UN 2922	102	Corrosive liquids, poisonous, n.o.s.
NA 2922	101	Dimethyl chlorothiophosphate
NA 2922	101	Sodium hydrosulfide, solution
UN 2923	102	Corrosive solids, poisonous, n.o.s.
NA 2923	101	Sodium hydrosulfide, solid
NA 2924	101	Dichlorobutene
UN 2924	101	Flammable liquid, corrosive, n.o.s.
UN 2924	102	Flammable liquids, corrosive, n.o.s.
UN 2925	101	Flammable solid, corrosive, n.o.s.
UN 2925	102	Flammable solids, corrosive, n.o.s.
UN 2926	101	Flammable solid, poisonous, n.o.s.
UN 2926	102	Flammable solids, poisonous, n.o.s.
UN 2927	102	Poisonous liquids, corrosive, n.o.s.
UN 2928	101	Poisonous solid, corrosive, n.o.s.
UN 2928	102	Poisonous solids, corrosive, n.o.s.
NA 2929	101	Chloropicrin mixture, flammable
UN 2929	102	Poisonous liquids, flammable, n.o.s.
UN 2930	102	Poisonous solids, flammable, n.o.s.
UN 2931	102	Vanadyl sulphate
UN 2933	102	Methyl-2-chloropropionate
UN 2934	102	Isopropyl-2-chloropropionate
UN 2935	102	Ethyl-2-chloropropionate
UN 2936	102	Thiolactic acid
UN 2937	102	alpha-Methylbenzyl alcohol
UN 2938	102	Methylbenzoate
UN 2940	102	9-Phosphabicyclononanes
UN 2941	102	2-Fluoroaniline
UN 2942	102	2-Trifluoromethyl aniline
UN 2943	102	Tetrahydrofurfurylamine
UN 2944	102	4-Fluoroaniline
UN 2945	102	N-Methylbutylamine
UN 2946	102	2-Amino-5-diethylaminopentane
UN 2947	102	Isopropyl chloroacetate
UN 2948	102	3-Trifluoromethyl aniline
UN 2949	102	Sodium hydrosulphide
UN 2950	101	Magnesium granules coated
UN 2950	102	Magnesium granules, coated
UN 2952	102	Azodiisobutyronitrile
UN 2953	102	2,2'-Azodi-(2,4-dimethylvaleronitrile)
UN 2954	102	Azodi-(1,1'-hexahydrobenzonitrile)
UN 2955	102	2,2'-Azodi-(2,4-dimethyl-4-methoxyvaleronitrile)
UN 2957	102	tert-Amyl peroxypivalate
UN 2958	102	Diperoxyazelaic acid
UN 2959	102	2,5-Dimethyl-2,5-di-(benzoylperoxy) hexane
UN 2960	102	Di-(2-ethylhexyl) peroxydicarbonate
UN 2961	102	2,4,4-Trimethylpentyl-2-peroxy phenoxy acetate
UN 2962	102	Disuccinic acid peroxide
UN 2963	102	Cumyl peroxyneodecanoate
UN 2964	102	Cumyl peroxypivalate
UN 2965	102	Boron trifluoride dimethyl etherate
UN 2966	102	Thioglycol
UN 2967	102	Sulphamic acid
UN 2968	102	Maneb, *or* Maneb preparation(s)
UN 2969	102	Castor beans, Castor meal, Castor pomace *or* Castor flake
UN 2974	101	Radioactive material, special form, n.o.s.
UN 2975	101	Thorium metal, pyrophoric
UN 2976	101	Thorium nitrate
UN 2977	101	Uranium hexafluoride, fissile
UN 2978	101	Uranium hexafluoride, low specific activity
UN 2979	101	Uranium metal, pyrophoric
UN 2980	101	Uranyl nitrate hexahydrate solution
UN 2981	101	Uranyl nitrate, solid
UN 2982	101	Radioactive material, n.o.s.
UN 2983	102	Ethylene oxide and propylene oxide mixtures
UN 2984	102	Hydrogen peroxide, aqueous solutions
UN 2989	102	Lead phosphite dibasic
UN 2991	102	Carbamate pesticides, liquid, toxic, flammable, n.o.s.
UN 2992	102	Carbamate pesticides, liquid, toxic, n.o.s.
UN 2993	102	Arsenical pesticides, liquid, toxic, flammable, n.o.s.
UN 2994	102	Arsenical pesticides, liquid, toxic, n.o.s.
UN 2995	102	Organochlorine pesticides, liquid, toxic, flammable, n.o.s.
UN 2996	102	Organochlorine pesticides, liquid, toxic, n.o.s.

(1)—Identification Number	(2)—Source 172.***	(3)—Description
UN 2997	102	Triazine pesticides, liquid, toxic, flammable, n.o.s.
UN 2998	102	Triazine pesticides, liquid, toxic, n.o.s.
UN 2999	102	Phenoxy pesticides, liquid, toxic, flammable, n.o.s.
UN 3000	102	Phenoxy pesticides, liquid, toxic, n.o.s.
UN 3001	102	Phenyl urea pesticides, liquid, toxic, flammable, n.o.s.
UN 3002	102	Phenyl urea pesticides, liquid, toxic, n.o.s.
UN 3003	102	Benzoic derivative pesticides, liquid, toxic, flammable, n.o.s.
UN 3004	102	Benzoic derivative pesticides liquid, toxic, n.o.s.
UN 3005	102	Dithiocarbamate pesticides, liquid, toxic, flammable, n.o.s.
UN 3006	102	Dithiocarbamate pesticides, liquid, toxic, n.o.s.
UN 3007	102	Phthalimide derivative pesticides, liquid, toxic, flammable, n.o.s.
UN 3008	102	Phthalimide derivative pesticides, liquid, toxic, n.o.s.
UN 3009	102	Copper based pesticides, liquid, toxic, flammable, n.o.s.
UN 3010	102	Copper based pesticides, liquid, toxic, n.o.s.
UN 3011	102	Mercury based pesticides, liquid, toxic, flammable, n.o.s.
UN 3012	102	Mercury based pesticides, liquid, toxic, n.o.s.
UN 3013	102	Substituted nitrophenol pesticides, liquid, toxic, flammable, n.o.s.
UN 3014	102	Substituted nitrophenol pesticides, liquid, toxic, n.o.s.
UN 3015	102	Bipyridilium pesticides, liquid, toxic, flammable, n.o.s.
UN 3016	102	Bipyridilium pesticides, liquid, toxic, n.o.s.
UN 3017	102	Organophosphorus pesticides, liquid, toxic, flammable, n.o.s.
UN 3018	102	Organophosphorus pesticides, liquid, toxic, n.o.s.
UN 3019	102	Organotin pesticides, liquid, toxic, flammable, n.o.s.
UN 3020	102	Organotin pesticides, liquid, toxic, n.o.s.
UN 3021	102	Pesticides, liquid, flammable, toxic, n.o.s.
NA 9011	101	Camphene
NA 9018	101	Dichlorodifluoroethylene
NA 9026	101	Dinitrocyclohexylphenol
NA 9035	101	Gas identification set
NA 9037	101	Hexachloroethane
NA 9053	101	Oiled material
NA 9069	101	Tetramethylmethylenediamine
NA 9077	101	Adipic acid
NA 9078	101	Aluminum sulfate, solid
NA 9079	101	Ammonium acetate
NA 9080	101	Ammonium benzoate
NA 9081	101	Ammonium bicarbonate
NA 9083	101	Ammonium carbamate
NA 9084	101	Ammonium carbonate
NA 9085	101	Ammonium chloride
NA 9086	101	Ammonium chromate
NA 9087	101	Ammonium citrate, dibasic
NA 9088	101	Ammonium fluoborate
NA 9089	101	Ammonium sulfamate
NA 9090	101	Ammonium sulfite
NA 9091	101	Ammonium tartrate
NA 9092	101	Ammonium thiocyanate
NA 9093	101	Ammonium thiosulfate
NA 9094	101	Benzoic acid
NA 9095	101	n-Butyl phthalate
NA 9096	101	Calcium chromate
NA 9097	101	Calcium dodecylbenzenesulfonate
NA 9099	101	Captan
NA 9100	101	Chromic sulfate
NA 9101	101	Chromic acetate
NA 9102	101	Chromous chloride
NA 9103	101	Cobaltous bromide
NA 9104	101	Cobaltous formate
NA 9105	101	Cobaltous sulfamate
NA 9106	101	Cupric acetate
NA 9109	101	Cupric sulfate
NA 9110	101	Cupric sulfate, ammoniated
NA 9111	101	Cupric tartrate
NA 9117	101	Ethylenediaminetetraacetic acid
NA 9118	101	Ferric ammonium citrate
NA 9119	101	Ferric ammonium oxalate
NA 9120	101	Ferric fluoride
NA 9121	101	Ferric sulfate
NA 9122	101	Ferrous ammonium sulfate
NA 9125	101	Ferrous sulfate
NA 9126	101	Fumaric acid
NA 9127	101	Isopropanolamine dodecylbenzenesulfonate
NA 9134	101	Lithium chromate
NA 9137	101	Naphthenic acid
NA 9138	101	Nickel ammonium sulfate
NA 9139	101	Nickel chloride
NA 9140	101	Nickel hydroxide
NA 9141	101	Nickel sulfate
NA 9142	101	Potassium chromate
NA 9145	101	Sodium chromate
NA 9146	101	Sodium dodecylbenzenesulfonate
NA 9147	101	Sodium phosphate, dibasic
NA 9148	101	Sodium phosphate, tribasic
NA 9149	101	Strontium chromate
NA 9151	101	Triethanolamine dodycylbenzenesulfonate
NA 9152	101	Vanadyl sulfate
NA 9153	101	Zinc acetate
NA 9154	101	Zinc ammonium chloride
NA 9155	101	Zinc borate
NA 9156	101	Zinc bromide
NA 9157	101	Zinc carbonate
NA 9158	101	Zinc fluoride
NA 9159	101	Zinc formate
NA 9160	101	Zinc phenolsulfonate
NA 9161	101	Zinc sulfate
NA 9162	101	Zirconium potassium fluoride
NA 9163	101	Zirconium sulfate
NA 9180	101	Uranyl acetate
NA 9183	101	Organic peroxide, liquid *or* solution, n.o.s.
NA 9184	101	Pyrethrins
NA 9187	101	Organic peroxide, solid, n.o.s.
NA 9188	101	Hazardous substance, liquid *or* solid, n.o.s.
NA 9189	101	Hazardous waste, liquid *or* solid, n.o.s.
NA 9190	101	Ammonium permanganate
NA 9191	101	Chlorine dioxide hydrate, frozen
NA 9193	101	Oxidizer, corrosive, liquid, n.o.s.
NA 9194	101	Oxidizer, corrosive, solid, n.o.s.
NA 9195	101	Metal alkyl, solution, n.o.s.
NA 9199	101	Oxidizer, poisonous, liquid, n.o.s.
NA 9200	101	Oxidizer, poisonous, solid, n.o.s.
NA 9201	101	Antimony trioxide
NA 9202	101	Carbon monoxide, cryogenic liquid
NA 9206	101	Methyl phosphonic dichloride

Shipping Paper And Manifest Requirements

General

1. A "shipping paper" is a document used to identify the freight being offered for transportation. This term covers any shipping order, bill of lading, waybill, manifest or other document serving a similar purpose. Except for the few exceptions outlined below, each shipment of hazardous material must be accompanied by a shipping paper. *(49 CFR, 171.8)*

2. A "manifest" is a shipping paper prepared on a prescribed form on which all hazardous wastes must be identified. The original manifest is required to accompany each shipment from the point of pickup to its final destination and appropriate copies of the document must be properly distributed. Copies are provided for the generator, transporter and treatment, storage or disposal (TSD) facility.

3. It is the responsibility of the shipper to properly prepare the shipping paper tendered with the shipment of hazardous materials to the initial carrier. *(49 CFR, 172.200)*. The carrier, in turn, must be certain that the shipping paper is properly prepared prior to accepting the shipment. *(49 CFR, 177.817)*. This shipping paper is required to accompany each shipment of hazardous materials during transportation. This requirement may be met by a photocopy of the shipping paper, or by a carrier prepared shipping paper such as a freight bill or waybill. For shipments of hazardous waste the original shipper prepared manifest is the only authorized documentation. *(40 CFR, 262.23)*

Hazardous Materials Description

1. The basic description of a hazardous material on a shipping paper must contain the following information: (Sections (a), (b) and (c) below must be presented in sequence) *(49 CFR, 172.202)*

a) The proper shipping name as presented in Column (2) of the Hazardous Material Table. *NOTE 1: Words shown in italics are not part of the shipping name, however their optional use to further describe the material, in addition to the proper shipping name, is permissible. NOTE 2: Qualifying words used in the description need not be shown in the order stated except when the material is a hazardous waste.*

b) The hazard class as shown in Column (3) of the Hazardous Material Table. *NOTE 1: Except for a proper shipping name that contains words describing more than one hazard class, inclusion of the hazard class is not required when the words of the proper shipping name contain the key word or words of the hazard class of the material, however for the sake of consistency it is recommended that they both be shown; e.g. "Radioactive Material, n.o.s. Radioactive Material." NOTE 2: The technical and chemical group names may be entered in parentheses between the proper shipping name and hazard class; e.g. "Corrosive liquid, n.o.s. (Caprylyl chloride) Corrosive Material." NOTE 3: If a material meets the definition of more than one hazard class the additional class(es) may be shown after the primary hazard class in the basic description.*

c) The identification number as specified for that material in Column (3A). The appropriate alpha prefex must also be included; e.g. "Pine oil, Combustible liquid, UN1272."

d) Except for empty packagings, the total quantity (by weight or volume) must be shown before or after, or both before and after, the basic description outlined in (a), (b) and (c) above. *NOTE 1: Abbreviations are only authorized when specifying the type of package and the weight or volume; e.g. "40 cyl. Nitrogen, Nonflammable gas, UN1066, 800 lbs." Abbreviations are not authorized for proper shipping name, or hazard class.*

e) A shipping paper must contain an emergency response telephone number as prescribed in Subpart G of Part 172 *(49 CFR 172.201(d))*.

f) If a shipment meets the criteria for a limited quantity exception as referenced in Column 5(a) of the Hazardous Materials Table, the notation, "Limited Quantity" or "Ltd. Qty." must follow the basic description.

g) For shipments of materials identified as Toxic by Inhalation, in containers over 1 liter capacity, the shipping paper description must include the words "Poison - Inhalation Hazard."

h) The shipping paper tendered to the initial carrier by the shipper must include the shipper's certification that the hazardous materials have been properly prepared for transportation and are in compliance with the regulations. See additional information under, "SHIPPER'S CERTIFICATION."

i) The required description of hazardous materials on a shipping paper, and on all copies used in connection with transportation, must be printed (manually or mechanically) in English.

j) A hazardous materials shipping paper may contain additional information not inconsistent with the required information. This information (e.g. a tariff description for rating or billing) must follow all the information required by the hazardous materials regulations.

k) When the description from the Optional Hazardous Materials Table in section 172.102 is used, the words "IMO" or "IMO CLASS" may be entered before or following the hazard class entry.

This Shipping Order must be legibly filled in, in ink, in indelible pencil, or in Carbon, and retained by the Agent.

PAGE 1 OF 1

SHIPMENT IDENTIFICATION NO.

FREIGHT BILL PRO NO CARRIER USE

CARRIER NAME Hazardous Haulers, Ltd.
CARRIER ADDRESS & ZIP 100 Granite Street, Rockville, USA 10000
SCAC DUNS

TO: Consignee Red, White and Blue Hardware, Inc.
On Collect on Delivery shipments the letters COD must appear before consignee's name or as otherwise provided in Item 430 Sec. 1
Street 770 Main Street
Destination (Code) Anytown, USA ZIP 20000

TRAILER / CAR NUMBER

ROUTE

FROM: Shipper Du-It-Urself Products
Street 402 Plant Street
Origin (Code) Somewhere, IN 60000 ZIP

SPECIAL INSTRUCTIONS
EMERGENCY RESPONSE CONTACT
John Smith
1-800-444-5555

FOR PAYMENT, SEND BILL TO:
Name
Street
City / State ZIP

SHIPPER'S INTERNAL DATA

SID NO.

NO. SHIPPING UNITS	H/M*	KIND OF PACKAGING, DESCRIPTION OF ARTICLES, SPECIAL MARKS & EXCEPTIONS	WEIGHT (SUBJECT TO CORR.)	RATE	CHARGES
10 cases		Scrapers	50 lbs.		
20 cases	X	Paint, Flammable liquid, UN1263	800 lbs.		
10 cases		Brushes for paint	50 lbs.		
20 drums	X	Acetone, Flammable liquid, UN1090	700 lbs.		
10 cases		Paint, waterbase	400 lbs.		

REMIT C.O.D. TO
ADDRESS

NOTE - Where the rate is dependent on value, shippers are required to state specifically in writing the agreed or declared value of the property.
The agreed or declared value of the property is hereby specifically stated by the shipper to be not exceeding
$ per

COD AMT: $

Subject to Section 7 of conditions, if this shipment is to be delivered to the consignee without recourse on the consignor, the consignor shall sign the following statement:
The carrier shall not make delivery of this shipment without payment of freight and all other lawful charges.

(Signature of Consignor)

C.O.D. FEE PREPAID ☐ COLLECT ☐ $

TOTAL CHARGES $

FREIGHT CHARGES ARE PREPAID UNLESS MARKED COLLECT

CHECK BOX IF COLLECT ☐

RECEIVED, subject to the classifications and tariffs in effect on the date of the issue of this Bill of Lading, the property described above in apparent good order, except as noted (contents and condition of contents of packages unknown), marked, consigned, and destined as indicated above which said carrier (the word carrier being understood throughout this contract as meaning any person or corporation in possession of the property under the contract) agrees to carry to its usual place of delivery at said destination, if on its route, otherwise to deliver to another carrier on the route to said destination. It is mutually agreed as to each carrier of all or any of said property, over all or any portion of said route to destination and as to each party at any time interested in all or any said property, that every service to be performed hereunder shall be subject to all the bill of lading terms and conditions in the governing classification on the date of shipment.
Shipper hereby certifies that he is familiar with all the bill of lading terms and conditions in the governing classification and the said terms and conditions are hereby agreed to by the shipper and accepted for himself and his assigns.

This is to certify that the above named materials are properly classified, described, packaged, marked and labeled and are in proper condition for transportation, according to the applicable regulations of the Department of Transportation.

SHIPPER Du-It-Urself Products
PER

CARRIER Hazardous Haulers, Ltd.
PER DATE

* Mark with "X" or "RQ" if appropriate to designate Hazardous Materials or Hazardous Substances as defined in the department of Transportation Regulations governing the transportation of hazardous materials. The use of this column is an optional method for identifying hazardous materials on bills of lading per Section 172.201 (a) (1) (iii) of Title 49, Code of Federal Regulations. Also, when shipping hazardous materials, the shipper's certification statement prescribed in Section 172.204 (a) of the Federal Regulations must be indicated on the bill of lading, unless a specific exception from this requirement is provided in the Regulations for a particular material.

NOTE: As an alternative to use of an H.M. Column, shippers and carriers are permitted to highlight hazardous materials freight by listing such freight on the shipping paper, waybill, manifest, etc. before listing of non-hazardous freight; or the hazardous listing can be entered in a color of ink that contrasts with other printing. If a carrier utilizes a reproduction of a shipping paper on which hazardous materials are described in a contrasting color, the hazardous materials listing must be highlighted.

Additional Description Requirements *(49 CFR, 172.-203)*

1. Hazardous Substances

a) When the proper shipping name of a mixture of solution that is a hazardous substance does not identify the component(s) making it a hazardous substance the name or names of such component(s) must be entered in parentheses in association with the basic description. For wastes or waste streams exhibiting the EPA characteristics of ignitability, corrosivity, reactivity, or EP toxicity, the identification number of the waste stream, or the notation "EPA' followed by the word "ignitability," "corrosivity," "reactivity," or "EP toxicity," shall be entered in a like manner. (See Appendix to §172.101)

Note: The proper shipping name for a hazardous substance or a hazardous waste that is not listed in the table as a hazardous materials is: "Hazardous Substance liquid or solid, n.o.s., NA9188," or "Hazardous Waste, liquid or solid, n.o.s., NA9189," and the hazard class is "ORM-E."

b) The letters "RQ" (Reportable Quantity) must be entered before or after the basic description, e.g. "RQ, Cresol, Corrosive material, UN 2076,"; or "Hazardous substance, liquid, n.o.s., ORM-E, NA9188 (Adipic acid), RQ." *NOTE: The "RQ" can be substituted for the "X" in the HM Column to designate the existance of a hazardous substance.*

c) "ORM-E" is shown as the hazard class only in conjunction with the proper shipping names "Hazardous substance, liquid *or* solid, n.o.s., NA9188," or "hazardous waste, liquid *or* solid, NA9189." With these shipping names, the component(s) making the commodity a hazardous substance must be shown, or the waste or waste stream must be properly identified as described in (a), above.

2. Hazardous Wastes - a hazardous waste manifest, EPA form 8700-22 and 8700-22A, when necessary, must be prepared and hand signed by the generator,

Please print or type. *(Form designed for use on elite (12-pitch) typewriter.)* *Form Approved. OMB No. 2050-0039. Expires 9-30-88*

UNIFORM HAZARDOUS WASTE MANIFEST

1. Generator's US EPA ID No. PAS001387924 Manifest Document No. 00103 | 2. Page 1 of 1 | Information in the shaded areas is not required by Federal law.

3. Generator's Name and Mailing Address: BENTLEY SCIENTIFIC INDUSTRIES, EDGECOMB POND RD., BOLTON, PA. 21640

4. Generator's Phone (212) 664-5225

A. State Manifest Document Number

B. State Generator's ID

5. Transporter 1 Company Name: BOOMER TRANSPORT | 6. US EPA ID Number: NYD001235674 | C. State Transporter's ID | D. Transporter's Phone

7. Transporter 2 Company Name | 8. US EPA ID Number | E. State Transporter's ID | F. Transporter's Phone

9. Designated Facility Name and Site Address: DUELL DISPOSAL SYSTEMS, MIDDLE Rd AT EASY ST., PADANARUM, VA., 21602 | 10. US EPA ID Number: VAD002468640 | G. State Facility's ID | H. Facility's Phone

GENERATOR

11. US DOT Description *(Including Proper Shipping Name, Hazard Class, and ID Number)*	12. Containers No.	Type	13. Total Quantity	14. Unit Wt/Vol	I. Waste No.
a. WASTE ACETONE, FLAMMABLE LIQUID, UN 1090	001	DM	00055	G	F003
b. HAZARDOUS WASTE, LIQUID, N.O.S. (KEPONE), ORM-E NA 9189 RQ	003	DM	00158	G	
c.					
d.					

J. Additional Descriptions for Materials Listed Above | K. Handling Codes for Wastes Listed Above

15. Special Handling Instructions and Additional Information

16. **GENERATOR'S CERTIFICATION:** I hereby declare that the contents of this consignment are fully and accurately described above by proper shipping name and are classified, packed, marked, and labeled, and are in all respects in proper condition for transport by highway according to applicable international and national government regulations.

If I am a large quantity generator, I certify that I have a program in place to reduce the volume and toxicity of waste generated to the degree I have determined to be economically practicable and that I have selected the practicable method of treatment, storage, or disposal currently available to me which minimizes the present and future threat to human health and the environment; **OR**, if I am a small quantity generator, I have made a good faith effort to minimize my waste generation and select the best waste management method that is available to me and that I can afford.

Printed/Typed Name: PETER H. BENTLEY, PRES. | Signature: Peter H Bentley | Month Day Year: 08 21 87

TRANSPORTER

17. Transporter 1 Acknowledgement of Receipt of Materials

Printed/Typed Name: JIM E. GREY | Signature: Jim E Grey | Month Day Year: 8 21 87

18. Transporter 2 Acknowledgement of Receipt of Materials

Printed/Typed Name | Signature | Month Day Year

FACILITY

19. Discrepancy Indication Space

20. Facility Owner or Operator: Certification of receipt of hazardous materials covered by this manifest except as noted in Item 19.

Printed/Typed Name: C. FRANKLIN GATES | Signature: C Frank Gates | Month Day Year: 8 23 87

EPA Form 8700-22 (Rev. 9-86) Previous editions are obsolete. AMERICAN TRUCKING ASSOCIATIONS FORM C1450 received hand signed and carried by the transporter and received, hand signed, and properly recorded by the TSD facility. Copies must be maintained by each of these facilities. In addition to the basic description for all hazardous materials, the waste manifest must also contain the following information: *(49 CFR, 172.205 & 40 CFR Part 262).*

a) A manifest document number;

b) The generator's name, mailing address, telephone number, and EPA identification number;

c) The name and EPA identification number of each transporter;

d) The name, address and EPA identification number of the designated facility and an alternative facility, if any;

e) Type and number of packages;

f) Handwritten signatures of both the shipper's representative and the carrier's representative; *(NOTE: For hazardous materials the signature may be handwritten or printed by typewritter or other mechanical means.)*

g) The generator's certification. See additional information under, "Shipper's Certification."

3. Radioactive Materials - In addition to the basic description required for all hazardous materials, radioactive materials must be further described by including: *(49 CFR, 172.203)*

a) The name of each radionuclide in the radioactive material as listed in section 173.435. The abreviations shown for the radionuclide are authorized and may be shown, for example, as ^{99}Mo or 99_{MO} for Molybdenum as an example.

b) A description of the physical and chemical form; of the materials, if the materials are not in special form.

c) The activity contained in each package of the shipment in terms of curies, millicuries cr microcuries and, for shipments of a highway route controlled quantity of R.A.M. the words "Highway Route Controlled Quantity" must be entered in association with the basic description. *(49 CFR, 173.403)*

d) The category of label applied to each package (e.g. Radioactive White - I);

e) The transport index of each package bearing Radioactive Yellow II or III labels;

f) For fissile radioactive materials:

1) The words "Fissile Exempt" if the package is exempt pursuant to section 173.-453;

2) The fissile class if not exempt pursuant to section 173.455;

3) The additional notation for fissile Class III shipments: "Warning - Fissile Class III Shipment. Do not Load More than (appropriate number) Packages per Vehicle. In loading and storage areas, keep at least 20 feet (6 meters) from other packages bearing Radioactive labels."

g) For a package approved by the U.S. Department of Energy (D.O.E.) or the U.S. Nuclear Regulatory Commission (USNRC) a notation of the package identification marking as prescribed in the approval for that package *(49 CFR, 173.471).*

h) For export shipments or foreign made packages, a notation of the package identification marking as prescribed in the applicable International Atomic Energy Agency (IAEA) Certificate of Competent Authority *(49 CFR, 173.473).*

NOTE: For additional information regarding the handling of radioactive materials shipments refer to the special section on Radioactive Materials on page 185.

4. Special Requirements *(49 CFR, 172.203)*

a) Dangerous When Wet - When the "Dangerous When Wet" label is applied to a package this notation must be entered on the shipping paper in association with the basic description. *(49 CFR, 172.203(j))*

b) N.O.S. Entries and other Generic Descriptions - Whenever a hazardous material is described on a shipping paper:

1) Where the proper shipping name is an n.o.s. entry as shown in 49 CFR 172.203(k)(3);

2) Where the hazardous material is a poisonous material subject to the requirements of 49 CFR 172.203(m); or

3) Where the hazardous material is a mixture or solution of two or more hazardous materials the name or names of the constituent(s) of those hazardous materials which give them the hazardous properties must be entered in parentheses in conjunction with the basic description. For example:

1) Corrosive liquid, n.o.s. (Caprylyl Chloride), UN1760

or

Corrosive liquid, n.o.s., UN1760 (contains Caprylyl Chloride)

2) Motor fuel antiknock compound (Tetraethyl Lead), Poison B, UN1649

or

Motor fuel antiknock compound, Poison B, UN1649 (Tetraethyl Lead)

3) Flammable liquid, corrosive, n.o.s. (contains Methanol, Potassium hydroxide), Flammable liquid, UN 2924

c) Poisonous Materials - Regardless of the class to which a material is assigned, if the name of the compound or principal constituent that causes the material to meet the definition of a poison is not included in the proper shipping name for the material, the name of that compound or constituent must be entered on the shipping paper in association with the basic description. *(49 CFR 172.203(m))*

When a material is a poison and it is not evident in the shipping name or class entry, the word "Poison" must be entered in the description. *(49 CFR 172.203(m))*

Highlighting Hazardous Materials

1. When a shipping paper covers both hazardous materials and non-hazardous materials, the regulations require that the hazardous materials be highlighted by one of the following means *(49 CFR, 172.201)*.

a) The hazardous materials shall be described first on the shipping paper.

b) The hazardous material description may be in a color that must clearly contrast with any description of a non-hazardous material. *NOTE: On a reproduction of a shipping paper, the entry for a hazardous material may be marked with a highlighter pen in lieu of being in a contrasting color as required for the original.*

c) The entry for a hazardous material may be noted by an "X" (or "RQ" for hazardous substances) placed in a vertical column on the shipping paper entitled, "HM", appearing before the proper shipping name.

2. The requirements for highlighting hazardous materials by one of the above options apply to original shipping papers prepared by the shipper, and to shipping papers prepared by the carrier.

Shipper's Certification

1. When a shipment is offered to the initial carrier, it must bear the prescribed shipper's certification that it is in compliance with the hazardous materials regulations *(49 CFR, 172.204)*:

"This is to certify that the above - named materials are properly classified, described, packaged, marked and labeled and are in proper condition for transportation according to the applicable regulations of the Department of Transportation."

2. For shipments of hazardous waste the generator's certification on the manifest must read as follows:

"I hereby declare that the contents of this consignment are fully and accurately described above by proper shipping names and are classified, packed, marked and labeled, and are in all respects in proper condition for transport by highway according to applicable international and national governmental regulations."

3. These certifications must be printed (manually or mechanically), and must be signed by an authorized representative of the shipper. This signature may be shown either manually, by typewriter, or by other mechanical means. *NOTE: For hazardous waste shipments the signature must be handwritten.*

4. For shipments moving by air the following alternative shipper's certification is authorized and may be encountered by motor carrier personnel handling such shipments. This certification is acceptable for motor carrier shipments and shipments moving by motor carrier and air carrier.

"I hereby certify that the contents of this consignment are fully and accurately described above by proper shipping name and are classified, packed, marked and labeled, and in proper condition for carriage by air according to applicable national governmental regulations."

On any shipment moving by air, the following additional certification must appear:

"This shipment is within the limitations prescribed for passenger aircraft cargo-only aircraft (delete non-applicable)."

5. Exception from shipper's certification - No Shipper's certification is requried for the highway shipment of a hazardous materials when:

a) Transported in a cargo tank furnished by the carrier.

b) By the shipper as a private carrier, unless the shipment is to be reshipped or transferred to another carrier.

Additional Notations Required on Shipping Papers in Special Cases *(49 CFR, 172.203)*

1. Exemptions - For a shipment of hazardous materials made under an "Exemption" the shipping paper must bear the notation "DOT-E" followed by the number issued for the exemption. The notation must clearly refer to the material covered by the exemption. *(49 CFR 172.203(a))*

2. Empty packagings - Except for hazardous substances, the shipping paper for an empty packaging containing the residue of a hazardous material may show the word "Empty", or the notation, "Empty - Last Contained..." before or after the basic description. *(49 CFR 172.203(e))*

3. Anhydrous Ammonia - The shipping paper must contain the notation "0.2 percent water" if the material is suitable for shipment in tanks of quenched and tempered steel. If the commodity does not contain 0.2 percent water, the notation, "Not for Q and T Tanks" must be shown. *(49 CFR 172.203(h)(1))*

4. Liquified Petroleum Gas - The word "Non-Corrosive" or the abbreviation "Non-Cor" must be on the shipping paper if the gas can be transported in a tank of quenched and tempered steel. *(See 49CFR 173.315(a)(a) NOTE 15)* If the gas is other than non-corrosive the words "NOT FOR Q AND T TANKS" must be entered. *(49 CFR 172.203(h)(2))*

5. Import/Export Shipments - Except for Class A and Class B explosives and radioactive materials, when a hazardous material is being imported into or exported from the U.S. it can be described, classified and labeled according to the international standards. However, international shipments of hazardous materials must conform to the requirements for emergency response information as prescribed in Subpart G of Part 172. When a material is transported under the International Maritime Organization (IMO) Class shown in the Optional Table (Column 3 of 172.202) the DOT Class from Part 173 that most closely corresponds to that IMO Class must also be shown; e.g. "Ethylene oxide, 2.1, Flammable gas, UN1040." *NOTE: The IMDG classes and their corresponding DOT classes are shown in the Introduction to the Optional Table at page 102 of this manual.*

6. Shipment by Rail - When a vehicle containing hazardous materials is delivered for rail transportation, the motor carrier's shipping paper must contain the notation "Placarded" followed by the name of the placard required for the rail car. *(49 CFR 172.203(g))*

7. Shipments by Air - When a hazardous material is prohibited from transportation aboard passenger - carrying aircraft, the words "Cargo-only aircraft" must be entered after the basic description. *(49 CFR 172.203(f))*

8. Shipments by Water - Employees of motor carriers who handle shipments by water should expect to see the following notations on shipping papers which are required for the water transportation:

a) The identification of the package type (barrels, drums, cylinders, boxes, etc.)

b) The number of each type of package, including those on a pallet, or in a container.

c) The gross weight of each type of package, or the weight of each package. *(49 CFR 172.203(i))*

9. Additional Documentation for Class A and B Explosives - In addition to properly prepared shipping papers, the Federal Motor Carrier Safety Regulations require that the driver of a vehicle transporting explosives be furnished with the following additional information which the driver must sign for and have with him during transportation: *(49 CFR, 397.19)*

a) A copy of Part 397, Federal Motor Carrier Safety Regulations;

b) A written route plan. This may be prepared by the driver if he/she is dispatched from a point other than a company terminal;

c) Information on the nature of the load;

d) Information on action to be taken in the event of an emergency or delay;

e) Shipper and carrier phone numbers to be contacted in the event of an emergency or delay.

Shipping Paper Accessibility *(49 CFR, 177.817)*

1. During transportation, shipping papers for hazardous materials must be accessible in case of accident, or for inspection by the authorities as follows:

a) Shipping papers for hazardous materials may be carried separately from all other shipping papers utilizing a Hazardous Materials envelope.

b) If hazardous materials shipping papers are carried with other shipping papers, they must be made conspicuous by appearing first, or by being distinctively tabbed. Markers used for tabbing may be of a distinct color and may also show the class of hazardous material involved.

c) When the driver is at the wheel, the hazardous materials shipping papers must be within his reach when restrained by a seatbelt, readily visible to a person entering the cab, or in a container affixed to the inside of the driver's door.

d) When the driver is not at the wheel, the hazardous materials shipping papers must be left on the driver's seat, or they must be in a container on the driver's door. *(NOTE: Manifests for hazardous waste shipments must be handled in a similar fashion.)*

Company Processing of Shipping Papers

1. As shipping papers are received, the rate clerk should review them for hazardous materials listings. A general review should be made to determine that the hazardous materials are properly highlighted and described. This should include a check against the Hazardous Materials Table to assure that the spelling and classification are correct.

2. When preparing freightbills, waybills, etc., the shipper's certification need not be carried forward. All other applicable descriptive information and notations described in this chapter must be entered.

3. Shipping documents for hazardous materials should be given to the dispatcher for special handling.

4. Because of the regulatory requirement that hazardous materials shipping documents be accessible and visible during transportation, the use of a special envelope for these documents is recommended.

PROHIBITED ENTRIES: *(49 CFR, 172.202)* No shipping paper entry shall identify a material as hazardous by hazard class or identification number unless the material described is a regulated hazardous material.

MARKINGS

General

It is the responsibility of the shipper to properly mark each package of hazardous materials for transportation. *(49 CFR, 172.300)*. The information on markings presented in this section is intended to provide assistance to motor carrier personnel in recognizing properly marked packages and to provide information on cross-checking of information to determine that packages are properly marked and in compliance.

Special marking requirements for portable tanks and cargo tanks are discussed below.

Information Required on Markings

1. The proper shipping name of the hazardous material as shown in the Hazardous Materials Table or the Optional Hazardous Materials Table (for an international shipment) must be shown on packagings of 110 gallons or less. These packagings must also be marked with the proper hazard identification number of the material as shown in column 3A of the Hazardous Materials Table. The appropriate "UN" or "NA" prefix should be included. The identification number is not required on packages containing Limited Quantities, or materials classed as ORM-D.

2. Each non-bulk package containing a hazardous material subject to the provisions in 49 CFR 172.203(k) - Technical names for n.o.s. and other Generic entries, must be marked with the technical name in parentheses immediately following the proper shipping name as prescribed by the rules and requirements for shipping papers.

3. The name and address of the shipper or the consignee must also be shown unless:

a) The package is transported by highway and will not be transferred from one motor carrier to another;

b) The package is part of a truckload or container load shipment from one shipper to one consignee; or

c) The shipment is being made in a portable tank or cargo tank.

4. On an export shipment moving by water for which the proper shipping name is an "N.O.S." designation, the technical name of the hazardous material must be shown in parentheses following the proper shipping name.

Additional Marking Requirements for Hazardous Substances and Hazardous Wastes in Packages of 110 Gallons or Less

1. If the proper shipping name does not identify the hazardous substance by name, one of the following descriptions shall be shown in parentheses in association with the proper shipping name (See Appendix to §172.101):

a) The name of the hazard substance;
b) For waste streams, the waste stream number;
c) For wastes meeting the EPA characteristics of ignitability, corrosivity, reactivity, or EP toxicity, the letters "EPA" followed by, "ignitability," "corrosivity," "reactivity," or "EP toxicity" as appropriate.

2. The letters "RQ" must be shown in association with the proper shipping name.

3. EPA regulations (40 CFR 262.32) require that a package of 110 gallons or less containing a hazardous waste be additionally marked:

"HAZARDOUS WASTE - Federal Law Prohibits Improper Disposal. If found, contact the nearest police or public safety authority or the U.S. Environmental Protection Agency.
Generator's Name and Address ____________.
Manifest Document No. ____________.

4. If the proper shipping name of a mixture or solution that is a hazardous substance does not indicate the component(s) that make it a hazardous substance, then the name of that component must be shown in parentheses in association with the proper shipping name. For a hazardous waste, the identification number of the waste or waste stream, or the notation, "EPA" followed by "corrosivity," "EP toxicity," "ignitability," or "reactivity," as appropriate, must be shown in parentheses in association with the shipping name.

5. Every package (except a portable tank or a cargo tank) containing a reportable quantity of a hazardous substance must display the letters "RQ" in association with the proper shipping name of the contents.

Specifications for Markings

1. Required markings must *(49 CFR, 172.304)*:

a) Be durable, in English, and printed on packages or on labels, tags or signs affixed to packages.

b) Be on a background of sharply contrasting color.

c) Be unobscured by labels or attachments.

d) Be located away from any other display, such as advertising, which would reduce its effectiveness.

2. The following abbreviations are authorized for markings *(49 CFR, 172.308)*:

a) The abbreviations "W" for "with" and "W/O" for "without"; e.g. "ammunition for cannons w/o projectiles."

b) The abbreviation "ORM" for "Other Regulated Materials."

3. Packages of liquid hazardous materials must be marked, "This Side Up," or "This End Up" to indicate the upward position of the closure of inside packages during transportation. This requirement does not apply to "Limited Quantity" shipments, cylinders of liquified compressed gas, or certain specified types of containers. The use of arrows to supplement the markings is recommended and the use of arrows for any other purpose is prohibited. It is essential that personnel handling these packagings heed the markings to avoid spills *(49 CFR, 172.312)*.

Other Regulated Materials (ORM)

1. Any package having a rated capacity of 110 gallons or less, containing material classed as ORM - A, B, C, D, or E must be marked with the appropriate ORM designation, enclosed within a rectangular border, following or below the proper shipping name. *(49 CFR, 172.316)*

2. An ORM which meets the definition of a hazardous substance must display the letters "RQ" along with other required markings. If the proper shipping name does not identify the ingredients which make it a hazardous substance the name of the substances must be shown with the shipping name.

3. An ORM - E which is a hazardous waste must be marked with additional markings as prescribed by EPA regulations.

4. Packagings of Hazardous Materials shipped as Consumer Commodity must always be marked with the proper shipping name "Consumer Commodity" and with one of the following markings:

a) ORM-D-AIR if the package meets the packaging requirements for air transportation.

b) ORM-D for packages meeting requirements for highway mode transportation.

Radioactive Materials

In addition to the specific markings outlined in Section 2 of this part, the following additional markings must appear on packages containing Radioactive materials *(49 CFR, 172.310)*:

a) The gross weight must be plainly and durably shown when the package of radioactive materials exceeds 110 gallons (50 kilograms);

b) The words "TYPE A" or "TYPE B" must be shown as appropriate in 1¼ inch letters on the exterior of packages conforming to specifications prescribed by DOT; and

c) The letters "USA" must be marked on packages destined for export.

NOTE: For additional information regarding the handling of radioactive materials shipments refer to the special section on Radioactive Materials at page 185.

Portable Tanks

1. Carrier personnel must be able to recognize the following required markings on portable tanks to minimize the likelihood of accepting shipments that are not in compliance, and to cross-check the information on shipping papers, markings, labels, placards, and identification numbers. Required markings include *(49 CFR, 172.326)*:

a) The name of the tank owner, or lessee, as appropriate;

b) The proper shipping name on two opposing sides in letters no less than two inches high. (The letters "RQ" are not required for hazardous substances transported in packages with a capacity of more than 110 gallons).

c) The specific identification number from column 3A of 172.101 on each end and each side of tanks with a capacity of 1000 gal. or more, or on two opposing sides in association with the proper shipping name for tanks with a capacity of less than 1000 gals., these number must be at least 4 inches high and may be displayed on orange panel or a placard for the designated class.

d) If a portable tank is transported in a vehicle or a container so that the identification numbers cannot be seen, the numbers must be displayed on each end and each side of the vehicle.

e) A portable tank marked for the transportation of one hazardous material cannot be used for the transportation of any other hazardous material, or any non-hazardous material, unless such markings are changed or removed as appropriate.

f) A marked portable tank must remain marked when empty unless filled with a material not subject to regulation, or it is completely cleaned and purged.

Cargo Tanks

1. The following special marking requirements pertain to cargo tanks transporting hazardous materials *(49 CFR, 172.328)*:.

a) A cargo tank must be marked on each side and each end with the identification number marking prescribed in the Hazardous Materials Table for the contents;

b) On a multi-compartment tank, with different hazardous materials in the compartments, the numbers for each of the materials need not be displayed on the ends of the tank if the numbers on each side are displayed in the same sequence as the compartments in which they are transported *(49 CFR, 172.336)*;

c) Numbers need not be displayed on a cargo tank transporting only fuel oil or gasoline if the tank is so marked in letters at least 2 inches high on each side and the rear;

d) If a cargo tank is transporting different distillate fuels, only the number for the fuel having the lowest flashpoint needs to be displayed;

e) If a multi-unit tank-car tank is transported by motor vehicle, identification number markings must be displayed on the front, rear, and both sides of the vehicle;

f) The shipper must provide the necessary identification numbers for the hazardous material being offered on placards or orange panels at the time of pickup, unless the proper numbers are already displayed. If the shipper elects to use the orange panel, he must affix the panel to the vehicle. If a cargo tank is loaded with hazardous materials for later pick-up by the carrier, the shipper must affix the proper identification numbers unless they are already on the tank;

g) On a cargo tank transporting flammable or nonflammable compressed gas, the proper shipping name of the gas, or an appropriate common name for it, must be shown in 2-inch letters in addition to the required identification number marking and the required placarding;

h) On an MC-330 OR MC-331 tank, the marking "QT" or "NQT", as appropriate, must be affixed near the specification plate to indicate whether or not the tank is constructed of quenched and tempered steel; and

i) If a cargo tank is marked with the name, or identification number, of a specific hazardous material, no other material may be transported in the tank unless the markings are removed or changed as appropriate.

Conflicting Markings

1. No other markings on packages may conflict with, obscure, or be subject to being confused with markings required by the regulations. A carrier cannot accept shipments which have conflicting or confusing markings.

Identification Number Markings

1. These markings are required on portable tanks and cargo tanks in addition to any other marking labeling or placarding that may be authorized or required. The identification number must be displayed in proximity to a required placard (or label).

a) Identification numbers are required to be on orange panels, 6¼" by 15¾" with a 9/16" black border. Numerals in 4-inch black Helvetica Medium are specified. The color specification for the orange panel is the same as that required for Explosiove labels and placards. The identification number panel must have the durability specified for placards;

b) On an identication number panel, the name and hazard class of the material may appear in the upper left border in letters not more than ¼" high.

c) Except when POISON GAS, RADIOACTIVE, or EXPLOSIVE placards are required, an identification number may be displayed on a placard in lieu of the number panel. When this alternative display of the identification number is used:

i) The number replaces the word(s) on the placard. If the number is affixed to a placard, the word(s) must be covered as completely as possible.

ii) Numbers in 3 1/2" black Alpine Gothic or Alternate Gothic No. 3 type on a 4" x 8½" white background are specified.

iii) When this Alternative display is used, the United Nations Class Number must be displayed in the lower portion of the placard (See introduction to Optional Hazardous Materials Table for UN Class numbers).

iv) The name and hazard class of the material may be shown in ¼" letters in the upper portion of the space on the placard where the hazard identification numbers are displayed.

v) If the alternative number display is used on a cargo tank containing a combustible liquid, the entire portion of the placard below the number may be white except for the UN Class number.

vi) The display of an identification number on a placard (except POISON GAS, RADIOACTIVE or EXPLOSIVE) is authorized at any time so long as the identification number applies to the hazardous materials on the vehicle.

vii) For hazardous materials for which placards are not authorized or required, identification numbers, alone, may be displayed on a plain white configuration having the same dimensions as a hazardous materials placard.

viii) If more than one of the identification number markings on the placards or orange panels are lost or destroyed during transportation, the carrier must replace all the missing identification numbers as soon as practicable.

Special Marking Requirements For The Transportation of PCB'S

1. Polychlorinated biphenyls (PCB's) are hazardous substances which must be transported in compliance with the hazardous materials regulations of the U.S. Department of Transportation and special additional requirements of the Environmental Protection Agency:

a) Under the hazardous materials regulations, the proper shipping description is, "Hazardous Substance, liquid, n.o.s. (PCB's) NA9188, RQ.," or "Hazardous Substance, solid, n.o.s. (PCB's) NA9188, RQ." which must appear on each package of 110 gallons or less.

b) Regulations of EPA require the package to be marked with the label illustrated below (minimum size 6" x 6").

c) A vehicle transporting a marked shipment of PCB's must be marked on the front, rear, and both sides with the marking illustrated below:

Required Markings For Transportation Of Liquids Which Are Toxic By Inhalation

If the inhalation toxicity of any liquid in a package, with a capacity over 1 liter, meets or exceeds the specific criteria in 49 CFR 173.3a, the package must be marked "Inhalation Hazard," in association with the required Poison label. *(49 CFR, 172.301)*

Labels

The labeling of packagings of hazardous materials is the responsibility of the shipper, and packagings must be properly labeled at the time they are picked up by the motor carrier. Labels provide a vital cross-reference with information on the shipping paper in determining that the shipment is in compliance with the regulations. It is required that motor carrier personnel have a working knowledge of the labels prescribed by the Department of Transportation *(49 CFR, 172.400)*:

1. The labels are a warning to carrier personnel to handle packagings with extra care because of the nature of their contents.

2. They provide guidance as to commodities which, under the regulations, cannot be stored next to each other in the terminal, or loaded into the same vehicle because of the likelihood of a hazardous reaction between the contents of the packagings.

3. They provide guidance as to the placarding that must be applied to vehicles - see "PLACARDING" Section.

4. Carrier personnel must be able to distinguish between those DOT-prescribed labels that apply to motor carrier transportation and those which are important only in other modes or which may be found on packages in compliance with the requirements of other agencies such as the Food and Drug Administration or the Environmental Protection Agency.

5. The carrier must replace lost or damaged labels based on information obtained from the shipping papers. Carriers must keep a supply of labels on hand for this purpose and personnel must be trained to determine the correct replacement label (Labels are available from ATA for this purpose).

7. When information is not available as to the nature of the contents of an unlabeled package of hazardous materials, the "Flammable Liquid" label must be applied.

General Labeling Requirements

1. In general, the prescribed label must be affixed to the packaging near the marked shipping name of the contents *(49 CFR, 172.406)*.

2. Except for packaging of radioactive materials, the required label may be on a securely attached tag if:

a) The packaging is smaller than the required label;

b) The packaging surface is so irregular that a label cannot be affixed to it; or

c) The packaging is a compressed gas cylinder.

3. Hazardous materials labels may not be affixed to packagings unless they contain hazardous materials and the label represents the hazard of the contents. The use of other marking or labels which resemble required labels, which could be confused with them, or which obscure them is prohibited and no carrier may accept such a packaging *(49 CFR, 172.401)*.

4. No label is required for any of the following:

a) A packaging which qualifies for a "Limited Quantity" exception; *(NOTE: A packaging of Class B Poison must always be labeled.)*

b) A compressed gas cylinder (Flammable or Nonflammable Gas), if it is not overpacked and if it bears markings prescribed by the standards of the Compressed Gas Association, when transported by private or contract carrier. This exception does not apply to cylinders transported by a common carrier. These cylinders must be properly labeled;

c) A packaging containing material classed as ORM-A, B, C, D, or E which does not contain any other hazardous material;

d) A packaging containing a combustible liquid;

e) A packaging containing a low specific activity radioactive material in a vehicle designated for the exclusive use of the consignor;

f) A placarded freight container, capacity 640 cubic feet or more;

g) A placarded portable tank;

h) A compressed gas cylinder permanently mounted on a vehicle;

i) A cargo tank;

j) A truckload shipment of military ammunition being transported for the Department of Defense which is loaded and unloaded by DOD or the shipper; or,

k) Any other packaging of hazardous materials transported for DOD, loaded and unloaded by or under the supervision of DOD and escorted by DOD personnel in a separate vehicle.

Multiple Label Requirements

1. When two or more labels are required on a packaging or overpack, the labels must be displayed next to each other *(49 CFR, 172.406)*.

2. When compatible hazardous materials of different hazard classes are in the same packaging, overpack or container, the label for each such class must be shown *(49 CFR, 172.404)*.

3. A material classed as an Explosive A, Poison A or Radioactive Material, that also meets the definition of another hazard class, must be labeled for each class *(49 CFR, 172.402)*.

4. A Poison B liquid that also meets the definition of a flammable liquid must be labeled "POISON" and "FLAMMABLE LIQUID." A Poison B liquid which also meets the definition of a corrosive liquid must be labeled "POISON" and "CORROSIVE."

5. A material which is classed as an oxidizer, corrosive, flammable solid or flammable liquid that also meets the definition of a poison must be labeled "POISON" in addition to the primary class label.

6. A liquid which is identified as TOXIC BY INHALATION, regardless of its primary hazard class, must be labeled "POISON" in addition to any other required label when in packages over 1 liter. It is not necessary to duplicate the poison label for packages already so labeled. *(49 CFR, 172.402)*

7. A flammable solid which also meets the definition of a water reactive material must be labeled, "FLAMMABLE SOLID" and "DANGEROUS WHEN WET." (See special placarding requirements.)

8. A flammable solid that also meets the definition of a corrosive material must be labeled "FLAMMABLE SOLID" and "CORROSIVE".

9. An Oxidizer that meets the definition of a corrosive material must be labeled "OXIDIZER" and "CORROSIVE".

10. A Flammable Liquid that also meets the definition of a corrosive must be labeled "FLAMMABLE LIQUID" and "CORROSIVE".

Determining the Hazard Class of Multiple-Labeled Packagings

1. In handling multiple-labeled packagings, it is necessary to determine the one hazard class of the shipment which will govern the placarding of the vehicle (if required) and which may require compliance with loading and storage restrictions in vehicles and terminals.

2. Check the shipping paper. Both the proper shipping name and the class must be shown. The class shown will govern placarding requirements and compliance with loading and storage restrictions. If further checking is necessary, follow the procedures outlined below.

a) If the commodity is specifically listed by name in the Hazardous Materials Table and packagings are multiple-labeled, the class shown on the shipping paper should agree with Column 3 of the table, and the multiple labeling on packagings should agree with that indicated in Column 4 of the table.

b) If the shipment is described as "n.o.s." the regulations do not require that the hazard class be repeated on the shipping paper, although the practice of repeating the hazard class is recommended. (See Shipping Paper Requirements - Rule 1 (b) Note 1.) To determine the class, or to verify the class shown, refer to the listing of priorities in (c), below, which is the order of hazards prescribed by D.O.T. in Section 173.2.

c) The order of hazards shown below has been prescribed by D.O.T. for the use of shippers in determining the hazard class of materials described as "n.o.s." for which multiple labeling is required. The order of priorities is useful to the motor carrier in verifying the proper classification of multiple-labeled packagings to determine proper placarding of vehicles and the applicability of loading and storage restrictions. In the case of any disagreement, carrier personnel should not reclassify these shipments without consulting the shipper.

1. Radioactive material (except Limited Quantities)
2. Poison A
3. Flammable gas
4. Nonflammable gas
5. Flammable liquid
6. Oxidizer
7. Flammable solid
8. Corrosive material (liquid)
9. Poison B
10. Corrosive material (solid)
11. Irritating material
12. Combustible liquid (containers over 110 gallons)
13. ORM-B
14. ORM-A
15. Combustible liquid (containers 110 gallons or less)
16. ORM-E

3. The hazards presented by the following materials should not be considered in determining priority hazards in Multi-Hazard Materials: a) An explosive which must be classed and approved as required by Section 173.86; b) Etiologic agents; c) Organic peroxides; and d) Blasting agents.

4. The above order of hazards does not apply when two or more compatible materials of different classes are in the same packaging which is labeled for each one. In this situation each material must be considered separately and the vehicle placarded accordingly.

Labeling for Special Situations

Described below, are other situations in which variations in labeling may be seen by motor carrier personnel. The significance in highway transportation of each variation is also outlined *(49 CFR, 172.402).*

1. **Cargo Aircraft Only** - Used in conjunction with a hazard class label indicates that no air shipment can be made on a passenger-carrying aircraft. No significance in highway transportation.

2. **Magnetized Material** - Denotes need for special treatment in air transportation. No significance in highway transportation.

3. **Bung** - Used in conjunction with the "Flammable Liquid" label on drums and barrels in which the flammable liquid is under a specified vapor pressure at 100 degrees F. (37.8 degrees C). Indicates need for special care in handling because of internal pressure of contents.

4. **Etiologic Agents** - Certain microorganisms may cause human disease and are thus considered hazardous materials. Packages containing these materials must bear the "etiologic agents" label. When required by the regulations of another country, a "Poison" label may be used in conjunction with the "Etiologic Agents" label. In this situation, the packagings may not be transported in the same vehicle with foodstuffs or animal feeds.

5. **Dangerous When Wet** - This label may be used in conjunction with any other hazard class label when the use is required by the regulations of another country. If this label is affixed to any packaging also bearing a "Flammable Solid" label, the commodity should be checked through the Hazardous Materials Table to determine whether or not the dual labeling is required by the U.S. regulations. If not required by U.S. regulations for the commodity, it is of no significance in highway transportation.

6. **Empty** - This label may be used to cover the prescribed hazard class label on an empty reusable container. As an alternative, the hazard class label may be removed or obliterated. Containers moving on a shipping paper which indicates empty containers should be checked for the presence of an "Empty" label or the removal of the old label.

7. **Spontaneously Combustible** - This label is used only in conjunction with a hazard class label when required by the regulations of another country. It has no significance in highway transportation.

8. **Oxygen** - A label bearing the word "Oxygen" may be used in lieu of the "Oxidizer" label prescribed in the Hazardous Materials Table on packagings of oxygen.

9. **Chlorine** - A label bearing the word "Chlorine" may be used in lieu of the dual labeling "Poison" and "Nonflammable Gas" prescribed in the Hazardous Materials Table for chlorine.

10. **Use of IMO Labels** - Except for Explosives A and B, labels meeting IMO specifications may be used so long as required wording is in English and the colors conform to U.S. regulations. Explosive A and B labels may bear additional inscriptions if required for export or import. A label may bear inscription required by the country of origin.

11. Packages labeled in conformance with Canadian "Transport Dangerous Goods" (TDG) regulations originating in Canada, may be transported within or through the U.S. to their final destination if all other TDG regulations are followed to the extent of reciprocity.

12. Freight containers containing hazardous materials having a capacity of more than 64 cubic feet and less than 640 cubic feet may either be labeled on two sides or two ends or be placarded with one placard. *(CFR 49, 172.406 and 172.512)*

NOTE: Placarding may not be used in place of labels on a container of radioactive material. Portable tanks of 1000 gallons or less may be labeled on two sides or two ends or placarded on two opposing sides.

Radioactive Materials *(49 CFR, 172.403)*

1. Labeling is not required of shipments of radioactive materials subject to Limited Quantity exception provisions and so noted on the shipping papers.

2. Labeling is not required for certain types of radioactive materials when being transported in a vehicle on an "exclusive use" basis.

3. When labeling is required, there must be two labels on opposite sides of the package (excluding the bottom).

4. A package containing a radioactive material which also meets the definition of another hazardous materials class must be additionally labeled for the other class: e.g. "Radioactive and Corrosive". Such multiple labels must be adjacent to each other.

5. The shipper must insert additional information on labels as specified below for each type of radioactive materials label in the spaces provided:

a) Radioactive White I - The name of the contents and the number of curies.

b) Radioactive Yellow II - The name of the contents, the number of curies and the transport index number.

c) Radioactive Yellow III - The name of the contents, the number of curies, and the transport index numbers.

NOTE 1: Drivers should check labels to be sure that this information has been filled in by the shipper. The information should be rechecked during handling at the terminal. Abbreviations for the name of the contents and the number of curies are authorized.

NOTE 2: For additional information regarding the handling of radioactive materials shipments refer to the special section on Radioactive Materials on page 185.

Label Specifications

The specifications for labels have been greatly expanded with attention given to colors of ink, durability, etc. Carriers intending to print their own labels should consult Section 172.407, and Appendix A. Comparison charts are available from the Materials Transportation Bureau, U.S. Department of Transportation, Washington, D.C. 20590 for labels and placards printed on ink surfaces and for those on metal lacquer, paint, plastic and similar surfaces. Contact the Materials Transportation Bureau for current prices of charts.

Placarding

1. Placarding requirements do not apply to: (a) Etiologic agents; (b) Hazardous materials classed as ORM-A, B, C, D or E; (c) Shipments being transported as "Limited Quantities" when so noted on shipping documents; (d) Radioactive materials bearing White I or Yellow II labels; (e) Shipments bearing only the Spontaneously Combustible label.

2. Placarding is prohibited unless: (a) The vehicle contains a hazardous material; and (b) The placard represents a hazard of the materials being transported.

3. As in the previous regulations, there are some classes of hazardous materials which require placarding of the vehicles when "any quantity" is aboard, and other classes which require placarding only when the aggregate gross weight is 1,000 pounds or more. The use of placards is outlined in the following tables showing: (a) Hazard class; (b) Prescribed label(s); (c) Prescribed placard(s).

TABLE 1 (Placarding Required For Any Quantity)

HAZARD CLASS	LABELS	PLACARD(S)
Class A Explosives	Explosive A	Explosive A
Class B explosives	Explosive B	Explosive B1
Poison A	Poison Gas	Poison Gas
Flammable Solid (Water Reactive Material)	Flammable Solid & Dangerous When Wet	Flammable Solid-W-2
Radioactive Material	Radioactive Material (Yellow III label only)	Radioactive3
Uranium Hexafluoride (fissile)	Radioactive & Corrosive	Radioactive4a & Corrosive4b
Uranium Hexafluoride (low specific activity)	Radioactive & Corrosive	Radioactive4a & Corrosive4b

NOTES

1. *When both Class A and B explosives are transported in a vehicle, only the Explosives A placards need be displayed.*
2. *The Flammable Solid -W- placard must be used for those commodities listed in the Hazardous Materials Table as requiring both the Flammable Solid and Dangerous When Wet labels.*
3. *Radioactive placards are also required for shipments of Low Specific Activity Radioactive Materials (marked "Radioactive - LSA") transported in a vehicle on an "exclusivie use" basis. This requirement does not apply to shipments of unconcentrated uranium or thorium ores.*
4. *The placarding of shipments of Uranium Hexafluoride is governed by the following provisions:*
   a) *The RADIOACTIVE placard is required for any quantity bearing Yellow Radioactive III label; and*
   b) *The CORROSIVE placard must be added if the gross weight of the shipment is 1,000 pounds or more.*

TABLE 2 (Placarding required when 1,000 pounds or more, aggregate gross weight of one or more of the classes below is loaded in a vehicle)

HAZARD CLASS	LABEL(S)	PLACARD(S)
Class C explosives	Explosive C	DANGEROUS 1,2
Blasting Agents	Blasting Agent	BLASTING AGENTS 2
Nonflammable Gas	Nonflammable Gas	NONFLAMMABLE GAS 7
Nonflammable Gas (Chlorine)	Chlorine or Non-Flammable Gas & Poison	CHLORINE or NON-FLAMMABLE GAS 3
Nonflammable Gas (Fluorine)	Poison & Oxidizer	POISON
Nonflammable Gas (Oxygen, cryogenic liquid)	Oxidizer	OXYGEN 4
Flammable Gas	Flammable Gas	FLAMMABLE GAS 7
Combustible Liquid	None	COMBUSTIBLE 5
Flammable Liquid	Flammable Liquid	FLAMMABLE 9
Flammable Solid (other than a water reactive material)	Flammable Solid	FLAMMABLE or FLAMMABLE SOLID 6
Oxidizer	Oxidizer	OXIDIZER 2, 8
Organic Peroxide	Organic Peroxide	ORGANIC PEROXIDE
Poison B	Poison	POISON 9
Corrosive Material	Corrosive	CORROSIVE 9
Irritating Material	Irritant	DANGEROUS

NOTES:

1. *No placarding required unless packages bear Explosive C labels.*
2. *The placards BLASTING AGENTS, OXIDIZER, and DANGEROUS not required on a freight container or a vehicle also transporting Explosives A or B which is placarded EXPLOSIVES A or EXPLOSIVES B as required.*
3. *NONFLAMMABLE GAS placard authorized for packagings of 110 gallons or less. CHLORINE placard required for larger packagings.*
4. *OXYGEN placard required for liquid oxygen. When oxygen is shipped as a gas, packagings must be labeled Oxidizer or Oxygen, and NONFLAMMABLE GAS placards are required.*
5. *COMBUSTIBLE placards only required for containers larger than 110 gallons. On cargo tanks or portable tanks, FLAMMABLE placards may be used in lieu of COMBUSTIBLE.*
6. *FLAMMABLE placards authorized except for transportation by water. FLAMMABLE SOLID placards required for transportation by water.*
7. *NONFLAMMABLE GAS placards not required on a vehicle displaying FLAMMABLE GAS placards.*
8. *Oxidizer placards are not required for highway transportation if the vehicle or freight container contains blasting agents and is placarded BLASTING AGENTS.*
9. *Liquids identified as Toxic By Inhalation must have a POISON Placard in addition to any other placard required for any package in excess of 1 liter, (1,000 lbs. exception does NOT apply).*

4. Use of the DANGEROUS placard for mixed loads - When two or more classes of hazardous materials requiring different placards, as specified in Table 2, are loaded into the same vehicle or freight container, and their aggregate gross weight is 1,000 pounds or more, DANGEROUS placards may be used in lieu of the specific class placards. However, if 5,000 pounds or more of one such class is loaded at one facility, the placard prescribed for that class must be displayed. The presence of smaller quantities of other classes may continue to be indicated by the presence of DANGEROUS placards. This "5,000 pound rule" applies to initial pick-ups, and to consolidations of shipments at local terminals and break-bulk terminals *(49 CFR, 172.504)*.

5. No placard required - No placard is required on a vehicle or freight container loaded with less than 1,000 pounds of one or more classes of hazardous materials listed in Table 2, or for any such vehicle or container loaded onto a rail car.

6. Provision of placards - The shipper is required to provide the placards applicable to the shipment being offered for transportation, unless the vehicle is already equipped with the proper placards. Because this provision of the regulations does not deal with how the vehicle might be equipped with the proper placards, the motor carrier will find that it is not necessarily relieved of the need to equip vehicles with placards. There will be no lessening of the need for the carrier to train drivers to make necessary changes of placards to meet changes in the nature and weight of hazardous materials classes included in any load *(49 CFR, 172.506)*.

7. Display of placards - Placards must be displayed on each end and each side of the vehicle *(49 CFR, 172.516)*.

a) Each placard must be visible from the direction in which it faces, unless the view in that direction is obstructed by another vehicle in a combination.

b) The front-facing placard on a motor vehicle or combination may be on the front of the power unit, the front of the cargo body, or both.

c) In a combination consisting of two or more cargo-carrying vehicles, each vehicle must be placarded for the class(es) of hazardous material(s) it contains.

d) Each placard must be securely attached or placed in a holder; located clear of accessories or appurtenances which would tend to obscure it; be so located as to minimize splash or spray from the vehicle's wheels; be located at least 3 inches from any other marking on the vehicle (further if the marking would reduce the effectiveness of the placard); be maintained so its legibility will not be substantially reduced during transportation; be displayed so that wording reads horizontally from left to right.

8. Moving unplacarded vehicle prohibited - A vehicle containing hazardous materials may not be moved unless properly placarded except under the following conditions *(49 CFR, 177.823)*:

a) The vehicle is escorted by a representative of a state or local government;

b) The carrier has permission from DOT;

c) Movement is necessary to protect life or property.

9. Special provisions for freight containers *(49 CFR, 172.512)*.

a) Requirements for the placarding of vehicles may be met by placarding of freight containers loaded on them.

b) Placarding requirements for freight containers having a volume of 640 cubic feet or more (8'x8'x10') are the same as those outlined above for vehicles.

c) Containers of less than 640 cubic feet may be placarded in lieu of being labeled as described under Special Labeling on page 167.

10. Special provisions for tank vehicles and portable tanks. *(49 CFR, 172.514)*

a) Tank vehicles must be placarded when transporting hazardous materials the same as other vehicles. In addition, those transporting compressed gases must be marked with the proper shipping name, or the common name for the gas. See rule (f) under Marking of Cargo Tanks on page 161.

b) A compartment tank in which different compartments contain hazardous materials of different classes must be placarded front, rear, and both sides to show the class of hazardous materials in each compartment. The use of the DANGEROUS placard for this purpose is not permitted. However, a mixed load of flammable liquid and combustible liquid would require only FLAMMABLE placards.

c) Portable tanks having a capacity of 1,000 gallons or more must be placarded in the manner prescribed for tank vehicles.

d) A portable tank of less than 1,000 gallons capacity need only have placards on two opposite sides.

e) Placards must be left on a cargo tank or portable tank after it has been emptied unless it is reloaded with material not requiring placarding, or it is cleaned and purged so that no hazard exists.

f) In addition to required placarding or labeling, tank vehicles and portable tanks must display identification number markings. See Page 161.

Emergency Response Communication Standards

General

In June, 1990, regulations regarding emergency response information needed to be displayed on shipping papers and packages of hazardous materials are effective. These standards were developed through a cooperative effort on the part of industry and the Federal Research and Special Programs Administration. Even though the mandatory date for compliance is June, 1990, you are permitted to prepare shipments of hazardous materials in accordance with the regulations prior to that date.

These regulations impact the preparation of hazardous materials shipping papers and information needed to accompany them. Primarily, it is the responsibility of the shipper to supply the information required. However, it is still incumbent upon the transporter of the material to make sure the information is present prior to shipment. In the event of a hazardous materials incident, it is imperative that the information required by the standard be made available to emergency response personnel immediately, whenever there is a hazardous material present.

Emergency response information is NOT required for hazardous materials which do not need shipping papers, such as small quantity shipments of hazardous materials.

The Emergency Response Communications Standards can be found in 49 CFR Part 172 Sub-Part G.

Emergency Response Information Required *(49 CFR 172.602)*

The Department of Transportation has identified seven specific areas of information which must accompany shipping papers for hazardous materials. This information must be present for each hazardous material being transported and must be immediately available to emergency responders whenever there is a hazardous material present. Regulations require that information concerning:

1) the proper description of the hazardous material(s);

2) the immediate hazards to health;

3) fire and explosion risks;

4) precautions to be taken immediately;

5) immediate methods for handling large or small fires;

6) initial methods for handling spills or leaks; and

7) preliminary first aid measures

be present. This information is similar to that which is provided in the Department of Transportation's Emergency Response Guidebook or on a manufacturer's Material Safety Data Sheet (MSDS).

The regulations permit several methods of supplying the required information; however, all methods are required to be:

1) printed in English; and

2) available for use away from the package containing the hazardous material.

This information can be printed:

1) directly on the shipping paper; or

2) on another document such as an MSDS or an emergency response guidance manual which cross-references the material to emergency response procedures.

During transportation, regulations guiding the availability of shipping papers *(49 CFR 177.817)* also apply to the emergency response information. At facilities, the emergency response information must be immediately accessible and maintained at the facility whenever a hazardous material is present.

Emergency Response Telephone Number *(49 CFR 172.604)*

Shippers of hazardous materials are required to provide an emergency response telephone number which must be entered onto the shipping papers. This telephone number must appear on the shipping paper:

1) immediately following a hazardous material entry; or

2) in a conspicuous place on the shipping paper, if the one number can be utilized for all hazardous materials listed on the paper.

There must be some indication that this number is to be used for emergency response. For example: EMERGENCY CONTACT: *** ****.

The telephone number provided by the shipper of the hazardous materials must be monitored at all times by:

1) a person offering the material who has knowledge of emergency response procedures for the product; or

2) an agency, such as CHEMTREC, which is capable of, and accepts responsibility for providing emergency response information.

It is the responsibility of the person providing the phone number of an agency to make sure that the agency has the latest emergency response information on file prior to shipment of the product.

Accepting Shipment

Prior to accepting a shipment of hazardous materials from a manufacturer or other customer, make sure that you have the information required by the Emergency Response Communications Standards. Although it is the responsibility of the shipper to supply the information, the carrier is still responsible for not accepting or transporting a shipment of hazardous materials unless that shipment is properly prepared and ready for transport. *(49 CFR 171.2)*

Hazardous Materials Handling Program

This hazardous materials handling program sets forth recommended company rules and procedures which are a combination of state and federal regulations, safe operating practices and general rules for the handling of freight. This will insure compliance with regulations and also the safety and efficiency necessary to control hazards of the dangerous commodity and essential for good customer service.

Any or all portions of this program should be established under the guidance of the carrier safety department after consultation with the traffic, sales and operations departments concerning rules, procedures and means by which the program will be integrated into each phase of the fleet's operations.

Key Person

In each terminal there is need for a key person who has an understanding of the hazardous materials program and who is given the responsibility for assuring that the program in his terminal functions smoothly. To carry out this responsibility he or she will need special instruction on federal and state regulations and company operating procedures as they relate to the transportation of hazardous materials. The key person may be a dispatcher, dock foreman, rate person or hold some other position at the terminal. He or she should be selected on the basis of ability to comprehend regulations and to take positive action when needed.

Duties of the key person will generally be as follows:

1. Make periodic inspections to determine that hazardous materials shipments are being handled properly in each phase of the operations. These inspections will include, but not be limited, to a review of shipping documents for accuracy and completeness, study of dispatch procedures, inspection of containers to determine adequacy and condition, check on vehicle placarding and driver handling of hazardous materials loads.

2. Instruct new employees in their responsibilities for the proper handling of hazardous materials.

3. When possible, meet with representatives of major hazardous materials shippers to determine that they know their responsibilities to comply with regulations and to determine the best handling procedures of the shipper's freight under routine circumstances and in the event of accident, damage or emergency conditions. If there is doubt about adequacy of preparation of shipments for transportation the key person should inspect shipper facilities as authorized by the D.O.T. regulations.

4. See that a record is developed which shows the most frequent shippers, common types of hazardous materials shipments handled, usual types of containers and routine and emergency handling procedures for hazardous materials frequently handled. The record should be available to all personnel handling hazardous materials.

5. Contact shipper or consignee when a hazardous materials shipment is not in good order (damaged containers, shipping papers not properly completed, improper labelling, etc.). The shipper should be advised that shipments in bad order will be returned if they are still at the original terminal, unless damaged containers are in very poor condition in which case the shipper should send a representative to the terminal to handle the shipment in the safest manner.

6. Discuss requests for pickup of hazardous materials with the shipper when there is any question about shipper responsibilities or about the shipment being in good order.

7. See that all required reports are filed on accidents involving hazardous materials shipments and when leaking, damaged or broken containers are found in transit. Federal regulations for completing such reports are provided at page 202. State and local reports required vary. For information regarding state reporting requirements refer to the ATA publication "How and Where".

8. Report to the fleet safety department any instance of unsafe conditions connected with the transportation of hazardous materials so that correc-

tive action can be taken throughout the system. Such conditions might relate to inadequte containers, poor handling procedures, minor and major incidents, offering of incompatible materials by shipper, shipper refusal to comply with regulations.

9. See that refused shipments which cannot be delivered with 48 hours are properly handled. Procedure to follow is:

a) Notify origin terminal to contact shipper for immediate instructions.

b) If shipper provides prompt instruction for delivery or return of shipment, load and dispatch as quickly as possible.

c) If instruction on delivery or return is not given, see that the commodity is placed in suitable storage, and contact fleet safety department for instructions.

d) If commodity is in bad order and presents a hazard, contact fire department as safety measure until instruction is received from fleet safety department. *(Note: Under Federal regulations refused shipments may be stored if suitable storage space is available, or they may be sold or destroyed if necessary for safety but, if possible, no action other than delivery or return should be taken without approval of safety department.)*

Electric storage batteries may be held 30 days pending delivery or disposition but other materials must be acted on promptly.

10. Develop recommendations, based on experience, for changes or addition to hazardous materials program and forward to fleet safety department.

Reference materials should be furnished to each terminal for use by the key person and others. Such material should include a copy of federal regulations governing the transportation of hazardous materials and a copy of federal safety regulations. Publications containing these regulations are available from American Trucking Associations, Inc.

Reference works which describe hazardous commodities, explain their hazards, and show how to cope with them are of value in every terminal. The Department of Transportation's Emergency Response Guidebook is recommended for this purpose. Copies are available through ATA.

Regulations of local and state governments pertinent to storing and transporting hazardous materials may be obtained from police and fire agencies. In areas where bridge, tunnel and turnpike authorities function there may be special regulations that pertain only to use of such facilities and copies of such regulations should be supplied to the terminals.

Instruction of Personnel

Hazardous materials shipments and shipping papers are handled by many different people within the carrier's operation. It is the objective of this program to provide training so such personnel can carry out responsibilities with maximum safety and efficiency.

From the time a request is made for pickup of a hazardous shipment to the time of delivery it will be handled by office workers, dockmen, drivers and dispatchers. Each must be trained on the following points:

- How to identify hazardous shipments
- Routine handling procedures
- Procedures to follow in case of emergency

Personnel responsibilities for handling of freight, shipping papers, customer requests, dispatching and other duties vary from carrier to carrier according to size and nature of operations. Thus the information set forth herein must be edited and grouped to fit the availability of personnel and the organization structure of a particular carrier.

Identification of Hazardous Shipments

The first and most important step in the proper handling of hazardous materials is its identification. This is the basis for carrying out all other phases of the program.

Identification will vary from shipment to shipment according to packaging, marking, labelling and information provided on shipping papers. Means of identifying shipments are:

1. Type of container: Containers such as cylinders, battery cases and drums are an indication that a shipment might be hazardous. However, non-hazardous materials may be shipped in similar containers so whether or not the shipment is hazardous must be verified by additional means.

2. Labels on containers: Presence of diamond shape labels on packages usually indicates a shipment is hazardous. Hazardous Materials labels must be read

in order to distinguish them from other shipping labels which might appear on the package. Absence of these hazardous materials labels does not necessarily mean that a shipment is not hazardous. Many hazardous shipments are exempt from labelling but not from other regulatory requirements.

3. Marking: Packages containing hazardous materials must be marked with the proper shipping name and identification number of the contents unless shipped under a "small quantity" or "limited quantity" exception or displaying the proper shipping name "Consumer Commodity." Unless a material is packaged in a portable tank or cargo tank, or in a truckload or freight container load lot from one consignee to one consignor, or is part of a shipment by one carrier only, the name and address of either the consignee or consignor must be displayed on the package. Recognition of a consignee or consignor name on a package may also be a basis for checking freight to identify hazardous materials shipments. Personnel should also be alert to recognize shipments to or from chemical companies, paint stores, plastics firms, photo refinishers, etc.

4. Hazardous Substances: If a material meets the definition of a hazardous substance, the letters "RQ" should appear on the package along with the proper shipping name. If the shipping name does not identify the ingredients in a mixture or solution which make it a hazardous substance, the names of these ingredients should be marked on the package along with the shipping name. *(See definition of Hazardous Substance on page 1.)*

5. Shipping Papers: Hazardous materials shipping papers may be readily reccgnizable by being distinctively tabbed, entries of hazardous materials shown first, highlighted, an "X" or "RQ" placed in a column marked HM, or, in the case of inter-modal shipments by air, a candy stripe red and white border on the document. Key information is available on properly prepared shipping papers including the proper shipping name, hazard class, identification number, total quantities, and additional handling information. Shipping paper entries and corresponding markings on packages such as "Dangerous When Wet," "Keep Away From Heat," "Do Not Stow With Foodstuffs" and "This End Up" will assist in identifying and safely handling hazardous materials. The shipping paper may include a signed certification by the shipper that the shipment is in compliance with the regulations.

6. Special Precautions: Personnel should be alert to recognize markings such as "Limited Quantity," "Consumer Commodity," "This Package Conforms to conditions and limitation specified in 49CFR 173.4" and marked flash points between 73°F and 100°F on drums or containers. These markings identify materials which, in larger quanitites would be more strictly regulated and which could cause problems if the containers were damaged or ruptured. Any damaged packages should be checked for leakage.

Handling the Request for Pickups

When the terminal receives a request to pick up freight it is routine procedure to inquire as to the nature of the shipment. Personnel taking calls for pickups should be instructed to also ask the shipper if the freight is classified as a hazardous material.

If the commodity is hazardous the employee should determine:

1. Proper shipping name.

2. Nature of the commodity. That is, whether it is classified as flammable, corrosive, oxidizing material, etc.

3. Weight or volume of shipment, and type and number of shipping containers.

4. That the shipment complies with DOT requirements for packaging, labelling and marking and that the shipper has certified on the shipping paper that it does comply (see certification shown under section titled 'SHIPPING PAPERS'). If the shipment is subject to a Limited Quantity Exception, the certification must still be shown. In addition the shipper must show on the shipping paper the notation "Limited Quantity" or "Ltd., Qty.". If the shipper does not know what is required under the regulations or if there is any question about shipment being in compliance, the pickup request should be turned over to the key person in the terminal.

5. Whether or not special precautions are to be taken in the routine handling of the shipment and any special action to be taken in case of emergency such as fire or spillage involving the shipment. If the shipper is willing to attach a summary of such information on the shipping paper such information should be passed through the key person so that he or she can record it for possible future reference.

The pickup request should be marked "Hazardous Materials" in bold letters to attract attention and should be given, with the above information, to the dispatcher.

DISPATCHER'S RESPONSIBILITIES

The dispatcher must be given an understanding of all phases of the hazardous materials program so that he or she can see the importance of his or her own activity and how it relates to that of other personnel. The dispatcher must have a copy of loading and unloading rules so he or she can check on compatibility of loading. The dispatcher will need to have a copy of driver rules so that he or she can properly instruct drivers on routine and emergency procedures they must follow.

The dispatcher should have information on special routing that might be required to avoid congested areas, tunnels and turnpikes where hazardous materials loads are not permitted or are subject to special restrictions. He or she will also need a listing of phone numbers for the key person, fire department and company officials who should be notifed in case of emergency. Phone numbers of regular shippers of hazardous materials may also be provided. He or she may be given the responsibility of making immediate phone reports to D.O.T. or to the fleet safety director as required when accidents or incidents occur. Reporting requirements are shown on page 203.

Pickup and Delivery

In Pickup and Delivery operations the dispatcher has the following responsibilities:

1. Be certain that any pickup and delivery driver who is assigned to handle hazardous materials is qualified and that his or her vehicle complies with the DOT equipment requirements. If a carrier has some vehicles and some drivers who do not meet such requirements the dispatcher should have a listing of them. *(Requirements for both drivers and vehicles are shown in the Motor Carrier Safety Regulations Booklet available from A.T.A.)*

2. Before the driver leaves the terminal or when he or she calls in to report that he or she is picking up hazardous materials, give him or her a brief resume of driving rules that apply to placarded vehicles and of his or her responsibilities for checking the vehicle, seeing that placards are properly applied, checking accuracy of shipping papers, determining condition of containers and labels and marking on shipments. *(The basis for this resume is the instruction for pickup and delivery drivers shown later in this guidebook).*

3. Placarding: See that pickup units have placards on board when assigned a hazardous materials pickup and that placards are removed after the hazardous materials have been unloaded at terminal. See that delivery units are properly placarded before leaving terminal and have necessary placards on board for enroute changes if necessary.

4. The dispatcher should advise the pickup and delivery driver of the presence of any hazardous materials in the shipment and should make the driver aware of any special routing he or she must follow to stay clear of tunnels, alleys, narrow streets or heavily populated areas. Operations in, on or near such places are prohibited unless necessary for freight operations and provided there is no practicable alternative.

5. Advise the dock foremen of which pickup vehicles will have hazardous materials on board when they return to the terminal so that the dock foreman can carry out prompt and correct procedures for unloading.

6. The key person must be notified when questions or problems arise. Calls from drivers concerning improperly prepared shipments or shipping papers, damaged freight or placarding problems should be among those handled by the key person.

A Hazardous Materials Guide, which sets forth placarding rules, and other driver responsibilities, is available from American Trucking Associations in large size (14" x 20") for use in dispatch offices, and in smaller size (8½ x 11") on adhesive stock for mounting on dashboards, cab doors, cargo bodies or clipboards.

If the carrier is interested in developing a record of hazardous materials shippers, containers, consignees and shipping information the dispatcher may be given the responsibility. All shipping papers should be routed by the dispatcher so that he or she can record the pertinent information on outbound and inbound shipments. Such a record can be expanded to show placards applied to vehicles so that the carrier can determine its placarding experience, needs and compliance of employees with placarding rules.

Over the Road Dispatching

In road operations the dispatcher must be responsible for overseeing the proper handling of both inbound and outbound loads which include hazardous materials.

The line dispatcher's duties on outbound loads are:

1. Find out from dock foreman which outbound units are carrying hazardous materials.

2. See that proper placards are applied to each unit as needed and that units are in a safe location in the terminal yard. In case of fire or other emergency in terminal area hazardous materials must be removed to a safe location immediately.

3. See that shipping documents accompany every hazardous materials shipment, including those which have the notation "Limited Quantity". Shipping papers must be in good order in accordance with the section on "SHIPPING PAPERS".

4. Hazardous materials shipping papers for each load should be placed on top of other shipping papers or distinctively tabbed. Special hazardous materials envelopes are recommended as a handy means of carrying out this procedure. Be certain that all shipments are compatible and allowed to be shipped on the same vehicle. Show placarding information on shipping paper envelopes in the space provided, if such envelopes are used.

5. See that the emergency response information is included in or attached to the shipping paper prior to movement of the shipment. Emergency Response information must conform to the requirements in the section on "Emergency Response Communications Standards."

6. Instruct line drivers on responsibilities (as noted under driver section). Show drivers each H.M. shipping paper so he or she will know what is being transported. Develop special routing away from tunnels, heavy populated areas, narrow streets and alleys unless there is not practicable alternative route available.

7. If Explosives A or B are being transported the driver must be given the following documents: Written Route Plan; Copy of Part 397 of the Motor Carrier Safety Regulations; Instructions on procedures to follow in event of accident or delay (including names and phone numbers of key shipper and carrier contacts); Instructions on nature of the explosives and action to take in emergency. The driver must sign a receipt for the foregoing materials and documents.

8. Notify intermediate and destination terminals that hazardous materials loads are en route so they will be alerted and can handle the loads properly.

Duties and responsibilities of the dispatcher for inbound loads are:

1. See that priority is given to unloading of hazardous materials units.

2. Have placards removed from vehicle after hazardous materials are unloaded.

3. See that close count and inspection is made of hazardous materials shipments and advise key person of any problems or discrepancies.

4. Have hazardous materials loaded on delivery unit as soon as possible or stored in safe location in terminal. Be certain that hazardous materials which are loaded or stored together are compatible and allowed together according to the loading and storage requirements.

5. Advise pickup and delivery driver or pickup and delivery dispatcher that hazardous materials have arrived for delivery and advise location in the terminal or in vehicle they have been moved to.

6. If the inbound vehicle is loaded with Explosives A or B it must be located so that it is always within unobstructed view while at the terminal facility. The only exception is when the terminal has been approved as a "safe haven" by governmental authorities in writing for parking of explosive vehicles.

Dock Handling

The following rules apply specifically to the supervisor of dock operations:

1. Notify dispatcher promptly of inbound shipments of hazardous materials on city or road vehicles. Provide information on destination, size and nature of shipment.

2. Unload hazardous materials shipments promptly and check count and condition of containers, completeness and accuracy of shipping documents labelling and package marking.

3. See that all loading and unloading rules outlined on page 194 are followed.

4. Reload shipments in outboud or delivery vehicle as soon as possible. Check loading and storage guide to see that all hazardous shipments are compatible. If hazardous materials cannot be reloaded promptly see that they are stored safely and not mixed with incompatible materials.

5. If the shipment is in good order note on the shipping papers "Hazardous Materials - Count Correct and Packages appear in proper condition." Initial and date, and take shipping papers to dispatcher. Point out that they are hazardous materials.

6. Advise dispatcher which units contain hazardous materials and where they are located in the terminal yard. If hazardous materials are stored on dock advise their location.

7. Advise key person of any discrepancies or problems with hazardous materials shipments.

8. See that vehicles used to transport Class A or B poisons, or radioactive material are inspected after each usage and not returned to service, if contaminated, until they have been decontaminated.

PICKUP & DELIVERY DRIVER RULES

General:

1. Only drivers who qualify under Federal Motor Carrier Safety Regulations can handle hazardous materials.

2. Vehicles carrying hazardous materials must meet federal requirements.

3. The driver must inspect his or her vehicle before loading hazardous materials or leaving the terminal to deliver hazardous materials. Inspection should follow the procedure set forth in ATA Form C0830, illustrated at page 177. The driver must check to see that the vehicle is equipped with emergency equipment - fire extinguisher, red flags, flares, fuses, electric lantern; or reflective devices. *(Note 1: Vehicles transporting hazardous materials must be equipped with a fire extinguisher having an Underwriter's Laboratories rating of 10 B:C or more. Note 2: Flame producing warning devices are prohibited on any motor vehicle transporting explosives, Class A or B; on any tank truck transporting flammable liquids or flammable compressed gas whether loaded or empty; or on any motor vehicle using compressed gas as a motor fuel.)*

4. Tire Inspections: The driver must inspect tires before leaving the terminal and upon his return. He must also inspect tires every two hours or 100 miles, whichever comes first, while the vehicle is placarded. If any tire is leaking or otherwise defective he must drive to the nearest safe place for repair. If a tire is "hot", the driver shall notify the terminal immediately and not move the vehicle until the defective tire has been removed.

5. Placarding: For deliveries placards must be displayed on the vehicle before it leaves the terminal. For pickups there should be a supply of assorted placards in the cab of the vehicle if it is not equipped with permanent placards. All placards must be removed when the hazardous materials are unloaded except for tank vehicles or portable tanks not yet cleaned and purged.

6. Smoking: Do not smoke when driving, loading or unloading hazardous materials, or when within 25 feet of a vehicle that is placarded.

7. Fires: Do not drive near open fires. Pass open fires only when it is positively safe to do so and there is no alternative route. Don't park within 300 feet of an open fire.

8. Routing: Avoid tunnels, narrow streets, alleys and heavily populated areas unless there is no practicable alternative route or unless pickups and deliveries must be made in the area. Ask the dispatcher if special routing is required.

9. Rail Crossings: Any cargo tank or placarded vehicle containing hazardous materials, or any vehicle carrying any amount of chlorine, must make a full stop not more than 50 feet or less than 15 feet from the nearest rail of a highway grade crossing. Stops are not required at crossings which are marked "abandoned" or "exempt", street car crossings, crossings used exclusively for industrial switching within a business district, or crossings directed by a flagman, police officer, or functioning green traffic signal. If no stop is required you must reduce speed and take due care in crossing.

10. Fueling: When fueling a placarded vehicle the engine must be off and the driver or attendant must stay at the nozzle until fueling is completed.

11. Parking: Vehicles transporting hazardous materials, cannot be parked on or within 5 feet of the traveled portion of a public street or highway except for brief periods when the necessities of operation require the vehicle to be parked, and it is impracticable to park the vehicle in any other place.

Vehicles containing Explosives A or B cannot be parked on or within 5 feet of the traveled portion of a street or highway, or on private property (including truck stops) without the property owners permission, or within 300 feet of a bridge, tunnel, house or any building or place where people congregate, except for brief periods when it is necessary and is inpractical to park any other place (such as loading and unloading).

12. Attendance: Vehicles containing hazardous materials, other than Class A or B explosives, that are

parked on a public street or highway or the shoulder thereof must be attended by the driver. However, the vehicle need not be attended while the driver is performing duties which are incident and necessary to his duties as operator.

A vehicle is attended when the person in charge of the vehicle is on the vehicle, awake, and not in the sleeper berth, or is within 100 feet of the vehicle and has it within unobstructed view.

Vehicles with explosives A or B must be attended by the driver or a qualified representative at truck stops, pick-up & delivery point and at all other times unless the vehicle is on property of the carrier, shipper or consignee.

When vehicles transporting explosives are on carrier, shipper or consignee property, the person caring for the vehicle must have it in unobstructed view, must know what's on the vehicle and must know what to do in an emergency. If the vehicle is parked on a lot which has been approved in writing by a governmental authority for parking of explosives vehicles, the vehicle does not have to be in unobstructed view.

If the driver or carrier selects a qualified representative to attend the vehicle that person must be authorized to move the vehicle, know what's on the vehicle and know what to do in an emergency.

13. Documents: Drivers of vehicles carrying Explosives A or B must carry a copy of Part 397 of the Motor Carrier Safety Regulations, Written route instructions, Instructions on what to do in an emergency or delay (Including names and phone numbers of carrier and shipper representatives to contact, information about the nature of the explosives and information about what to do in case of fire, accident, leakage or other emergencies).

The driver must sign a receipt for such documents and the carrier must retain the receipt for one year. A driver must be familiar with the contents of such documents and keep them in his or her possession.

Deliveries:

1. Determine what types of hazardous materials are on the vehicle, where they are located and how to handle them.

2. Follow loading and unloading rules in handling hazardous shipments. These are shown on pages 194 to 196.

3. If shipment is damaged or refused by consignee, contact the terminal for instructions.

4. Remove placards when hazardous materials have been unloaded, or change placards to show hazardous materials still on vehicle.

Pickups:

1. If hazardous materials are to be picked up determine nature of shipment and methods of handling before leaving the terminal.

2. Do not make unassigned pickups of hazardous materials without checking the terminal for instructions.

3. Make a careful check and count of hazardous materials packages. Inspect containers for leakage or damage. Do not accept shipments if containers, shipping papers, labels or markings do not appear to be in proper condition. Call the terminal about these if there is any question about the shipment, loading or placarding.

Shipping papers must show:

a) proper shipping name for the material as shown in the Hazardous Materials Table §172.101, e.g. Acetaldehyde; followed by

b) the hazard class of the commodity, e.g. Flammable Liquid; followed by

c) the identification number for the product, e.g. UN1089;

NOTE: Entries a, b, and c must be entered in sequence.

d) an emergency response telephone number for each hazardous material, or one number for all materials if appropriate.

e) the letters 'RQ' before or after the shipping name, or in the HM column, if the material is a hazardous substance (e.g. 1000 lbs. or more of acetaldehyde in one package)

f) the total quantity by weight, volume or other neasure

g) any additional pertinent information applicable to the commodity, packaging, or special handling instructions

h) certification, if required, with prescribed wording and signature of the shipper or his agent

i) "Limited Quantity" entry in association with the shipping name to show that the shipment need not be packaged in specification packages or labeled if the Limited Quantity exception is used.

4. If both hazardous materials and other freight are on the same shipping paper, the hazardous materials must be: (a) Listed ahead of all other items; or (b) Entered in a different color and otherwise highlighted on photocopies; or (c) Indicated by an "X" in an HM column. *(Note: The letters "RQ" can be substituted for the "X" when a reportable quantity of the hazardous substance described is contained in any package in the shipment.)*

5. Before loading pickups of hazardous materials check the loading and storage guide to be sure that the shipments are allowed on the same vehicle together. Hazardous materials must be blocked and braced as necessary to prevent movement and overturning during transportation.

6. Placard, or change placards as necessary when additional hazardous materials are loaded on the vehicle.

7. Upon returning to the terminal notify the dispatcher immediately of the nature and location of the load and of the location of the vehicle in the terminal yard.

ROAD DRIVER RULES

The following rules apply to drivers of line haul units carrying hazardous materials:

1. Documentation: Every driver must determine the nature of the load and check freight bills and manifest to see that they are properly made out.

Drivers of vehicles carrying explosives A or B must carry a copy of Part 397 of the Federal Motor Carrier Safety Regulations, a written route plan, instructions on what to do in an emergency or delay (including names and phone numbers of carrier and shipper representatives to contact, information about the nature of the explosives and what to do in case of fire, accident, leakage, or other emergency.)

The driver must sign a receipt for such documents, and must become familiar with their contents and keep them in his possession. The carrier will retain the receipt in its files for one year.

2. Inspection of vehicle: Check equipment on the vehicle for completeness and good working condition as required in Part 392 of the Federal Motor Carrier Safety Regulations. Be certain emergency equipment is complete and in good working order. Use ATA Form C0830 to insure a thorough inspection.

Tires must be inspected at beginning and end of each trip. Also inspect tires every 2 hours or 100 miles, whichever comes first. If any tire is low, leaking or otherwise defective the vehicle can be driven to a safe place for repair. If a tire is overheated the vehicle cannot be moved until the "hot" tire has been replaced.

3. Placards: Be certain that unit is properly placarded according to the Hazardous Materials Guide, shipping papers and the hazardous materials tables shown on page 167.

4. Routing: All drivers must know what routing to take to avoid tunnels, crowds, narrow roads & streets, etc. Drivers of vehicles carrying Explosives A or B must receive a written route plan from the company and must carry the routing with them.

5. Rail crossings: Make a full stop when placarded or when carrying any quantity of chlorine. Stop within 15 to 50 feet of the nearest rail. Stops are not required for crossings which are abandoned or exempt, streetcar crossings within a town, crossings protected by a police officer, crossing guard or a stop and go traffic light (not a railroad flashing signal), which displays a green light.

6. Smoking: Do not smoke when driving, loading or unloading hazardous materials, or when within 25 feet of a vehicle that is placarded.

7. Fires: Pass open fires when it is absolutely safe to do so and there is no alternative route. Don't park within 300 feet of an open fire.

8. Fueling: When fueling a placarded vehicle the engine must be off and the driver or pump attendant must stay at the nozzle.

9. Parking: Vehicles transporting hazardous materials, cannot be parked on or within 5 feet of the traveled portion of a public street or highway except for brief periods when the necessities of operation require the vehicle to be parked, and it is impracticable to park the vehicle in any other place.

DRIVER'S INSPECTION REPORT

(SEE INSTRUCTIONS ON REVERSE SIDE)

MAINTENANCE **CHECK DEFECTS ONLY. Explain under REMARKS** **C 338501**

COMPLETION OF THIS REPORT REQUIRED BY FEDERAL LAW, 49CFR 396.11 & 396.13.

Truck or Tractor No. ____________ Mileage (No Tenths) |_|_|_|_|_|_| Trailer No. ____________

Dolly No. ____________ Trailer No. ____________ Location: ____________

ATA/VMRS System Code Numbers for Shop Use Only

POWER UNIT

GENERAL CONDITION	IN-CAB	EXTERIOR
☐ 02 Cab/Doors/Windows	☐ 03 Gauges/Warning Indicators	☐ 34 Lights
☐ 02 Body/Doors	☐ 02 Windshield Wipers/Washers	☐ 34 Reflectors
☐ __ Oil Leak ______	☐ 54 Horn(s)	☐ 16 Suspension
☐ __ Grease Leak ______	☐ 01 Heater/Defroster	☐ 17 Tires
☐ 42 Coolant Leak	☐ 02 Mirrors	☐ 18 Wheels/Rims/Lugs
☐ 44 Fuel Leak	☐ 15 Steering	☐ 32 Battery
☐ __ Other ______	☐ 23 Clutch	☐ 43 Exhaust
______ (IDENTIFY)	☐ 13 Service Brakes	☐ 13 Brakes
	☐ 13 Parking Brake	☐ 13 Air Lines
ENGINE COMPARTMENT	☐ 13 Emergency Brakes	☐ 34 Light Line
☐ 45 Oil Level	☐ 53 Triangles	☐ 49 Fifth-Wheel
☐ 42 Coolant Level	☐ 53 Fire Extinguisher	☐ 49 Other Coupling
☐ __ Belts ______	☐ 53 Other Safety Equipment	☐ 71 Tie-Downs
☐ __ Other ______	☐ 34 Spare Fuses	☐ 14 Rear-End Protection
______ (IDENTIFY)	☐ 02 Seat Belts	☐ __ Other ______
	☐ __ Other ______	______ (IDENTIFY)
	______ (IDENTIFY)	☐ NO DEFECTS

TOWED UNIT(S)

☐ 71 Body/Doors	☐ 16 Suspension	☐ 77 Landing Gear	☐ 79 Rear-End Protection
☐ 71 Tie-Downs	☐ 17 Tires	☐ 59 Kingpin/Upper Plate	☐ __ Other ______
☐ 34 Lights	☐ 18 Wheels/Rims/Lugs	☐ 59 Fifth-Wheel (Dolly)	______ (IDENTIFY)
☐ 34 Reflectors	☐ 13 Brakes	☐ 59 Other Coupling Devices	☐ NO DEFECTS

REMARKS: ______________________________

____

____

____

REPORTING DRIVER: Date ______	**MAINTENANCE ACTION:** Date ______
Name ______ Emp. No. ______	Repairs Made ☐ No Repairs Needed ☐
REVIEWING DRIVER: Date ______	R.O.#'s ______
	Certified By: ______
Name ______ Emp. No. ______	Location: ______

SHOP REMARKS: ______________________________

Reorder From: American Trucking Associations, 2200 Mill Road, Alexandria, VA 22314 1 800 ATA LINE

ATA FORM C0830 (REV. 1-80)

Vehicles carrying Explosives A or B cannot be parked on or within 5 ft. of the traveled portion of the highway or on private property (including truck stops) without the permission of the owner, or within 300 feet of bridges, tunnels, houses, or buildings or places where people congregate except when necessary and it is not practical to park anywhere else (such as loading or unloading).

10. Attendance: Vehicles containing hazardous materials, other than Class A or B explosives, that are parked on a public street or highway or the shoulder thereof must be attended by the driver. However, the vehicle need not be attended while the driver is performing duties which are incident and necessary to his or her duties as operator.

A vehicle is attended when the person in charge of the vehicle is on the vehicle, awake, and not in the sleeper berth, or is within 100 feet of the vehicle and has it within unobstructed view.

Vehicles with Explosives A or B must be attended by the driver or a qualified representative at truck stops, pick-up & delivery points and at all other times unless the vehicle is on property of the carrier, shipper or consignee.

When explosives vehicles are on carrier, shipper or consignee property, the person caring for the vehicle must have it in unobstructed view, must know what's on the vehicle and must know what to do in an emergency. If the vehicle is parked on a lot which has been approved in writing by a governmental authority for parking of explosives vehicles, the vehicle does not have to be in unobstructed view.

11. Emergencies or delays: Review the section of this manual entitled "Emergency Procedures" for the proper handling of damaged packages and the disposition or repair thereof. In case of fire, accident, breakdown or delay notify the nearest terminal and follow procedure outlined in the section titled "Emergency Handling Hazardous Materials." If transporting Explosives A or B follow the procedure in document provided which lists names of persons to contact and procedure to follow.

12. Upon arrival at destination notify dispatcher or other authority and advise of the nature of the load and location of the vehicle on the property.

Loading & Unloading

The following procedures and precautions must be observed by the party (shipper or carrier) loading and unloading hazardous materials:

1. Guard against hazards. Commodities may offer several hazards under varying conditions. For example, the primary hazard of a flammable liquid is fire but in certain situations vapors of the liquid may be toxic and leaking shipments of flammable liquids can cause damage to skin tissues or to other freight.

The following precautions should be taken with all hazardous materials shipments:

a) Do not drop, jar or bump. Impact can cause breakage and release of hazardous materials, fumes or vapors.

b) Keep from heat. Excessive heat, such as from storing next to heating units or loading into trailers and leaving the unit in the hot sun can increase the possibility of hazard from hazardous materials.

c) Protect from moisture. Wet packages will break open more easily and moisture can cause a reaction in certain hazardous materials and result in fire, toxic fumes or contamination of other materials.

d) Keep well ventilated. Air circulating between shipments prevents fumes and materials from becoming mixed and contaminating each other.

e) Avoid friction. Don't slide hazardous materials shipments as heat generated through friction may create a hazard.

f) Check the loading guide for hazardous materials to be certain that shipments which will react with each other are not loaded together.

2. Check shipping papers and packages to be certain they are in proper condition. Information required on shipping papers is shown in the shipping paper section of this guidebook. Packages must be in good condition and show required labels and markings. Report any discrepancies to the dock foreman or key person in the terminal.

3. No smoking or open flame is permitted around any unit being loaded or unloaded.

4. The parking brake on the vehicle must be set and the engine must not be running while the vehicle is being loaded or unloaded.

5. Tools which might damage the shipment must not be used on any packages.

6. When flammable solids, oxidizing materials or corrosive liquids are transported on a unit with other commodities not forbidden in the loading and storage charts, they shall be so loaded as to provide ready accessibility thereto for shifting or removal.

7. All hazardous materials should be kept free from extreme changes in temperature.

8. Tanks, barrels, drums and cylinders shall be reasonably secured against movement within the motor vehicle during transport and against relative movement between each other. Such motion might create friction or cause damage to valves or fittings and result in dangerous leakage.

9. Heaters may be used when transporting flammable liquid or flammable gas provided:

a) It is a catalytic heater;

b) The heater's surface temperature cannot exceed 130° F. (54° C.);

c) The heater is not ignited in a loaded vehicle;

d) There is no flame, either on the catalyst or anywhere in the heater;

e) The heater bears the manufacturers' certification "Meets DOT requirements for catalytic heaters used with flammable liquid and gas." and,

f) The heater is also marked "Do Not Load Into or use in cargo compartments containing flammable liquids or gas if flame is visible on catalyst or in heater."

10. Vehicles equipped with heaters not meeting the specifications outlined above may be used to transport flammable liquid or flammable gas only if the device is first rendered inoperable by emptying or removing the heater fuel tank, closing its discharge valve and disconnecting its fuel feed line.

11. Damaged or leaking packages of hazardous materials may be reloaded into a recovery drum for shipment to destination or return to the shipper. A recovery drum must be of metal construction. Cushioning must be provided if necessary, and there must be a means of absorbing any free liquid. The drum, the cushioning and the absorptive material must be compatible with the hazardous material. The drum must be marked with the proper shipping name of the hazardous material, the name and address of the consignee, labeled as required, and bear the additional marking "Recovery Drum". The capacity of a recovery drum must not exceed 110 gallons.

12. Transfer of Flammable Liquids:

a) Transfer of flammable liquids between containers not in metallic contact may be made only if metallic bonds or ground conductor are provided. Such bonding shall be made by first connecting an electrical conductor to the container to be filled and then to the container from which the liquid is to be emptied. The latter connection must be at a point away from the opening from which the liquid is to discharge.

b) No flammable liquid shall be unloaded or loaded with the engine running unless the vehicle engine is used for the operation of a pump.

13. Flammable Solids and Oxidizing Materials shall be handled as follows:

a) Lading shall be entirely within and covered by the body of the vehicle or by tarpaulins or by other suitable means and the tailgate or doors shall be securely closed and fastened.

b) Commodities which are susceptible to heat or spontaneous combustion shall be loaded in such a manner as to insure adequate ventilation between packages.

c) Extra precautions shall be taken to prevent wetting of flammable solids and oxidizing materials which become more hazardous when wet. Examples of such materials are charcoal screenings, hair, felt, jute and other fibrous materials.

d) Matches of any type shall not be loaded near other flammable articles. "Strike-anywhere" matches may be loaded only in vehicles with smooth interiors unless the matches are contained in wooden outside boxes. Care should be taken to prevent shifting or jamming of match shipments in transit.

e) Reloading of damaged packages of matches after examination and repair is permitted. Loose matches shall be disposed of.

f) Loose or baled Nitrate of Soda Bags which are empty and unwashed may be transported in truck load lots only and only in closed or covered bodies lined with paper. Loading and unloading must be handled by the shipper and consignee.

14. Corrosive liquids shipments shall be loaded and unloaded as follows:

a) Carboys and frangible containers of acids and charged electric storage batteries shall be individually handled when loaded or unloaded by hand.

b) Carboys and frangible containers may be loaded in tiers if the weight of the upper tiers is upon the crates or boxes enclosing the carboys and frangible containers and not upon the carboys and the carboys and frangible containers themselves. Means must be provided to prevent shifting of the tiers and individual carboys and frangible containers in the tiers.

c) Corrosive liquids may be hauled with other lading only as permitted by the loading chart. Ready access must be provided to corrosive liquids to permit their shifting or removal.

d) Nitric acid loaded in the same motor vehicle with other acids or corrosive liquids in carboys must be separated from the other carboys by a 2 by 6 inch plank, set on edge, nailed across the vehicle floor. The board shall be placed a distance of at least 12 inches from the nitric acid carboys and the space between shall be filled with an incombustible absorbent material such as sand, sifted ashes, etc.

e) Electric storage batteries containing electrolyte must be protected against short circuits and against other lading falling against them. Batteries may be loaded higher than one tier provided none of the weight from the upper tier rests on the lower tiers.

15. Compressed gas loading should be conducted as follows:

Cylinders shall be securely lashed in an upright position; or loaded into racks securely attached to the motor vehicle; or packed in boxes or crates which will prevent their overturn; or loaded in horizontal position. Spec. DOT - 4L cylinders must be loaded upright and securely braced. Cylinders of liquified hydrogen cannot be transported by common carrier. Other carriers must comply with special loading rules of DOT.

16. Loading and Unloading of Poisons

a) Class A Poison or Irritating Materials may not be transported if there are any inter connecting means of any character between containers.

b) No Class A Poison or Irritating Substance may be loaded into any cargo tank.

c) Class A or B Poisons cannot be loaded on the same vehicle with foodstuffs, feeds or any edible material intended for consumption by humans or animals unless they are overpacked in a specification container which is liquid tight and dust proof.

d) Vehicles used for hauling general freight must be inspected for contamination after transporting Class A or B poisons. If contaminated they must be removed from service and the shipper contacted for instruction on decontamination.

17. Loading and Storage of Radioactive Materials

Radioactive materials cannot be loaded with Class A Explosives, with minor exceptions as shown in the loading charts. No more than 50 radiation units of radioactive materials bearing Radioactive Yellow II or Yellow III labels can be loaded on one vehicle or stored in one location. The total number of radiation units in a shipment is determined by adding together the transport index numbers shown on the label of each package. Exceptions to this requirement are granted for "exclusive use" vehicles.

Packages of Radioactive material bearing a Yellow II or Yellow III label must be loaded or stored incidental to transportation in accordance with the requirements of 49CFR 177.842.

Persons should not remain unnecessarily in a vehicle or storage area containing radioactive materials.

18. Loading and Unloading of Asbestos

Asbestos, and any contamination in vehicles, must be handled in a way which will minimize the spread of airborne particles.

Reorder # C1060 from:
American Trucking Associations
2200 Mill Road • Alexandria, Virginia 22314-4677
1 800 ATA-LINE

American Trucking Associations

HAZARDOUS MATERIALS GUIDE

CHECKLIST FOR PICKUP DRIVERS

For detailed information refer to 49 CFR Part 172 (Hazardous Materials Regulations).

Shipping Papers:

- Hazardous materials must be listed first OR highlighted by color OR be indicated by an "X" in an HM Column.
- Each description must show in order proper shipping name; hazard class; hazard identification number; quantity by weight/volume. Abbreviations are prohibited except for packaging type, quantity, or "Ltd. Qty." notation.
- Certification must be shown and signed by shipper manually or by mechanical means.
- Hazardous substances - letters "RQ" required before or after basic description. "RQ" can be substituted for "X" in HM Column.
- **Additional requirements - POISONS 1) Word "POISON" must be shown if not part of proper shipping name or hazardous class entry; 2) Certain materials require note: "POISON-INHALATION HAZARD." "DANGEROUS WHEN WET" notation for flammable solid requiring "Dangerous When Wet" label.**

Hazardous Wastes:

- Uniform Hazardous Waste Manifest (EPA Form 8700-22 (9-86)) must be prepared. All signatures must be manual. Driver must sign manifest at time of pick-up and must obtain signature of receiver on carrier's copy.

Packaging Marking: Shipper Responsibility

- Shipper must show proper shipping name; hazard identification number; name and address of shipper or consignee. Liquids - "This End Up," "Inhalation Hazard."
- Must be durable; in English; printed or affixed to packaging; in contrasting color; unobscured; away from other markings.

Labels: Shipper Responsibility

- Shipper must securely affix to package; near proper shipping name; unobscured by any other markings.
- Represent hazard class (check against hazard class on shipping paper). Multiple labels are required in special cases including Toxic By Inhalation.
- No labels required for "Limited Quantity" shipments except Class B Poisons.

Placards:

- Provided by shipper for proper hazard class offered, unless already on unit.
- Represent hazard class (See table to right and special requirements).
- On 4 sides of vehicle. Front placard may be on power unit or cargo body. Wording must read horizontally, left to right.
- Unobstructed view, at least 3 inches from other marking, lettering, or display. Multiple placards, must be near each other.
- Must be changed if HM cargo changes, and removed if HM not on board.
- Identification numbers required on cargo tanks portable tanks, or vans carrying portable tanks.

In Transit:

- HM shipping papers must be in reach with seat belt on; clearly distinguishable from other documentation.
- No smoking when transporting explosives, flammables, or oxidizers.
- Before leaving vehicle secure against movement. Place HM shipping papers on your seat or in the pouch on your door.

MIXED LOADS

DANGEROUS

1. When the total weight of two or more materials from the blue shaded area is 1000 lbs. or more, a DANGEROUS placard may be used. If 5,000 pounds or more of any of these materials are loaded at one location use its class placard as shown to the right along with either a dangerous placard or the class placard(s) for the other materials.
2. The class placard for each "Any Quantity" material (Red shaded area) in a mixed load must be displayed regardless of any other placards displayed.

LABEL	WEIGHT	PLACARD	REMARKS
EXPLOSIVE A	ANY QUANTITY	EXPLOSIVES A	Mixed Loads of EXPLOSIVES A & B, use EXPLOSIVES A placard.
EXPLOSIVE B	ANY QUANTITY	EXPLOSIVES B	
RADIOACTIVE	ANY QUANTITY	RADIOACTIVE	Uranium Hexafluoride must be placarded RADIOACTIVE (any quantity) and CORROSIVE (1000 lbs. or more).
POISON GAS	ANY QUANTITY	POISON GAS	Do **not** load with food, animal feed, or medicines for internal use.
FLAMMABLE SOLID & DANGEROUS WHEN WET	ANY QUANTITY	FLAMMABLE SOLID (W)	Keep dry in transit. Do not use water for fire or other emergencies.
FLAMMABLE LIQUID	1000 Pounds or More	FLAMMABLE	If a flammable liquid meets the definition of Toxic By Inhalation a POISON placard is also required if the quantity is over 1 liter.
FLAMMABLE SOLID	1000 Pounds or More	FLAMMABLE SOLID	FLAMMABLE placard may be used in place of the FLAMMABLE SOLID placard.
NON-FLAMMABLE GAS	1000 Pounds or More	NON-FLAMMABLE GAS	A **NON-FLAMMABLE GAS** placard is not required when a **FLAMMABLE GAS** or **OXYGEN** placard is displayed.
OXYGEN	1000 Pounds or More	OXYGEN [1] OR NON-FLAMMABLE GAS [2]	[1] OXYGEN placard is required for Oxygen Cryogenic Liquid. [2] NON-FLAMMABLE GAS placard required for all other shipments of oxygen.
CHLORINE	1000 Pounds or More	CHLORINE OR NON-FLAMMABLE GAS	CHLORINE placard is required only for packagings of more than 110 gallons.
CORROSIVE	1000 Pounds or More	CORROSIVE	If a corrosive liquid meets the definition of Toxic By Inhalation a POISON placard is also required if the quantity is over 1 liter.
POISON	1000 Pounds or More	POISON	Do **not** load with food, animal feed, or medicines for internal use, unless properly overpacked. POISON placard required for Fluorine.
OXIDIZER /3	1000 Pounds or More	OXIDIZER	Do **not** use OXIDIZER placard for shipments of oxygen. [3] OXYGEN label is allowed for packages containing oxygen in place of the OXIDIZER label.
EXPLOSIVE C	1000 Pounds or More	DANGEROUS	
ORGANIC PEROXIDE	1000 Pounds or More	ORGANIC PEROXIDE	
IRRITANT	1000 Pounds or More	DANGEROUS	
FLAMMABLE GAS	1000 Pounds or More	FLAMMABLE GAS	
BLASTING AGENT	1000 Pounds or More	BLASTING AGENTS	
Combustible Liquid	More Than 110 Gallons	COMBUSTIBLE	No Placard Required if Containers are 110 Gallons or Less.
Materials with the following labels do **not** require placarding: Radioactive I, Radioactive Yellow II, Etiological Agents, O.R.M.s. Combustible liquids in containers less than 110 gallons.		NO Placard Required	Any Hazardous Material identified as a "Limited Quantity" on the shipping paper does **not** need to be placarded.

TOXIC BY INHALATION

POISON

Liquids which have been identified as Toxic by Inhalation, in packages of over 1 liter, and the shipping paper states "Poison - Inhalation Hazard," must be marked "Inhalation Hazard" and labeled "POISON." The vehicle must be placarded "POISON," in addition to the primary hazard requirements.

Note: 1000 lb exception does not apply to vehicles carrying these materials.

CHEMICAL EMERGENCY-CONTACT CHEMTREC (800) 424-9300

4/89
C1060

AMERICAN TRUCKING ASSOCIATIONS

HAZARDOUS MATERIALS LOADING & STORAGE GUIDE

INSTRUCTIONS An "X" at the intersection of any two columns means that, unless there is an exception indicated by a numbered footnote, hazardous materials of the classes indicated may not be loaded into the same vehicle, nor placed adjacent to each other on a dock or in a terminal.

FOOTNOTES

1. Except as prescribed in §173.81(c) or 177.835(g) loading and transportation of detonators or detonating primers with materials named in rows b, c, e, f, 3, 8a or 10 is prohibited.
2. Corrosive liquids must not be loaded above or adjacent to flammable solids, oxidizing materials, ammunition for cannon with or without projectiles, or propellant explosives, except that shippers loading truckload shipments of corrosive liquids and flammable solids or oxidizing materials packages and who have obtained prior approval from the Department may load such materials together when it is known that the mixture of contents would not cause a dangerous evolution of heat or gas.
3. Explosives, class A, and explosives, class B must not be loaded or stored with chemical ammunition containing incendiary charges or white phosphorus either with or without bursting charges.
4. Bursters (explosive), boosters (explosive), or supplementary charges (explosive) without detonators when shipped by, to, or for the Departments of the Army, Navy, and Air Force of the United States Government may be loaded with any of the articles named except those in columns c, d, 3, 9, 11, 12, 13, 14, 15 and 16.
5. Does not includes ammonium nitrate, fertilizer grade, which may be loaded, transported or stored with high explosives, or with detonators containing no more than 1 gram of explosive each, excluding ignition and delay charges.
6. Normal uranium, depleted uranium, and thorium metal in solid form may also be loaded and transported with articles named in columns a, b, c, d, e, f, and g.

§177.848 Segregation and separation chart of hazardous materials.

(a) Charged electric storage batteries must not be loaded in the same vehicle with explosives, class A.
(b) Cyanides or cyanide mixtures must not be loaded or stored with acids or any other acidic materials which could release hydrocyanic acid from cyanides.
(C) Gas identification sets may be loaded and transported with all articles named in the segregation and separation chart, except those in column c.
(d) Nitric acid, when loaded in the same motor vehicle with other corrosive liquids in carboys, must be separated from the other carboys. A 2 by 6 inch plank set on edge, should be nailed across the motor vehicle floor at least 12 inches from the nitric acid carboys, and the space between the plank and the carboys or nitric acid should be filled with sand, sifted ashes, or other incombustible absorbent material.
(e) Smokeless powder for small arms in quantities not exceeding 100 pounds net weight in one motor vehicle shall be classed as a flammable solid for purposes of transportation when examined for this classification by the Bureau of Explosives and approved by the Director, OHMT.
(f) Hazardous materials must not be loaded, transported, or stored together, except as provided in the following table:

Class	Row	Material	16	15	14	13	12	11	10	9	8a	8	7a	7	6	5	4	3	2	1	g	f	e	d	c	b	a
CLASS A EXPLOSIVES	a	Low explosives or black powder.																									
CLASS A EXPLOSIVES	b	High explosives, propellant explosives or detonating cord.																									
CLASS A EXPLOSIVES	c	Initiating or priming explosives, wet: Diazodinitrophenol, fulminate of mercury, guanyl nitrosamino guanylidene hydrazine, lead azide, lead styphnate, nitro mannite, nitrosoguanidine, pentaerythrite tetranitrate, tetrazene, lead mononitroresorcinate.																								X	X
CLASS A EXPLOSIVES	d	Detonators, detonating primers.																							[1]X	[1]X	
CLASS A EXPLOSIVES	e	Ammunition for cannon with explosive projectiles, gas projectiles, smoke projectiles, incendiary projectiles, illuminating projectiles, ammunition for small arms with explosive projectiles, incendiary projectiles; rocket ammunition with explosive projectiles, gas projectiles, smoke projectiles, incendiary projectiles, illuminating projectiles; boosters (explosive); bursters (explosive); and supplementary charges (explosive) without detonators, 3/, 4/.																						[1]X	X		
CLASS A EXPLOSIVES	f	Explosive projectiles; bombs; torpedoes; mines; rifle or hand grenades (explosive); jet thrust units (jato), igniters, jet thrust, rocket motors, igniters rocket motor, 3/.																						[1]X	X		
CLASS A EXPLOSIVES	g	Detonating fuzes, class A, with or without radioactive components.																				X	X		X	X	
CLASS B EXPLOSIVES	1	Ammunition for cannon with empty, inert-loaded or solid projectiles, or without projectiles; or rocket ammunition with empty projectiles, inert-loaded or solid projectiles or without projectiles.																							X		
CLASS B EXPLOSIVES	2	Propellant explosives, jet thrust units (jato), igniters, jet thrust, rocket motors, rocket engines (liquid), igniters, rocket motor, starter cartridges.																							X		
CLASS B EXPLOSIVES	3	Fireworks, special or railway torpedoes.																			X	X	X	[1]X	X	X	X
CLASS C EXPLOSIVES	4	Small arms ammunition, or cartridges, practice ammunition.																							X		
CLASS C EXPLOSIVES	5	Primers for cannon or small arms, empty cartridge bags—black powder igniters, empty cartridge cases, primed, empty grenades, primed, combination primers or percussion caps, toy caps, explosive cable cutters, explosive rivets.																							X		
CLASS C EXPLOSIVES	6	Percussion fuzes, tracer fuzes or tracers.																							X		
CLASS C EXPLOSIVES	7	Time, combination or detonating fuzes.																							X		
CLASS C EXPLOSIVES	7a	Detonators, detonating primers.																[1]X				[1]X	[1]X		[1]X	[1]X	
CLASS C EXPLOSIVES	8	Safety squibs, fuse lighters, fuse igniters, delay electric igniters, electric squibs, instantaneous fuse or igniter cord.																							X		
CLASS C EXPLOSIVES	8a	Detonating cord.											[1]X					X			X			[1]X	X		
CLASS C EXPLOSIVES	9	Fireworks, common.																			X	X	X	X	X	X	X
OTHER HAZARDOUS MATERIALS	10	Blasting agents, n.o.s., or ammonium nitrate fuel oil mixture; blasing agent label.											[1]X					X						[1]X	X		
OTHER HAZARDOUS MATERIALS	11	Flammable liquids or flammable gases; flammable liquid or flammable gas label.																			X	X	X	X	X	X	X
OTHER HAZARDOUS MATERIALS	12	Flammable solids; flammable solid label. Oxidizer; oxidizer label. Organic peroxide; organic peroxide label.																			X	X	X	[5]X	X	[5]X	X
OTHER HAZARDOUS MATERIALS	13	Corrosive liquids; corrosive label.					[2]X												[2]X	[2]X	X	X	[2]X	X	X	X	X
OTHER HAZARDOUS MATERIALS	14	Nonflammable gases; N.F.G. label.																			X	X	X	X	X	X	X
OTHER HAZARDOUS MATERIALS	15	Poisonous gases or liquids in tank car tanks, cylinders, projectiles or bombs; poison gas labels.				X	X	X		X								X	X	X	X	X	X	X	X	X	X
OTHER HAZARDOUS MATERIALS	16	Radioactive materials, radioactive label.																			[6]X	[6]X	[6]X	[6]X	[6]X	[6]X	[6]X

C1010 Reorder From:
American Trucking Associations
2200 Mill Road
Alexandria, VA 22314-4677

Call Toll Free 1 800 ATA-LINE

Handling Radioactive Materials

The provisions set forth in this guidebook for the handling of other hazardous materials also pertain to the handling of radioactive materials and must be referred to for guidance on labeling, placarding, marking and other regulatory requirements.

The following specific rules and procedures are to be followed in the handling of radioactive materials shipments. They should be studied with care each time such shipments are to be transported.

1. Compliance with the Transport Index Rule - A major difference in handling and loading radioactive materials from other hazardous materials is the need to limit the quantity of radioactive materials that is assembled in one place or vehicle. A simple method of controlling this has been established by Federal Regulations and is known as the "50" Unit rule. First, this rule applies only to those packages bearing a Radioactive Yellow II or Radioactive Yellow III label. Second, these labels have a space which must be filled in by the shipper designating the number of radiation "units" in the package. The total number of packages that can be assembled in one place or vehicle is determined by simply adding together the total "Units" stated on the label on each package. Not more than "50" such units may be loaded on one vehicle or stored in one location. For example, a driver may pick up 50 packages marked with one unit each, or 5 packages of 10 units each, or any other combination not exceeding 50 units. Any number of packages bearing the Radioactive White I label may be loaded or stored together without regard for the "50 Unit" rule.

2. Drivers should not make pick-ups of radioactive materials without first contacting their dispatchers. The driver must give the dispatcher the information shown on shipping papers and should tell him or her the types of labels attached to packages, total radiation "Units" as shown on Radioactive Yellow II and III labels, markings on packages and the general condition of all packages.

3. Dispatchers are required to instruct drivers, dock personnel and others on how the radioactive materials should be handled in their particular phase of the movement. The instructions should be based on information received from the pick-up driver, or the shipper, and upon the following rules. Terminal storage and dispatch of Radioactive Yellow II and III materials must be planned to comply with the "50 Unit" rule. The dispatcher should have a working knowledge of the regulations in the ATA Hazardous Materials Tariff on which this instruction is based.

4. Dispatchers must be notified immediately if any radioactive materials shipment is found to be in poor condition. All personnel are responsible for checking shipping papers, labels, packages, etc. on radioactive materials movements to see that they are in proper condition for transportation. Those that are not in good order shall not be accepted by the pick-up driver or, if the condition is discovered in transit, they shall not be loaded on over-the-road units.

5. For specific details regarding the requirements for shipping papers, labels, package marking and placarding for radioactive materials shipments refer to the respective sections beginning at page 137.

6. In case of accident, breakdown or other emergency (including those at terminal locations) which result in breakage of packages of radioactive material, personnel should keep at least 50 feet away when the spillage is from containers of roughly 200 pounds, and 10 feet or more when the broken container weighs less. Personnel may move closer to the spillage to carry out necessary emergency activity but they should not come in contact with the material and they should stay in the area for as brief a time as possible.

The following steps should be taken, when pertinent, in case of accident, breakdown or other emergency which results in spillage or broken packages of radioactive materials:

a) Place warning signals as required in Part 392 of the DOT Safety Regulations and take other steps necessary to protect the vehicle.

b) Move injured persons only if they are located in a hazardous area.

c) Keep traffic and spectators moving. Do not let vehicles stop in the area of spilled materials or pass through the spillage.

d) If there is a hazard of radioactive materials being carried in the air, stay upwind from the spillage. Do not allow smoking, eating or drinking in the area. A handkerchief over the mouth and nose will minimize the possibility of breathing radioactive materials carried in the air.

e) Spilled liquids should be dammed to keep them from spreading to sewers, streams, etc. and tipped containers that might be leaking should be set upright, if possible. Powdered materials should be covered with a blanket, dirt or other material to prevent blowing.

f) In case of fire, summon the fire department. Personnel should fight fire if it is readily extinguishable, if there is no hazard from materials being carried in smoke and fumes and if no one person engaged in fire fighting stays in the vicinity of spillage for a prolonged period.

g) Be prepared to give cargo information to police and firefighters. Do not give information to unauthorized persons and avoid statements or action likely to panic uninformed spectators or passersby.

h) Take names and addresses of persons who might have breathed or been in contact with radioactive materials and ask that they stay on the scene until released by responsible authorities.

i) Personnel shall immediately contact the company by phone in case of accident, breakdown or other emergency in which radioactive materials containers are broken. The shipper and the Department of Transportation must be notified. In the absence of instructions from the shipper, call a U.S. Department of Energy (D.O.E.) Office or when the assistance of radiation hazard experts is wanted, call the closest D.O.E. Office. The offices shown on the map reproduced in this guide will respond to requests for assistance on an emergency basis with the action appropriate to the nature of the emergency. A request for assistance can be made at any time of the day or night by a collect telephone call if necessary. In some states, it may be necessary to call a designated state agency, such as the Health Department, in the event of an incident involving radioactive materials. In some cases, instructions on this matter may be given when the regional office of the Department of Energy is contacted.

j) When a shipment of radioactive materials is unduly delayed or found to be improperly packaged, a report shall be made immediately to the shipper.

k) When containers of radioactive materials are found to be leaking, broken or seriously damaged or when fire or explosions occur in connection with the transportation of radioactive materials a prompt report must be made to the Department of Transportation. Use ATA form C0670 (Hazardous Materials Incident Report) prescribed in section 171.16 of 49 CFR to report such incidents.

l) Contaminated vehicles cannot be returned to service until approved after inspection by an individual qualified to determine radiation levels in the vehicle.

DEPARTMENT OF ENERGY

REGIONAL COORDINATING OFFICES
FOR
RADIOLOGICAL ASSISTANCE
AND
GEOGRAPHICAL AREAS OF RESPONSIBILITY

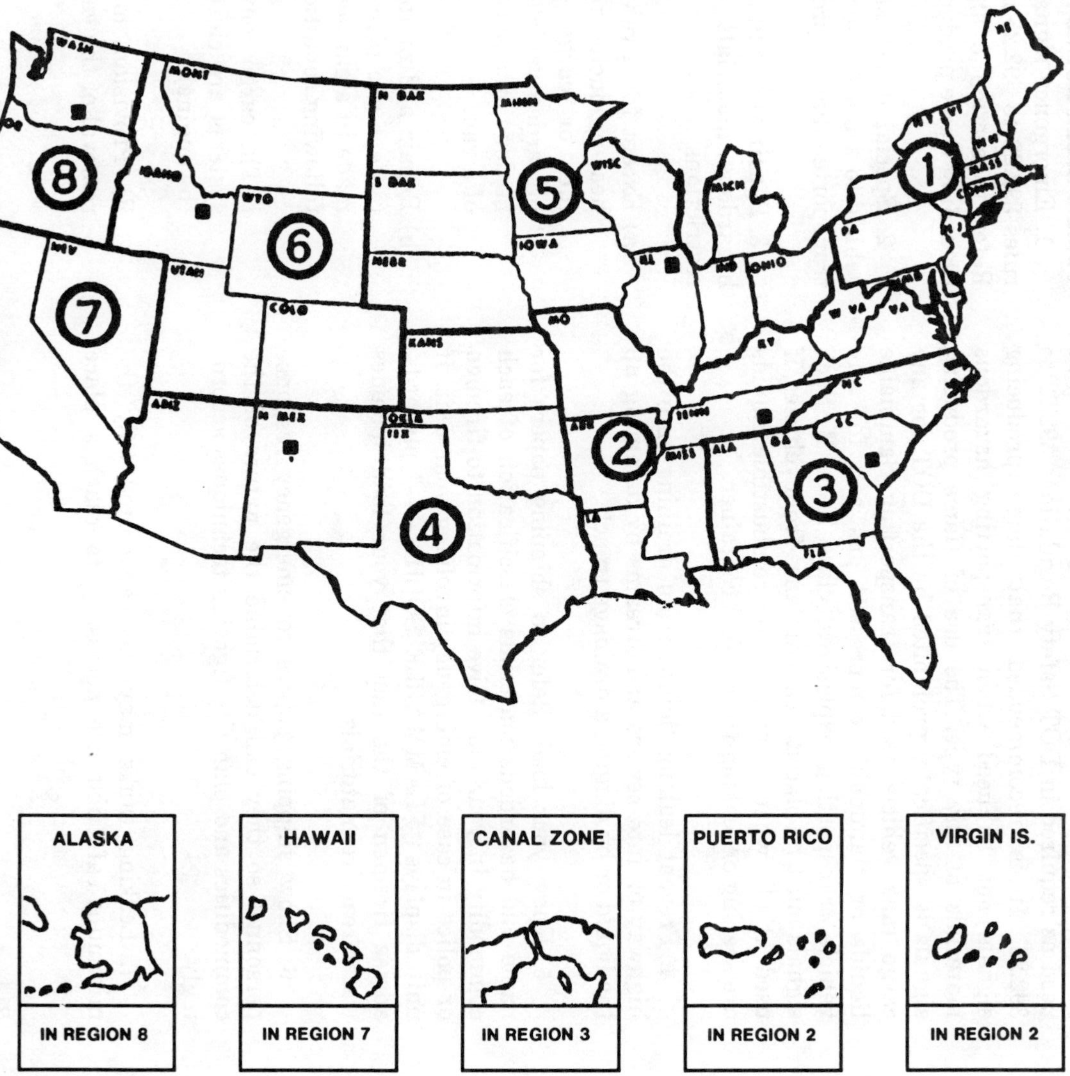

	REGIONAL COORDINATING OFFICE	POST OFFICE ADDRESS	TELEPHONE for ASSISTANCE
1	BROOKHAVEN AREA OFFICE	UPTON, L. I. NEW YORK 11973	(516) 282-2200
2	OAK RIDGE OPERATIONS OFFICE	P. O. BOX E OAK RIDGE, TENNESSEE 37830	(615) 576-1005 or (615) 525-7885
3	SAVANNAH RIVER OPERATIONS OFFICE	P. O. BOX A AIKEN, S. C. 29801	(303) 725-3333
4	ALBBUQUERQUE OPERATIONS OFFICE	P. O. BOX 5400 ALBUQUERQUE, NEW MEXICO 87115	(505) 844-4667
5	CHICAGO OPERATIONS OFFICE	9700 S. CASS AVE. ARGONNE, ILLINOIS 60439	Duty Hrs. (312) 972-4800 Off. Hrs. 972-5731
6	IDAHO OPERATIONS OFFICE	785 DOE DRIVE IDAHO FALLS, IDAHO 83401	(208) 526-1515
7	SAN FRANCISCO OPERATIONS OFFICE	1333 BROADWAY OAKLAND CALIFORNIA 94612	(416) 273-4237
8	RICHLAND OPERATIONS OFFICE	P. O. BOX 550 RICHLAND WASHINGTON 99352	(509) 378-7381

Correct as of June 4, 1986

Emergency Procedures

In the event of accident, breakdown or other emergency the driver must follow established company procedures for protecting the scene, obtaining assistance, notifying the fleet and gathering information necessary for making reports. In addition, the following procedures should be followed when hazardous materials are involved:

On The Highway

1. Do not let people congregate in the vicinity of the accident or fire unless they are authorized to engage in combatting the fire or helping to handle the accident.

2. Keep fires, flame and lighted cigarettes, cigars and pipes away from the scene.

3. Set up warning signals on the highway to prevent further accident. Emergency signals shall be used as required in DOT Safety Regulations 392.22 to 392.25. It is recommended that flame producing signals not be used when transporting hazardous materials of any type. The use of flame producing signals is specifically prohibited by the DOT for any cargo tank vehicle used for transporting flammable liquids or flammable compressed gas and for any vehicle transporting explosives, class A or B. Such signals must be placed a safe distance from vehicles, if used at all, when flammable or combustible liquids are seeping or leaking from a fuel container.

4. Prevent leaking liquids from draining onto the highway or into sewers and streams by damming up the liquid or by digging a drainage trench.

5. Know your load. Refer to shipping papers for name and hazardous materials classification of each commodity in your unit. Give information to firemen or police in case of emergency involving your unit. If unit displays "FLAMMABLE SOLID -W-" placards, advise firemen of the fact that your load includes water-reactive materials.

6. Show shipping papers to emergency response personnel so they can determine the nature of the commodities and plan fire fighting techniques accordingly.

7. Leaking tanks may be transported only the minimum safe distance necessary to reach a place where safe transfer of the hazardous material to another tank may be made.

8. Flammable liquid may be transferred from one container to another or from one vehicle to another on public roads in emergencies only. Containers must be bonded or grounded before and during transfer of the liquid from one container to another. Emergency signals (non-flame producing) must be set up during this operation.

9. Flammable solids and oxidizing materials should be unloaded if it will decrease the fire and other hazards. Broken boxes and loose materials should be gathered and disposed. Smoking boxes of matches and loose matches should be disposed of at a safe distance from the vehicle.

Vehicle Break Downs

1. Emergency signals shall be used as required in rules 392.22 to 392.25 of the DOT Safety Regulations. Refer to section 3 above for additional procedures regarding flame producing signals.

2. Special effort should be made to remove the vehicle to a place where the hazards of the load being transported may be protected against exposure.

3. Repair and maintenance of vehicles transporting hazardous materials are subject to the following restrictions:

a) Except for a vehicle placarded COMBUSTIBLE, heat-producing repairs may not be made on the fuel or cargo containment systems of any placarded vehicle; e.g. welding or cutting could be done on a spring shackle, but not on the door of a van.

b) Repair and maintenance on a placarded vehicle (except one placarded COMBUSTIBLE) may be made in a garage or closed space only under the following conditions:

i) The cargo containment and fuel containment systems are closed (except as necessary to run the engine);

ii) Provision must be made for immediate removal of the vehicle in case of emergency;

iii) The vehicle must be removed from the building on completion of the work;

iv) If the vehicle is transporting class A or B explosives, flammable liquid or flammable gas, all sources of spark, flame or glowing heat in the enclosure (including a heat system drawing air from the enclosure) must be shut down or made explosion-proof by an acceptable method. This does not apply to the electrical system of the vehicle if its use is necessary in connection with the maintenance or repair.

c) No heat-producing repair may be made on a cargo tank used for flammable liquids or gases, or poisonous liquids, or on a fuel tank of any type unless the tank, or any compartment has first been made gas-free.

Emergencies In Terminals

1. In case of fire, give priority to moving vehicles containing hazardous materials away from terminal. Whenever possible a prearranged safe location should be used.

2. Check all inbound units for signs of leakage before starting to handle cargo.

3. If leakage is noted at any time, stop operations and determine source of leak and nature of material. Look for HM labels on packagings and note type (if present). Check shipping papers for proper shipping name and hazard class of material.

4. Personnel must avoid direct contact with leaking hazardous materials.

5. If strong fumes are present, or if personnel become ill, remove affected personnel from scene, close up unit (if possible) and move it as far from all terminal buildings as possible. Avoid leaving the unit where leakage will adversely affect other nearby persons or property.

6. Notify fire, police, ambulance, health, pollution control, or other authorities as appropriate.

7. Contact shipper and/or Chemtrec to determine what measures can be taken, or what assistance is available, to alleviate the problem.

Emergency Assistance, Accidents And Incidents

1. If hazardous materials on a vehicle are found to be leaking, damaged, on fire or have any appearance of being in unsafe condition the driver should notify the fleet safety director or a responsible individual (dispatcher, key person or terminal manager) at the nearest terminal as quickly as possible under the circumstances. He or she must not leave his or her vehicle to make such notification unless there is no danger to the general public or unless the vehicle is under guard by a responsible individual. It may be necessary to have another person make such notification while the driver stays with the vehicle.

2. If a vehicle transporting hazardous materials is involved in an accident, the driver should ascertain if the hazardous materials have been damaged or if it appears that such materials may become involved for any other reason such as fire in the wreckage. After he or she has taken the necessary steps to protect the scene and the public he or she should contact his or her fleet as outlined in rule No. 1 above.

3. If hazardous materials at the terminal are found to be leaking or damaged immediate notification should be made to the terminal manager, key person or other company official on duty at the terminal.

Emergency Assistance

1. Shipper Liaison - Terminal key person should have the responsibility of contacting each shipper of hazardous materials to ascertain whether or not the shipper has a capability of providing advice and assistance to carrier personnel, and the authorities, in the event of an emergency involving its hazardous materials. The telephone number provided by each such shipper should be recorded and sent to the carrier's Safety Department for the development of a comprehensive listing of such shipper telephone numbers for distribution to all terminals.

2. Chemtrec - The Chemical Manufacturers Association, Inc. has developed the Chemical Transportation Emergency Center (Chemtrec) to provide immediate advice and assistance in dealing with hazardous materials emergencies to any concerned party, including any motor carrier. This service operates 24 hours a day, 7 days a week.

a) Chemtrec phone numbers:

General - (800) 424-9300 (add long-distance access number where required)

The above is a toll-free WATS number.

Calls within the District of Columbia - 483-7616.

fire or law enforcement should call

Calls from outside the continental U.S. - (202) 483-7616. Applicable Chemtrec numbers should be posted in each terminal. At least one copy should be posted where it can be located by a responsible employee at any hour of operation. It is recommended that all Chemtrec contacts be made by responsible supervisory personnel at carrier headquarters or terminals.

b) User Guidance - Chemtrec can usually provide hazard information warnings and guidance when given: (a) Name of the Product; (b) Nature of the Problem.

For more detailed information and/or assistance, or if the product is unknown, attempt to provide as much of the following information as possible:

Name of caller and call-back number
Carrier's name
Location of problem
Consignee
Shipper or manufacturer
Local conditions

c) Chemtrec also has contact with emergency assistance services of certain other shipper organizations.

Required notification for Hazardous Material, Hazardous Waste or Hazardous Substances Incidents (spills, leaks, discharges, etc.)

Whenever there is a spillage or leakage of hazardous materials, including hazardous wastes and hazardous substances in a vehicle or terminal, or there is direct involvement of the hazardous materials arising out of a motor vehicle accident, special reports must be made in accordance with the following:

a) IMMEDIATE NOTICE OF CERTAIN HAZARDOUS MATERIALS INCIDENTS *(49 CFR, 171.15)*

(1) A call to the DOT National Response Center must be made as promptly as possible if an incident involves hazardous materials during transportation, loading, unloading or storage and results in one or more of the following:

(a) Death of any person(s).

(b) Injury requiring hospitalization of any person(s).

(c) Estimated damage to carrier or other property exceeding $50,000.

(d) An evacuation of the general public occurs lasting one or more hours.

(e) One or more major transportation arteries or facilities are closed or shut down for one hour or more.

(f) Fire, breakage, spillage or suspected radioactive contamination involving a shipment of radioactive materials.

(g) Fire, breakage, spillage or suspected contamination involving a shipment of etiologic agents.

NOTE: Incidents involving etiologic agents may be reported by phone to the Center for Disease Control, (404) 633-5313, instead of the Department of Transportation number. The CDC phone number appears on the Etiologic Agents label.

(h) In the judgement of the carrier there exists a situation which should be reported even though it does not meet one of the specific criteria listed above.

SPECIAL DOT PHONE NUMBERS (National Response Center): Nationwide, 1-800-424-8802; Washington, D.C. Metropolitan Area, 202-426-2675.

(2) In making a telephone report, the following information must be provided to the extent that it is available:

(a) Name of reporter

(b) Name and address of carrier represented by reporter

(c) Phone number where reporter can be contacted

(d) Date, time and location of incident

(e) Extent of injuries, if any

(f) Classification, name and quantity of hazardous materials involved

(g) Type of incident, nature of hazardous materials involvement, and whether or not there is a continuing danger to life at the scene.

b) DETAILED HAZARDOUS MATERIALS INCIDENT REPORTS *(49 CFR, 171.16)*

(1) A written report must be made within 15 days of discovery of an incident arising out of the transportation, loading, unloading or storage of hazardous materials as follows:

(a) As a follow-up to any such incident reported by phone in accordance with the criteria set forth in A-1, above.

(b) As the result of an unintentional release of hazardous materials from any packaging including a cargo tank

(c) As the result of release of any amount of hazardous waste.

NOTE: Written incident reports are not required for the following hazardous materials: 1) consumer commodities; 2) batteries, electric storage, wet; 3) paint or paint related materials in containers of five gallons or less, unless a condition from (a) (1) of this section exists.

(2) The report in duplicate shall be sent to:

Information Systems Manager
Research & Special Programs Administration
U. S. Dept. of Transportation
Washington, D.C. 20590

The report shall be made on the prescribed form DOT F-5800.1 available from American Trucking Associations, as form C0670.

(3) The filing of a Hazardous Materials Incident Report does not relieve the interstate carrier from the necessity of also filing an MCS 50T report with the Regional Motor Carrier Safety Office, if the hazardous materials incident arises out of a motor vehicle accident.

(4) In filing a Hazardous Materials Incident Report, the carrier should endeavor to provide all of the information required to complete the report and to provide as much information as possible about the incident. These reports are analyzed by DOT to discover unsafe or inefficient packagings of hazardous materials, and to make such changes in the hazardous materials regulations as experience indicates are necessary for safety.

(5) For reports of hazardous waste discharges a copy of the hazardous waste manifest must be attached to the written report.

(6) The filing of the reports mentioned above does not relieve a carrier from the responsibility to report spillages of hazardous materials, or spillages of other materials that may create a pollution problem to local, state and federal agencies concerned with environmental controls as required in the area where any such spillage occurs.

c) HAZARDOUS SUBSTANCE DISCHARGE NOTIFICATION *(49 CFR, 171.17)*

(1) Any release of hazardous substances or hazardous wastes must be reported to the U.S. Coast Guard National Reponse Center at (toll free) 800-424-8802 or 202-426-2675. *(Note: Only spills of hazardous substances which equal or exceed the designated reportable quantity are required to be reported, however, it is recommeded that all spills be reported.)*

(2) The report must be made by either the driver or a company official as soon as that person has knowledge of the discharge.

(3) The discharge notification should include:

(a) The information listed in (a)(2) above;

(b) The name of the shipper; and

(c) The quantity of hazardous substance discharged, if known.

There are severe penalties for failure to make these reports.

(4) Releases of reportable quantities of an extremely hazardous waste or CERCLA hazardous substance listed in Title 40, Part 355, section 355.40, when they occur in transportation or storage incidental to transportation must be immediately reported to the local emergency planning committee, or if none available to the state emergency planning commission, or as a last resort by dialing 911 or calling the operateor. (See appendix A)

§ 355.40 Emergency release notification.

(a) *Applicability.* (1) The requirements of this section apply to any facility: (i) at which a hazardous chemical is produced, used or stored and (ii) at which there is release of a reportable quantity of any extremely hazardous substance or CERCLA hazardous substance.

(2) This section does not apply to: (i) Any release which results in exposure to persons solely within the boundaries of the facility. (ii) Any release which is a "federally permitted release" as defined in section 101 (10) of CERCLA, (iii) any release which is "continuous," as defined under section 103 (f) of CERCLA (except for "statistically significant increases" as defined under section 103(e) of CERCLA and (v) any release exempt from CERCLA section 103(a) reporting under section 101(22) of CERCLA.

Note to paragraph (a): Releases of CERCLA hazardous substances are subject to the release reporting requirements of CERCLA section 103, codified at 40 CFR Part 302, in addition to the requirements of this Part.]

(b) *Notice requirements.* (1) The owner or operator of a facility subject to this section shall immediately notify the community emergency coordinator for the local emergency planning committee of any area likely to be affected by the release and the State emergency response commission of any State likely to be affected by the release. If there is no local emergency planning committee, notification shall be provided under this section to relevant local emergency response personnel.

(2) The notice required under this section shall include the following to the extent known at the time of notice and so long as no delay in notice or emergency response results:

(i) The chemical name or identity of any substance involved in the release.

(ii) An indication of whether the substance is an extremely hazardous substance.

(iii) An estimate of the quantity of any such substance that was released into the environment.

(iv) The time and duration of the release.

(v) The medium or media into which the release occurred.

(vi) Any known or anticipated acute or chronic health risks associated with the emergency and, where appropriate, advice regarding medical attention necessary for exposed individuals.

(vii) Proper precautions to take as a result of the release, including evacuation (unless such information is readily available to the community emergency coordination pursuant to the emergency plan).

(viii) The names and telephone number of the person or persons to be contacted for further information.

(3) As soon as practicable after a release which requires notice under (b)(1) of this section, such owner or operator shall provide a written follow-up emergency notice (or notices, as more information becomes available) setting forth and updating the information required under paragraph (b)(2) of this section, and including additional information with respect to:

(i) Actions taken to respond to and contain the release,

(ii) Any known or anticipated acute or chronic health risks associated with the release, and,

(iii) Where appropriate, advice regarding medical attention necessary for exposed individuals.

(4) Exceptions. (i) Until April 30, 1988, in lieu of the notice specified in paragraph (b)(2) of this section, any owner or operator of a facility subject to this section from which there is a release of a CERCLA hazardous substance which is not an extremely hazardous substance and has a statutory reportable quantity may provide the same notice required under CERCLA section 103(a) to the local emergency planning committee.

(ii) An owner or operator of a facility from which there is a transportation-related release may meet the requirements of this section by providing the information indicated in paragraph (b)(2) to the 911 operator, or in the absence of a 911 emergency telephone number, to the operator. For purposes of this paragraph, a "transportation-related release" means a release during transportation, or storage incident to transportation if the stored substance is moving under active shipping papers and has not reached the ultimate consignee.

(Approved by the Office of Management and Budget under the control number 2050–0046)

§ 355.50 Penalties.

(a) *Civil penalties.* Any person who fails to comply with the requirements of § 355.40 shall be subject to civil penalties of up to $25,000 for each violation in accordance with section 325(b)(1) of the Act.

(b) *Civil penalties for continuing violations.* Any person who fails to comply with the requirements of § 355.40 shall be subject to civil penalties of up to $25,000 for each day during which the violation continues, in accordance with section 325(b)(2) of the Act. In the case of a second or subsequent violation, any such person may be subject to civil penalties of up to $75,000 for each day the violation continues, in accordance with section 325(b)(2) of the Act.

(c) *Criminal penalties.* Any person who knowingly and willfully fails to provide notice in accordance with § 355.40 shall, upon conviction, be fined not more than $25,000 or imprisoned for not more than two (2) years, or both (or, in the case of a second or subsequent conviction, shall be fined not more than $50,000 or imprisoned for not more than five (5) years, or both) in accordance with section 325(b)(4) of the Act.

Appendix A—The List of Extremely Hazardous Substances and their Threshold Planning Quantities

[Alphabetical Order]

CAS No.	Chemical name	Notes	Reportable quantity* (pounds)	Threshold planning quantity (pounds)
75-86-5	Acetone Cyanohydrin		10	1,000
1752-30-3	Acetone Thiosemicarbazide	e	1	1,000/10,000
107-02-8	Acrolein		1	500
79-06-1	Acrylamide	d, l	5,000	1,000/10,000
107-13-1	Acrylonitrile	d, l	100	10,000
814-68-6	Acrylyl Chloride	e, h	1	100
111-69-3	Adiponitrile	e, l	1	1,000
116-06-3	Aldicarb	c	1	100/10,000
309-00-2	Aldrin	d	1	500/10,000
107-18-6	Allyl Alcohol		100	1,000
107-11-9	Allylamine	e	1	500
20859-73-8	Aluminum Phosphide	b	100	500
54-62-6	Aminopterin	e	1	500/10,000
78-53-5	Amiton	e	1	500
3734-97-2	Amiton Oxalate	e	1	100/10,000
7664-41-7	Ammonia	l	100	500
16919-58-7	Ammonium Chloroplatinate	a, e	1	10,000
300-62-9	Amphetamine	e	1	1,000
62-53-3	Aniline	d, l	5,000	1,000
88-05-1	Aniline, 2,4,6-Trimethyl-	e	1	500
7783-70-2	Antimony Pentafluoride	e	1	500
1397-94-0	Antimycin A	c, e	1	1,000/10,000
86-88-4	ANTU		100	500/10,000
1303-28-2	Arsenic Pentoxide	d	5,000	100/10,000
1327-53-3	Arsenous Oxide	d, h	5,000	100/10,000
7784-34-1	Arsenous Trichloride	d	5,000	500
7784-42-1	Arsine	e	1	100
2642-71-9	Azinphos-Ethyl	e	1	100/10,000
86-50-0	Azinphos-Methyl		1	10/10,000
1405-87-4	Bacitracin	a, e	1	10,000
98-87-3	Benzal Chloride	d	5,000	500
98-16-8	Benzenamine, 3-(Trifluoromethyl)-	e	1	500
100-14-1	Benzene, 1-(Chloromethyl)-4-Nitro-	e	1	500/10,000
98-05-5	Benzenearsonic Acid	e	1	10/10,000
98-09-9	Benzenesulfonyl Chloride	a	100	10,000
3615-21-2	Benzimidazole, 4,5-Dichloro-2-(Trifluoromethyl)-	e, g	1	500/10,000
98-07-7	Benzotrichloride	d	1	100
100-44-7	Benzyl Chloride	d	100	500
140-29-4	Benzyl Cyanide	e, h	1	500
15271-41-7	Bicyclo[2.2.1]Heptane-2-Carbonitrile, 5-Chloro-6-((((Methylamino)Carbonyl)Oxy)Imino)-, (1s-(1-alpha, 2-beta, 4-alpha, 5-alpha, 6E))-	e	1	500/10,000

APPENDIX A—THE LIST OF EXTREMELY HAZARDOUS SUBSTANCES AND THEIR THRESHOLD PLANNING QUANTITIES—Continued

[Alphabetical Order]

CAS No.	Chemical name	Notes	Reportable quantity* (pounds)	Threshold planning quantity (pounds)
534-07-6	Bis(Chloromethyl) Ketone	e	1	10/10,000
4044-65-9	Bitoscanate	e	1	500/10,000
10294-34-5	Boron Trichloride	e	1	500
7637-07-2	Boron Trifluoride	e	1	500
353-42-4	Boron Trifluoride Compound With Methyl Ether (1:1)	e	1	1,000
28772-56-7	Bromadiolone	e	1	100/10,000
7726-95-6	Bromine	e. l	1	500
106-99-0	Butadiene	a, e	1	10,000
109-19-3	Butyl Isovalerate	a, e	1	10,000
111-34-2	Butyl Vinyl Ether	a, e	1	10,000
1306-19-0	Cadmium Oxide	e	1	100/10,000
2223-93-0	Cadmium Stearate	c, e	1	1,000/10,000
7778-44-1	Calcium Arsenate	d	1,000	500/10,000
8001-35-2	Camphechlor	d	1	500/10,000
56-25-7	Cantharidin	e	1	100/10,000
51-83-2	Carbachol Chloride	e	1	500/10,000
26419-73-8	Carbamic Acid, Methyl-, 0-(((2,4-Dimethyl-1, 3-Dithiolan-2-yl)Methylene)Amino)-	e	1	100/10,000
1563-66-2	Carbofuran		10	10/10,000
75-15-0	Carbon Disulfide	l	100	10,000
786-19-6	Carbophenothion	e	1	500
2244-16-8	Carvone	a, e	1	10,000
57-74-9	Chlordane	d	1	1,000
470-90-6	Chlorfenvinfos	e	1	500
7782-50-5	Chlorine		10	100
24934-91-6	Chlormephos	e	1	500
999-81-5	Chlormequat Chloride	e, h	1	100/10,000
107-20-0	Chloroacetaldehyde	a	1,000	10,000
79-11-8	Chloroacetic Acid	e	1	100/10,000
107-07-3	Chloroethanol	e	1	500
627-11-2	Chloroethyl Chloroformate	e	1	1,000
67-66-3	Chloroform	d, l	5,000	10,000
542-88-1	Chloromethyl Ether	d, h	1	100
107-30-2	Chloromethyl Methyl Ether	c, d	1	100
3691-35-8	Chlorophacinone	e	1	100/10,000
1982-47-4	Chloroxuron	e	1	500/10,000
21923-23-9	Chlorthiophos	e, h	1	500
10025-73-7	Chromic Chloride	e	1	1/10,000
7440-48-4	Cobalt	a, e	1	10,000
62207-76-5	Cobalt, ((2,2'-(1,2-Ethanediylbis (Nitrilomethylidyne))Bis(6-Fluorophenolato))(2-)-N,N',O,O')-,	e	1	100/10,000
10210-68-1	Cobalt Carbonyl	e, h	1	10/10,000
64-86-8	Colchicine	e, h	1	10/10,000
117-52-2	Coumafuryl	a, e	1	10,000
56-72-4	Coumaphos		10	100/10,000
5836-29-3	Coumatetralyl	e	1	500/10,000
95-48-7	Cresol, o-	d	1,000	1,000/10,000
535-89-7	Crimidine	e	1	100/10,000
4170-30-3	Crotonaldehyde		100	1,000
123-73-9	Crotonaldehyde, (E)-		100	1,000
506-68-3	Cyanogen Bromide		1,000	500/10,000
506-78-5	Cyanogen Iodide	e	1	1,000/10,000
2636-26-2	Cyanophos	e	1	1,000
675-14-9	Cyanuric Fluoride	e	1	100
66-81-9	Cycloheximide	e	1	100/10,000
108-91-8	Cyclohexylamine	e, l	1	10,000
287-92-3	Cyclopentane	a, e	1	10,000
633-03-4	C. I. Basic Green 1	a, e	1	10,000
17702-41-9	Decaborane(14)	e	1	500/10,000
8065-48-3	Demeton	e	1	500
919-86-8	Demeton-S-Methyl	e	1	500
10311-84-9	Dialifor	e	1	100/10,000
19287-45-7	Diborane	e	1	100
84-74-2	Dibutyl Phthalate	a	10	10,000
8023-53-8	Dichlorobenzalkonium Chloride	a, e	1	10,000
111-44-4	Dichloroethyl Ether	d	1	10,000
149-74-6	Dichloromethylphenylsilane	e	1	1,000
62-73-7	Dichlorvos		10	1,000
141-66-2	Dicrotophos	e	1	100
1464-53-5	Diepoxybutane	d	1	500
814-49-3	Diethyl Chlorophospate	e, h	1	500
1642-54-2	Diethylcarbamazine Citrate	e	1	100/10,000
93-05-0	Diethyl-p-Phenylenediamine	a,e	1	10,000
71-63-6	Digitoxin	c, e	1	100/10,000
2238-07-5	Diglycidyl Ether	e	1	1,000
20830-75-5	Digoxin	e, h	1	10/10,000
115-26-4	Dimefox	e	1	500
60-51-5	Dimethoate		10	500/10,000
2524-03-0	Dimethyl Phosphorochloridothioate	e	1	500
131-11-3	Dimethyl Phthalate	a	5,000	10,000
77-78-1	Dimethyl Sulfate	d	1	500
75-18-3	Dimethyl Sulfide	e	1	100
75-78-5	Dimethyldichlorosilane	e, h	1	500
57-14-7	Dimethylhydrazine	d	1	1,000
99-98-9	Dimethyl-p-Phenylenediamine	e	1	10/10,000
644-64-4	Dimetilan	e	1	500/10,000
534-52-1	Dinitrocresol		10	10/10,000
88-85-7	Dinoseb		1,000	100/10,000
1420-07-1	Dinoterb	e	1	500/10,000
117-84-0	Dioctyl Phthalate	a	5,000	10,000

Appendix A—The List of Extremely Hazardous Substances and their Threshold Planning Quantities—Continued

[Alphabetical Order]

CAS No.	Chemical name	Notes	Reportable quantity* (pounds)	Threshold planning quantity (pounds)
78-34-2	Dioxathion	e	1	500
646-06-0	Dioxolane	a, e	1	10,000
82-66-6	Diphacinone	e	1	10/10,000
152-16-9	Diphosphoramide, Octamethyl-		100	100
298-04-4	Disulfoton		1	500
514-73-8	Dithiazanine Iodide	e	1	500/10,000
541-53-7	Dithiobiuret		100	100/10,000
316-42-7	Emetine, Dihydrochloride	e, h	1	1/10,000
115-29-7	Endosulfan		1	10/10,000
2778-04-3	Endothion	e	1	500/10,000
72-20-8	Endrin		1	500/10,000
106-89-8	Epichlorohydrin	d, l	1,000	1,000
2104-64-5	EPN	e	1	100/10,000
50-14-6	Ergocalciferol	c, e	1	1,000/10,000
379-79-3	Ergotamine Tartrate	e	1	500/10,000
1622-32-8	Ethanesulfonyl Chloride, 2-Chloro-	e	1	500
10140-87-1	Ethanol, 1,2-Dichloro-, Acetate	e	1	1,000
563-12-2	Ethion		10	1,000
13194-48-4	Ethoprophos	e	1	1,000
538-07-8	Ethylbis(2-Chloroethyl)Amine	e, h	1	500
371-62-0	Ethylene Fluorohydrin	c, e, h	1	10
75-21-8	Ethylene Oxide	d, l	1	1,000
107-15-3	Ethylenediamine		5,000	10,000
151-56-4	Ethyleneimine	d	1	500
2235-25-8	Ethylmercuric Phosphate	a, e	1	10,000
542-90-5	Ethylthiocyanate	e	1	10,000
22224-92-6	Fenamiphos	e	1	10/10,000
122-14-5	Fenitrothion	e	1	500
115-90-2	Fensulfothion	e, h	1	500
4301-50-2	Fluenetil	e	1	100/10,000
7782-41-4	Fluorine	k	10	500
640-19-7	Fluoroacetamide	j	100	100/10,000
144-49-0	Fluoroacetic Acid	e	1	10/10,000
359-06-8	Fluoroacetyl Chloride	c, e	1	10
51-21-8	Fluorouracil	e	1	500/10,000
944-22-9	Fonofos	e	1	500
50-00-0	Formaldehyde	d,l	1,000	500
107-16-4	Formaldehyde Cyanohydrin	e, h	1	1,000
23422-53-9	Formetanate Hydrochloride	e,h	1	500/10,000
2540-82-1	Formothion	e	1	100
17702-57-7	Formparanate	e	1	100/10,000
21548-32-3	Fosthietan	e	1	500
3878-19-1	Fuberidazole	e	1	100/10,000
110-00-9	Furan		100	500
13450-90-3	Gallium Trichloride	e	1	500/10,000
77-47-4	Hexachlorocyclopentadiene	d, h	1	100
1335-87-1	Hexachloronaphthalene	a, e	1	10,000
4835-11-4	Hexamethylenediamine, N,N'-Dibutyl-	e	1	500
302-01-2	Hydrazine	d	1	1,000
74-90-8	Hydrocyanic Acid		10	100
7647-01-0	Hydrogen Chloride (Gas Only)	e, l	1	500
7664-39-3	Hydrogen Fluoride		100	100
7722-84-1	Hydrogen Peroxide (Conc >52%)	e, l	1	1,000
7783-07-5	Hydrogen Selenide	e	1	10
7783-06-4	Hydrogen Sulfide	l	100	500
123-31-9	Hydroquinone	l	1	500/10,000
53-86-1	Indomethacin	a, e	1	10,000
10025-97-5	Iridium Tetrachloride	a, e	1	10,000
13463-40-6	Iron, Pentacarbonyl-	e	1	100
297-78-9	Isobenzan	e	1	100/10,000
78-82-0	Isobutyronitrile	e, h	1	1,000
102-36-3	Isocyanic Acid, 3,4-Dichlorophenyl Ester	e	1	500/10,000
465-73-6	Isodrin		1	100/10,000
55-91-4	Isofluorphate	c	100	100
4098-71-9	Isophorone Diisocyanate	b, e	1	100
108-23-6	Isopropyl Chloroformate	e	1	1,000
625-55-8	Isopropyl Formate	e	1	500
119-38-0	Isopropylmethylpyrazolyl Dimethylcarbamate	e	1	500
78-97-7	Lactonitrile	e	1	1,000
21609-90-5	Leptophos	e	1	500/10,000
541-25-3	Lewisite	c, e, h	1	10
58-89-9	Lindane	d	1	1,000/10,000
7580-67-8	Lithium Hydride	b, e	1	100
109-77-3	Malononitrile		1,000	500/10,000
12108-13-3	Manganese, Tricarbonyl Methylcyclopentadienyl	e, h	1	100
51-75-2	Mechlorethamine	c, e	1	10
950-10-7	Mephosfolan	e	1	500
1600-27-7	Mercuric Acetate	e	1	500/10,000
7487-94-7	Mercuric Chloride	e	1	500/10,000
21908-53-2	Mercuric Oxide	e	1	500/10,000
108-67-8	Mesitylene	a, e	1	10,000
10476-95-6	Methacrolein Diacetate	e	1	1,000
760-93-0	Methacrylic Anhydride	e	1	500
126-98-7	Methacrylonitrile	h	1	500
920-46-7	Methacryloyl Chloride	e	1	100
30674-80-7	Methacryloyloxyethyl Isocyanate	e, h	1	100
10265-92-6	Methamidophos	e	1	100/10,000
558-25-8	Methanesulfonyl Fluoride	e	1	1,000

[Alphabetical Order]

CAS No.	Chemical name	Notes	Reportable quantity* (pounds)	Threshold planning quantity (pounds)
950-37-8	Methidathion	e	1	500/10,000
2032-65-7	Methiocarb		10	500/10,000
16752-77-5	Methomyl	h	100	500/10,000
151-38-2	Methoxyethylmercuric Acetate	e	1	500/10,000
80-63-7	Methyl 2-Chloroacrylate	e	1	500
74-83-9	Methyl Bromide	l	1,000	1,000
79-22-1	Methyl Chloroformate	d, h	1,000	500
624-92-0	Methyl Disulfide	e	1	100
60-34-4	Methyl Hydrazine		10	500
624-83-9	Methyl Isocyanate	f	1	500
556-61-6	Methyl Isothiocyanate	b, e	1	500
74-93-1	Methyl Mercaptan		100	500
3735-23-7	Methyl Phenkapton	e	1	500
676-97-1	Methyl Phosphonic Dichloride	b, e	1	100
556-64-9	Methyl Thiocyanate	e	1	10,000
78-94-4	Methyl Vinyl Ketone	e	1	10
502-39-6	Methylmercuric Dicyanamide	e	1	500/10,000
75-79-6	Methyltrichlorosilane	e, h	1	500
1129-41-5	Metolcarb	e	1	100/10,000
7786-34-7	Mevinphos		10	500
315-18-4	Mexacarbate		1,000	500/10,000
50-07-7	Mitomycin C	d	1	500/10,000
6923-22-4	Monocrotophos	e	1	10/10,000
2763-96-4	Muscimol	a, h	1,000	10,000
505-60-2	Mustard Gas	e, h	1	500
7440-02-0	Nickel	a, d	1	10,000
13463-39-3	Nickel Carbonyl	d	1	1
54-11-5	Nicotine	c	100	100
65-30-5	Nicotine Sulfate	e	1	100/10,000
7697-37-2	Nitric Acid		1,000	1,000
10102-43-9	Nitric Oxide	c	10	100
98-95-3	Nitrobenzene	l	1,000	10,000
1122-60-7	Nitrocyclohexane	e	1	500
10102-44-0	Nitrogen Dioxide		10	100
62-75-9	Nitrosodimethylamine	d, h	1	1,000
991-42-4	Norbormide	e	1	100/10,000
0	Organorhodium Complex (PMN-82-147)	e	1	10/10,000
65-86-1	Orotic Acid	a, e	1	10,000
20816-12-0	Osmium Tetroxide	a	1,000	10,000
630-60-4	Ouabain	c. e	1	100/10,000
23135-22-0	Oxamyl	e	1	100/10,000
78-71-7	Oxetane, 3,3-Bis(Chloromethyl)-	1	e	500
2497-07-6	Oxydisulfoton	e, h	1	500
10028-15-6	Ozone	e	1	100
1910-42-5	Paraquat	e	1	10/10,000
2074-50-2	Paraquat Methosulfate	e	1	10/10,000
56-38-2	Parathion	c, d	1	100
298-00-0	Parathion-Methyl	c	100	100/10,000
12002-03-8	Paris Green	d	100	500/10,000
19624-22-7	Pentaborane	e	1	500
76-01-7	Pentachloroethane	a, d	1	10,000
87-86-5	Pentachlorophenol	a, d	10	10,000
2570-26-5	Pentadecylamine	e	1	100/10,000
79-21-0	Peracetic Acid	e	1	500
594-42-3	Perchloromethylmercaptan		100	500
108-95-2	Phenol		1,000	500/10,000
97-18-7	Phenol, 2,2'-Thiobis(4,6-Dichloro-	e	1	100/10,000
4418-66-0	Phenol, 2,2'-Thiobis(4-Chloro-6-Methyl)-	e	1	100/10,000
64-00-6	Phenol, 3-(1-Methylethyl)-, Methylcarbamate	e	1	500/10,000
58-36-6	Phenoxarsine, 10,10'-Oxydi-	e	1	500/10,000
696-28-6	Phenyl Dichloroarsine	d, h	1	500
59-88-1	Phenylhydrazine Hydrochloride	e	1	1,000/10,000
62-38-4	Phenylmercury Acetate		100	500/10,000
2097-19-0	Phenylsilatrane	e, h	1	100/10,000
103-85-5	Phenylthiourma		108	100/10,000
298-02-2	Phorate		10	10
4104-14-7	Phosacetim	e	1	100/10,000
947-02-4	Phosfolan	e	1	100/10,000
75-44-5	Phosgene	l	10	10
732-11-6	Phosmet	e	1	10/10,000
13171-21-6	Phosphamidon	e	1	100
7803-51-2	Phosphine		100	500
2703-13-1	Phosphonothioic Acid, Methyl-, O-Ethyl O-(4-(Methylthio)Phenyl) Ester	e	1	500
50782-69-9	Phosphonothioic Acid, Methyl-, S-(2-(Bis(1-Methylethyl)Amino)Ethyl O-Ethyl Ester	e	1	100
2665-30-7	Phosphonothioic Acid, Methyl-, O-(4-Nitrophenyl) O-Phenyl Ester	e	1	500
3254-63-5	Phosphoric Acid, Dimethyl 4-(Methylthio) Phenyl Ester	e	1	500
2587-90-8	Phosphorothioic Acid, O,O-Dimethyl-S-(2-Methylthio) Ethyl Ester	c, e, g	1	500
7723-14-0	Phosphorus	b, h	1	100
10025-87-3	Phosphorus Oxychloride	d	1,000	500
10026-13-8	Phosphorus Pentachloride	b, e	1	500
1314-56-3	Phosphorus Pentoxide	b, e	1	10
7719-12-2	Phosphorus Trichloride		1,000	1,000
84-80-0	Phylloquinone	a, e	1	10,000
57-47-6	Physostigmine	e	1	100/10,000
57-64-7	Physostigmine, Salicylate (1:1)	e	1	100/10,000
124-87-8	Picrotoxin	e	1	500/10,000
110-89-4	Piperidine	e	1	1,000
5281-13-0	Piprotal	e	1	100/10,000

APPENDIX A—THE LIST OF EXTREMELY HAZARDOUS SUBSTANCES AND THEIR THRESHOLD PLANNING QUANTITIES—Continued

[Alphabetical Order]

CAS No.	Chemical name	Notes	Reportable quantity* (pounds)	Threshold planning quantity (pounds)
23505-41-1	Pirimifos-Ethyl	e	1	1,000
10025-65-7	Platinous Chloride	a, e	1	10,000
13454-96-1	Platinum Tetrachloride	a, e	1	10,000
10124-50-2	Potassium Arsenite	d	1,000	500/10,000
151-50-8	Potassium Cyanide	b	10	100
506-61-6	Potassium Silver Cyanide	b	1	500
2631-37-0	Promecarb	e, h	1	500/10,000
106-96-7	Propargyl Bromide	e	1	10
57-57-8	Propiolactone, Beta-	e	1	500
107-12-0	Propionitrile		10	500
542-76-7	Propionitrile, 3-Chloro-		1,000	1,000
70-69-9	Propiophenone, 4-Amino-	e, g	1	100/10,000
109-61-5	Propyl Chloroformate	e	1	500
1331-17-5	Propylene Glycol, Allyl Ether	a, e	1	10,000
75-56-9	Propylene Oxide	l	100	10,000
75-55-8	Propyleneimine	d	1	10,000
2275-18-5	Prothoate	e	1	100/10,000
95-63-6	Pseudocumene	a, e	1	10,000
129-00-0	Pyrene	c	5,000	1,000/10,000
140-76-1	Pyridine, 2-Methyl-5-Vinyl-	e	1	500
504-24-5	Pyridine, 4-Amino-	h	1,000	500/10,000
1124-33-0	Pyridine, 4-Nitro-, 1-Oxide	e	1	500/10,000
53558-25-1	Pyriminil	e, h	1	100/10,000
10049-07-7	Rhodium Trichloride	a, e	1	10,000
14167-18-1	Salcomine	e	1	500/10,000
107-44-8	Sarin	e, h	1	10
7783-00-8	Selenious Acid		10	1,000/10,000
7791-23-3	Selenium Oxychloride	e	1	500
563-41-7	Semicarbazide Hydrochloride	e	1	1,000/10,000
3037-72-7	Silane, (4-Aminobutyl)Diethoxymethyl-	e	1	1,000
128-56-3	Sodium Anthraquinone-1-Sulfonate	a, e	1	10,000
7631-89-2	Sodium Arsenate	d	1,000	1,000/10,000
7784-46-5	Sodium Arsenite	d	1,000	500/10,000
26628-22-8	Sodium Azide (Na(N3))	b	1,000	500
124-65-2	Sodium Cacodylate	e	1	100/10,000
143-33-9	Sodium Cyanide (Na(CN))	b	10	100
62-74-8	Sodium Fluoroacetate		10	10/10,000
131-52-2	Sodium Pentachlorophenate	e	1	100/10,000
13410-01-0	Sodium Selenate	e	1	100/10,000
10102-18-8	Sodium Selenite	h	100	100/10,000
10102-20-2	Sodium Tellurite	e	1	500/10,000
900-95-8	Stannane, Acetoxytriphenyl-	e, g	1	500/10,000
57-24-9	Strychnine	c	10	100/10,000
60-41-3	Strychnine, Sulfate	e	1	100/10,000
3689-24-5	Sulfotep		100	500
3569-57-1	Sulfoxide, 3-Chloropropyl Octyl	e	1	500
7446-09-5	Sulfur Dioxide	e, l	1	500
7783-60-0	Sulfur Tetrafluoride	e	1	100
7446-11-9	Sulfur Trioxide	b, e	1	100
7664-93-9	Sulfuric Acid		1,000	1,000
77-81-6	Tabun	c, e, h	1	10
13494-80-9	Tellurium	e	1	500/10,000
7783-80-4	Tellurium Hexafluoride	e, k	1	100
107-49-3	TEPP		10	100
13071-79-9	Terbufos	e, h	1	100
78-00-2	Tetraethyllead	c, d	10	100
597-64-8	Tetraethyltin	c, e	1	100
75-74-1	Tetramethyllead	c, e, l	1	100
509-14-8	Tetranitromethane		10	500
1314-32-5	Thallic Oxide	a	100	10,000
10031-59-1	Thallium Sulfate	h	100	100/10,000
6533-73-9	Thallous Carbonate	c, h	100	100/10,000
7791-12-0	Thallous Chloride	c, h	100	100/10,000
2757-18-8	Thallous Malonate	c, e, h	1	100/10,000
7446-18-6	Thallous Sulfate		100	100/10,000
2231-57-4	Thiocarbazide	e	1	1,000/10,000
21564-17-0	Thiocyanic Acid, 2-(Benzothiazolylthio)Methyl Ester	a, e	1	10,000
39196-18-4	Thiofanox		100	100/10,000
640-15-3	Thiometon	a, e	1	10,000
297-97-2	Thionazin		100	500
108-98-5	Thiophenol		100	500
79-19-6	Thiosemicarbazide		100	100/10,000
5344-82-1	Thiourea, (2-Chlorophenyl)-		100	100/10,000
614-78-8	Thiourea, (2-Methylphenyl)-	e	1	500/10,000
7550-45-0	Titanium Tetrachloride	e	1	100
584-84-9	Toluene 2,4-Diisocyanate		100	500
91-08-7	Toluene 2,6-Diisocyanate		100	100
110-57-6	Trans-1,4-Dichlorobutene	e	1	500
1031-47-6	Triamiphos	e	1	500/10,000
24017-47-8	Triazofos	e	1	500
76-02-8	Trichloroacety Chloride	e	1	500
115-21-9	Trichloroethylsilane	e, h	1	500
327-98-0	Trichloronate	e, k	1	500
98-13-5	Trichlorophenylsilane	e, h	1	500
52-68-6	Trichlorophon	a	100	10,000
1558-25-4	Trichloro(Chloromethyl)Silane	e	1	100
27137-85-5	Trichloro(Dichlorophenyl)Silane	e	1	500
998-30-1	Triethoxysilane	e	1	500

Appendix A—The List of Extremely Hazardous Substances and their Threshold Planning Quantities—Continued

[Alphabetical Order]

CAS No.	Chemical name	Notes	Reportable quantity* (pounds)	Threshold planning quantity (pounds)
75-77-4	Trimethylchlorosilane	e	1	1,000
824-11-3	Trimethylolpropane Phosphite	e, h	1	100/10,000
1066-45-1	Trimethyltin Chloride	e	1	500/10,000
639-58-7	Triphenyltin Chloride	e	1	500/10,000
555-77-1	Tris(2-Chloroethyl)Amine	e, h	1	100
2001-95-8	Valinomycin	c, e	1	1,000/10,000
1314-62-1	Vanadium Pentoxide		1,000	100/10,000
108-05-4	Vinyl Acetate Monomer	d, l	5,000	1,000
3048-64-4	Vinylnorbornene	a, e	1	10,000
81-81-2	Warfarin		100	500/10,000
129-06-6	Warfarin Sodium	e, h	1	100/10,000
28347-13-9	Xylylene Dichloride	e	1	100/10,000
58270-08-9	Zinc, Dichloro(4,4-Dimethyl-5((((Methylamino) Carbonyl)Oxy)Imino)Pentanenitrile)-,(T-4)-	e	1	100/10,000
1314-84-7	Zinc Phosphide	b	100	500

*Only the statutory or final RQ is shown. For more information, see 40 CFR Table 302.4

Notes:

a This chemical does not meet acute toxicity criteria. Its TPQ is set at 10,000 pounds.
b This material is a reactive solid. The TPQ does not default to 10,000 pounds for non-powder, non-molten, non-solution form.
c The calculated TPQ changed after technical review as described in the technical support document.
d Indicates that the RQ is subject to change when the assessment of potential carcinogenicity and/or other toxicity is completed.
e Statutory reportable quantity for purposes of notification under SARA sect 304(a)(2).
f The statutory 1 pound reportable quantity for methyl isocyanate may be adjusted in a future rulemaking action.
g New chemicals added that were not part of the original list of 402 substances.
h Revised TPQ based on new or re-evaluated toxicity data.
j TPQ is revised to its calculated value and does not change due to technical review as in proposed rule.
k The TPQ was revised after proposal due to calculation error.
l Chemicals on the original list that do not meet toxicity criteria but because of their high production volume and recognized toxicity are considered chemicals of concern ("Other chemicals").

Public Relations In Emergencies

Introduction

The transportation of hazardous materials is becoming increasingly complex with the introduction of new products and increased transportation demands. As a result the regulations which govern the transportation of these goods are constantly undergoing amendment and new packaging designs and specifications are being developed. Recent technology has provided more economical and efficient methods to ship goods such as the introduction of cryogenic liquids which reduce volumes of gases by up to 600 times through refrigeration. The nuclear industry throughout the world has burgeoned with the development of nuclear power and nuclear pharmaceuticals to provide a better way of life for our citizens. As a direct result of the use of nuclear technology, we now have increased tonnages of radioactive waste to safely transport and store. A combination of all these factors results in more hazardous materials than ever before on our highways at any given time.

Accidents have been few but increased interest by enforcement agencies, news media and the public in the potential dangers involved in the transportation of hazardous materials has resulted in some citations of carriers for violations of the regulations and in bad publicity for these carriers and this industry.

This section explains how public relations planning can be made an integral part of a trucking company's action program. It is intended to assist trucking employees in handling emergencies and communicating necessary information to each other and to the public. It suggests plans of action which can be used to help insure against bad publicity, exaggerations, distortions and misunderstandings.

A Bad Example

Several years ago, one company learned the hard way what can happen when public relations planning is left undone. There was an emergency - explosion and fire - at one of the largest of the company's terminals. Within two hours, a crowd had gathered outside the gates - news media representatives, curious and alarmed neighbors and employees, and officials representing several federal, state and local investigative organizations.

Terminal employees handled the fire control and injury treatment with the aid of local firemen because there was an explicit emergency plan but no thought had been given to including public relations instructions. Terminal management personnel and other employees found themselves being harrassed by the press while they were trying to devote full attention to the emergency. The information they provided under pressure resulted in poor and inaccurate news stories, rampant rumors and bitter criticism of the company.

Responsibility

When an emergency occurs, there are three major areas of responsibility for the company's management and employees:

Personnel (care for the injured and protect against further injury)

Property (confine the disaster to the smallest possible area and protect property and equipment)

Public (the basic facts must be given to an interested and anxious public as quickly and accurately and as completely as possible)

With regard to Personnel, one public relations responsibility is to find out who is involved and get full information on them. Contact relatives before releasing names to the press. Where Property is concerned, such as equipment involved in an incident, refer to specific models in reporting on the incident.

In getting the story to the Public, be sure to include safety measures taken, the previous good safety record of your firm and of the employees involved. Be sure to note in your disaster reporting any uncontrollable factors which might have contributed to the cause of the emergency (poor visibility, gas leak in terminal area, etc.). And follow up as more information becomes available with calls to the news media regarding your company's efforts to clean up after the incident.

A Written Plan

Don't rely on anyone to remember what to do in a hazardous materials emergency. Put the plans in writing. Give copies to everyone who might be involved, directly or indirectly, in the plan's execution (don't overlook telephone operators).

Have on hand an alphabetical listing of hazardous materials with instructions on how to deal with them in emergencies. (This information is contained in the DOT Emergency Response Manual which is available from ATA.) These instructions, besides telling you what to do, can provide the factual basis for your press releases. This is a very important point since it is virtually impossible for an individual to know about all the dangerous commodities he handles. More than 15,000 are listed and classified as hazardous materials and there are thousands of shipments of these materials made daily in the United States.

Drivers should have a brief outline of actions to take in an emergency.

For The Driver

Drivers should have brief instructions about appropriate actions to take in an emergency. Such instructions should recognize that, except for reporting the incident, the driver(s) may be essentially out of touch with company personnel, especially if the incident occurs at a distant location. Instructions to drivers should be based on the rules for conduct at the scene of an accident as set forth in the driver handbook, or listed in the accident report kit.

The driver should advise emergency service personnel of the names and classes of hazardous materials in the cargo and show the shipping papers to them. He or she should not speculate on the nature of the material or the cause of the incident. The driver should politely decline to answer questions at the scene and refer questioners to company officials.

Emergency Information Chief

Include in the emergency instructions the suggestion that each terminal appoint an emergency chief for each shift (this should be the key person as covered in the Hazardous Materials Handling Program). It would be this person's responsibility to pass on information to reporters as fast as he can get accurate data. During periods of delay in assembling facts, the information chief should reassure the press and electronic media representatives that information is coming as quickly as possible.

Reporters and photographers may want to visit the scene of a fire or explosion. The decision as to when to permit them to do so must be based fundamentally on safety considerations. As soon as the key person determines it is safe for a small number of reporters to visit the scene, the information chief should arrange to escort them on a tour, pointing out action the company has taken to contain the disaster.

Importance of Attitude

Chances are that newspaper reporters, photographers and electronic media representatives will arrive at the scene of the disaster expecting to be met with denials, flat refusals of information, thin excuses for non-cooperation or an arbitrary prohibition against photographs. They have come to have that attitude after bitter experiences.

The attitude of the company representative who deals with the media will make a big difference in the handling of news about your emergency. To quote one public relations manual, "Never try to cover up, no matter what the circumstances. It's regrettable but not a public disgrace to have an accident. To have one in print after you have denied it is highly embarrassing to a company...Remember the newspapers have free access to hospitals, coroners, fire and police departments and to your own employees. They will get the news. See that they get it correctly."

Just as bad as covering up is speculating, spreading rumor unfounded in fact. Only established facts should be given. Never make guesses. It is much better to say, "the cause has not been determined" or "due to safety and security reasons, we cannot allow visitors to approach the scene of the accident as yet."

For the best possible treatment by the media, be forthrightly honest, openly reasonable, willing to cooperate and ready to explain delays. Recognition and proper handling of these public relations responsibilities will enhance the reputation of your company as a responsible organization in the community where the emergency occurs.

One Company's Procedure

Many companies already have a specific order of action for emergency procedures. Here is one firm's recommended course of action for the emergency information chief following an incident. (First, division management and other specified company personnel are notified.)

1. Attend first to the three P's - personnel, property and public.

2. Inform the public through the news media.

3. Report all available facts concerning the incident and assure newsmen that additional information will be forthcoming as soon as possible.

4. Allow newsmen to visit the scene as soon as the area is safe. Be sure they are accompanied by a company representative.

5. Make sure the newsmen understand the scope of the emergency as it is common for them to exaggerate the seriousness of it.

6. Inform families of injured personnel first, before releasing their names to the press. Newsmen understand this procedure and should cooperate.

7. Make telephone facilities available to the newsmen when possible.

8. Explain safety procedures and safeguards that have been established to protect against emergencies. Be sure that terminal personnel are briefed from time to time on what to do in case of an emergency or disaster.

9. Government officials will usually investigate. Cooperate fully and be sure they are assisted in every way.

A Common Problem

Perhaps the most common problem the trucking industry faces in emergencies is the futile attempt of employees to "cover-up", fearing repercussions from management if they tell too much. They usually are driven to such measures out of loyalty and uncertainty. It is not enough to draft a good public relations plan for emergency situations. The company must see that all employees, especially supervisors, are aware of just how far you want them to go in providing facts, quickly and accurately, to the public.

Positive Actions in Between

Work with firefighters and police departments in your community before emergencies force it. Not all policemen and firemen are familiar with how to handle hazardous materials. For many years leading motor carriers have spoken out encouraging trucking companies to familiarize firemen with how to handle the equipment (hooking and unhooking trailers, for instance). Go over your emergency action plans with them and with policemen, and have them tour the terminal so they will be familiar with it should an emergency ever occur. This action will also make them aware of your company's desire to cooperate, to be a good citizen, and will help when your reputation is on the line in an emergency.

In addition, a carrier should be aware that in an emergency, immediate help may be obtained from the Chemical Manufacturers Association, Chemical Transportation Emergency Center (CHEMTREC), telephone number (800) 424-9300.

EPA Regional Hazardous Waste Offices

For specific information regarding the federal hazardous waste requirements contact the Office of the Environmental Protection Agency in your region.

EPA Region 1:
(Connecticut, Maine, New Hampshire, Rhode Island, Vermont.)

Waste Management Division
State Waste Programs Branch
John F. Kennedy Federal Building
Boston, Massachusetts 02203
(617) 223-1922

EPA Region 2
(New Jersey, New York, Puerto Rico, Virgin Islands)

Air and Waste Management Division
Hazardous Waste Facilities Branch
26 Federal Plaza
New York, New York 10278
(212) 264-0504

EPA Region 3:
(Delaware, District of Columbia, Maryland, Pennsylvania, Virginia, West Virginia)

Hazardous Waste Management Division
Waste Management Branch
841 Chestnut Street
Philadelphia, PA 19107
(215) 597-0980

EPA Region 4:
(Alabama, Florida, Georgia, Kentucky, Mississippi, North Carolina, South Carolina, Tennessee)

Waste Management Division
RCRA Branch
345 Courtland Street, NE
Atlanta, Georgia 30365
(404) 347-3016

EPA Region 5:
(Illinois, Indiana, Michigan, Minnesota, Ohio, Wisconsin)

Waste Management Division
Solid Waste Branch
230 South Dearborn Street
Chicago, IL 60604
(312) 886-7435

EPA Region 6:
(Arkansas, Louisiana, New Mexico, Oklahoma, Texas)

Allied Bank Tower
1445 Ross Ave.
Dallas, TX 75203-2733
(214) 655-6700

EPA Region 7:
(Iowa, Kansas, Missouri, Nebraska)

Waste Management Division
726 Minnesota Avenue
Kansas City, KS 66101
(913) 236-2850

EPA Region 8:
(Colorado, Montana, North Dakota, South Dakota, Utah, Wyoming)

Waste Management Division
Hazardous Waste Branch (8HWM-ON)
One Denver Place, Suite 1300
999 18th Street
Denver, CO 80202-2413
(303) 293-1502

EPA Region 9:
(Arizona, California, Hawaii, Nevada, Guam, Marianas)

Toxics and Waste Management Division
Waste Programs Branch
215 Fremont Street
San Francisco, CA 94105
(415) 974-8119

EPA Region 10:
(Alaska, Idaho, Oregon, Washington)

Hazardous Waste Division
Waste Management Branch
1200 6th Avenue
Seattle, WA 98101
(206) 442-2782

State Hazardous Waste Agencies & Requirements

The American Trucking Associations publication "Trasporting Hazardous Waste" tells what each state requires of transporters of hazardous waste, and the name and location of the agency in each state that regulates such transportation.

The publication also summarizes Federal requirements of the Environmental Protection Agency.

It may be obtained from the ATA Customer Services Department: 1-800-ATA-LINE.

Alabama
Department of Environmental Management
Land Division
1751 Congressman W.L. Dickinson Drive
Montgomery, Alabama 36130
(205) 271-7735

Alaska
Department of Environmental Conservation
Hazardous and Solid Waste Management
PO Box 0
Juneau, Alaska 99811-1800
(907) 465-2671

American Samoa
Department of Public Works
Pago Pago, American Samoa 96799
Overseas Operator (Commercial Call 633-4141)

Arizona
Department of Environmental Quality
Compliance Section
2005 N Central, Room 302
Phoenix, Arizona 85004
(602) 257-2211

Arkansas
Department of Pollution Control and Ecology
Hazardous Waste Division
8001 National Drive
Little Rock, Arkansas 72209
(501) 562-7444

California
Department of Health Services
Toxic Substances Control Division
P.O. Box 942732
Sacramento, California 94234-7320
(916) 324-2430

Colorado
Department of Health/Hazardous Materials
and Waste Management Division
4210 East Eleventh Avenue
Denver, Colorado 80220
(303) 320-8333 or (303) 331-4830

Commonwealth of North Mariana Islands
Department of Health Services
Environmental Protection Board
Trust Territory of the Pacific Islands
Saipan, Mariana Islands 96950
Overseas Operator (Commercial Call 9370-9433)

Connecticut
Department of Environmental Protection
Hazardous Materials Management Section
State Office Building
165 Capitol Avenue
Hartford, Connecticut 06106
(203) 566-8843

Delaware
Department of Natural Resources
and Environmental Control
Solid Waste Management Branch
Division of Air & Waste Management
P.O. Box 1401
89 Kings Highway
Dover, Delaware 19903
(302) 736-4781

District of Columbia
Department of Environmental Services
Division of Pecticides and Hazardous Waste
614 H St., N.W., Room 605
Washington, D.C. 20032-5397
(202) 767-8422

Florida
Department of Environmental Regulation
Hazardous Waste Section
Twin Towers Office Building
2600 Blair Stone Road
Tallahassee, Florida 32399-2400
(904) 488-0300

Georgia
Department of Natural Resources
Environmental Protection Division
Hazardous Waste Management Program
205 Butler Street, SE
Floyd Towers East
Suite 1154
Atlanta, Georgia 30334
(404) 656-2833

Guam
EPA, Government of Guam
P.O. Box 2999
Agana, Guam 96910
Overseas Operator (Commercial Call 646-8863)

Hawaii
Department of Health
Environmental Protection & Health
Services Division
Hazardous Waste Program
645 Halekauwila Street
Honolulu, Hawaii 96813
(808) 548-2270

Idaho
Hazardous Materials Bureau
Division of Environmental Quality
Statehouse
Boise, Idaho 83720
(208) 334-5879

Illinois
Environmental Protection Agency
Division of Land Pollution Control
2200 Churchill Road
Springfield, Illinois 62794-9276
(217) 782-6761 or (217) 524-6175

Indiana
Dept. of Environmental Management
Hazardous Waste Management Branch
Office of Solid & Hazardous Waste Management
P.O. Box 6015
Indianapolis, Indiana 46206-6015
(317) 232-4535

Iowa
U.S. EPA Region VII
726 Minnesota Ave
Kansas City, KS 66101
(913) 236-2887

Kansas
Department of Health and Environment
Hazardous Waste Section
Bureau of Waste Management Sanitation
Forbes Field, Bldg. 321
Topeka, Kansas 66620-0001
(913) 296-1607 or (913) 296-1500

Kentucky
Natural Resources and Environmental
Protection Cabinet
Division of Waste Management
Frankfort Office Park
18 Reilly Road
Frankfort, Kentucky 40601
(502) 564-6716

Louisiana
Department of Environmental Quality
Office of Solid & Hazardous Waste
Hazardous Waste Management Division
P.O. Box 44307
Baton Rouge, Louisiana 70804
(504) 342-1354

Maine
Department of Environmental Protection
Bureau of Oil and Hazardous
Materials Control
State Health Station 17
Augusta, Maine 04333
(207) 289-2651

Maryland
Department of the Environment
Hazardous and Solid Waste Management
Administration
Hazardous Waste Program
2500 Broening Highway
Building 40, Second Floor
Baltimore, Maryland 21224
(301) 631-3343

Massachusetts
Department of Environmental Quality
Engineering
Division of Hazardous Waste
One Winter Street, 5th Floor
Boston, Massachusetts 02108
(617) 292-5580

Michigan
Department of Natural Resources
Waste Management Division
PO Box 30038
Lansing, Michigan 48909
(517) 373-2730

Minnesota
Pollution Control Agency
Hazardous Waste Division
520 Lafayette
St. Paul, Minnesota 55155
(612) 297-2760

Mississippi
Department of Natural Resources
Bureau of Pollution Control
Hazardous Waste Division
P.O. Box 10385
Jackson, Mississippi 39289
(601) 961-5171

Missouri
Department of Natural Resources
Waste Management Program
205 Jefferson Street
P.O. Box 176
Jefferson City, Missouri 65102
(314) 751-9428

Montana
Department of Health and Environmental Sciences
Solid & Hazardous Waste Bureau
Room B201, Cogswell Bldg.
Helena, Montana 59620
(406) 444-2821

Nebraska
Department of Environmental Control
Hazardous Waste Section
Land Quality Division
P.O. Box 94877
State House Station
Lincoln, Nebraska 68509-4877
(402) 471-4217

Nevada
Department of Conservation and Natural Resources
Division of Environmental Protection
201 S. Fall Street, Room 221
Carson City, Nevada 89710
(702) 885-5872

New Hampshire
Department of Environmental Services
Waste Management Division
6 Hazen Drive
Concord, New Hampshire 03301
(603) 271-2900

New Jersey
Department of Environmental Quality
Bureau of Registration and Permits Administration
Division of Solid Waste Management
401 E. State St.
Trenton, New Jersey 08625
(609) 292-5196

New Mexico
Health and Environment Department
Hazardous Waste Bureau
Harold Runnels Building
Santa Fe, New Mexico 87503
(505) 827-2929

New York
Department of Environmental Conservation
Division of Hazardous Substances Regulation
Bureau of Hazardous Waste Operations
50 Wolf Road, Room 205
Albany, New York 12233-7252
(518) 457-3254 or (518) 457-6858

North Carolina
Department of Human Resources
Hazardous Waste Branch
Division of Health Services
P.O. Box 2091
Raleigh, North Carolina 27602-2091
(919) 733-2178

North Dakota
Department of Health
Environmental Waste and Special Studies
1200 Missouri Avenue
P.O. Box 5520
Bismarck, North Dakota 58502-5520
(701) 224-2366

Ohio
Environmental Protection Agency
Division of Solid and Hazardous Waste Management
P.O. Box 1049
1800 Watermark Dr
Columbus, Ohio 43216
(614) 644-2934

Oklahoma
State Department of Health
Waste Management Service
P.O. Box 53551
1000 NE 10th Street
Oklahoma City, Oklahoma 73152
(405) 271-7054

Oregon
Department of Environmental Quality
Hazardous and Solid Waste Division
811 SW Sixth Avenue
Portland, Oregon 97204
(503) 229-5913

Pennsylvania
Department of Environmental Resources
Bureau of Waste Management
8th Floor, Fulton Building
P.O. Box 2063
Harrisburg, Pennsylvania 17120
(717) 787-7381

Puerto Rico
Environmental Quality Board
Office of the Governor
P.O. Box 11488
Santurce, Puerto Rico 00910
(809) 725-5140

Rhode Island
Department of Environmental Management
Division of Air and Hazardous Materials
291 Promenade Street
Providence, Rhode Island 02908
(401) 277-2797

South Carolina
Department of Health and Environmental Control
Bureau of Solid and Hazardous Waste Management
R.J. Aycock Building
2600 Bull Street
Columbia, South Carolina 29201
(803) 734-5200

South Dakota
Department of Water and Natural Resources
Division of Air Quality and Solid Waste
Joe Foss Building - Room 416
Pierre, South Dakota 57501
(605) 773-3153

Tennessee
Department of Health and Environment
Division of Solid Waste Management
701 Broadway
4th Floor - Customs House
Nashville, Tennessee 37219-5403
(615) 741-3424

Texas
Texas Water Commission
Hazardous and Solid Waste Division
P.O. Box 13087
Capitol Station
Austin, Texas 78711
(512) 463-8175

Utah
Department of Health
Bureau of Solid and Hazardous Waste
288 North 1460 West
P.O. Box 16690
Salt Lake City, Utah 84116-0690
(801) 538-6170

Vermont
Agency of Natural Resources
State Office Complex
Waterbury, Vermont 05676
(802) 244-8702 or (802) 828-2070

Virgin Islands
Department of Public Works
Solid Waste Planning Office
Government of the Virgin Islands
Charlotte Amalie
St. Thomas, Virgin Islands 00801
(809) 776-7971

Virginia
Department of Waste Management
101 N. 14th Street
11th Floor, Monroe Building
Richmond, Virginia 23219
(804) 225-2667

Washington
Department of Ecology
Solid and Hazardous Waste Program
Mail Stop PV-11
Olympia, Washington 98504-8711
(206) 459-6322

Washington, D.C.
Department of Consumer and Regulatory Affairs
Environmental Control Division
Hazardous Waste Section
5010 Overlook Avenue, SW
Washington, D.C. 20032-5397
(202) 783-3194

West Virginia
Department of Highways
1900 Washington Street, East
Charleston, West Virginia 25305
(304) 348-3338

Wisconsin
Department of Natural Resources
Bureau of Solid and Hazardous Waste Management
Box 7921
Madison, Wisconsin 53707-7921
(608) 266-2111

Wyoming
Department of Environmental Quality
Solid Waste Management Program
State of Wyoming
122 W 25th Street
Herschler Bldg.
Cheyenne, Wyoming 82002
(307) 777-7752

Hazardous Materials Supplies

The materials described below are essential for the safe handling of hazardous materials and to assure compliance with applicable regulations. For more information or for your free copy of the American Trucking Associations' Catalog of Publications and Services call toll free: 1 800 ATA-LINE. In Virginia call collect: 0 703 838-1754. Local Washington, DC area: 838-1754, 2200 Mill Road, Alexandria, VA 22314.

HAZARDOUS MATERIALS EMERGENCY RESPONSE GUIDE

The complete listing of hazardous materials and procedures that must be taken in the event of an emergency. Hazardous materials are listed both alphabetically and numerically for quick reference, and easy step-by-step emergency procedures are provided to minimize the hazards associated with a release of hazardous materials.

FEDERAL MOTOR CARRIER SAFETY REGULATIONS

When transporting hazardous materials, both city and line-haul drivers as well as vehicles must be in full compliance with these regulations. This is the only pocket-sized printing of the federal regulations for handy use by drivers, supervisors, company officials, and others. Included is Part 397 — Transportation of Explosives and Other Hazardous Materials by Motor Vehicle, plus a digest of the hazardous materials regulations.

PLACARDS

All required placards are available on plastic coated tagboard or plastic film adhesive material and meet DOT requirements for size, durability and color.

LABELS

The ATA label assortment is in a small package to be kept at each terminal. The package contains all DOT-prescribed types and enables carriers to comply with the regulation requiring replacement of labels lost or damaged in transit.

HAZARDOUS MATERIALS SHIPPING PAPERS ENVELOPES

The easy way to keep these papers together and meet the requiremment of DOT to have them handy for inspection or in case of emergency. Printed in red, the 9" x 12" envelopes include routine and emergency instructions for drivers and dispatchers and provide space to note placarding of vehicles for the trip.

HAZARDOUS MATERIALS GUIDE FOR TRUCK TRANSPORTATION

Summarizes requirements for Shipping papers and the Marking and Labeling of packages. Shows prescribed labels and indicates placards required for each class of hazardous materials. Includes instructions for placarding mixed loads. Available in two sizes: 8½" x 11" for vehicles and clipboards; 14" x 20" for desktops and wall display in terminals. Vehicle and clipboard sizes are on pressure-sensitive stock.

LOADING AND STORAGE GUIDES

Uses full-color presentations of hazard class labels to show which classes cannot be transported together in the same vehicle and cannot be stored adjacent to each other in terminals. Available in 8½" x 11" adhesive and 14" x 20."

HAZARDOUS MATERIALS INCIDENT REPORT

The DOT form for reporting an accident or incident involving the transportation or storage of hazardous materials including any accidental release of hazardous materials. The size of the form is 8½" x 11," 50 sheets per pad.

TITLE 49, CFR

The latest edition of the Code of Federal Regulations avilable. Title 49 includes all federal regulations for marking, labeling, placarding, packaging and transporting hazardous materials. Every terminal should have Title 49 on hand.

DRIVER'S GUIDE TO HAZARDOUS MATERIALS

An easy to read and understand pocket size book that covers all federal regulations governing hazardous materials. Designed to provide drivers with a good working knowledge to hazardous materials regulations. Meets the training requirements of DOT and the National Master Freight Agreement. Features: instructions on the use of table 172.101, 49 CFR; shipping papers, packaging, loading, unloading, driving and parking regulations; emergency procedures, phone numbers and contacts; marking, labeling and placarding requirements displayed on a 4 color chart.

For those carriers involved in the bulk transportation of hazardous materials, the following are available from the National Tank Truck Carriers Conference, 2200 Mill Road, Alexandria, VA 22314, 703 838-1960.

CARGO TANK HAZARDOUS MATERIALS REGULATIONS

A comprehensive listing of the federal regulations applicable to the bulk transportation of hazardous materials. Includes specifications for cargo tanks and portable tanks.

HAZARDOUS COMMODITY HANDBOOK

A compilation of commodities suitable for transportation in bulk. Provides information regarding applicable regulations, chemical formulas, specification cargo tanks and other special requirements of DOT for tank trucks.